Parametric Array Loudspeakers

Jun Yang · Peifeng Ji

Parametric Array Loudspeakers

From Theory to Application

Springer

Jun Yang
Institute of Acoustics
Chinese Academy of Sciences
Beijing, China

Peifeng Ji
Institute of Acoustics
Chinese Academy of Sciences
Beijing, China

ISBN 978-981-96-3547-4   ISBN 978-981-96-3548-1 (eBook)
https://doi.org/10.1007/978-981-96-3548-1

Translation from the Chinese Simplified language edition: "Parametric Array Loudspeakers: From Theory to Application" by Jun Yang and Peifeng Ji, © Science Press 2022. Published by Science Press. All Rights Reserved.

The original submitted manuscript has been translated into English. The translation was done using artificial intelligence. A subsequent revision was performed by the author(s) to further refine the work and to ensure that the translation is appropriate concerning content and scientific correctness. It may, however, read stylistically different from a conventional translation.

This Springer imprint is published by the registered company Springer Nature Singapore Pte Ltd.
The registered company address is: 152 Beach Road, #21-01/04 Gateway East, Singapore 189721, Singapore

Preface

"Sound Beaming", a unique skill in martial arts novels, refers to the loudspeaker being able to condense the sound into a line and directly deliver it to the listener's ears, achieving directional sound propagation, while others cannot hear it. In real-life applications, when a highly directional sound beam is required, the conventional approach is to employ loudspeaker arrays, but they suffer from poor low-frequency directivity, the presence of sidelobes, and large size, which limit their practical applications. The parametric array loudspeaker, as a new concept sound source, has the advantages of narrow beam, strong directivity, and very small sidelobes. Different from traditional loudspeakers, the sound production principle of the parametric array loudspeaker is based on the parametric array effect, modulating the audio signal with an ultrasonic carrier, and then amplifying it to obtain a modulated sound wave signal with a finite amplitude. After being emitted by the ultrasonic transducer, the modulated audio signal can self-demodulate due to the nonlinearity in the air during the propagation process, thereby obtaining a highly directional audio frequency signal.

Since the concept of the parametric array was proposed by Westervelt in 1963, it has received widespread attention and has continued to develop, from the initial underwater parametric array to the parametric array loudspeaker in the air. At present, the parametric array loudspeaker has become a research hotspot in the field of audio engineering and has received extensive research in both theory and application. Many domestic and international companies have also launched their own parametric array loudspeaker products, which have been increasingly used in exhibition halls, supermarkets, transportation, offices, and public places, bringing certain economic benefits. At the same time, the rapid development and application of parametric array loudspeakers can provide ideas for solving social problems caused by noise pollution and have certain social benefits.

The parametric array loudspeaker, as a typical application of nonlinear acoustics, involves a very wide range of knowledge, including modeling simulation, computational science, transducer technology, signal processing technology, and psychoacoustic technology, among many other disciplines. For newcomers, it is not easy to grasp so much relevant knowledge, and the existing related books are basically comprehensive books on nonlinear acoustics. There is currently no monograph on parametric array loudspeakers both domestic and international. This book focuses on parametric array loudspeakers, involves research in all major aspects, and details a large amount of related research content in basic theory, key technology development, and system practicality. This book mainly includes theoretical calculations, measurements, signal processing, beamsteering methods, implementation, and applications, and can provide comprehensive and valuable insights for researchers in the field of acoustics. This book can assist acoustic researchers in swiftly delving into the research areas related to parametric array loudspeakers, and is an indispensable reference book in the research of parametric array loudspeakers and even nonlinear acoustics.

This book is a summary and refinement of our research group's many years of work on parametric array loudspeakers. We would like to thank our research collaborators, Profs. Meng Hwa Er, Woon-Seng Gan, Tomoo Kamakura, and our research group members, Profs. Jing Tian, Ming Wu, and Associate Professor Kan Sha for their guidance and assistance in the related research process of this book. We also appreciate the extensive work done by students Wei Liu, Chao Ye, Zheng Kuang, Shuaibing Wu, Yongsheng Mu, Chenxi Huang, Wei Zhang, Wei Ji, Dengyong Ma, Yew Hin Liew, Furi Andi Karnapi, Khim-Sia Tan, and Kelvin Chee-Mun Lee in data collection, modeling calculations, and implementation. We are grateful to Dr. Chuang Shi for his assistance in the publication process of this book. We thank the responsible editor of the Science Press for the extensive work done for this book. The writing and publication of this book have received the care and support of the leaders and researchers of the Key Laboratory of Noise and Vibration of the Chinese Academy of Sciences, and have also received the support and funding of the National Natural Science Foundation of China (Nos. 11004217, 11674348). Here, we express our heartfelt thanks.

Due to the limitations of the authors' expertise, the book may contain shortcomings. We sincerely invite our colleagues to offer criticism and corrections.

Beijing, China
2021

Jun Yang
Peifeng Ji

Introduction

The parametric array loudspeaker, which can achieve directional sound radiation, has received widespread attention from researchers due to its advantages of narrow beam, highly directivity, and very small sidelobes. It has developed rapidly in theory and practical applications, and has formed several commercial products. At present, the parametric array loudspeaker has become a research hotspot in the field of audio engineering. This book mainly introduces the current research status of parametric array loudspeakers, including modeling calculations, measurements, signal processing, beamsteering, implementation, and applications. The introduction of the above aspects helps to deeply understand the parametric array loudspeaker and the scientific problems involved, which is of great guiding significance for further research on parametric array loudspeakers and provides a reference for other acoustic research.

This book can be used as a reference book for graduate students in acoustics and acoustic researchers. We hope this book can help readers in their scientific research work.

Contents

Chapter 1
Overview

1.1 History and Current Status of Nonlinear Acoustics in Parametric Arrays

1.1.1 History of Nonlinear Acoustics

Acoustics mainly studies the emission, propagation, reception of sound, the properties of sound, and the interaction between sound and other substances. Acoustics is one of the oldest disciplines in natural science, with descriptions of acoustics dating back to ancient times. With the invention and application of radio technology, acoustics has moved from classical acoustics to the development period of modern acoustics. Modern acoustics has a strong permeability and intersects with many other disciplines, engineering technologies, and artistic fields. It plays an important and unique role in these fields, and further develops the corresponding theories and technologies, and gradually forms independent branches of acoustics, such as physical acoustics, nonlinear acoustics, quantum acoustics, molecular acoustics, ultrasonics, optoacoustics, electroacoustics, architectural acoustics, environmental acoustics, language acoustics, bioacoustics, underwater acoustics, atmospheric acoustics, geosonics, physiological acoustics, psychoacoustics, musical acoustics, and sonochemistry, etc.

In early acoustic research, the motion equation of acoustics could satisfy people's basic needs by solving it on a linear basis, thus ignoring the nonlinearity of motion and medium. Therefore, the assumption of linear acoustics is mainly based on small amplitude sound waves. In linear acoustics, two sound waves propagating at the same time will not interact with each other, and the vibration of the medium caused by them is equal to the linear superposition of the vibration of the medium caused by them when they exist separately, satisfying the superposition principle. However, if the acoustic equation retains nonlinear terms, the superposition principle is no longer

J. Yang and P. Ji, *Parametric Array Loudspeakers*,
https://doi.org/10.1007/978-981-96-3548-1_1

applicable, and it no longer follows the rules of linear acoustics, thus giving rise to a new branch of discipline—nonlinear acoustics.

Nonlinear acoustics is a science that studies intensive sound. In nonlinear acoustics, intensive sound is usually referred to as finite amplitude sound waves or large amplitude sound waves, which is a phenomenon between small amplitude sound waves and weak shock waves. The main research object of nonlinear acoustics is phenomena related to the propagation of finite amplitude sound waves, such as shock formation, harmonic distortion, non-constant propagation speed, etc. These phenomena cannot be explained by general linear acoustics. They are caused by the nonlinear effects of the medium on finite amplitude sound waves [1].

The field of nonlinear acoustics is a discipline with a long history, especially in the past sixty years, it has made great progress. Reference [2] gives a detailed introduction to the development of nonlinear theory before the 1930s. In 1755, Euler first gave the nonlinear wave equation describing finite amplitude acoustics, but did not give a strict solution to this equation. Soon after, Lagrange derived a general solution, but he believed that the changing sound wave speed would destroy the theory of sound propagation. In 1808, Poisson obtained the exact solution of the propagation of finite amplitude plane waves. Stokes [3] explained the meaning of the non-constant propagation speed that troubled Lagrange and Poisson, and he published the first explanation about waveform distortion in 1848. Stokes was also the first to give an analysis of shock waves, and proposed that viscosity and heat conduction are the inevitable loss causes of shock front. Two important papers published around 1860 [2, 4] can be regarded as the beginning of nonlinear acoustic research, that is, the simple wave theory of Riemann-Earnshaw, which is the strict solution form of the nonlinear one-dimensional wave equation independently published by Riemann and Earnshaw. Beyer compiled some early papers into the famous collection of nonlinear acoustics [6].

In addition to the analysis of shock waves by Rankine [7], Hugoniot [1], Rayleigh [8] and Taylor [9], the publication of three papers in the 1930s marked a significant progress in the understanding of finite amplitude sound. One paper by Fubini [10] is applicable to waves in lossless fluids before shock formation, and another by Fay [11] provides an asymptotic solution for finite amplitude waves with viscous losses. These two papers first provided explicit models for harmonic generation in sound waves, and the third paper by Thuras et al. [12] first provided experimental research on this phenomenon.

Eckart [13], Lighthill [14] and Mendousse [15] derived the wave equation around 1950, marking the beginning of the modern era of nonlinear acoustics. The equations of Eckart and Lighthill allow the study of non-planar finite amplitude waves, while Mendousse proved that the Burgers equation can be used to simulate plane waves in viscous fluids. Khokhlov and his collaborators demonstrated that the generalization of the Burgers equation can simulate cylindrical waves [16]. Blackstock [17] studied the detailed solution of the Burgers equation in detail, using the intermediate values of nonlinear wave solutions to smoothly connect the Fubini solution and the Fay solution, solving the problem of nonlinear propagation of single-frequency large amplitude waves. Fenlon [18] started from the lossless Burgers equation, using

Fourier expansion to represent the sound waves of each frequency component after interaction in the form of a spectrum. So far, in the most famous practical applications, the most noteworthy contribution is Westervelt's work on sound of scattering by sound [19, 20], which eventually formed the theory of acoustic parametric arrays in the 1960s [21, 22]. In the parametric array, the nonlinear interaction of two high-frequency sound beams produces a narrow low-frequency sound beam with almost no sidelobes. This process allows high directivity sound to be radiated from a relatively small transducer, with the additional benefit of being able to transmit a wide frequency range of sound. The proposal of the acoustic parametric array has brought broad prospects for the application of nonlinear acoustics. At a conference of the Acoustical Society of America in 1962, papers following Westervelt reported the experimental verification of the parametric array by Bellin and Beyer [23].

The parametric array was first conceived and tested in the United States, followed by many related works in the UK and Norway. Berktay [24–26] studied various possible application examples of using the parametric array, but later found that some of his theoretical predictions [7, 27, 28] were somewhat overly optimistic. When it was discovered from Berktay's work that the attractive features of the parametric array often exceeded the inherent low efficiency of the phenomenon, more related research work began to appear. By the early 1970s, the parametric array had been applied in civilian and military sonar systems, and the number of papers on the parametric array published at the Acoustical Society meetings and seminars each year had exceeded 100. The next turning point in the practical application of nonlinear effects in sound beams came from the theoretical work carried out inthe former Soviet Union around 1970 by Zabolotskaya and Khokhlov [29] and Kuznetsov [30]. Their research resulted in a parabolic nonlinear wave equation, known as the Khokhlov-Zabolotskaya-Kuznetsov (KZK) equation, which provides an excellent model for the combined effects of nonlinearity, diffraction, and dissipation in directional sound beams. Throughout the 1970s, numerous works based on the KZK equation appeared in the former Soviet Union, most of which were aimed at Gaussian beams. However, the most authoritative subsequent research on the nonlinear effects of sound beams was a series of important works published by Norwegian mathematicians Jacqueline Naze Tjøtta, Sigve Tjøtta and their collaborators [31–36], who explained the complex diffraction effects present in high-intensity sound fields radiated from circular sound sources, which characterize the geometric characteristics of most ultrasonic transducers. The book written by Beyer [37] emphasizes many achievements in the field of nonlinear acoustics, including experimental work. Rudenko and Soluyan [38] reported the latest progress in the former Soviet Union. Hamilton and Blackstock's [1] work on nonlinear acoustics is a good introductory textbook on nonlinear acoustic theory and applications. Qian systematically introduced the propagation of finite amplitude sound waves in unbounded and bounded spaces [39], he carried out a lot of work in many areas such as broadband parametric array research, interaction between sound waves research [40–52], greatly promoting the development of nonlinear acoustics both domestic and international. Based on Fenlon theory, Du et al. [53] mainly conducted theoretical research on the interaction between finite amplitude sound waves and small amplitude plane sound waves and sound energy suppression, and conducted related experimental verification in the sound wave tube. Based on the Burgers equation, Yang's research group mainly used the spectral decomposition method to derive the mechanism of the interaction process between low-frequency signal waves and high-frequency pump waves, and studied the energy transfer problem after the nonlinear action of variable parameter sound waves [54–57].

In the past few decades, the development of high-power sound sources and the improvement of digital signal processing technology have increasing supported practical applications using high-power sound sources. Parametric arrays with unique technical advantages such as "wideband, high directivity, small size" have been widely used in the field of underwater acoustics engineering [28, 43, 58–63]. With the continuous progress of science and technology, the application fields and scope of parametric arrays are constantly expanding.

1.1.2 Multi-Beam Sound Field Research

From the perspective of nonlinear acoustics, when there is a sound wave propagating in space, the original medium in space is disturbed by this sound wave. If the medium was originally uniform, the result of the disturbance makes it non-uniform in space. If there is another sound wave in this disturbed medium, then it will be scattered by the non-uniform medium, or it can be said that one sound wave scatters another sound wave, this phenomenon is called sound of scattering by sound [64–70]. The sound field formed in this way is called a multi-beam sound field. All about the nonlinear interaction of two beams (i.e., the phenomenon of sound of scattering by sound) started with Lighthill's theory [14, 71]. Lighthill gave the form of the wave equation

$$\frac{\partial p}{\partial t} + \frac{\partial(\rho u_j)}{\partial x_i} = 0 \tag{1.1}$$

$$\frac{\partial}{\partial t}(\rho u_j) + c_0^2 \frac{\partial p}{\partial x_i} = -\frac{\partial T_{ij}}{\partial x_j} \tag{1.2}$$

$$\frac{\partial^2 p}{\partial t^2} - c_0^2 \nabla^2 \rho = \frac{\partial^2 T_{ij}}{\partial x_i \partial x_j} \tag{1.3}$$

where T_{ij} is the stress tensor, ρ. is the fluid density, ∇^2 is the Laplace operator. In 1956, Ingard and Pridmore-Brown [72] first proposed the concept of sound of scattering by sound. Ignoring viscosity, they derived the expression for the scattered pressure field based on Lighthill's far-field expression. The result suggests that when two orthogonal quasi-straight beams interact, the scattered sound occurs outside the common region, and their experimental results also support their theoretical results. In 1957, Westervelt [19] published two papers expressing different views on the results of Ingard et al. He argued that the first-order field in the above theory does not satisfy the homogeneous wave equation required by the Lighthill equation. He chose a plane wave as the first-order field, and derived from the Lighthill equation that there is no second-order scattered wave outside the common region of two non-parallel waves. In 1960, Bellin and Beyer [73] experimentally verified Westervelt's theory. Dean [74] considered the interaction of two concentric cylindrical waves and spherical waves. Garrett et al. [34] used parabolic approximation to explain the interaction of two wave fields at small angles. Darvennes et al. [64] theoretically derived the interaction of two Gaussian beams at small angles. There are many related studies on this issue, and the books by Beyer [6] and Rudenko [38] specifically include a chapter on sound of scattering by sound. From the above literature, there is a dispute about whether there is scattered sound outside the common region. A new theory has been used to reinterpret the interaction of two beams, and a series of articles by Tjøtta et al. [75–80] have been published to derive the sound scattering in the case of arbitrary crossing angles, pointing out the reasons why related research may reach different conclusions. They illustrate this using the simplest model of rectangular sound source interaction. From the continuity equation and equation of state of a uniform non-viscous fluid, the equation for density ρ is obtained

$$\Box^2\rho = \left(\nabla^2 - \frac{1}{c_0^2}\frac{\partial^2}{\partial t^2}\right)\rho = -\frac{1}{c_0^2}\sum_{i,j}\frac{\partial^2 T_{ij}}{\partial x_i \partial x_j} \tag{1.4}$$

When the two interacting sound waves are plane waves, the above equation can be transformed into $\Box^2(\rho + f) = 0$ form, where f is the quadratic function of the primary wave field, $\Box^2$ is the D'Alembert operator, that is, $\Box^2 = \nabla^2 - \frac{1}{c_0^2}\frac{\partial^2}{\partial t^2}$. For collimated beam plane waves, this formula is valid in the common region, with $\rho_F = -f$ as a particular solution. Westervelt used this feature to believe that there is no scattered field outside the common region. However, ρ_F is not the general solution of the equation, because any ρ_H can always be added such that $\Box^2\rho_H = 0$ holds. Thus, $\rho = \rho_F + \rho_H$ provides another solution for the density equation in the common region. Outside the common region, the sound field must satisfy $\Box^2\rho_H = 0$, which depends on the boundary conditions. If the solution in the common region is not uniquely determined, it is impossible to determine the boundary conditions of the collimated plane wave. Tjøtta et al. believed that the reason why Westervelt did not find scattered sound outside the common region is that he ignored the effect of diffraction and used the approximation of a narrow beam. They also considered the effects of boundary conditions and sound absorption on the results of multi-beam nonlinear effects. Tjøtta's method can be seen as a generalization of small-angle multi-beam interaction. When in small-angle situations, this method is approximately consistent with the results obtained by Garrett et al. [34] Tjøtta's method is a general solution of quasi-linear approximation, which is valid at any distance, crossing angle, and frequency ratio, but their results contain multiple integrals, which are not easy to solve. Although they provided an approximate solution in the derivation process, this solution is only applicable to the far field and contains two integrals, which takes a lot of time. Yang's research group provided a fast algorithm for simplifying the calculation of the small-angle offset model using the Gaussian beam expansion method [81–83], and studied the experimental phenomenon of the difference frequency sound generated by two ultrasonic beams in the presence of an artificial head [84]. Qian studied the interaction between plane waves and plane pulse waves in a non-dispersive medium, pointing out that only second-order scattered waves exist in the common region of the two parallel waves [47]. Garner [85] used cascaded second-order nonlinear effects to simulate the third-order nonlinear action of non-collimated ultrasonic beams produced by two displaced piston sound sources.

1.1.3 Research on Parametric Arrays

The basic principle of parametric array research is to modulate the audio frequency signal to be emitted with a high-frequency (ultrasonic frequency) carrier signal, and then emit it with ultrasonic transducers. Due to the nonlinear acoustic effect of the medium, the audio frequency signal in the sound wave will demodulate itself during propagation. Due to the super directivity of ultrasonic propagation, it is possible to achieve directional radiation of audio frequency signals, as shown in Fig. 1.1.

Figure 1.2 describes the schematic diagram during propagation. In the figure, f_1 and f_2 are two ultrasonic signals with a small difference in frequency, called primary waves, assuming $f_1 > f_2, f_1 - f_2$ is the audio frequency signal. These two signals are emitted using the ultrasonic transducer. Due to the nonlinear effect of the medium, if only considering the second order, the difference frequency signal $f_1 - f_2$, the sum frequency signal $f_1 + f_2$, plus the two primary signals and their second harmonic signals will be obtained. According to the principle of sound absorption in the medium, the absorption is roughly proportional to the square of the frequency value of the sound wave, so the sum frequency signal $f_1 + f_2$ and f_1, f_2 and their harmonic signals will decay quickly, leaving mainly the audio frequency signal $f_1 - f_2$.

Generally speaking, parametric arrays are divided into three types of sound sources [62, 86, 87]: absorption-limited sound sources, diffraction-limited sound sources, and saturation-limited sound sources. The types of these sources are determined by three distances, namely the Rayleigh distance, the absorption distance, and

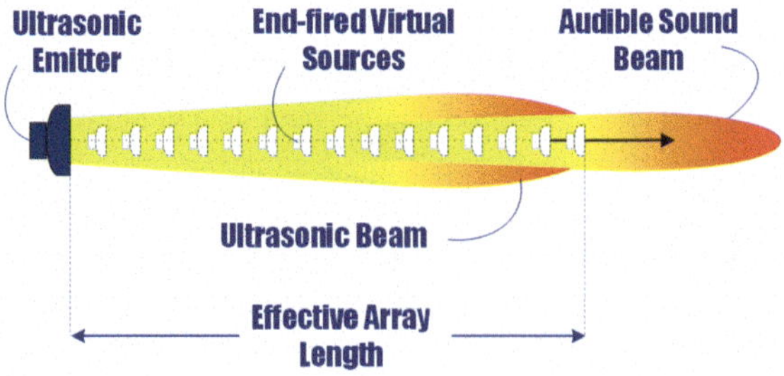

Fig. 1.1 Schematic diagram of directional sound source formation by parametric array

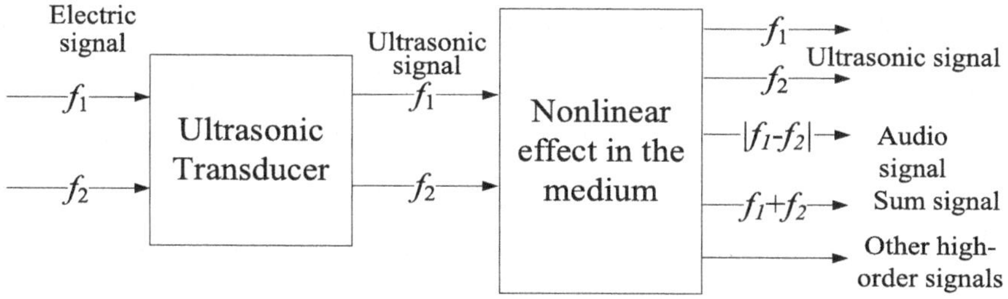

Fig. 1.2 Schematic diagram of propagation in medium

the saturation distance. The Rayleigh distance is defined as the distance from the transducer to the transition between the near-field region and the far-field region. Therefore, the wavefront of the primary wave is a plane wave within the Rayleigh distance, but when it propagates beyond the Rayleigh distance, it will propagate spherically. The absorption distance determines the length of the primary wave before it stops generating secondary waves. Therefore, the absorption distance represents the length of the virtual source array. The saturation distance represents the area of saturation effect. When the amplitude of the primary wave is high enough, saturation effects will occur. In this case, harmonics and shock formations are generated in the primary wave beam.

When the absorption distance is shorter than the Rayleigh distance, it is a situation where absorption is limited, and the nonlinear effect terminates within the near field. In this case, the difference frequency sound is generated by the collimated beam of the primary wave. When the absorption distance is longer than the Rayleigh distance, the virtual source array extends to the far field, forming a diffraction-limited situation. In addition, if the absorption distance is much longer than the Rayleigh distance, the secondary waves generated in the near field can be ignored. Although the primary wave propagates spherically in the far field, as long as the primary wave beam is very narrow, the primary wave will still produce parametric array effect similar to the Westervelt equation. When the sound pressure level of the primary wave exceeds a certain limit, saturation effects occur within the near field of the source. This results in a saturation-limited situation, where the primary wave in the near field can be considered as a plane wave.

The parametric array can be seen as a special case of a multi-beam sound field, that is, two sound sources emit in the same center and direction. The study of parametric arrays is a multidisciplinary subject, which not only involves mechanics, acoustics, psychology, but also includes knowledge of circuits and signal processing. Therefore, the research on parametric arrays has been greatly developed with the progress of various disciplines and technologies, and has gradually been widely used in different aspects.

The initial understanding of the parametric array phenomenon was the Tartini tone discovered by Italian pianist Tartini in the mid-eighteenth century [88]. Since this is a form of beat, it was initially thought to be caused by the vibration of the human ear when perceiving sound. However, Helmholtz doubted it, and later he confirmed that this was caused by the nonlinear effect of air. He believed that this effect simultaneously produced the sum (sum frequency) and the absolute difference (difference frequency) of two frequency signals. But he did not give the theoretical analysis results. In the 1960s, Westervelt [21, 22], based on Lighthill's theory [14, 71], gave the relationship and specific theoretical derivation of the second-order field generated when two plane waves propagate in a non-uniform medium.

In 1965, Berktay [24] proposed a more accurate and complete explanation of the parametric array. He did not limit the analysis to the primary wave being a form of two frequencies, but further used the concept of envelope in modulation. This is very practical, because parametric arrays usually do not just emit a single frequency. The final demodulated signal (the signal of interest) is determined by this envelope, that is, the signal after the parametric array is demodulated is proportional to the second derivative of the square of the envelope with respect to time, that is $p_2(t) \propto \frac{\partial}{\partial t^2} E^2(t)$, where p_2 is the demodulated audio frequency signal, $E(t)$ is any envelope function. Since Berktay assumed that this solution was obtained in the case of being far from the sound source, ignoring the influence of ultrasound, this solution is called "Berktay's farfield solution". Although the primary wave signal, the sum frequency signal and the demodulated difference frequency signal coexist in the near-field, this solution is also applicable. This is the basic expression of the output of the parametric array. This conclusion was quickly verified in underwater acoustics [89, 90]. In 1981, Qian believed that Berktay ignored the amplitude characteristics of the near-field source, re-established the theoretical model of the nonlinear sound field calculation of the parametric array, and conducted experimental verification [42]. In the same year, Qian proposed a new method for calculating Huygens' integration, and used it to study the radiation field of the line source parametric array and the truncated parametric array [41]. In 2008, Zheng et al. improved the Berktay's model, obtained the difference frequency sound field in the axial and off-axis directions, and obtained the distribution of the difference frequency sound field more accurately [91].

If the Berktay far-field solution is used to derive an output of a single-frequency double-sideband modulation method with an input of 2 kHz (assuming the modulation index m is 1), two single-frequency signal outputs with the same amplitude are obtained, which is equivalent to 100% total harmonic distortion (THD). This shows that the double-sideband modulation method cannot provide the desired single-frequency output, so it is necessary to preprocess the audio frequency signal [92].

The parametric array was first widely used in underwater acoustics. Because the absorption coefficient of air is large and its nonlinear parameter is small, it is relatively difficult to produce this nonlinear effect in the air. It was not until 1975 that Bennett and Blackstock [93] realized the parametric array in the air.

Because the audio frequency obtained by the parametric array demodulation has very highly directivity, and this audio frequency decays much slower when propagating in the air than the sound emitted by traditional loudspeakers, making the development of directional sound sources possible. In recent years, with the development of parametric array technology research, the application of parametric arrays in the air, namely parametric array loudspeakers, is becoming more and more widespread.

The earliest recorded was Yoneyama et al. in Japan [94]. They formed a hexagonal array with 547 ultrasonic transducers in 1983, with a center frequency of 40 kHz. To overcome the frequency response slope of 12dB/octave proposed by Berktay, they suggested the input signal should be preprocessed before double-sideband modulation. In actual experiments, the audio frequency signal is simply processed with the transducer frequency response curve. They obtained a relatively flat frequency response between 1.5~7 kHz. At the International Nonlinear Acoustics Conference in 1984, an article on directional speakers [95] mentioned three issues: the optimal carrier frequency, distortion calibration, and the impact of ultrasound on listeners. Their view is that the smaller the carrier frequency, the better, but it cannot be too small, otherwise the beam will diverge, and a range of recommended carrier frequency values is given. They made an array with 581 ultrasonic transducers and found that the distortion was significantly improved after preprocessing. During this period, in-depth discussions were held on issues such as preprocessing methods, sound source shapes, conversion efficiency, etc. [96–102]. Steer's research group [103, 104] also calculated the third-order nonlinear sound field, and it was simplified using the Gaussian beam expansion method. The issue of low conversion efficiency of the parametric array has also attracted the attention of scholars [105], who have conducted research from the perspective of transient signal technology [97, 106–110] or transducer technology [111–113], to some extent alleviating the problem. For this issue, some scholars have also tried to solve it by using multiple parametric array loudspeakers or combining various modulation methods [114–117].

Similar to the end-fire array effect [21, 118], the parametric array can be regarded as a virtual source array composed of countless virtual sources. The parametric array is a cumulative array, and it needs a certain propagation distance to fully form. It is the aforementioned absorption distance, which can be referred to as the array length [119], where the primary waves no longer interact with each other. For some reasons, if the observation point is less than the array length before the propagation has reached the array length, this situation is the truncation of the parametric array, and the formed part of the parametric array sound source is called the truncated parametric array [41, 58, 59, 120]. This research is often used in practical applications in many fields such as detection and material acoustic properties.

1.1.4 Research on Parametric Array Loudspeakers

Since the 21st century, with the launch of commercial products and its emergence as a research focus in audio engineering, parametric array loudspeakers have achieved substantial academic progress in both foundational theories and key technologies [119, 121–129], mainly reflected in the following directions.

1.1.4.1 Preprocessing Algorithms

The parametric array loudspeaker involves nonlinear effects, so in order to achieve sound field reproduction, the input signal of the parametric array loudspeaker must be preprocessed to reduce harmonic distortion. Based on the Berktay far-field solution, various signal processing algorithms are applied to the parametric array loudspeaker. Yoneyama et al. [94] used the ordinary double sideband amplitude modulation (DSBAM) method to generate the envelope signal $E(t) = 1 + mg(t)$, where $g(t)$ is the input audio signal, m is the modulation index. However, this simple processing method has caused significant harmonic distortion. In 1984, Kamakura et al. [95] proposed the square root double sideband modulation method (Square Root AM, SRAM), which generated the envelope signal $E(t) = \sqrt{1 + mg(t)}$. Compared with ordinary double sideband modulation, this method can effectively reduce harmonic distortion, but it requires an ultrasonic transducer with infinite bandwidth to generate harmonic signals formed by square root processing. Subsequently, they proposed the single-sideband amplitude modulation method (SSB-AM) [130], which only requires half the bandwidth compared to ordinary double sideband modulation. However, for complex input signals, single-sideband modulation is prone to errors. Therefore, Croft et al. [131, 132] proposed the recursive single-sideband modulation method (Recursive SSB-AM) to approximate the results of square root double sideband modulation, but this requires a high-speed processor to achieve real-time calculation. Wu et al. [133] adopted multi-rate processing technology and proposed an implementation method for recursive single-sideband modulation technology. Its feasibility was verified by experiments. Tan et al. [134] proposed adjustable single-sideband modulation (Modified AM, abbreviated as MAM) using the interaction of orthogonal signals, which can effectively reduce harmonic distortion and reduce computational complexity. Sakai and Kamakura [135] proposed a single-sideband modulation method with dynamic carrier (Dynamic SSB-AM), which dynamically adjusts the amplitude of the carrier signal according to the amplitude of the modulation signal to improve the linear relationship between the output signal and the modulation signal, thereby reducing harmonic distortion. Shi et al. [136] provided the experimental results of four preprocessing methods: DSBAM, SRAM, Lower sideband-AM (LSB-AM) and Upper sideband-AM (USB-AM), and based on DSB-AM and SRAM, respectively, they proposed two new improved preprocessing methods. Ikefuji et al. [137] combined the DSB-AM and the SSB-AM methods, using different preprocessing methods for different frequency bands, and proved the effectiveness of this method through subjective testing. Since the Berktay far-field solution is only an approximate solution of the parametric loudspeaker, it cannot truly reflect the actual situation of the sound field. Ji et al. [138] summarized various preprocessing algorithms and compared the above DSB-AM, SSB-AM, and MAM preprocessing algorithms based on the time-domain solutions of the KZK equation. All of the above are based on amplitude modulation, and frequency modulation methods have also been studied by scholars due to their low implementation costs [139, 140], some progress has been made, but the high harmonic distortion has not yet been well resolved [105], which is worth further in-depth study.

Due to the limitations of the Berktay far-field solution, the preprocessing methods based on the Berktay far-field solution cannot completely eliminate the nonlinear distortion in the parametric array loudspeaker system, and in recent years, the preprocessing method based on the Volterra filter has been introduced. In 2002, Lee et al. introduced the Volterra filter (VF) into the research of parametric array loudspeakers [141], based on the Berktay far-field solution model. They established a 2nd-order VF model, which simulated the nonlinearity of the parametric array loudspeaker well. Later, based on the KZK equation, Ji et al. obtained a more accurate 2nd-order VF model, and on the basis of the VF model, designed a 2nd-order inverse filter, which can effectively reduce the harmonic distortion of the parametric array loudspeaker [142]. Yang's research group introduced the diagonal Volterra filter with lower complexity into the research of parametric array loudspeakers, and theoretically demonstrated the feasibility of using a one-dimensional Volterra filter (ODVF) to compensate for the nonlinear distortion of the loudspeaker instead of the traditional VF. They proposed a solution based on ODVF to compensate for the harmonic and intermodulation distortion of the parametric array loudspeaker [143–146]. Shi et al. proposed an ultrasonic-ultrasonic VF [147], and applied the sparse normalized least mean square algorithm to the identification of the first-order Volterra filter and acoustic delay [148]. For the parametric array loudspeaker system based on frequency modulation, Hatano et al. designed a parallel cascade structure of VF and its inverse system [140].

The above-mentioned preprocessing methods are mainly to solve the harmonic distortion problem generated in the parametric array loudspeaker system, and designers can therefore choose the appropriate preprocessing technology according to the specific system requirements, such as analog or digital implementation, available computing resources, bandwidth limitations of ultrasonic transmitters, etc. To solve the problem of poor low-frequency response of the parametric array loudspeaker, Shi et al. [149] proposed the "missing fundamental" concept widely used in traditional speaker virtual bass enhancement to improve the bass quality of the parametric array loudspeaker. This psychoacoustic preprocessing technique splits the audio input with a low-pass and a high-pass filter under a cutoff frequency of 500 Hz. The high-pass audio content uses general preprocessing techniques, while the low-pass bass is processed by the function $F(x)$, which serves as the harmonic generator of the parametric array loudspeaker. Subjective tests showed that this preprocessing method can promote the enhancement of low volume while maintaining a good auditory experience.

1.1.4.2 Beam Control

Beam control, based on the principle of parametric array effect, the difference frequency sound of parametric array loudspeakers has high directivity. This high directivity can make the difference frequency sound only emitted to the target area, without polluting the sound environment outside the target area. However, to ensure that the beam of the difference frequency sound to follow the target area for mobile tracking, Woodward et al. [150] in 1994 applied phased array technology to underwater parametric arrays for seabed mapping, which can meet the beamsteering requirements within $\pm 20°$. In 2005, Olszewski et al. [151] used a mechanical motor to rotate each channel's transducer, and then compensated the phase of each channel by delaying, eliminating the sound path difference caused by rotation, thereby achieving the purpose of beamsteering. However, this method has a slow beamsteering speed and requires motor assistance, which adds extra equipment. The use of digital signal processing to achieve beamsteering is low cost, energy saving, fast, and does not require additional auxiliary equipment, which is the trend of future development. The delay and sum algorithm is a common method for array beamforming. This method is simple and practical, and can conveniently and quickly realize the beamsteering of the parametric array loudspeaker. In 2006, Yang et al. [152] proposed to use the zero-order Bessel function weighting method for parametric array difference frequency sound beamsteering. In addition, Yang et al. [153–155] used the delay and sum method, used the Chebyshev weighting method combined with SSB modulation method, weighted the carrier and sideband frequencies separately, and obtained a sound beam with a constant beam width. In actual implementation, the angle of the parametric array beamsteering is limited by the sampling rate of the digital-to-analog converter. In the case of low sampling rate, if integer sampling period delay is used, although the parametric array beam can be deflected, the beam has a minimum steering angle, and the steering angle of the beam is only distributed at a few discrete angles. In 2006, Gan et al. [156] pointed out that at a sampling rate of 192 kHz, the minimum steering angle is about 26 degrees. For this reason, Gan et al. achieved a very small steering angle with the smallest amount of calculation by delaying the carrier and sideband signals separately without increasing the sampling frequency, but this method requires the generation of an omnidirectional carrier beam, so the demodulated difference frequency sound energy is severely reduced. In 2010, Takeoka et al. [157] first used $\Sigma - \Delta$ high-pass filter to upsample and then integer delay, and then replaced the power amplifier to drive the ultrasonic transducer to emit sound waves with an inverter. Although this method solves the problem of delay accuracy, the signal-to-noise ratio of this method depends heavily on the stability of the clock, and secondly, the output dynamic range is small, resulting in a small sound pressure level of the difference frequency sound. The research group led by Gan at Nanyang Technological University has done a lot of theoretical and experimental work in this area [158–163]. The research group led by Yang [164] applied the Farrow structure fractional delay algorithm to the parametric array loudspeaker, and designed a phased parametric array loudspeaker system to achieve arbitrary beamsteering with fixed filter coefficients.

1.1.4.3 New Transducers

The new transducer, the ultrasonic transducer, as the emission source of the parametric array loudspeaker, has a significant impact on its performance. Factors such as the frequency response, phase, and amplitude consistency of the transducer directly affect the radiation power, output sound pressure level, and distortion of the parametric array loudspeaker. Early parametric array loudspeakers used traditional piezoelectric ceramic ultrasonic transducers (PZT), generally using a parallel method to form a transducer array sound source with multiple ultrasonic transducer units. This method is simple in structure and easy to implement. However, PZT has a large mechanical impedance, resulting in a small bandwidth and low radiation efficiency [165]. Polyvinylidene Difluoride (PVDF), as a new type of piezoelectric high polymer material, has a high piezoelectric constant, light weight, good flexibility, can be bent into any shape, and has a flat frequency response in a wide frequency band. Toda conducted a detailed research on PVDF film, using various different structures (cylindrical, arc, and corrugated, etc.), and carried out a theoretical and experimental research on its resonance frequency, directivity, radiation sound pressure, driving voltage, etc. [166–171], and used PVDF film to make a parametric array loudspeaker for actual measurement [172]. However, the dielectric constant of PVDF film is small, it is difficult to produce high-intensity ultrasonic signals, and it requires a high driving voltage, which puts higher requirements on the power amplifier. In recent years, with the development of MEMS technology, new types of ultrasonic transducers have appeared. Capacitive Micromachined Ultrasonic Transducers (CMUTs), fabricated utilizing CMOS technology, exhibit high sensitivity and a broad bandwidth. They are characterized by their compact form factor and low intrinsic noise. The compatibility of their manufacturing process with CMOS technology enables the integration of the pre-stage driving circuit, preamplifier, signal processing circuit onto a single silicon chip, thereby achieving electronic integration and minimizing parasitic effects. These attributes render CMUTs highly suitable for a range of applications, including ultrasonic non-destructive testing and medical imaging, etc. [173–177]. Wygant et al. [112] designed CMUT transducers with two different membrane thicknesses (40 and 60 μm), with resonant frequencies of 46 kHz and 55 kHz, respectively. When supplied with a 200 VDC bias and 380 V and 350 VAC signals, the transducers were capable of generating 135 dB and 129 dB ultrasonic signals at their surfaces. When used to produce a 5 kHz difference frequency signal, the sound pressure level measured at a distance of 3 m was 58 dB, with a 6 dB bandwidth of 8.7°. Zhang et al. [178] designed and manufactured an air-coupled hexagonal CMUT array for the application of the parametric array loudspeaker. The piezoelectric Micro-machined Ultrasonic Transducer (pMUT) is composed of a vibrating membrane made of a certain material and a piezoelectric thin film with upper and lower electrodes. The mechanical vibration of the vibrating membrane and the piezoelectric effect of the piezoelectric thin film can interact to realize the mutual conversion of sound energy and electrical energy. Since the reception and transmission of silicon micro piezoelectric ultrasonic transducers are both in the bending vibration mode, their working frequency is mainly determined by the first-order vibration frequency of the bending vibration of the vibrating membrane. Therefore, the resonance frequency of the transducer can be changed by changing the thickness and area of its vibrating membrane. In addition, silicon micro piezoelectric ultrasonic transducers also have the advantages of low power consumption, small volume, and easy integration [179–183]. Lee et al. [184] developed a pMUT transducer with a size of $35 \times 33\ mm^2$, with primary wave frequencies of 95 and 135 kHz, producing a difference frequency sound of 40 kHz. The experimental measurement at 0.8 m showed a 6 dB beam angle of 5° and a maximum axial sound pressure level of 85.4 dB. The development of new ultrasonic transducers has greatly promoted the development of parametric array loudspeakers. Je et al. [185] designed a more efficient (up to 70%) and wider bandwidth (15 kHz) pMUT array. Ahn et al. [186] used a pMUT array to design a compact parametric array loudspeaker.

1.1.4.4 Measurement Issue

Measurement problems arise because parametric array loudspeakers are different from ordinary loudspeakers, and ultrasonic primary waves coexist with the demodulated audio frequency wave, which puts higher requirements on the accurate measurement of parametric array loudspeakers. When two ultrasonic waves with frequencies f_1 and f_2 ($f_1 > f_2$) are emitted by the ultrasonic transducer, due to the nonlinear effect in the air, multiple different frequency sound waves can be reproduced, that is, the difference of these two frequencies ($f_1 - f_2$) and the sum of the two frequencies ($f_1 + f_2$), as well as higher-order harmonics. The sound wave attenuation coefficient is almost directly proportional to the square of the frequency, causing the high-frequency components to attenuate faster during propagation. Therefore, after a certain distance of propagation, only the difference frequency sound wave remains, forming the audio frequency. However, in the near-field of the parametric loudspeaker, there are both ultrasonic waves p_1 and p_2 and the audio frequency p_d. Because the sound pressure level of the ultrasonic waves p_1 and p_2 is high, especially in the near field, it is generally above 120 dB, so when measuring directly with a loudspeaker in this area, spurious sound will be also generated. The spurious sound comes from two aspects, one is due to the nonlinear effect of the microphone pre-amplification circuit, and the other is due to the radiated sound pressure on the surface of the loudspeaker. The frequency values of the above spurious sound generated is the same as the demodulated audio frequency wave. The amplitude of the spurious sound is proportional to the product of the amplitudes of the ultrasound waves p_1 and p_2, that is, $p_n \propto p_1 p_2$. It is apparent that the spurious sound greatly interferes with the measurement of the actual demodulated audio frequency. Especially in the near field area, the spurious sound is significantly larger than the sound frequency generated by the parametric loudspeaker. Therefore, it is necessary to design a sound filter to filter out the spurious sound and obtain the amplitude of the actual demodulated audio frequency.

In 1975, Bennett and Blackstock [93] first realized the experiment of a parametric loudspeaker in the air, using a hemispherical sound filter made of 0.06 mm thick plastic glass paper, placed at the front end of the measurement transducer, attenuating the difference frequency sound (5 kHz) by 3.5 dB, but attenuating the ultrasonic primary wave (18.6 and 23.5 kHz) by 20 dB, achieving low-pass characteristics, thus effectively reducing spurious sound. Moffett et al. [187] studied the nonlinear problem of measuring hydrophones in water in 1982. Toda [172] designed a tubular sound filter with a four-layer polymer film structure, where the spacing between the four films is half the carrier wavelength (carrier frequency is 34 kHz), and the maximum attenuation of the ultrasonic carrier is up to 30 dB. However, this sound filter structure is too complex and not conducive to practical application. Wygant et al. [112] used a layer of Celen synthetic fiber as a sound filter when using CMUT transducers as the emission source for parametric array loudspeakers, but did not provide a detailed description of the structure and performance of the filter. Kamakura et al. [188, 189] used two sets of primary waves with opposite phases as the driving signals for the parametric array loudspeaker to reduce the sound pressure level of the primary wave, and it has small effect on the generated difference frequency sound. This method can also be used to reduce spurious sound. Yang's research group used this phase cancellation method to measure the sound pressure on the axis [190]. Cylindrical filters [191] and filters based on phononic crystals [192, 193] were used to reduce the impact of spurious sound on measurements. On this basis, the effects of four main parameters on spurious sound, namely the intensity of the primary wave, the difference frequency, the radius of the sound source, and the observation distance, were investigated [194].

1.1.4.5 Research on Applications

As a new type of directional sound source, parametric array loudspeakers have a wide range of application prospects, and there are already many parametric array loudspeaker products and their related applications on the market. Nakashima et al. [195, 196] installed two small parametric array loudspeakers on a mobile phone to form a dual-channel system. The audio frequency at 50 cm reached 70 dB in experimental measurements, and the sound pressure level difference at both ears was 15 dB using a head and torso simulator (HATS), indicating a good directivity. Nakadai and Tsujino [197] installed a parametric array loudspeaker in the mouth of a humanoid robot, achieving human–robot interactive voice communication. When a person and the humanoid robot speak at the same time, the robot can pick up voice signals through the loudspeaker without being disturbed by its own voice signals, realizing the function of "listening while speaking". Johannes et al. [198] used a parametric array loudspeaker to achieve 3D sound reproduction, reducing interaural crosstalk. Phanomchoeng et al. used the high-directivity characteristics of the parametric array at low-frequency to warn specific lanes on highways without affecting the normal driving of vehicles in other directions, ensuring the smooth progress of construction on that lane [199]. Castagnede et al. [200,201] used a parametric array loudspeaker as a sound source to measure the absorption coefficient of materials using a single transducer pulse reflection method. Haupt and Rolt from the Lincoln Laboratory in the United States [202] used a parametric array loudspeaker from Holosonics to detect landmines buried at different depths below the ground, and discussed the indicators that a parametric array loudspeaker suitable for landmine detection should have from aspects such as power, frequency, and volume. Calicchia et al. [203] used a parametric array loudspeaker to achieve non-destructive testing of Italian Renaissance murals.

Gibson et al. [204] designed and built a parametric array loudspeaker consisting of an ultrasonic transducer array, forming a high-intensity, collimated low-frequency sound to achieve sound penetration of small areas/small targets and induce their vibration. Achanta et al. [205] and Kaduchak et al. [206] used a parametric array loudspeaker to achieve long-distance portable hidden weapon detection. Sayin et al. took a different approach and explored the feasibility of using a parametric array loudspeaker with 750 ultrasonic transducer units to construct an omnidirectional sound source [207]. They found that its performance at high frequencies is better than traditional dodecahedron sound sources, and it is smaller and lighter. Arnela et al. further improved the omnidirectional parametric array loudspeaker based on the theoretical model, making it easier to manufacture and install [208].

Compared to foreign countries, domestic research on parametric array loudspeakers started relatively late, and is mostly focused on the development and experimental research of parametric array loudspeakers [110, 209–219], with relatively fewer reports on application research [220–222]. The research group led by Yang at Institute of Acoustics, Chinese Academy of Sciences, has done a lot of research work on the theoretical foundation, measurement, engineering development, and application research of parametric array loudspeakers, and has achieved fruitful results. In theoretical research, a new calculation method of Gaussian beam expansion coefficients was proposed [223], and the sound field of multiple beams crossing at small angles was calculated using the Gaussian beam expansion method [82, 83], and various preprocessing algorithms for parametric array loudspeakers were discussed in detail [133, 138, 143–146, 224, 225]. In terms of measurement, several methods to remove or reduce spurious sound were proposed [190–192], and the parameters affecting spurious sound were systematically studied [194]. In terms of engineering development, research was conducted on the new ultrasonic transducer PVDF film [226], multi-path parametric array loudspeakers [227], and impedance matching of transducer arrays [228, 229], and several prototypes were successively introduced. In terms of application research, field measurement research on the sound absorption coefficient of materials [230] and active noise reduction control research [231, 232] were conducted using parametric array loudspeakers. The research group led by Xu at the University of Electronic Science and Technology conducted research on parametric array loudspeakers from the aspects of preprocessing algorithms [233], ultrasonic transducers [234], and hardware implementation [235, 236]. In addition, the research group led by Zhao at Shandong University of Science and Technology provided a design scheme for implementing parametric array loudspeakers using DSP [237, 238]. Yi et al. at the Jiaxing Engineering Center of Institute of Acoustics, Chinese Academy of Sciences [239] also conducted system design and testing of parametric array loudspeakers.

1.2 Fundamentals of Nonlinear Acoustics

The basics of nonlinear acoustics will be discussed in the following sections, focusing on the equations used in nonlinear acoustics. The review will start with the propagation of acoustic waves in lossless fluids, which provides the basic principles of nonlinear acoustics.

1.2.1 Propagation of Finite Amplitude Waves in Non-Attenuating Fluids

This section first introduces the basic part of nonlinear acoustics, that is, the propagation of finite amplitude waves in non-attenuating fluids.

Currently, there are four main equations used to describe the general motion of viscous heat-conducting fluids: ① mass conservation equation, ② momentum conservation equation, ③ entropy balance equation, and ④ thermodynamic state equation. To derive the wave equation for sound propagation in fluids, it is assumed that the fluid is homogeneous and the effects of viscous and thermal conductivity coefficients caused by sound wave disturbances are ignored.

The mass conservation equation is as follows

$$\frac{D\rho}{Dt} + \rho\nabla \cdot u = 0 \tag{1.5}$$

where ρ is the medium density, u is the fluid velocity vector, ∇ is the gradient operator, and $D/Dt = \partial/\partial t + u \cdot \nabla$.

The momentum conservation equation can be written as

$$\rho\frac{Du}{Dt} + \nabla P = \mu\nabla^2 u + \left(u_B + \frac{1}{3}\mu\right)\nabla(\nabla \cdot u) \tag{1.6}$$

where P is the thermodynamic pressure, μ is the shear viscosity coefficient, ∇^2 is the Laplace operator, u_B is the bulk viscosity coefficient. The shear viscosity coefficient refers to the momentum diffusion between adjacent fluid elements with different speeds. The bulk viscosity coefficient provides an effective approximation at low frequencies, that is, the imbalance deviation between the actual local pressure and the thermodynamic pressure. Here, for convenience, it is assumed that all relaxation times are much smaller than the time scale of sound disturbances.

The entropy equation can be expressed as

$$\rho T\frac{Ds}{Dt} = \kappa\nabla^2 T + \mu_B(\nabla \cdot u)^2 + \frac{1}{2}\mu\left(\frac{\partial u_i}{\partial x_j} + \frac{\partial u_j}{\partial x_i} - \frac{1}{2}\delta_{ij}\frac{\partial u_k}{\partial x_k}\right)^2 \tag{1.7}$$

where T is the absolute time, s is the specific entropy, κ is the thermal conductivity, δ_{ij} is the Kronecker delta function, that is, when $i = j$, it is 1, and 0 in all other cases.

The equation of state is

$$P = P(\rho, s) \tag{1.8}$$

For an ideal gas, $P/\rho T$ and the heat capacity ratio are constants, the above equation can be written as

$$\frac{P}{P_0} = \left(\frac{\rho}{\rho_0}\right)^{\gamma} \exp\left(\frac{s - s_0}{c_v}\right) \tag{1.9}$$

where P_0, ρ_0 and s_0 are the corresponding equilibrium values, γ is the ratio of specific heat at constant pressure c_p to that at constant volume c_v. To derive the sound propagation formula, the above equation uses the Taylor formula to expand at the point (ρ_0, s_0).

If the fluid under consideration is lossless, μ, μ_B and κ are all zero. This allows us to obtain the exact solutions of the above formulas, which are applicable to the propagation of plane waves in lossless fluids. Specifically, for the problem of plane wave propagation in an ideal isentropic gas, a simple formula can be derived as follows

$$\frac{\partial u}{\partial t} + (c_0 + \beta u)\frac{\partial u}{\partial x} = 0 \tag{1.10}$$

where $u(x, t)$ is the velocity of the gas, x is the spatial variable, t is the time, $c_0 = (\gamma P_0/\rho)^{1/2}$ is the speed of small signal sound, β is a nonlinear coefficient.

To obtain the sound field under a specific excitation at the sound source $u = f(x)$, that is, the signal at the origin is

$$u(0, t) = f(t) \tag{1.11}$$

The implicit solution satisfying the above equation is obtained by Poisson as follows

$$u = f(t - x/(c_0 + \beta u)) \tag{1.12}$$

The above equation is known as the Poisson solution. Another implicit solution is obtained by Earnshaw. Assume that the sound wave is excited by a piston with the finite amplitude. If the piston displacement is specified as $X(t)$, the sound source condition is: when $x = X(t)$, $u = \dot{X}(t)$, where $\dot{X}(t) = \frac{dX}{dt}$. At the moment ϕ, the piston is at $x = X(\phi)$, and the speed is $\dot{X}(\phi)$. This speed is applied to the fluid causing disturbance, that is, the fluid particle speed at this wave point is $u = \dot{X}(\phi)$. Thus, the Earnshaw solution is obtained

$$u = \dot{X}(\phi), \quad \phi = t - \frac{x - X(\phi)}{c_0 + \beta \dot{X}(\phi)} \tag{1.13}$$

For example, let the sound source excitation be a sine wave, i.e., $u_0 \sin\omega t$, where u_0 is the amplitude, ω is the angular frequency. If a sinusoidal excitation is applied at the origin, $u(0, t) = u_0 \sin\omega t$, then the Poisson solution is

$$u = u_0 \sin\omega\left(t - \frac{x}{c_0 + \beta u}\right) \tag{1.14}$$

When the sound source is a vibrating piston, the boundary condition is: $u = u_0$, at $x = \frac{u_0}{\omega}(1 - \cos\omega t)$

As solved by Earnshaw

$$u = u_0 \sin\omega\phi, \quad \phi = t - \frac{x - u_0/[\omega(1 - \cos\omega\phi)]}{c_0 + \beta u_0 \sin\omega\phi} \tag{1.15}$$

Once the particle velocity is obtained, the sound pressure p can be calculated using the characteristic impedance relationship as follows

$$\frac{dP}{du} = \frac{dp}{du} = \rho c \tag{1.16}$$

where P is the total pressure of the fluid. For small signal waves, the right side of the above equation has a constant value $\rho_0 c_0$, and $p = \rho_0 c_0 u$ is obtained. For finite amplitude waves, the relationship is not simple. For adiabatic gas, after integration and a series expansion, the following equation can be obtained

$$p = \rho_0 c_0^2 \left[\frac{u}{c_0} + \frac{\beta}{2}\left(\frac{u}{c_0}\right)^2 + \frac{\beta}{6}\left(\frac{u}{c_0}\right)^3 + \ldots\right] \tag{1.17}$$

1.2.2 Approximation of Thermoviscous Fluids

1.2.2.1 Second-Order Approximation Theory

The second-order approximation theory of the thermal viscous fluid approximation finite amplitude is affected by two different nonlinear effects: cumulative and local effects. In the case of traveling waves, the cumulative effect is generally dominant, that is, it accumulates with the increase of propagation distance, leading to the steepening of the waveform. The distortion that occurs at a location is based on all previous distortions and continues to expand. The local nonlinear effect, on the other hand, produces distortion that does not increase with the propagation distance. An example of the latter is the nonlinearity of the characteristic impedance relationship. If the particle velocity waveform has been calculated, the pressure waveform is obtained by Eq. (1.17). The linear term in the formula can obtain a pressure waveform with the same cumulative distortion as the particle velocity waveform. Since the remaining terms in Eq. (1.17) are at least smaller than the linear term by $O(\varepsilon)$ (when a term is determined to be $O(\varepsilon^n)$ or smaller to ε^n, for very small ε, its size is $\alpha\varepsilon^n$, where α is a finite positive number), so the additional distortion it introduces is very small. Moreover, since the additional distortion only depends on the local waveform, it often remains unchanged and does not increase with the increase of propagation distance. Therefore, the only place where local distortion dominates is near the sound source, where the cumulative distortion is still very small.

Local effects make the analysis of traveling wave propagation more complex, but their impact on the solution is generally negligible because the distortion they cause is usually much smaller than the distortion resulting from cumulative effects. Therefore, many simplifications can be achieved at a small cost by ignoring local effects. When dissipation must be considered, a certain small value of the second-order approximation is usually introduced to obtain simplified wave equations and solutions. Two small order parameters are usually used in the approximation. One is the acoustic Mach number $\varepsilon = u_0/c_0$ (where u_0 is the typical sound speed amplitude). This value is 154 dB in the air (reference value 2×10^{-5} Pa), and 264 dB in water (reference value 1×10^{-6} Pa). The other small parameter is $\eta = \mu\omega/\rho_0 c_0^2$. In physics, η is an important measure of the significance of viscous stress in plane traveling sound waves, relative to the fluctuating pressure. Under standard conditions, $\eta = 10^{-6}$ for 1 kHz in the air and 1 MHz in water can be obtained.

In the analysis of the second-order approximation, expand Eqs. (1.5) to (1.8), and discard all $O(\eta^2\varepsilon)$, $O(\eta\varepsilon^2)$ and $O(\varepsilon^3)$ terms. The resulting model retains the order of ε, $\eta\varepsilon$ and ε^2 terms, describing the small signal sound as the leading order of η. More importantly, it is expected to explain the combined effects of nonlinearity and dissipation on weak nonlinear three-dimensional sound waves.

The basis of the second-order approximation theory is two assumptions. First, the intensity of the wave is not large, that is, $|u| \ll c_0$ (or $\varepsilon \ll 1$). On this basis, although the $O(\varepsilon^2)$ term must be retained, the $O(\varepsilon^3)$ term can be discarded. Secondly, distortion is mainly cumulative. That is to say, the observation point is not close to the sound source. For periodic waves, "not close to the source" means $x \gg \lambda/(2\pi\beta)$, where λ is the wavelength of the fundamental frequency component. For most fluids, this means that x is greater than about one wavelength. The direct consequences of these two assumptions are as follows.

The relationship of linear characteristic impedance can be used. The finite displacement of the source from its rest position may be ignored, that is, for a sinusoidal source, the equation is sufficient to describe the sound source conditions. The difference between material characterization (Lagrangian) and spatial characterization (Eulerian) can be ignored. $O(\varepsilon^2)$ term and $O(\varepsilon)$ factors can be replaced by equivalent traveling waves $O(\varepsilon)$, which will introduce error term $O(\varepsilon^3)$. Substitute $\rho = \rho_0 + \rho'$ into Eq. (1.5), move $O(\varepsilon)$ term to the left side of the equation, and move $O(\varepsilon^2)$ term to the right side of the equation, the equation can be rewritten as

$$\frac{\partial \rho'}{\partial t} + \rho_0 \nabla \cdot u = -\rho' \nabla \cdot u - u \nabla \cdot \rho' \tag{1.18}$$

Equation (1.18) is still an exact solution. Introduce the vector identifier into Eq. (1.6)

$$\nabla(\nabla \cdot u) = \nabla^2 u + \nabla \times \nabla \times u \tag{1.19}$$

and

$$(u \cdot \nabla) \cdot u = \frac{1}{2} \nabla u^2 - u \times \nabla \times u \tag{1.20}$$

where $u^2 = u \cdot u$. Let $P = P_0 + p$, discard $O(\varepsilon^3)$ term, and collect $O(\varepsilon)$ term and $O(\varepsilon^2)$ term on the left and right, respectively. The momentum conservation equation can be written as

$$\rho_0 \frac{\partial u}{\partial t} + \nabla p = \left(\mu_B + \frac{3}{4}\mu\right) \nabla^2 u - \frac{1}{2} \rho_0 \nabla u^2 - \rho' \frac{\partial u}{\partial t} \tag{1.21}$$

The entropy equation can be written as

$$\rho_0 T_0 \frac{\partial s'}{\partial t} = \kappa \nabla^2 T' \tag{1.22}$$

Finally, using the Taylor series to expand the state equation at equilibrium state (ρ_0, s_0) and ignoring $O(\varepsilon^3)$ terms to obtain

$$p = c_0^2 \rho' + \frac{c_0^2}{\rho_0} \frac{B}{2A} \rho'^2 + \left(\frac{\partial P}{\partial s}\right)_{\rho,0} s' \tag{1.23}$$

where $B/A = (\rho_0/c_0^2)(\partial^2 P/\partial p^2)_{s,0}$ is the nonlinear parameter.

The Eqs. (1.18), (1.21), (1.22) and (1.23) represent consistent $O(\varepsilon^2)$ approximations of the mass conservation equation, momentum conservation equation, entropy equation, and state equation from Eqs. (1.5) to (1.8).

1.2.2.2 Second-Order Sound Wave Equation

The second-order sound wave equation integrates the Eqs. (1.18), (1.21), (1.22) and (1.23) into a single second-order sound wave equation. The first term on the right side of Eq. (1.18) can be written as

$$-\rho' \nabla \cdot u = -\left(\frac{p}{c_0^2}\right)\left(-\frac{1}{\rho_0 c_0^4} \frac{\partial p'}{\partial t}\right) = \frac{p}{\rho_0 c_0^4} \frac{\partial p}{\partial t} \tag{1.24}$$

The conservation of mass equation can be rewritten as

$$\frac{\partial \rho'}{\partial t} + \rho_0 \nabla \cdot u = \frac{1}{\rho_0 c_0^4} \frac{\partial p^2}{\partial t^2} + \frac{1}{c_0^2} \frac{\partial \mathcal{L}}{\partial t} \tag{1.25}$$

where $\mathcal{L} = \frac{1}{2}\rho_0 u^2 - \frac{p^2}{2\rho_0 c_0^2}$ is the second-order Lagrangian density.

The momentum conservation equation can be written as

$$\rho_0 \frac{\partial u}{\partial t} + \nabla p = -\frac{1}{\rho_0 c_0^2}\left(\mu_B + \frac{4}{3}\mu\right)\nabla \frac{\partial p}{\partial t} - \nabla \mathcal{L} \tag{1.26}$$

For a propagating plane wave $\mathcal{L} = 0$, that is, the momentum conservation equation is linear, so it does not contribute to finite amplitude effects on $O(\varepsilon^2)$.

The entropy equation and the state equation can be integrated into a formula as follows.

$$\rho' = \frac{p}{c_0^2} - \frac{1}{\rho_0 c_0^4}\frac{B}{2A}p^2 - \frac{\kappa}{\rho_0 c_0^4}\left(\frac{1}{c_v} - \frac{1}{c_p}\right)\frac{\partial p}{\partial t} \tag{1.27}$$

After certain mathematical operations, the desired second-order wave equation is obtained.

$$\square^2 p + \frac{\delta}{c_0^4}\frac{\partial^3 p}{\partial t^3} = -\frac{\beta}{\rho_0 c_0^4}\frac{\partial^2 p^2}{\partial t^2} - \left(\nabla^2 + \frac{1}{c_0^2}\frac{\partial^2}{\partial t^2}\right)\mathcal{L} \tag{1.28}$$

where $\beta = 1 + B/(2A)$ is the nonlinear coefficient, δ is the diffusion coefficient of the thermal viscous medium, defined as follows.

$$\delta = \frac{1}{\rho_0}\left(\frac{4}{3} + \mu + \mu_B\right) + \frac{\kappa}{\rho_0}\left(\frac{1}{c_v} - \frac{1}{c_p}\right) = v\left(\frac{4}{3} + \frac{\mu_B}{\mu} + \frac{\gamma - 1}{\mathrm{Pr}}\right) \tag{1.29}$$

where $v = \mu/\rho_0$ is the kinematic viscosity.

Equation (1.28) applies to the second-order wave equation of weak thermal viscous fluids where $\eta \ll 1$. It accounts for the mutual interaction of nonlinear and dissipative processes in modifying sound propagation in three-dimensional space.

1.3 Theoretical Basis of Parametric Arrays

The foundation of parametric array theory is the directional sound source based on the principle of sound wave interaction, that is, when finite amplitude sound waves of different frequencies propagate in a medium, new sound waves with different frequencies will be generated. The propagating signal is called the primary wave. If only two frequencies and second-order effects are considered, the final sound waves will be the sum frequency sound wave, the difference frequency sound wave and higher-order harmonics. The sum frequency wave and higher-order harmonics will be attenuated quickly, and the main consideration is the difference frequency sound. The difference frequency sound will propagate forward with the primary wave along the sound axis, and continuously generate new difference frequency sounds superimposed with the primary wave during the propagation process. Due to the phase change of the difference frequency sound along the sound axis, the axial sound field is the strongest, so this array is also called the virtual source end-fire array. It can be expected that the difference frequency sound will reach a larger value in the sound axis direction, and in the non-axis direction, due to non-coherent superposition, the amplitude will be very small. This will form a high directivity. If the primary wave is a broadband signal, in general, it can always be represented by a Fourier integral, and the function to be integrated is a single-frequency sine wave. The interaction of two frequencies of infinite many frequency components generates a broadband difference frequency signal, thus forming the so-called broadband parametric array.

Early research on acoustic parametric arrays was basically based on underwater systems, so when applying parametric array loudspeakers in acoustic engineering, it is necessary to consider the impact of water and air as different media on the parametric array model. The parametric array loudspeaker, as a new type of sound source, produces audio frequencies with highly directivity, which is one of its major advantages. It is necessary to study the directivity of the parametric array loudspeaker, so this part further studies the directivity of the parametric array. First, the Westervelt model of the parametric array [22], the KZK equation, and the Berktay model [24] are discussed. An analysis was conducted; then, through simulation experiments, the Berktay model was studied in depth for different results both in water and in the air.

1.3.1 Westervelt Model

When Eq. (1.28) discards the terms containing $\mathcal{L}$, it results in the Westervelt equation, which was discussed by Westervelt [22] in 1963 in relation to the sound emission problem of two coaxial intensive sound beams. The parametric array proposed by Westervelt is a line source parametric array, assuming that the source is distributed on a straight line. In the theoretical derivation process, it is assumed that the medium is an ideal fluid, and the attenuation of the medium to sound waves is only introduced in a "presumptuous" way when solving the wave equation. In addition, the finite amplitude effect was not considered in his study, that is, it was assumed that only small amplitude absorption was considered when discussing the attenuation in the medium, and it was assumed that both the primary wave and the difference frequency sound propagated with small amplitude, and the whole theory actually only took second order approximation.

Lighthill derived the wave equation of fluid mechanics sound emission from the general fluid mechanics equations. Starting from the famous Lighthill wave equation

$$\nabla^2\rho - \frac{1}{c_0^2}\frac{\partial^2\rho}{\partial t^2} = -\frac{1}{c_0^2}\frac{\partial^2 T_{ij}}{\partial x_i \partial x_j} \tag{1.30}$$

Ignore viscous stress to obtain

$$T_{ij} = \rho \upsilon_i \upsilon_j + \left(P - \rho c_0^2\right)\delta_{ij} \tag{1.31}$$

Substitute T_{ij} into Eq. (1.30), and change it to the wave equation represented by P, that is

$$\nabla^2 P - \frac{1}{c_0^2}\frac{\partial^2 P}{\partial t^2} = \frac{1}{c_0^2}\frac{\partial^2}{\partial t^2}\left(\rho c_0^2 - P\right) - \frac{\partial^2}{\partial x_i \partial x_j}\left(\rho \upsilon_i \upsilon_j\right) \tag{1.32}$$

For the fluid undergoing irrotational motion, $\nabla \times V = 0$, accurate to second order approximation, it is obtained

$$\frac{\partial^2}{\partial x_i \partial x_j}\left(\rho \upsilon_i \upsilon_j\right) = \rho_0\left[(\nabla \cdot V)^2 + V \cdot \nabla^2 V + \nabla\left(\frac{1}{2}V^2\right)\right] \tag{1.33}$$

Applying the equation satisfied by the first-order field, the above equation can be transformed into

$$\frac{\partial^2}{\partial x_i \partial x_j}\left(\rho \upsilon_i \upsilon_j\right) = \rho_0^{-1} c_0^{-2}\left[-\Box^2\left(\frac{1}{2}\rho^2 + \frac{1}{2}\rho_0^2 c_0^{-2} V^2\right) + \nabla^2\left(\rho_0^2 c_0^{-2} V^2\right)\right] \tag{1.34}$$

In addition, the expansion is approximated to second order.

$$\begin{aligned}
\rho - c_0^{-2}P &\approx \rho - c_0^{-2}\left[P_0 + \left(\frac{\partial P}{\partial \rho}\right)_{\rho=\rho_0}(\rho - \rho_0) + \frac{1}{2}\left(\frac{\partial^2 P}{\partial \rho^2}\right)_{\rho=\rho_0}(\rho - \rho_0)^2\right] \\
&= -\frac{1}{2}c_0^{-2}\left(\frac{\partial^2 P}{\partial \rho^2}\right)_{\rho=\rho_0}(\rho - \rho_0)^2 c_0^{-2} \\
&= \frac{1}{2}c_0^{-6}\left(\frac{\partial^2 P}{\partial \rho^2}\right)_{\rho=\rho_0} p^2
\end{aligned} \tag{1.35}$$

Here

$$P = P_0 + p + p_s$$

Substituting these relationships into the equation, note that the first term within the curly brackets on the right side of the equation does not constitute a source function, because the quantity operated by the D'Alembert operator $\Box^2$ is equivalent to a solution of the homogeneous wave equation. Therefore, it is discarded or merged into the homogeneous solution. As a result, the sum and difference frequency sound pressure p_s should satisfy

$$\Box^2 p_s = -\frac{1}{2} c_0^{-6} \left(\frac{\partial^2 P}{\partial \rho^2} \right)_{\rho=\rho_0} \frac{\partial^2 p^2}{\partial t^2} - \rho_0 \nabla^2 V_1^2 \tag{1.36}$$

where p, V_1 are respectively the first-order pressure (sound pressure) and the first-order particle velocity vector. To eliminate V_1, assume that V_1 and p satisfy the plane wave impedance relationship, that is

$$V_1^2 = \frac{p^2}{\rho_0^2 c_0^2} \tag{1.37}$$

Substitute it into the wave equation to yield

$$\Box^2 p_s = -\rho_0 \frac{\partial q}{\partial t} \tag{1.38}$$

where

$$q = \rho_0^{-2} c_0^{-4} \left[1 + \frac{1}{2} \rho_0 c_0^{-2} \left(\frac{\partial^2 P}{\partial \rho^2} \right)_{\rho=\rho_0} \right] \frac{\partial p^2}{\partial t^2} \tag{1.39}$$

Under the quasi-linear approximation (only considering one scattering), the amplitude the difference frequency sound p_- satisfies the following inhomogeneous wave equation.

$$\nabla^2 p_- - \frac{1}{c_0^2} \frac{\partial^2 p_-}{\partial t^2} = -\rho_0 \frac{\partial q}{\partial t} \tag{1.40}$$

where

$$q = \frac{\beta}{\rho_0^2 c_0^4} \frac{\partial p_i^2}{\partial t} \tag{1.41}$$

$$\beta = 1 + \frac{1}{2} \rho_0 c_0^{-2} \left(\frac{d^2 p}{d \rho^2} \right)_{\rho=\rho_0} \approx 1 + \frac{B}{2A} \tag{1.42}$$

p_i is the instantaneous sound pressure of the primary wave, q is the source strength per unit volume generated by the nonlinear interaction of two primary waves, known as the source strength density. The Westervelt equation is an accurate approximation of the full second-order wave equation, where cumulative nonlinear effects dominate local nonlinear effects. Generally, except within a wavelength from the sound source,

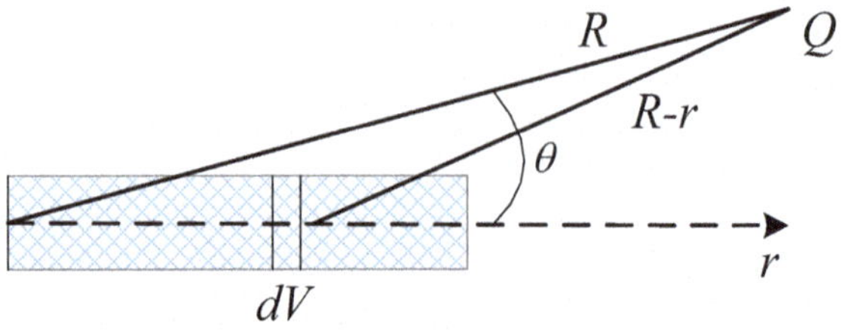

Fig. 1.3 Schematic diagram of parametric array

local effects can be ignored for problems involving quasi-plane waves such as sound beams. The general solution of the inhomogeneous wave equation can be written in the form of a volume integral,

$$p_-(\mathbf{R}, t) = -\frac{\rho_0}{4\pi} \int_V \frac{\partial q}{\partial t} \frac{\exp\left(ik_- \left|\mathbf{R} - \mathbf{r}'\right|\right)}{\left|\mathbf{R} - \mathbf{r}'\right|} dV \tag{1.43}$$

$\mathbf{R}$ is the position vector from the origin to the field point, $\mathbf{r}$ is the position vector from the origin to the volume element in the volume integral region, dV, and $\mathbf{R} - \mathbf{r}'$ is the position vector from the integral volume element dV to the field point, as shown in Fig. 1.3 [241].

Assume the primary wave p_i is the superposition of two finite amplitude waves with the same propagation direction and frequencies of ω_1 and ω_2, its sound pressure is

$$p_i = P_0 e^{-\alpha x}[\cos(\omega_1 t - k_1 x) + \cos(\omega_2 t - k_2 x)] \tag{1.44}$$

where $\omega_1 \approx \omega_2$, $|\omega_1 - \omega_2| << \omega_1$ or ω_2, $\alpha = \frac{1}{2}(\alpha_1 + \alpha_2)$, $k_{1,2} = \omega_{1,2}/c_0$, x is the distance along the sound axis r direction. This formula represents the superposition of two equal amplitude plane waves with equal sound absorption coefficient α. Substitute it into Eq. (1.41) to yield

$$\begin{aligned} q = & \frac{\beta}{\rho_0^2 c_0^4} P_0^2 e^{-2\alpha x} \frac{\partial}{\partial t}\left[\cos^2(\omega_1 t - k_1 x) + \cos^2(\omega_2 t - k_2 x)\right. \\ & \left. + 2\cos(\omega_1 t - k_1 x)\cos(\omega_2 t - k_2 x)\right] \end{aligned} \tag{1.45}$$

Only consider the difference frequency sound, so in q only retain the terms related to the difference frequency $\omega_- = |\omega_1 - \omega_2|$.

$$q = -i\omega_- \frac{\beta}{\rho_0^2 c_0^4} P_0^2 e^{-2\alpha x} \exp(-i\omega_- t + ik_- x) \tag{1.46}$$

Substitute the above equation into Eq. (1.43) to obtain the sound pressure of the difference frequency sound.

$$p_-(\mathbf{R},t) = -\frac{\rho_0}{4\pi}\int\limits_V \frac{\partial q}{\partial t}\frac{\exp(ik_-|\mathbf{R}-\mathbf{r}'|)}{|\mathbf{R}-\mathbf{r}'|}dV$$
$$= \frac{\omega_-^2 P_0^2 \beta}{4\pi\rho_0 c_0^4}\int\limits_V \exp(ik_- x - 2\alpha x)\frac{\exp(ik_-|\mathbf{R}-\mathbf{r}'|)}{|\mathbf{R}-\mathbf{r}'|}dVe^{-i\omega_- t} \quad (1.47)$$

Doing so actually assumes that the difference frequency sound is only generated by the difference frequency term in the source intensity density, that is, the second order scattering effect is ignored. If it is assumed that the beams of the two high-frequency plane primary waves are sufficiently narrow, the volume distribution of the wave source can be approximated by a line distribution along the axis of the primary wave, taking the integral volume element $dV = S_0 dx$, where, S_0 is the cross-sectional area of the primary wave beam. The above integral formula shows that the difference frequency sound field is generated by the radiation of the continuously distributed array source along the entire interaction area. This area is bounded by the cross-sectional area S_0, and the distance extended in the sound axis direction (i.e., the upper limit of integration) depends on the small amplitude attenuation coefficient of the primary wave α. Therefore, this virtual source array is called an end-fire parametric array [21, 242]. Considering that the field point Q is in the far field of the parametric array

$$R >> \frac{S_0}{\lambda_-} \quad (1.48)$$

λ_- is the wavelength of the difference frequency sound, $\lambda_- = 2\pi/k_-$. The vector $\mathbf{R}$ is approximately parallel to $\mathbf{R}-\mathbf{r}'$, therefore

$$|\mathbf{R}-\mathbf{r}'| \approx R - x\cos\theta \quad (1.49)$$

Substitute it into the exponent factor of the numerator of the integral. Further approximate in the denominator, and let $|\mathbf{R}-\mathbf{r}'| \approx R$ to yield

$$p_-(R,\theta) = \frac{\omega_-^2 P_0^2 \beta}{4\pi\rho_0 c_0^4}\int\limits_0^l \frac{S_0}{R}e^{ik_- R}\exp\left[i(k_- x - k_- x\cos\theta) - 2\alpha x\right]dx \quad (1.50)$$

The upper limit of the integral is taken as a certain distance l, at which the integrand decreases to a negligible level. In fact, when $l >> 1/(2\alpha)$, the integrand is very small, and l is taken to be infinitely large. Also, $1 - \cos\theta = 2\sin^2\frac{\theta}{2}$, and then the difference frequency sound pressure of the coaxial sound beam is

$$p_-(R,\theta) = \frac{\omega_-^2 P_0^2 \beta S_0}{4\pi\rho_0 c_0^4 R}\int_0^\infty \exp\left[i\left(2k_- \sin^2\frac{\theta}{2} + 2i\alpha\right)x\right]dx$$
$$= \frac{\omega_-^2 P_0^2 \beta S_0}{4\pi\rho_0 c_0^4 R\alpha_T}\frac{\exp[-(\alpha_- + ik_-)R]}{1 + i(2k_-/\alpha_T)\sin^2(\theta/2)} \quad (1.51)$$

where $\alpha_T = \alpha_1 + \alpha_2 - \alpha_-$, and P_0 is the amplitude of primary waves with equal amplitude. If the amplitude values of the two primary waves are not equal, then the above formula is rewritten as

$$p_-(R,\theta) = \frac{\omega_-^2 p_{01} p_{02} \beta S_0}{4\pi\rho_0 c_0^4 R\alpha_T}\frac{\exp[-(\alpha_- + ik_-)R]}{1 + i(2k_-/\alpha_T)\sin^2(\theta/2)}$$
$$= p_W(R,0)D_W(\theta) \quad (1.52)$$

The maximum sound pressure amplitude in the axial direction in the formula is

$$p_W(R,0) = \frac{\omega_-^2 p_{01} p_{02} \beta S_0}{4\pi\rho_0 c_0^4 R\alpha_T} \quad (1.53)$$

The directional function is

$$D_W(\theta) = \frac{1}{1 + i(2k_-/\alpha_T)\sin^2(\theta/2)} \quad (1.54)$$

It is known as the Westervelt directional function. The −3 dB beam width is

$$\theta_W \cong \sqrt{2\alpha_T/k_-} \quad (1.55)$$

From Eq. (1.54), when θ increases, the difference frequency sound pressure monotonically decreases, and sidelobes will not appear. Therefore, compared with ordinary piston transducers, the parametric array not only has sharp directivity but also can avoid the appearance of sidelobes. The above-mentioned directivity function $D_W(\theta)$ is obtained in cylindrical coordinates. In the general coordinate system shown in Fig. 1.4, (R, θ, η) and (r, γ, ϕ) are the coordinates of the field point and the source point, respectively. Then the difference frequency sound directivity function is

$$D_W(\theta,\eta) = \frac{1}{1 + i\left(\frac{k_-}{2\alpha_T}\right)(\theta^2 + \eta^2)} \quad (1.56)$$

The Burgers equation is a one-dimensional sound wave equation, which only considers the propagation of plane waves under the effects of nonlinearity and absorption. When the Westervelt equation considers sound absorption, the equation becomes

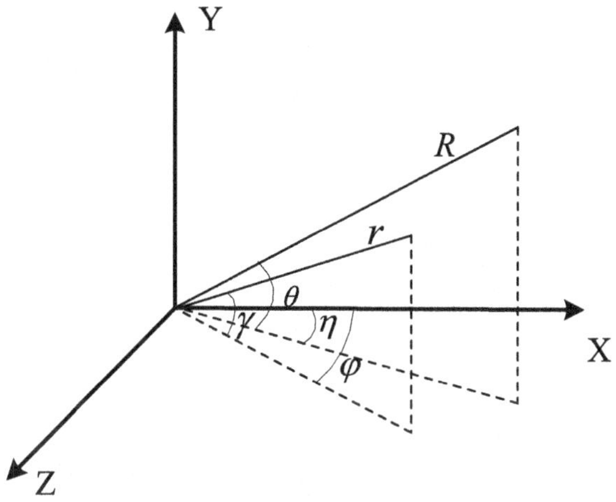

Fig. 1.4 Schematic diagram of sound source coordinate system

$$\Box^2 p + \frac{\delta}{c_0^4}\frac{\partial^3 p}{\partial t^3} = -\frac{\beta}{\rho_0 c_0^4}\frac{\partial^2 p^2}{\partial t^2}$$

If only the one-dimensional Westervelt equation is considered, the above equation becomes

$$\left(\frac{\partial^2}{\partial x^2} - \frac{1}{c_0^2}\frac{\partial^2}{\partial t^2}\right)p + \frac{\delta}{c_0^4}\frac{\partial^3 p}{\partial t^3} = -\frac{\beta}{\rho_0 c_0^4}\frac{\partial^2 p^2}{\partial t^2}$$

To obtain the Burgers equation, a new coordinate system is introduced. The new variables are

$$p = p(x_1, \tau), x_1 = \varepsilon x, \tau = t - x/c_0$$

where τ is the delay time. Nonlinearity and absorption produce only slow variables of the distance function, respectively. Then, substitute the new variables into the above formula to obtain

$$\varepsilon^2\frac{\partial^2 p}{\partial x_1^2} - \varepsilon\frac{2}{c_0}\frac{\partial^2 p}{\partial x_1 \partial \tau} + \frac{\delta}{c_0^4}\frac{\partial^3 p}{\partial \tau^3} = -\frac{\beta}{\rho_0 c_0^4}\frac{\partial^2 p^2}{\partial \tau^2}$$

Integrate with respect to τ and discard the relatively small first term to obtain

$$\varepsilon\frac{\partial p}{\partial x_1} - \frac{\delta}{2c_0^3}\frac{\partial^2 p}{\partial \tau^2} = \frac{\beta}{2\rho_0 c_0^3}\frac{\partial p^2}{\partial \tau}$$

Finally, substitute x, integrate with respect to τ, and discard the relatively small third term.

$$\frac{\partial p}{\partial x} - \frac{\delta}{2c_0^3}\frac{\partial^2 p}{\partial \tau^2} = \frac{\beta}{\rho_0 c_0^3}\frac{\partial p^2}{\partial \tau} \tag{1.57}$$

This leads to the Burgers equation. The Burgers equation is a one-dimensional model that can illustrate the effects of dissipation and nonlinearity on propagating plane waves.

1.3.2 KZK Equation and Berktay Model

1.3.2.1 KZK Equation

In the 1970s, the Khokhlov-Zabolotskaya-Kuznetsov (KZK) equation was proposed by Khokhlov, Zabolotskaya, and Kuznetsov from the former Soviet Union [235, 239]. The KZK equation is an extension of the Burgers equation [22, 243], which includes the effects of diffraction, absorption, and nonlinearity in directional sound sources. It can be derived by introducing a slow scale and operating partial derivatives in a manner similar to the derivation of the Burgers equation. For a detailed derivation process, see reference [1]. It is defined in the plane $z = 0$, it has a characteristic radius a, and it radiates at frequencies that satisfies the relationship $ka >> 1$, where k is the wavenumber, $k = \omega/c_0$. So the beam has reasonable directivity. The KZK equation is described as

$$\frac{\partial^2 p}{\partial z \partial \tau} = \frac{c_0}{2}\nabla_\perp^2 p + \frac{\delta}{2c_0^3}\frac{\partial^3 p}{\partial \tau^3} + \frac{\beta}{2\rho_0 c_0^3}\frac{\partial^2 p^2}{\partial \tau^2} \tag{1.58}$$

where p represents the sound pressure; the operator $\nabla_\perp^2$ is the Laplacian in the plane perpendicular to the axis of the beam. For non-axisymmetric beams, there exists $\nabla_\perp^2 = \partial^2/\partial x^2 + \partial^2/\partial y^2$. The first term on the right side of the equation represents the diffraction effect, the second term represents the sound absorption effect, and the third term represents the nonlinear effect. The equation includes three aspects: sound diffraction, sound absorption, and nonlinear effects. If the diffraction effect is not considered, the equation can be transformed into the Burgers equation. The KZK equation is used for the theoretical prediction of parametric array sound fields. It can accurately describe the self-demodulation process of the primary wave from the near field to the far field, including the situations at various points on the sound beam axis and outside the axis. The KZK equation is also applicable to the sound field situation where two beams intersect at a small angle. The original KZK equation does not include relaxation effects, so it is only applicable to thermoviscous absorption media. However, actual media (such as air, seawater [244], etc.) often contain relaxation effects, so the extended version of the KZK equation is generally used

$$\frac{\partial p}{\partial z} = \frac{c_0}{2}\int_{-\infty}^{t'}\left(\frac{\partial^2 p}{\partial x^2} + \frac{\partial^2 p}{\partial y^2}\right)dt'' + \frac{\delta}{2c_0^3}\frac{\partial^2 p}{\partial t'^2} + \frac{\beta}{2\rho_0 c_0^3}\frac{\partial p^2}{\partial t'}$$

$$+\sum_{\nu}\frac{c'_{\nu}}{c_0^2}\int_{-\infty}^{t'}\frac{\partial^2 p}{\partial t''^2}\exp\left(-\frac{t'-t''}{t_{\nu}}\right)dt'' \quad (1.59)$$

where c'_{ν} and t_{ν} are the small signal sound speed increment and relaxation time of the relaxation process $\nu(\nu = 1, 2, ...)$, respectively. The right side of Eq. (1.59) represents the diffraction effect, thermoviscous absorption effect, nonlinear effect, and relaxation effect in turn.

The sound absorption in fluids [245–248] mainly comes from the energy loss in the continuous collision and energy exchange of molecular motion. Molecular motion is divided into three types: translation, rotation, and vibration. All molecules have translational motion, with three degrees of freedom. Multi-atom molecules also have rotation and vibration, with two more degrees of freedom (diatomic molecules) or more, and single-atom molecules only have translation. According to the law of equal partition of energy, the average energy of each degree of freedom is the same. Molecular translation and rotation are manifested as viscosity and heat conduction on a macro scale, and this absorption is called classical absorption, which is proportional to the square of the frequency. Each vibration has its own natural frequency, and the sound absorption caused by these vibrations is called relaxation absorption. For single-atom molecules (such as argon, helium, etc.), some viscous liquids (such as glycerin) and liquid metals (such as mercury), the absorption is only classical absorption, and the experimental results are completely consistent with the theory. For multi-atom molecule gases and most liquids, it is much larger than classical absorption. The sound attenuation coefficient measured in the air is tens or hundreds of times larger than the classical value (especially in the audio frequency range), and the relaxation attenuation is considerable. In the measured results, there are two relaxation frequencies that are quite far apart. The two relaxation frequencies in the air are: oxygen relaxation frequency f_{rO} and nitrogen relaxation frequency f_{rN}. f_{rO} and f_{rN}, both in Hz, are calculated by

$$f_{rO} = \frac{p_a}{p_r}\left(24 + 4.04 \times 10^4 h\frac{0.02 + h}{0.391 + h}\right) \quad (1.60)$$

and

$$f_{rN} = \frac{p_a}{p_r}(T/T_0)^{-1/2}\{9 + 280h \times \exp\{-4.170[(T/T_0)^{-1/3} - 1]\}\} \quad (1.61)$$

where p_a, p_r are the environmental atmospheric pressure and the reference environmental atmospheric pressure respectively, T, T_0 are the environmental atmospheric temperature and the reference atmospheric temperature respectively, and h is the absolute humidity (molar concentration of water vapor) in %.

The air absorption attenuation coefficient is measured in Np/m, which can be expressed as

$$\alpha(f) = f^2 \left\{ \left[1.84 \times 10^{-11} \left(\frac{p_a}{p_r} \right)^{-1} (T/T_0)^{1/2} \right] + (T/T_0)^{-5/2} \left[0.01275 e^{-2239.1/T} \times \left(f_{rO} + f^2/f_{rO} \right)^{-1} + 0.1068 e^{-3352.0/T} \left(f_{rN} + f^2/f_{rN} \right)^{-1} \right] \right\} \tag{1.62}$$

For the sound wave with frequency f the total sound absorption coefficient [248] can also be expressed as

$$\alpha(f) = \alpha_0(f) + \sum_\nu \alpha_\nu(f) = \frac{\delta}{2c_0^3}(2\pi f)^2 + \sum_\nu \frac{c'_\nu}{c_0^2} \frac{2\pi f^2 f_\nu}{f^2 + f_\nu^2} \tag{1.63}$$

where $f_\nu = 1/(2\pi t_\nu)$ is the relaxation frequency, α_0 is the thermal viscous absorption coefficient, and α_ν is the absorption coefficient of the νth relaxation process.

1.3.2.2 Berktay Model

For many practical application of parameter arrays, the interaction process of the primary waves may occur in the far field, where the directivity of the primary waves in the far field must be considered [243], and the primary waves cannot be simply approximated as collimated beam sound waves. Assume that the transducer is placed on the yOz plane, and the sound axis coincides with the x axis, as shown in Fig. 1.4. The expressions of the primary wave signals with frequencies ω_1 and ω_2 are

$$p_{1,2} = \frac{P_{1,2}}{r} D_{1,2}(\gamma, \varphi) \exp\left[-(\alpha_{1,2} + ik_{1,2}) r \right] \tag{1.64}$$

where $\alpha_{1,2}$ and $k_{1,2}$ are the attenuation coefficients and wave numbers at the two primary wave frequencies, $P_{1,2}$ is the sound pressure amplitude at unit distance on two frequencies, and $D_{1,2}(\gamma, \varphi)$ is the normalized directivity function. In Fig. 1.4, (R, θ, η) and (r, γ, ϕ) are the spherical coordinates of the field point and source point, respectively.

The total sound pressure of the primary wave $p_i = p_1 + p_2$, the total source strength density is $\frac{\beta}{\rho_0^2 c_0^4} \frac{\partial}{\partial t} (p_1 + p_2)^2$. Only considering the source strength density q related to the difference frequency sound $\omega_- = |\omega_1 - \omega_2|$, then

$$q = \frac{i\beta P_1 P_2 D_1(\gamma,\varphi) D_2(\gamma,\varphi)\omega_-}{\rho_0^2 c_0^4 r^2} \exp\left[-(\alpha_1+\alpha_2+ik_-)r\right] \tag{1.65}$$

Substitute q into the solution in the form of volume integral, and consider the exponential decay factor of the difference frequency sound $e^{-\alpha_-|R-r|}$

$$p_-(R,\theta,\eta) = \frac{\omega_-^2 P_1 P_2 \beta S_0}{4\pi\rho_0 c_0^4} \int \frac{D_1(\gamma,\varphi) D_2(\gamma,\varphi)}{r^2|\mathbf{R}-\mathbf{r}|}$$
$$\times \exp\left[-(\alpha_1+\alpha_2+ik_-)r - (\alpha_- + ik_-)|\mathbf{R}-\mathbf{r}|\right] dV \tag{1.66}$$

The integral element, as can be seen from Fig. 1.4, provided that the field point is far away from the interaction area ($R >> r$), then

$$|\mathbf{R}-\mathbf{r}| = R - ru \tag{1.67}$$

where $u = \cos\gamma\cos\theta\cos(\varphi-\eta) + \sin\gamma\sin\theta$.

If both the primary wave and the difference frequency sound have sharp directivity, the angles contributing to the sound field are taken to be small, accurate to the second order

$$u \approx 1 - \frac{1}{2}(\gamma-\theta)^2 - \frac{1}{2}(\varphi-\eta)^2 \tag{1.68}$$

Substitute the above expressions into Eq. (1.66) to yield

$$p_-(R,\theta,\eta) = \frac{\omega_-^2 P_1 P_2 \beta S_0}{4\pi\rho_0 c_0^4 R} \exp\left[-(\alpha_- + ik_-)R\right]$$
$$\int_0^\infty \int_{-\pi/2}^{\pi/2} \int_{-\pi/2}^{\pi/2} D_1(\gamma,\varphi) D_2(\gamma,\varphi)$$
$$\times \exp\left\{-\left[(\alpha_1+\alpha_2-u\alpha_-) + ik_-(1-u)\right]r\right\}\cos\gamma\, d\gamma\, d\varphi\, dr \tag{1.69}$$

Since $\alpha_- \ll \alpha_1 + \alpha_2$

$$\alpha_1 + \alpha_2 - u\alpha_- \approx \alpha_1 + \alpha_2 - \alpha_- = \alpha_T \tag{1.70}$$

Integrate the above formula with respect to r to yield

$$p_-(R,\theta,\eta) = \frac{\omega_-^2 P_1 P_2 \beta S_0}{4\pi\rho_0 c_0^4 R\alpha_T} \exp\left[-(\alpha_- + ik_-)R\right]$$
$$\times \int_{-\pi/2}^{\pi/2} \int_{-\pi/2}^{\pi/2} \frac{D_1(\gamma,\varphi) D_2(\gamma,\varphi)\cos\gamma}{1 + i(k_-/2\alpha_T)\left[(\gamma-\theta)^2 + (\varphi-\eta)^2\right]} d\gamma\, d\varphi \tag{1.71}$$

The directivity of the difference frequency sound is determined by the integral part of the above formula. Substitute $D_W(\theta, \eta)$ into this formula, and the directivity function of the difference frequency sound can be obtained.

$$\begin{aligned} D_-(\theta, \eta) &= \int_{-\pi/2}^{\pi/2} \int_{-\pi/2}^{\pi/2} \frac{D_1(\gamma, \varphi) D_2(\gamma, \varphi) cos\gamma}{1 + i(k_-/2\alpha_T)\left[(\gamma - \theta)^2 + (\varphi - \eta)^2\right]} d\gamma d\varphi \\ &= D_1(\theta, \eta) D_2(\theta, \eta) cos\theta \otimes D_W(\theta, \eta) \end{aligned} \quad (1.72)$$

As can be seen, the directivity of the difference frequency sound is the convolution of the product of the directivity functions of the two primary waves $D_1(\theta, \eta)D_2(\theta, \eta)\cos\theta$ and $D_W(\theta, \eta)$.

According to Eq. (1.72), consider two extreme cases.

(1) When $D_1(\theta, \eta) \approx D_2(\theta, \eta)$ and $D_1(\theta, \eta)D_2(\theta, \eta)$ is very sharp, which is close to the impulse function, then

$$D_-(\theta, \eta) = D_1(\theta, \eta) D_2(\theta, \eta) cos\theta \otimes D_W(\theta, \eta) = D_W(\theta, \eta) \quad (1.73)$$

This leads to the Westervelt model.

(B) When $D_W(\theta, \eta)$ is very sharp and it is close to the impulse function, that is $\theta_W \cong \sqrt{2\alpha_T/k_-} \to 0$, then

$$D_-(\theta, \eta) = D_1(\theta, \eta) D_2(\theta, \eta) cos\theta \otimes D_W(\theta, \eta) = D_1(\theta, \eta) D_2(\theta, \eta) \quad (1.74)$$

The directivity of the difference frequency sound equals the product of the directivity functions of the two primary waves.

This conclusion is cited in many documents [154–156]. In practical applications, the product of the directivity functions of the primary waves is often used to estimate the directivity of the difference frequency sound, but the premise of this conclusion cannot be ignored, i.e., $\theta_W \cong \sqrt{2\alpha_T/k_-} \to 0$. At this time $D_W(\theta, \eta)$ can be considered as an impulse function. This condition is easily met underwater because although the sound absorption coefficient is directly proportional to the square of the frequency, the underwater sound absorption coefficients are very small, thus θ_W is relatively small. However, in the air, the sound absorption coefficient is significantly larger than underwater, and the variation with frequency is much greater. Therefore, the condition $\theta_W \cong \sqrt{2\alpha_T/k_-} \to 0$ is not easily met. In the commonly used parametric array loudspeakers, the primary wave frequency is 40 and 38 kHz, and $D_W(\theta, \eta)$ cannot be approximated to an impulse function. It cannot simply be concluded that $D_-(\theta, \eta) = D_1(\theta, \eta), \eta)$ is the conclusion.

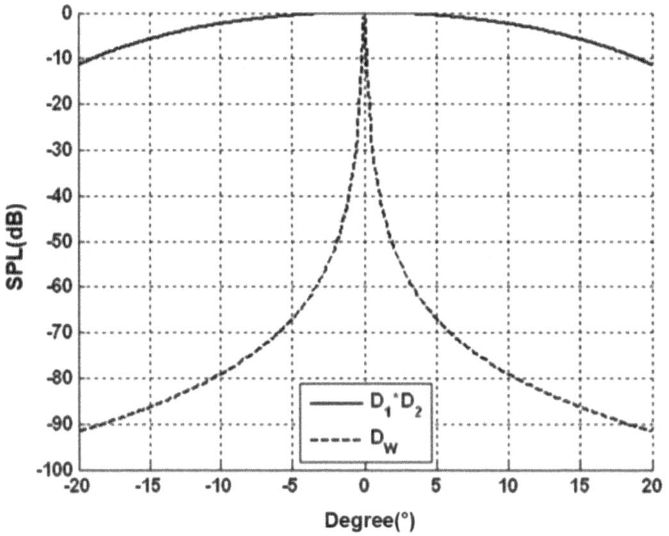

Fig. 1.5 Comparison of $D_W(\theta, \eta)$ and $D_1(\theta, \eta)D_2(\theta, \eta)$ in water (Primary wave frequency 40 kHz and 38 kHz)

1.3.3 Simulation Experiment of Parametric Array Directivity

The following is a simulation experiment on the directivity of the parametric array under the Berktay model, using water and air as different media, and calculating the directivity function of the primary wave and the difference frequency sound respectively (Fig. 1.5).

1.3.3.1 Underwater Parametric Transmitting Array

Assumes that the primary wave frequency is 40 and 38 kHz, the sound source is a circular transducer with a radius of 5 cm, and the directivity of the primary wave and the difference frequency sound is compared at a far field at 3 m. First, calculate the directivity of the primary wave according to the following formula [1]

$$D(\theta) = \frac{2J_1(ka\tan\theta)}{ka\tan\theta} \tag{1.75}$$

The results are shown in Fig. 1.5 by using Eq. (1.54) to calculate $D_W(\theta)$. It is evident it is quite sharp and it is close to the impulse function. So it can be concluded that $D_-(\theta, \eta) = D_1(\theta, \eta)D_2(\theta, \eta)$.

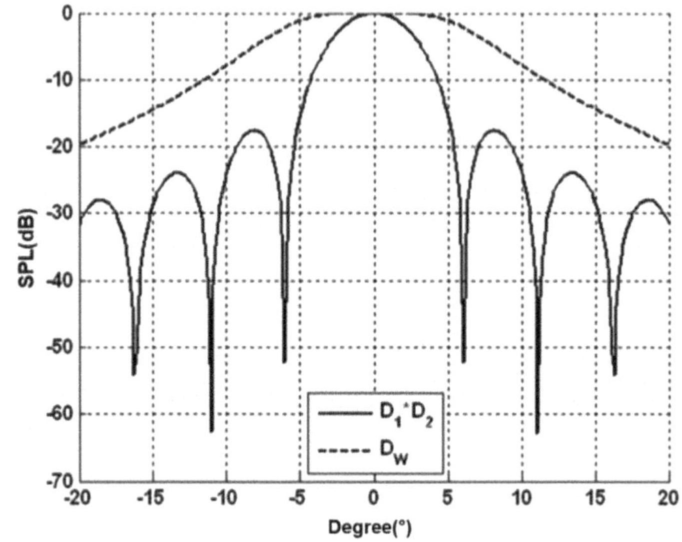

Fig. 1.6 Comparison of $D_W(\theta, \eta)$ and $D_1(\theta, \eta)D_2(\theta, \eta)$ in the air (Primary wave frequency 40 kHz and 38 kHz)

1.3.3.2 $D_W(\theta)$of Parametric Array Loudspeaker

The initial conditions of the parametric array loudspeaker in the air $D_W(\theta)$ are exactly the same as those of the underwater parametric array: the primary wave frequency is 40 and 38 kHz, the sound source is a circular transducer with a radius of 5 cm, and the directivity of the primary wave and the difference frequency sound is compared at a far field of 3 *m*. First, use Eqs. (1.75) and (1.54) to calculate the directivity of the primary wave and $D_W(\theta)$, as shown in Fig. 1.6. The main lobe is wide and cannot be approximated to the impulse function, so it cannot be concluded that $D_-(\theta, \eta) = D_1(\theta, \eta)D_2(\theta, \eta)$.

Further, to obtain a conclusion consistent with the Berktay model, the primary wave frequency is changed to 6 and 5 kHz, with other conditions unchanged. Fig. 1.7 is the calculation result using Eqs. (1.75) and (1.54), and it becomes clear that $D_W(\theta)$ is very sharp, so directional difference frequency sound satisfies $D_-(\theta, \eta) = D_1(\theta, \eta)D_2(\theta, \eta)$. However, in commonly used parametric array loudspeakers, ultrasonic waves are used as the primary wave to produce directional difference frequency sound. Therefore, the primary wave frequencies of 6 and 5 kHz have no practical significance in engineering applications.

Combining the above simulation results, the conclusion of applying the Berktay model $D_-(\theta, \eta) = D_1(\theta, \eta)D_2(\theta, \eta)$ is conditional, and it is summarized as follows.

1. The sound absorption distance is greater than the Rayleigh distance, that is, the nonlinear effect is completed in the far field.

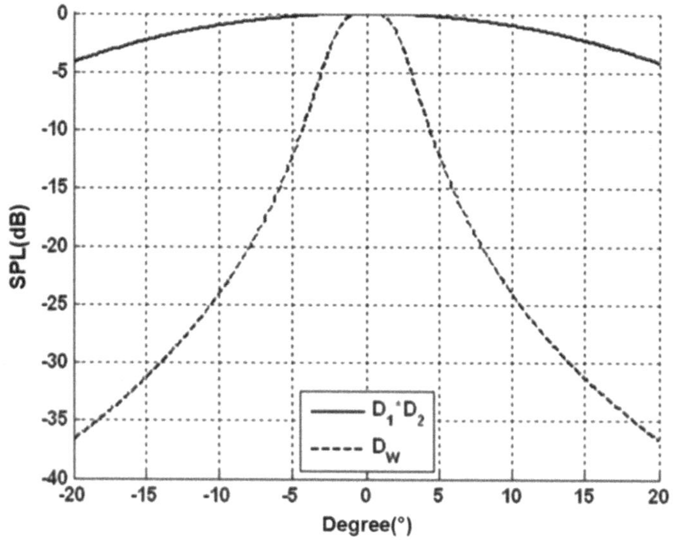

Fig. 1.7 Comparison $D_W(\theta, \eta)$ and in the air (Primary wave frequency 6 kHz and 5 kHz)

2. $\theta_W = \sqrt{2\alpha_T/k_-} \to 0$, that is, $D_W(\theta)$ is very sharp, and it is close to the impulse function, which means that the Westervelt directivity function is sharper than the product of the primary wave directivity functions.

For parametric array loudspeakers, due to the large sound absorption coefficient, the second condition is difficult to meet. Therefore, the directivity of the difference frequency sound is wider than that of the primary wave. In underwater systems, because the sound absorption coefficient is very small, and its impact can almost be ignored, it is easy to meet the second condition, which means that the directivity of the difference frequency sound can be very sharp.

1.4 Summary

This chapter first reviews the history and current state of research related to parametric arrays. Starting from the four equations used to describe the general motion of viscous heat-conducting fluids, it provides the propagation formulas for finite amplitude waves in non-attenuating fluids and thermoviscous fluids. Based on the Westervelt model and Berktay model of parametric arrays, the relationship between the directivity of the primary wave and the difference frequency sound is analyzed. Through simulation experiments, it is pointed out that due to the difference in sound absorption coefficients, the directivity of the difference frequency sound of underwater parametric arrays can be very sharp, which is approximately equal to the product of the primary wave directivities. However, for parametric array loudspeakers, due to the larger sound absorption coefficient of the air, the directivity of the difference frequency sound is wider than that of the primary waves.

Chapter 2
Directional Beam Calculation

Currently, the numerical calculation of the sound propagation problem of directional beams is generally based on the parabolic approximation of the Khokhlov-Zabolotskaya-Kuznetsov (KZK) equation. The mathematical representation of the KZK equation is as follows,

$$\frac{\partial^2 p}{\partial z \partial \tau} = \frac{c_0}{2}\nabla_\perp^2 p + \frac{\delta}{2c_0^3}\frac{\partial^3 p}{\partial \tau^3} + \frac{\beta}{2\rho_0 c_0^3}\frac{\partial^2 p^2}{\partial \tau^2} \tag{2.1}$$

where p is the sound pressure, z is the direction of sound wave propagation, $\tau = t - z/c_0$ is the delay time, c_0 is the small signal sound speed, δ is the diffusion coefficient of the viscous heat medium, β is the non-linear coefficient of the medium, ρ_0 is the density of the medium. $\nabla_\perp^2$ is the Laplace operator, in the axisymmetric (two-dimensional) coordinate system.

$$\nabla_\perp^2 p = \frac{\partial^2 p}{\partial r^2} + \frac{1}{r}\frac{\partial p}{\partial r} \tag{2.2}$$

where r is the radial coordinate. In the non-axisymmetric (three-dimensional) coordinate system

$$\nabla_\perp^2 p = \frac{\partial^2 p}{\partial x^2} + \frac{\partial^2 p}{\partial y^2} \tag{2.3}$$

The first term on the right side of the formula represents the diffraction effect (under the parabolic approximation condition), the second term represents the thermal viscous absorption effect, and the third term represents the non-linear effect.

The KZK equation was proposed by the former Soviet scientists Khokhlov, Zabolotskaya, and Kuznetsov in the 1970s. Zabolotskaya and Khokhlov [29] first derived the lossless version of the KZK equation (sometimes referred to as the KZ

J. Yang and P. Ji, *Parametric Array Loudspeakers*,
https://doi.org/10.1007/978-981-96-3548-1_2

equation), and in their derivation $\delta = 0$, that is, it does not include the effect of sound absorption. Kuznetsov [30] subsequently took into account the effect of thermal viscous absorption. By removing the diffraction term from the KZK equation, the equation becomes the Burgers equation [2]. The Burgers equation is a one-dimensional sound wave equation, which describes a plane wave with nonlinearity. The KZK equation can be used to describe the directional finite amplitude wave emitted by a finite aperture transducer precisely because the diffraction term is taken into account.

The KZK equation is limited by the parabolic approximation, requiring $ka >> 1$, where k is the wavenumber, a is the characteristic size of the sound source. The physical meaning of $ka >> 1$ is that the size of the sound source should be much larger than the wavelength of the sound wave emitted. Far from the sound axis (more than about 20° off the sound axis) and near the sound source (for example, within a few radii of the circular sound source), the parabolic approximation will introduce errors into the sound field.

The KZK equation has no analytical solution and can only be solved by numerical calculation methods. But under quasi-linear conditions, the sound field is simplified into the sum of the first and second order sound fields, which can be solved by integration [32, 33, 36, 249]. The usual method is to solve the KZK equation by finite difference in the frequency domain or time domain.

Since the KZK equation is derived from the Westervelt equation under the parabolic approximation, it inevitably introduces errors in places far from the axis (>20°) or near the sound source [1]. To overcome this limitation, scholars have proposed various methods, most of which are based on the Westervelt equation. Huijssen et al. obtained the axial and lateral beam diagrams of the harmonic components by applying the explicit finite difference time domain (FDTD) scheme to the Westervelt equation [250]. Based on the Neumann iterative solution, Huijssen and Verweij [251] proposed a new method to solve the lossless Westervelt equation, in which the nonlinear term is regarded as a nonlinear contrast source. Kamakura et al. [252] solved the Westervelt equation by introducing the step-by-step Padé method, extending its propagation angle to ±40°, which is twice the applicable angle of the KZK equation. Recently, the Angular Spectrum Approach (ASA) has been widely used in ultrasound simulation due to its fast calculation time [253, 254]. ASA was first introduced into the field of nonlinear acoustics by Alais and Hennion [255], used to analyze the sum frequency and difference frequency generation in the parametric array. Jing et al. proposed the Wave-Vector Frequency-Domain method to describe the second-order sound field characteristics in anisotropic nonlinear media based on ASA, to describe the sound field of a strongly focused sound source or a phased parametric array. This method can effectively consider the sound propagation status in all directions and can effectively calculate the backscattering sound field in dispersive media [256]. Landsberger and Hamilton [257] studied the generation of second harmonic sound beams in immersed elastic solids. Frijlink et al. [258] provided a simulation package called Abersim for solving the directional Westervelt equation in Cartesian two-dimensional and three-dimensional grids. Although it uses ASA to calculate diffraction effects, if the nonlinear term is considered, its calculation time will increase dramatically. Du and Jensen [259] used ASA in conjunction with sources generated by Field II [260] to solve the nonlinear wave equation for second harmonic generation. Its calculation speed is much faster than Abersim [258] because ASA is an analytical solution. Based on the Gaussian beam expansion method in the quasi-linear approximation, Cervenka and Bednarik [261] also obtained the formula for calculating the difference frequency sound of the baffle axisymmetric piston by the continuous approximation method. Haigh [262] improved the efficiency of sound field calculation by dividing the calculation area into several modules for parallel calculation by solving the Westervelt equation. The research group led by Qiu obtained a simple and rigorous expression of the quasi-linear solution of the Westervelt equation through the spherical harmonic expansion method [121].

2.1 Solution of KZK Model

The solution of the KZK model is necessary because the KZK equation can describe the propagation process of sound waves more accurately. At present, it can be divided into two types from the applicable range: one is to directly use finite difference to solve term by term in the time domain or frequency domain, and then superimpose the results. This method is numerical solution, and this result is applicable to any parameters; the other is to use the successive approximation method to obtain the result in the weak nonlinearity situation. This method is used to derive the expression of the nonlinear effect result, which is only applicable to the weak nonlinearity situation. The following are briefly introduced respectively.

2.1.1 Finite Difference Method

2.1.1.1 Combined Time-Frequency Domain Algorithm

Bakhvalov et al. [263–267] conducted pioneering numerical research on the KZK equation in a series of articles. Their algorithm calculates the effects of absorption and diffraction in the frequency domain and uses the Godunov method [268] to calculate the nonlinear effects in the time domain. In their algorithm, to ensure the stability of numerical calculation, they consider relatively smooth sound source distribution functions. The near field is the main area of concern for Bakhvalov et al. Zhileikin et al. [269] used this algorithm to study the high-frequency sound propagation in rigid cylindrical tubes. In addition, McKendree [270] and Frøysa et al. [271] also developed their own combined time-frequency domain algorithms.

2.1.1.2 Frequency Domain Algorithm

The first complete set of frequency domain algorithms was developed by Aanonsen et al. [35] to calculate the near field of directional finite amplitude waves called the Bergen code. The core of this algorithm is to expand the sound wave signal into a Fourier series and solve for each frequency component separately. However, the initial code is not efficient in calculating the far field. To reduce the time to calculate the far field, Hamilton et al. [272] proposed a special coordinate transformation method applied to the Bergen code, which ensures the stability of the numerical algorithm and allows the calculation step in the far field to be increased, thereby saving the time to calculate the far field. Subsequently, the frequency domain algorithm was extended to calculate dual-frequency sound sources [273–275] plane sound sources [276], focused sound beam [277, 278] and non-axisymmetric sound sources [98, 279, 280].

Because the subsequent frequency domain algorithms are influenced by the Bergen code, a general description of the Bergen code is given here.

First, perform the following variable transformation on Eq. (2.1)

$$\begin{cases} \sigma = z/z_0 \\ X = x/(a(1+\sigma)) \\ Y = y/(a(1+\sigma)) \\ \tau = \omega_0 t' - \dfrac{\omega_0\left(x^2+y^2\right)}{2c_0 z_0(1+\sigma)} \\ P = p(1+\sigma)/p_0 \end{cases} \tag{2.4}$$

It is obtained that

$$\begin{aligned} \frac{\partial^2 P}{\partial\tau\partial\sigma} = & \frac{1}{4(1+\sigma)^2}\left(\frac{\partial^2 P}{\partial X^2}+\frac{\partial^2 P}{\partial Y^2}\right) \\ & + A\frac{\partial^3 P}{\partial\tau^3} + \frac{N}{2(1+\sigma)}\frac{\partial^2 P^2}{\partial\tau^2} \end{aligned} \tag{2.5}$$

where $N = z_0/\bar{z}$, $A = z_0\alpha$, $\bar{z} = \rho_0 c_0^3/(\beta\omega_0 p_0)$, $z_0 = \omega_0 a^2/(2c_0)$. Let P be expanded into a Fourier series form

$$P = \sum_{n=1}^{+\infty}\left[g_n \sin(n\tau) + h_n \cos(n\tau)\right] \tag{2.6}$$

Substituting Eq. (2.6) into Eq. (2.5), the following n set of equations can be obtained

$$\frac{\partial g_n}{\partial \sigma} = \frac{1}{4n(1+\sigma)^2}\left(\frac{\partial^2}{\partial X^2} + \frac{\partial^2}{\partial Y^2}\right)h_n - n^2 A g_n$$
$$+ \frac{nN}{2(1+\sigma)}\left(\frac{1}{2}\sum_{m=1}^{n-1}(g_m g_{n-m} - h_m h_{n-m}) - \sum_{m=n+1}^{\infty}(g_{m-n} g_m + h_{m-n} h_m)\right) \quad (2.7)$$

$$\frac{\partial h_n}{\partial \sigma} = -\frac{1}{4n(1+\sigma)^2}\left(\frac{\partial^2}{\partial X^2} + \frac{\partial^2}{\partial Y^2}\right)g_n - n^2 A h_n$$
$$+ \frac{nN}{2(1+\sigma)}\left(\frac{1}{2}\sum_{m=1}^{n-1}(h_m g_{n-m} - g_m h_{n-m}) + \sum_{m=n+1}^{\infty}(h_{m-n} g_m - g_{m-n} h_m)\right) \quad (2.8)$$

These equations can be solved using the finite difference method.

2.1.1.3 Time Domain Algorithm

Lee and Hamilton [281] proposed a two-dimensional time-domain algorithm (the Texas code), which avoids the truncation of the Fourier series in the frequency domain algorithm. However, the initial time-domain algorithm was only for thermal viscous media and did not include relaxation effects. The algorithm for relaxation effects was completed by Cleveland et al. [5] The Texas code has been used to study lithotripters [282], self-demodulation phenomena of amplitude and frequency modulated pulses [283], nonlinear distortion of short pulses [284], and parametric arrays in the air [285], etc. Recently, Yang and Cleveland [286] extended the Texas code to three dimensions to study harmonic imaging of rectangular sound sources.

2.1.1.4 Other Algorithms

Christopher and Parker [287, 288] proposed a new frequency domain algorithm that is not limited by the parabolic approximation [289], i.e., it can accurately calculate the effect of diffraction. This method has been used to study lithotripters [290], harmonic imaging [291], etc. Khokhlova et al. [292, 293] also proposed an algorithm similar to Christopher's algorithm, but they used a second-order operator splitting method [294, 295].

There are many institutions in China that conduct research on nonlinear acoustics, such as Nanjing University, University of Science and Technology of China, National University of Defense Technology, and Institute of Acoustics of the Chinese Academy of Sciences, most of which commonly used are frequency-domain algorithms [296–299], while time-domain algorithms are relatively less frequently used [300].

2.1.2 Gaussian Beam Expansion Method

When the sound pressure level of the sound source is moderate and all harmonics higher than the second can be ignored, results in a quasi-linear approximation [1], which also meets the safety considerations of parametric array loudspeakers [125, 301, 302]. At this time, for an axisymmetric sound source, the quasi-linear solution of the KZK equation can be obtained using a successive approximation method. Using this method, the integral expressions of the pump wave, the harmonics, and the difference frequency can be obtained, and for special sound source forms such as Gaussian sound sources, their analytical expressions can be obtained. Under the quasi-linear approximation condition, the sound field of a directional finite amplitude wave can be simplified into a sum of a first-order sound wave (also known as a fundamental wave or pump wave) and a second-order sound wave (second harmonics, sum frequency wave, difference-frequency sound). The first-order sound wave can be represented by a double integral, and the second-order sound wave can be represented by a quintuple integral, so the related sound field calculation requires a lot of time. The Gaussian beam expansion method to be introduced below is a further development based on the integral method. When the sound source distribution function is represented in the form of a sum of Gaussian functions, using the integrability of the Gaussian function, the sound field of the first-order sound wave can be obtained in an analytical form, and the sound field of the second-order sound wave can be simplified into a single integral, thereby greatly reducing the time for sound field calculations.

The Gaussian beam expansion method was first used by Wen and Breazeale [101] in 1988 to calculate the linear sound field of a circular piston transducer in water. Compared with the method proposed by Cavanagh and Cook [303, 304] to expand the sound source distribution function into a sum of orthogonal Gauss-Lagueere or Gauss-Hermite functions, the method of Wen and Breazeale can achieve similar accuracy with fewer terms. The Gaussian beam expansion method has high computational efficiency and is widely used in the calculation and modeling of directional beams. For example, Huang and Breazeale [305] combined the Gaussian beam expansion method with the finite element method to model the diffracted sound field when the medium is non-uniformly distributed or when there is reflection. The key to the Gaussian beam expansion method is how to determine the Gaussian beam expansion coefficients. Wen and Breazeale used the method of minimizing the mean square error in the spatial domain to obtain the Gaussian beam expansion coefficients. Because of the non-orthogonality of the Gaussian function, this process can only be completed in a nonlinear optimization way, which is exactly why the number of terms of the Gaussian function can be small. In reference [101], the Gaussian function with 10 terms can accurately simulate the acoustic field distribution. In reference [306], Wen and Breazeale improved the algorithm and obtained a set of 15 coefficients, which can more accurately simulate the near field of the sound wave.

A group led by Ding has done a lot of theoretical work in the field of Gaussian beam expansion [307–316]. The Gaussian beam expansion method is extended to calculate the radiation sound field of rectangular, elliptical or even arbitrary sound source distributions [313]. The analysis shifts from the calculation of the first-order sound field to the calculation of the second-order sound field [310–312]. The Gaussian expansion of the rectangular function is extended to the Gaussian expansion of the zero-order Bessel function [314]. They summarized the reciprocity relationship between the Gaussian beam in the spatial domain and *k* spatial domain [307]. Ding et al. [316] initially proposed a method called "Gaussian function polynomial" in the spatial domain to calculate the Gaussian beam expansion coefficients. This method is characterized by expanding the sound source distribution function with orthogonal Gaussian function polynomials. The algorithm in the later article [308] made some improvements to this method. In general, Ding's method is to solve a set of linear equations to calculate the Gaussian beam expansion coefficients, which is simple. However, this method uses real Gaussian functions, and the exponential coefficients of the Gaussian functions are predetermined, which leads to the linear equations to be solved becoming ill-conditioned when the number of Gaussian functions is large, so the application of this method is limited.

Sha et al. [100, 317, 318] extended the Gaussian beam expansion method from the spatial domain to the *k* spatial domain. Kim et al. [319] reviewed the generation of Gaussian beam expansion coefficients, summarized the existing methods for calculating Gaussian beam expansion coefficients, and it was the first time that a transformation relationship between the Gaussian beam expansion coefficients in the spatial domain and the spatial domain was proposed. In addition, they simulated the sound field using a set of 25 Gaussian beam expansion coefficients that they calculated themselves, and concluded that the accuracy of the Gaussian beam expansion method would be limited in near-field simulations.

The team at Iowa State University led by Schmerr [320, 321] has done a lot of theoretical and experimental work in the field of nondestructive testing using the Gaussian beam expansion method. They theoretically established Gaussian beam models for complex problems such as curved surfaces [322] or anisotropic solids [323, 324] and verified the correctness of the theory through experiments. In addition, they simulated phased array ultrasound by using a new method called the "multi-Gaussian array beam model" [325]. The principle of this method is to transform the sound axis of the Gaussian beam model to the phased direction, so that the phased direction is exactly within the limit of the near-axis approximation of the usual Gaussian beam model, and then consider the directivity of the beam and the equivalent area of the sound source in the phased direction, which can accurately simulate large-angle phased and focused ultrasound beams.

Spies [326, 327] also adopt the Gaussian beam expansion method to model in the field of nondestructive testing. The difference lies in that Spies used the method for calculating Gaussian beam expansion coefficients introduced by Prange and Shenoy [328] from signal processing, which is the Prony algorithm. Schmerr et al. [329] leveraged the characteristic that the sound source function in the spatial domain is infinitely continuous, extending the Prony method to the spatial domain for calculating Gaussian beam expansion coefficients, and achieved good results. Yu et al. [330] used the Gaussian coefficient superposition method to approximately represent the distribution function of the conical focusing sound source, combined with the near-axis approximation of the KZK equation, and obtained the analytical solution of the basic and second harmonic sound field of the conical focusing transducer in the lossy medium. Liu et al. [331] obtained a theoretical model for the temperature increase caused by piston-focused ultrasound using the Gaussian beam expansion method.

In terms of simulating phased arrays, Zhao et al. [332, 333] used a method different from Schemrr's, and they proposed a Gaussian beam model based on spherical waves, fundamentally eliminating the near-axis limitation of the traditional Gaussian beam model based on plane waves. Jeong [334] used the Gaussian beam method for modeling time reversal in solids.

2.1.2.1 Quasi-Linear Theory

When the amplitude of the finite amplitude wave is small, the nonlinearity of the sound field is weak, and the integral expression of the quasi-linear solution of the formula can be derived using the continuous approximation method. The following briefly describes its derivation process. The form of the quasi-linear solution of the formula is

$$p = p_1 + p_2 \tag{2.9}$$

where p_1 is the first-order sound field (also known as the pump wave or fundamental wave), p_2 is the second-order sound field (including sum frequency waves, difference frequency waves and second harmonics), and other higher-order sound fields are ignored. p_1 is the linear solution (first approximation) of the formula, p_2 is a small correction to p_1 (second approximation), $|p_2| \ll |p_1|$. Assume that the first-order sound field of a dual-frequency finite amplitude wave is represented as follows

$$p_1 = Re\left[q_a \exp(-i\omega_a t' - \alpha_a z) + q_b \exp(-i\omega_b t' - \alpha_b z)\right] \tag{2.10}$$

The second-order sound field is represented as follows

$$p_2 = \mathrm{Re}\begin{bmatrix} q_{2a}\exp(-i2\omega_a t' - \alpha_{2a} z) + q_{2b}\exp(-i2\omega_b t' - \alpha_{2b} z) \\ +q_s \exp(-i\omega_s t' - \alpha_s z) + q_d \exp(-i\omega_d t' - \alpha_d z) \end{bmatrix} \tag{2.11}$$

where ω_a, ω_b, ω_{2b}, ω_d correspond to the angular frequencies of primary waves, the sum frequency wave and difference frequency wave, respectively, α_a, α_b, α_{2a}, α_{2b}, α_s, α_d represent the thermal viscous absorption coefficients at ω_a, ω_b, ω_{2a}, ω_{2b}, ω_s, ω_d, respectively and the complex amplitudes of each frequency are q_a, q_b, q_{2a}, q_{2b}, q_s, q_d. Equations (2.9), (2.10), and (2.11) into Eq. (2.1), the following quasi-linear equation group can be obtained

$$\nabla_\perp^2 q_a + i2k_a \frac{\partial q_a}{\partial z} = 0 \tag{2.12}$$

$$\nabla_\perp^2 q_b + i2k_b \frac{\partial q_b}{\partial z} = 0 \tag{2.13}$$

$$\nabla_\perp^2 q_{2a} + i2k_{2a} \frac{\partial q_{2a}}{\partial z} = \frac{2\beta k_a^2}{\rho_0 c_0^2} q_a^2 e^{(-2\alpha_a + \alpha_{2a})z} \tag{2.14}$$

$$\nabla_\perp^2 q_{2b} + i2k_{2b} \frac{\partial q_{2b}}{\partial z} = \frac{2\beta k_b^2}{\rho_0 c_0^2} q_b^2 e^{(-2\alpha_b + \alpha_{2b})z} \tag{2.15}$$

$$\nabla_\perp^2 q_s + i2k_s \frac{\partial q_s}{\partial z} = \frac{\beta k_s^2}{\rho_0 c_0^2} q_a q_b e^{(-\alpha_a - \alpha_b + \alpha_s)z} \tag{2.16}$$

$$\nabla_\perp^2 q_d + i2k_d \frac{\partial q_d}{\partial z} = \frac{\beta k_d^2}{\rho_0 c_0^2} q_a q_b^* e^{(-\alpha_a - \alpha_b + \alpha_d)z} \tag{2.17}$$

where $k_a, k_b, k_{2a}, k_{2b}, k_s, k_d$ are the wavenumbers of each frequency. By solving these equations using the Green's function method to yield

$$q_a(x, y, z) = \frac{p_{0a} k_a}{i2\pi z} \int_{-\infty}^{+\infty} \int_{-\infty}^{+\infty} q_a(x', y') \exp\left(ik_a \frac{(x - x')^2 + (y - y')^2}{2z} \right) dx' dy' \tag{2.18}$$

$$q_b(x, y, z) = \frac{p_{0b} k_b}{i2\pi z} \int_{-\infty}^{+\infty} \int_{-\infty}^{+\infty} q_b(x', y') \exp\left(ik_b \frac{(x - x')^2 + (y - y')^2}{2z} \right) dx' dy' \tag{2.19}$$

$$\begin{aligned} q_{2a}(x, y, z) = & \frac{-\beta k_a^2}{2\pi \rho_0 c_0^2} \int_0^z \int_{-\infty}^{+\infty} \int_{-\infty}^{+\infty} \frac{q_a^2(x', y', z')}{(z - z')} e^{(-2\alpha_a + \alpha_{2a})z'} \\ & \times \exp\left(ik_a \frac{(x - x')^2 + (y - y')^2}{(z - z')} \right) dx' dy' dz' \end{aligned} \tag{2.20}$$

$$\begin{aligned} q_{2b}(x, y, z) = & \frac{-\beta k_b^2}{2\pi \rho_0 c_0^2} \int_0^z \int_{-\infty}^{+\infty} \int_{-\infty}^{+\infty} \frac{q_b^2(x', y', z')}{(z - z')} e^{(-2\alpha_b + \alpha_{2b})z'} \\ & \times \exp\left(ik_b \frac{(x - x')^2 + (y - y')^2}{(z - z')} \right) dx' dy' dz' \end{aligned} \tag{2.21}$$

$$\begin{aligned} q_s(x, y, z) = & -\frac{\beta k_s^2}{2\pi \rho_0 c_0^2} \int_0^z \int_{-\infty}^{+\infty} \int_{-\infty}^{+\infty} \frac{q_a(x', y', z') q_b(x', y', z')}{2(z - z')} e^{(-\alpha_a - \alpha_b + \alpha_s)z'} \\ & \times \exp\left(ik_s \frac{(x - x')^2 + (y - y')^2}{2(z - z')} \right) dx' dy' dz' \end{aligned} \tag{2.22}$$

$$q_d(x,y,z) = \frac{-\beta k_d^2}{2\pi \rho_0 c_0^2} \int_0^z \int_{-\infty}^{+\infty} \int_{-\infty}^{+\infty} \frac{q_a(x',y',z') q_b^*(x',y',z')}{2(z-z')} e^{(-\alpha_a-\alpha_b+\alpha_d)z'} \times \exp\left(ik_d \frac{(x-x')^2+(y-y')^2}{2(z-z')} \right) dx'dy'dz' \tag{2.23}$$

where, p_{0a} and p_{0b} are the sound pressure at the sound source of angular frequencies ω_a and ω_b, respectively, $q_a(x',y')$ and $q_b(x',y')$ are the source distribution functions. Equations (2.18)–(2.23) is the quasi-linear integral solution of Eq. (2.1). It should be noted that the sound absorption in the above formula only includes thermal viscous absorption, so it is only suitable for thermal viscous media. To describe the sound absorption coefficient with any frequency-dependent relationship, one simply needs to replace the thermal viscous absorption coefficient in the aforementioned formula with the actual absorption coefficient.

2.1.2.2 Directional Beam Calculation Based on Gaussian Beam Expansion Method

The directional beam calculation based on the Gaussian beam expansion method can simplify the calculation of the second-order sound field from Eqs. (2.20) to (2.23), reducing the calculation time. Ding [312] and Yang et al. [318] have completed the modeling of the second-order sound field using the Gaussian beam in the spatial domain and k space domain respectively. The following gives the derivation process of the second-order sound field Gaussian beam in the spatial domain.

The sound source function of the array composed of N_C transducers is represented as

$$q_a(x',y') = q_b(x',y') = \sum_{c=1}^{N_C} \sum_{n=1}^{N_G} A_{n,xy} \exp\left(-B_{n,x} \frac{(x'-x_c)^2}{a_x^2} - B_{n,y} \frac{(y'-y_c)^2}{a_y^2} \right) \tag{2.24}$$

where (x_c, y_c) is the center coordinate of the transducer unit, a_x and a_y are the characteristic lengths in the x and y directions respectively, $A_{n,xy}$, $B_{n,x}$ and $B_{n,y}$ is the Gaussian beam coefficient (related to the shape of the transducer unit, the sound pressure distribution, a_x and a_y, for details refer to reference [313]).

Substitute Eq. (2.24) into Eqs. (2.18) and (2.19), and use the following formula

$$\int_{-\infty}^{+\infty} \exp(-\psi t^2 - \phi t) dt = \sqrt{\frac{\pi}{\psi}} \exp\left(\frac{\phi^2}{4\psi} \right) \tag{2.25}$$

thus yield

$$q_a(x, y, z) = p_{0a} \sum_{c=1}^{N_C} \sum_{n=1}^{N_G} A_{n,xy} \frac{\exp\left(-\frac{B_{n,x}(x-x_c)^2/l_x^2}{1+iB_{n,x}z/z_{a,x}} - \frac{B_{n,y}(y-y_c)^2/l_y^2}{1+iB_{n,y}z/z_{a,y}}\right)}{\sqrt{1+iB_{n,x}z/z_{a,x}}\sqrt{1+iB_{n,y}z/z_{a,y}}} \tag{2.26}$$

$$q_b(x, y, z) = p_{0b} \sum_{c=1}^{N_C} \sum_{n=1}^{N_G} A_{n,xy} \frac{\exp\left(-\frac{B_{n,x}(x-x_c)^2/l_x^2}{1+iB_{n,x}z/z_{b,x}} - \frac{B_{n,y}(y-y_c)^2/l_y^2}{1+iB_{n,y}z/z_{b,y}}\right)}{\sqrt{1+iB_{n,x}z/z_{b,x}}\sqrt{1+iB_{n,y}z/z_{b,y}}} \tag{2.27}$$

where $z_{a,x} = k_a a_x^2/2$, $z_{a,y} = k_a a_y^2/2$, $z_{b,x} = k_b a_x^2/2$, $z_{b,y} = k_b a_y^2/2$. Substitute these Eqs. (2.26) and (2.27) into Eqs. (2.23) and (2.23), respectively, and integrate with respect to x' and y' to obtain the calculation formula for the complex amplitude of the second-order sound field. For example, the calculation formula for the complex amplitude of the difference frequency sound can be derived from the equation as follows

$$q_d(x, y, z) = \frac{-\beta k_d^2 p_{0a} p_{0b}}{2\pi \rho_0 c_0^2} \int_0^z \frac{e^{(\alpha_d - \alpha_a - \alpha_b)z'}}{2(z-z')} \sum_{c1=1}^{N_C} \sum_{c2=1}^{N_C} \sum_{n=1}^{N_G} \sum_{j=1}^{N_G} A_{n,xy} A_{j,xy}^* D_x D_y dz' \tag{2.28}$$

where

$$D_x = \frac{\int_{-\infty}^{+\infty} \exp\left[-\frac{B_{n,x}(x'-x_{c1})^2/l_x^2}{1+iB_{n,x}z'/z_{a,x}} - \left(\frac{B_{j,x}(x'-x_{c2})^2/l_x^2}{1+iB_{j,x}z'/z_{b,x}}\right)^* + ik_d \frac{(x-x')^2}{2(z-z')}\right] dx'}{\sqrt{1+iB_{n,x}z'/z_{a,x}}\left(\sqrt{1+iB_{j,x}z'/z_{b,x}}\right)^*}$$
$$= \frac{\frac{\sqrt{\pi l_x^2}}{\sqrt{1+iB_{n,x}z'/z_{a,x}}\left(\sqrt{1+iB_{j,x}z'/z_{b,x}}\right)^*} \exp\left(\frac{r_3 z'/z_{0,x} + r_4}{r_1 z'/z_{0,x} + r_2}\right)}{\sqrt{\frac{B_{n,x}}{1+iB_{n,x}z'/z_{a,x}} + \left(\frac{B_{j,x}}{1+iB_{j,x}z'/z_{b,x}}\right)^* - i\frac{(l+m)}{(z-z')/z_{0,x}}}} \tag{2.29}$$

The following parameters are used in the above equation

$$k = (k_a + k_b)/2, \; z_{0,x} = k a_x^2/2, \; l = k_a/k, \; m = -k_b/k$$

$$r_1 = i\left[m^2 B_{n,x} + l^2 B_{j,x}^* + i(l+m) B_{n,x} B_{j,x}^* \frac{z}{z_{0,x}}\right]$$

$$r_2 = lm\left[l + m + i\left(B_{n,x} + B_{j,x}^*\right)\frac{z}{z_{0,x}}\right]$$

$$r_3 = -i\frac{[l(x - x_{c1}) + m(x - x_{c2})]^2}{l_x^2} B_{n,x} B_{j,x}^*$$

$$r_4 = -lm(l+m)\left(\frac{(x-x_{c1})^2}{l_x^2}B_{n,x} + \frac{(x-x_{c2})^2}{l_x^2}B_{j,x}^*\right) - ilm\frac{z}{z_{0,x}}\frac{(x_{c1}-x_{c2})^2}{l_x^2}B_{n,x}B_{j,x}^*$$

Replace all the xs in Eq. (2.29) with ys to obtain the calculation formula for D_y.

The calculation formula for the amplitude of the sum frequency wave can be derived from Eq. (2.22), which is

$$q_s(x,y,z) = \frac{-\beta k_s^2 p_{0a} p_{0b}}{2\pi\rho_0 c_0^2}\int_0^z \frac{e^{(\alpha_s-\alpha_a-\alpha_b)z'}}{2(z-z')} \sum_{c1=1}^{N_C}\sum_{c2=1}^{N_C}\sum_{n=1}^{N_G}\sum_{j=1}^{N_G} A_{n,xy}A_{j,xy}S_{ab,x}S_{ab,y}dz' \tag{2.30}$$

By removing all the conjugates in Eq. (2.29) and defining $m = k_b/k$, the calculation formula for $S_{ab,x}$ in Eq. (2.30) is obtained. By replacing all instances of x in the calculation formula of $S_{ab,x}$ with y, the calculation formula for $S_{ab,y}$ is obtained.

As for the calculation formula for the complex amplitude of the second harmonics, it can be derived from Eqs. (2.20) and (2.21), which are as follows

$$q_{2a}(x,y,z) = \frac{-\beta k_a^2 p_{0a}^2}{2\pi\rho_0 c_0^2}\int_0^z \frac{e^{(\alpha_{2a}-2\alpha_a)z'}}{(z-z')} \sum_{c1=1}^{N_C}\sum_{c2=1}^{N_C}\sum_{n=1}^{N_G}\sum_{j=1}^{N_G} A_{n,xy}A_{j,xy}S_{aa,x}S_{aa,y}dz' \tag{2.31}$$

$$q_{2b}(x,y,z) = \frac{-\beta k_b^2 p_{0b}^2}{2\pi\rho_0 c_0^2}\int_0^z \frac{e^{(\alpha_{2b}-2\alpha_b)z'}}{(z-z')} \sum_{c1=1}^{N_C}\sum_{c2=1}^{N_C}\sum_{n=1}^{N_G}\sum_{j=1}^{N_G} A_{n,xy}A_{j,xy}S_{bb,x}S_{bb,y}dz' \tag{2.32}$$

By replacing all instances of b in the calculation formula for $S_{ab,x}$ in Eq. (2.30) with a, the calculation formula for $S_{aa,x}$ is obtained. Similarly, by replacing all instances of a with b, the calculation formula for $S_{bb,x}$ is obtained. By replacing all instances of x in the calculation formulas, for both $S_{aa,x}$ and $S_{bb,x}$ with y, the calculation formulas for $S_{aa,y}$ and $S_{bb,y}$ are obtained.

2.2 Calculation of Gaussian Beam Expansion Coefficients

The basic principles and representation of the Gaussian beam expansion method have been introduced earlier. This section introduces the core of the Gaussian beam expansion method, which is how to represent the sound source distribution function in the form of a superposition of Gaussian functions, specifically, the calculation of Gaussian beam expansion coefficients [335].

First, the existing work on Gaussian beam expansion coefficients will be summarized, their advantages and disadvantages will be analyzed. On this basis, this section presents a new method for calculating Gaussian beam expansion coefficients. This method is simple, fast, highly accurate, and widely applicable, suitable for both spatial domain and k space domain.

The methods for calculating Gaussian beam expansion coefficients can be roughly divided into spatial domain methods and k space domain methods. Wen and Breazeale [101] first proposed the first spatial domain method, they obtained the Gaussian beam expansion coefficients by minimizing the mean square error between the sound source distribution function and the superposition of Gaussian functions. Since they did not provide detailed content of the algorithm in their paper, it is impossible to implement their algorithm specifically, but according to the description in related literature, this algorithm based on nonlinear optimization is complex and time-consuming. However, several sets of Gaussian beam expansion coefficients are given in references [20, 24], which can be used as a standard for comparison and reference. The k space domain method was proposed by Sha et al. [100]. They used a probe search algorithm called A^* to obtain the Gaussian beam expansion coefficients of the rectangular transducer. Their method is carried out in a two-dimensional space Fourier transform domain (also known as k space domain or angular spectrum domain), so it is called k space domain method.

There is a reciprocal relationship between the Gaussian beam expansion coefficients in the spatial domain and the k space domain, which can be converted to each other through a simple formula. Kim et al. reviewed the spatial domain method and k space domain method in reference [319], and obtained the reciprocal relationship between the Gaussian beam expansion coefficients. Ding et al. discussed these two methods and the reciprocal relationship of Gaussian beam expansion coefficients in more detail in reference [307]. However, both methods are implemented through nonlinear optimization algorithms, which are complex and time-consuming. Therefore, a simple method is needed to calculate the Gaussian beam expansion coefficients.

Ding et al. [308] proposed a linear algorithm in the spatial domain to calculate the Gaussian beam expansion coefficients. This method calculates the Gaussian beam expansion coefficients by solving a set of linear equations, which is simple and fast. However, Ding's method requires pre-determining some of the Gaussian beam expansion coefficients, and the method discusses only real Gaussian functions, so its application range is limited.

Prange and Shenoy [328] used the Prony method, commonly used in the field of signal processing, to solve for the coefficients of Gaussian beam expansion. Their method is a type of spatial domain method. However, unlike the methods of Wen and Breazeale, Prange and Shenoy's algorithm does not calculate the coefficients of Gaussian beam expansion by approximating the sound source distribution function, but by approximating some sampling points in the spatial sound field, which leads to the need to use data from the spatial sound field (i.e., the need to use another sound field calculation model) in the process of calculating the coefficients of Gaussian beam expansion. One drawback of the Prony method is that it is not suitable for rectangular functions with very steep edges (ill-conditioned problems will occur in the calculation), which is why Prange and Shenoy did not use this method to approximate the sound source distribution function in the spatial domain. Recently, Schmerr et al. have applied the Prony method to the k spatial domain, first approximating the sound source distribution function in the k spatial domain (this method cleverly utilizes the feature that the sound source distribution function in the k spatial domain is infinitely untruncated), obtaining the Gaussian beam expansion coefficients in the k spatial domain, and then using the reciprocity relationship between the spatial domain and the k spatial domain of the Gaussian beam expansion coefficients, transforming the Gaussian beam expansion coefficients in the k spatial domain to the spatial domain. The advantage of the Prony method is its fast calculation speed, but it should be noted that during the use of the Prony method, there may be situations where the real part of B_n is less than 0, which is not allowed.

2.2.1 *Simple, Fast, High-Precision Method*

The general Gaussian series expansion form of the sound source distribution function is

$$q(x', y') = \sum_{n=1}^{N} A_{n,xy}\exp\left(-B_{n,x}\frac{x'^2}{a_x^2} - B_{,y}\frac{y'^2}{a_y^2}\right) \tag{2.33}$$

For a one-dimensional sound source (i.e., a circular axisymmetric sound source), the sound source distribution function is represented by a Gaussian series as follows

$$q(x', y') = \sum_{n=1}^{N} A_n\exp\left(-B_n\frac{x'^2 + y'^2}{a^2}\right) \tag{2.34}$$

where a is the radius. For a general two-dimensional sound source (such as a rectangular sound source), the sound source distribution function has symmetry in the x axis direction and y axis direction, and the two-dimensional sound source distribution function can be decomposed into the product of two one-dimensional functions, that is

$$q(x', y') = q(x')q(y') \tag{2.35}$$

Perform a Gaussian series expansion on the two one-dimensional functions separately, to yield

$$q(x', y') = \sum_{n=1}^{N} A_{n,x}\exp\left(-B_{n,x}\frac{x'^2}{a_x^2}\right)\sum_{m=1}^{M} A_{m,y}\exp\left(-B_{m,y}\frac{y'^2}{a_y^2}\right) \tag{2.36}$$

Both Eqs. (2.34) and (2.36) can be written in the form of Eq. (2.33).

As can be seen from the above, the acquisition of any Gaussian beam expansion coefficient can actually be reduced to the calculation of a one-dimensional Gaussian beam expansion coefficient. Therefore, this part will limit the discussion to the one-dimensional case, that is, how to express any function $v(\xi)$ in the form shown below

$$v(\xi) = \sum_{n=1}^{N} A_n\exp(-B_n\xi^2) \tag{2.37}$$

where A_n and B_n are the coefficients of the Gaussian beam expansion to be solved, and ξ is the normalized spatial coordinate variable.

2.2.1.1 Spatial Domain Algorithm

Because the Gaussian function is non-orthogonal, the Gaussian beam expansion coefficients A_n and B_n are generally obtained by nonlinear optimization methods. This section presents a quasi-linear algorithm to solve this problem. Define a target function Q

$$Q(A_n, B_n, N) = \int_{-\infty}^{+\infty}\left|v(\xi) - \sum_{n=1}^{N} A_n \exp(-B_n\xi^2)\right|^2 d\xi \tag{2.38}$$

The desired Gaussian beam expansion coefficients A_n and B_n are obtained by minimizing the objective function Q, while requiring N to be as small as possible. From Eq. (2.38), it is evident that A_n is the amplitude coefficient of the Gaussian series, and B_n is the exponential coefficient of the Gaussian series. These two types of coefficients have different positions and functions, and should be determined by different methods.

(1) Determination of A_n

When determining Q at its minimum, it must satisfy the condition

$$\frac{\partial Q}{\partial A_j^*} = 0 \quad (j = 1, 2, \ldots, N) \tag{2.39}$$

where * represents complex conjugate. From Eq. (2.39), it can be deduced that

$$\mathbf{A} = \mathbf{C}^{-1}\mathbf{D} \tag{2.40}$$

where

$$\mathbf{A} = \begin{pmatrix} A_1 \\ \vdots \\ A_N \end{pmatrix}, \mathbf{C} = \begin{pmatrix} C_{11} & \dots & C_{1N} \\ \vdots & \ddots & \vdots \\ C_{N1} & \cdots & C_{NN} \end{pmatrix}, \mathbf{D} = \begin{pmatrix} D_1 \\ \vdots \\ D_N \end{pmatrix}$$

$$C_{jn} = \int_{-\infty}^{+\infty} \exp\left[-\left(B_j^* + B_n\right)\xi^2\right]d\xi = \sqrt{\frac{\pi}{\left(B_j^* + B_n\right)}}$$

$$D_j = \int_{-\infty}^{+\infty} v(\xi)\exp\left(-B_j^*\xi^2\right)d\xi$$

Therefore, when B_n is already determined, A_n can be obtained through Eq. (2.40).

(B) Determination of B_n

By substituting Eq. (2.40) into Eq. (2.38) to yield.

$$Q(B_n, N) = \int_{-\infty}^{+\infty} |v(\xi)|^2 d\xi - \mathbf{D}^T\mathbf{C}^{-1}\mathbf{D}^* \tag{2.41}$$

As can be seen, Q has become a function only related to B_n and N, and is independent of A_n. However, using the condition $\partial Q/\partial B_j^* = 0$ cannot derive a result similar to that concerning B_n in the Eq. (2.40). Inspired by the Fourier series, this section uses another method to determine the value of B_n.

The Fourier series is as follows

$$f(x) = \sum_{m=-\infty}^{+\infty} F(mk_0)\exp(imk_0x) \tag{2.42}$$

where $k_0 = 2\pi/L$, L is the period of $f(x)$, and $f(x)$ is continuous within $(-L/2, L/2)$. Replace x with ξ^2, and both sides of Eq. (2.42) are multiplied by $\exp\left(-\delta\xi^2\right)$, a function can be constructed

$$h(\xi) = f\left(\xi^2\right)\exp\left(-\delta\xi^2\right) = \sum_{m=-\infty}^{+\infty} F(mk_0)\exp\left[-(\delta - imk_0)\xi^2\right] \tag{2.43}$$

$\delta > 0$ ensures $h(\xi)$ converges. For $h(\xi)$, it can be represented in the similar form of Eq. (2.37) using the Gaussian series expansion

$$h(\xi) = \sum_{n=1}^{N} A_n \exp\left(-B_n \xi^2\right) \tag{2.44}$$

Comparing Eqs. (2.44) and (2.43), it is evident that B_n in Eq. (2.44) is similar to the position of $(\delta - imk_0)$ in Eq. (2.43). Therefore, B_n is replaced by the following formula

$$B_n = \delta - imk_0 \tag{2.45}$$

where $\delta > 0$, and $m = n - (N+1)/2$. N unknown complex coefficients B_n in Eq. (2.45) are reduced to be determined by two real numbers δ and k_0, and the function $Q(B_n, N)$ is simplified to $Q(\delta, k_0, N)$, which can be obtained through a simple two-dimensional search for δ and k_0.

In the following examples, it will be discussed in detail about how to choose δ and k_0, here only the main idea of this algorithm is explained. The complex exponential functions of different frequencies $\exp(imk_0x)$ have orthogonality, so they can be used as the basis functions of the Fourier series. Although the complex Gaussian function $\exp\left(-B_n\xi^2\right)$ is not orthogonal, but it has a great similarity with the complex exponential function. B_n The imaginary part of it is actually very similar to the meaning of frequency in the Fourier series, and it is defined as the "frequency" (in quotes to distinguish from the real frequency) of the complex Gaussian function. Just as $f(x)$ is expanded into the sum of complex exponential functions of different frequencies $\exp(imk_0x)$, $h(\xi)$ The Gaussian series expansion also needs to use complex Gaussian functions of different "frequencies" $\exp\left(-B_n\xi^2\right)$. This is the starting point of this part of the algorithm, by searching for the "frequency" of the complex Gaussian function $\exp\left(-B_n\xi^2\right)$ to determine the basis function set of the Gaussian series. The Gaussian function used in Ding's method [308] is a real Gaussian function, that is, the imaginary part of B_n is all 0, completely cutting off the connection with the complex exponential function, which is the limitation of their algorithm.

The real part of B_n controls the decay characteristics of the complex Gaussian function. Relatively speaking, its importance is not as important as the imaginary part, so it is simply set to the same value δ. In the following "further optimization", the real part of B_n. will be automatically optimized.

It should be noted that, B_n does not necessarily have to be selected as in Eq. (2.45). The reason for using Eq. (2.45) in this section is that, B_n will exhibit the property of complex conjugate pairs (i.e., $B_n = B^*_{N-n+1}$), which will be seen in the examples given below. In fact, the only principle for selecting B_n is that the "frequency" components should be rich and the intervals between "frequencies" should be appropriate, such as $B_n = \delta - i(n - N/2)k_0$ is a strategy that can be used as a substitute for Eq. (2.45).

(C) Further optimization

Based on the above searched δ and k_0, and using Eqs. (2.45) and (2.40), a set of A_n and B_n values can be calculated. Generally speaking, this set of values already has a high precision, but it can be further improved through optimization. By substituting this set of coefficient values as the initial values into the iterative calculation of the steepest descent method below, more optimal coefficient values can be obtained.

$$(A_n)_{next} = A_n - \varepsilon \frac{\partial Q}{\partial A_n^*}, \quad (B_n)_{next} = B_n - \varepsilon \frac{\partial Q}{\partial B_n^*} \tag{2.46}$$

where ε is the step length factor. Stop the iteration when Q reaches a local minimum. Another optional iteration strategy is

$$(B_n)_{next} = B_n - \varepsilon \frac{\partial Q}{\partial B_n^*}, \ (\mathbf{A})_{next} = \mathbf{C}^{-1}\mathbf{D} \tag{2.47}$$

2.2.1.2 Corresponding k-Space Domain Algorithm

It has been introduced in the previous section on Gaussian beam expansion: there is a Fourier transform pair relationship between the spatial domain and k space domain. Since this section focuses on the calculation of one-dimensional Gaussian beam expansion coefficients, the following is the one-dimensional Fourier transform pair

$$\begin{cases} v(k_\xi) = \int_{-\infty}^{+\infty} v(\xi)\exp(-ik_\xi \xi)d\xi \\ v(\xi) = \frac{1}{2\pi}\int_{-\infty}^{+\infty} v(k_\xi)\exp(ik_\xi \xi)dk_\xi \end{cases} \tag{2.48}$$

$v(k_\xi)$ and $v(\xi)$ are in the following Gaussian series form

$$\begin{cases} v(k_\xi) = \sum_{n=1}^{N} A_n^k \exp(-B_n^k k_\xi^2) \\ v(\xi) = \sum_{n=1}^{N} A_n \exp(-B_n \xi^2) \end{cases} \tag{2.49}$$

Substitute Eq. (2.49) into Eq. (2.48), and obtain the following relationship

$$B_n^k = \frac{1}{4B_n}, \ A_n^k = A_n \sqrt{\frac{\pi}{B_n}} \tag{2.50}$$

This reciprocal relationship in Eq. (2.50) is used to extend the algorithm from the spatial domain to the k spatial domain.

The basic algorithm is similar to that in the spatial domain. First, define the objective function of the k spatial domain

$$Q^k\left(A_n^k, B_n^k, N\right) = \int_{-\infty}^{+\infty} \left| v\left(k_\xi\right) - \sum_{n=1}^{N} A_n^k \exp\left(-B_n^k k_\xi^2\right) \right|^2 dk_\xi \tag{2.51}$$

Note the superscript k represents k spatial domain. A_n^k is determined by the condition

$$\frac{\partial Q^k}{\partial \left(A_j^k\right)^*} = 0, \quad j = 1, 2, \ldots, N \tag{2.52}$$

From Eq. (2.52), it can be inferred that

$$\mathbf{A}^k = \left(\mathbf{C}^k\right)^{-1} \mathbf{D}^k \tag{2.53}$$

where

$$\mathbf{A}^k = \begin{pmatrix} A_1^k \\ \vdots \\ A_N^k \end{pmatrix}, \mathbf{C}^k = \begin{pmatrix} C_{11}^k & \cdots & C_{1N}^k \\ \vdots & \ddots & \vdots \\ C_{N1}^k & \cdots & C_{NN}^k \end{pmatrix}, \mathbf{D}^k = \begin{pmatrix} D_1^k \\ \vdots \\ D_N^k \end{pmatrix}$$

$$C_{jn}^k = \int_{-\infty}^{+\infty} \exp\left[-\left(\left(B_j^k\right)^* + B_n^k\right) k_\xi^2\right] dk_\xi = \sqrt{\frac{\pi}{\left(\left(B_j^k\right)^* + B_n^k\right)}}$$

$$D_j^k = \int_{-\infty}^{+\infty} v\left(k_\xi\right) \exp\left(-\left(B_j^k\right)^* k_\xi^2\right) dk_\xi$$

B_n^k is still determined by δ and k_0, but unlike Eq. (2.45), the reciprocal relationship of Eq. (2.50) is used to define B_n^k as

$$B_n^k = \frac{1}{4B_n} = \frac{1}{4(\delta - imk_0)} \tag{2.54}$$

δ and k_0 are still obtained through a two-dimensional search, and their meaning is the same as in Eq. (2.45). Further optimization of A_n^k and B_n^k can be achieved through the following iterative calculation

$$\left(B_n^k\right)_{next} = B_n^k - \varepsilon \frac{\partial Q}{\partial \left(B_n^k\right)^*} \left(\mathbf{A}^k\right)_{next} = \left(\mathbf{C}^k\right)^{-1} \mathbf{D}^k \tag{2.55}$$

From the above description, it becomes clear that the algorithm in the k space domain is basically consistent with the algorithm in the space domain. The only difference is that it uses the reciprocal relationship of Eq. (2.50) to express B_n^k in terms of Eq. (2.54).

2.2.1.3 On-Axis Near-Field Approximation

The following introduces another extension of this algorithm—"on-axis near field matching". It is known that the Rayleigh integral is an exact solution of the linear sound field. When the transducer radius is much larger than the sound wave wavelength (i.e., $ka >> 1$), the Fresnel integral is a good approximation to the Rayleigh integral. The Gaussian beam is derived from the Fresnel integral. Therefore, when the Fresnel integral cannot approximate the near field of the sound field well, the Gaussian beam cannot either. To improve the accuracy of the Gaussian beam in the near field, Wen and Breazeale [306] introduced the dependence of ka into the calculation of the Gaussian beam expansion coefficient by using the on-axis near field solution of the Rayleigh integral as a constraint. This optimization method is called "on-axis near field matching". For this method, this section provides the following algorithm.

Define the target function as

$$Q_M = \int_{-\infty}^{+\infty} \left| v(\xi) - \sum_{n=1}^{N} A_n \exp\left(-B_n \xi^2\right) \right|^2 d\xi + w \int_{z_1}^{z_2} |p_R(z) - p_G(z)|^2 dz \tag{2.56}$$

$p_R(z)$ is the on-axis solution of the Rayleigh integral, which is generally still in integral form, but for a circular transducer with uniform surface sound pressure distribution, it becomes

$$p_R(z) = p_0 e^{-i\omega t + ikz}\left[1 - \exp\left(ik\sqrt{a^2 + z^2} - ikz\right)\right] \tag{2.57}$$

$p_G(z)$ is the on-axis solution of the Gaussian beam.

$$p_G(z) = p_0 e^{-i\omega t + ikz} \sum_{n=1}^{N} \frac{A_n}{1 + iB_n z / z_R} \tag{2.58}$$

where $z_R = ka^2/2$ is the Rayleigh distance, w is the weight coefficient, used to adjust between the approximation of the source distribution function and the approximation of the on-axis sound field, z_1 and z_2 defines the interval that needs to be approximated for the on-axis sound field. Generally speaking, z_2 should be less than z_R. Because when $z > z_R$, $p_G(z)$ and $p_R(z)$ match very well, there is no need to approximate axially.

Using the condition $\partial Q_M/\partial A_j^* = 0$, it can be derived that

$$\mathbf{A} = (\mathbf{C} + w\mathbf{E})^{-1}(\mathbf{D} + w\mathbf{F}) \tag{2.59}$$

where **A**, **C**, **D** is the same as in Eq. (2.40), and

$$\mathbf{E} = \begin{pmatrix} E_{11} & \ldots & E_{1N} \\ \vdots & \ddots & \vdots \\ E_{N1} & \cdots & E_{NN} \end{pmatrix}, \mathbf{F} = \begin{pmatrix} F_1 \\ \vdots \\ F_N \end{pmatrix}$$

$$E_{jn} = \int_{z_1}^{z_2} \frac{1}{1 + iB_n z/z_R} \frac{1}{1 - iB_j^* z/z_R} dz$$

$$F_j = \int_{z_1}^{z_2} p_R(z) \frac{1}{1 - iB_j^* z/z_R} dz.$$

Therefore, the coefficient A_n of the on-axis near-field approximation can be quickly calculated by Eq. (2.59). As for the coefficient B_n of the on-axis near-field approximation, it can be obtained by using the following iterative calculation

$$(B_n)_{next} = B_n - \varepsilon \partial Q_M/\partial B_n^*, \ (\mathbf{A})_{next} = (\mathbf{C} + w\mathbf{E})^{-1}(\mathbf{D} + w\mathbf{F}).$$

The initial value can be attempted using the result of Eq. (2.47).

2.2.2 *Comparison of Numerical Results*

To prove the effectiveness of the algorithm, the Gaussian beam expansion coefficients of various different sound source distribution functions were calculated, and these coefficients were used for the simulation of the sound field. In addition, it is compared with other existing methods to illustrate the advantages of this algorithm.

2.2.2.1 Numerical Results in the Spatial Domain

The numerical results of the spatial domain use the spatial domain algorithm in this section to deal with the most classic and difficult situation, where the sound source distribution function is

$$v(\xi) = \begin{cases} 1 & |\xi| \le 1 \\ 0 & \text{others} \end{cases} \tag{2.60}$$

It represents a circular piston sound source with a uniform sound pressure distribution on the transducer surface. Because the edge of this sound source distribution function is truncated and very steep, it is not easy to fit with a continuous function.

In the algorithm part introduced earlier, the search process of δ and k_0 was not specifically described. This section illustrates this process through examples. Table 2.1 gives the results of the local minimum of $Q(\delta, k_0, N)$ through searching. A total of 63 local minima were found with the range of N from 1 to 26. To illustrate these data more intuitively, they are represented graphically. Figure 2.1a–d respectively show the relationship graphs of N and δ, N and k_0, δ and k_0, and N and Q. From the figures, it is obvious that the positions where the local minima of Q occur are very regular. For a certain value of N, δ and k_0 are confined to a very small area, and as N increases, δ and k_0 both show an increasing trend. Therefore, the area where the local minima of Q occur is relatively concentrated, which provides certain guiding significance for searching for the optimal values of δ and k_0. From Fig. 2.1d it is evident that Q gradually decreases as N increases, but this decreasing trend is constantly weakening, that is, as N increases, the reduction of Q is also constantly decreasing. It indicates that although the source function can be better approximated by increasing the number of Gaussian functions, when N is large enough, the effect of function approximation that can be achieved is very limited.

Figure 2.2 shows the fitting results of the source function obtained by different combinations of δ and k_0 when $N = 10$, to illustrate the impact of these two parameters on the Gaussian series. In Fig. 2.2, the result corresponding to $\delta = 2.98$ and $k_0 = 4.69$ is a local minimum of Q, and it can be observed that the Gaussian series is already in good agreement with the source function. By fixing $k_0 = 4.69$ and varying δ to observe the fitting results, it can be evident that when δ is too small (such as $\delta = 1.3$), the error between the Gaussian series and the source function is in the sidelobe. When δ is too large (such as $\delta = 9$), the side lobes of the Gaussian series are well suppressed, but the main lobe shows obvious oscillations, and the error between it and the source function mainly occurs in the main lobe. When δ is close to the local minimum (such as $\delta = 2$ or $\delta = 5$), the error between the Gaussian series and the source function will be balanced between the main lobe and the side lobes. By fixing $\delta = 2.98$ and varying k_0 to observe the fitting results, it can be apparent that when k_0 is too small (such as $k_0 = 1$), the transition zone of the Gaussian series is very smooth, and the error between it and the source function is mainly concentrated in this area. As k_0 increases, the transition zone becomes steeper and the error decreases. When k_0 is too large (such as $k_0 = 7$), the fitted curve has completely deviated from the steep edge of the sound source function, and the error begins to increase again. In summary, as the decay parameter of the Gaussian series basis function, δ controls the error size of the main lobe and side lobes, and cannot be too large or too small. The appropriate δ can balance the error between the main lobe and the side lobes. Too small δ makes the side lobes very large, while too large δ increases the amplitude of each basis function, leading to oscillation in the main lobe.

Table 2.1 $Q(\delta, k_0, N)$ local minimum

Corresponding to the ○ in Fig. 2.1				Corresponding to the □ in Fig. 2.1				Corresponding to the * in Fig. 2.1			
N	δ	k_0	Q	N	δ	k_0	Q	N	δ	k_0	Q
1	0.98	0	0.2198								
2	1.25	3.330	0.0952								
3	2.00	4.021	0.0672	3	1.00	2.324	0.0858				
4	1.99	4.377	0.0443								
5	2.57	4.665	0.0375	5	1.73	3.471	0.0397				
6	2.43	4.875	0.0283								
7	2.96	5.016	0.0256	7	2.24	4.105	0.0252				
8	2.75	5.161	0.0207	8	2.57	4.416	0.0217				
9	3.21	5.238	0.0193	9	2.60	4.500	0.0183				
10	2.98	5.340	0.0163	10	2.90	4.691	0.0164	10	2.42	4.042	0.0175
11	3.42	5.397	0.0155	11	2.87	4.769	0.0144				
12	3.17	5.472	0.0134	12	3.16	4.900	0.0132	12	2.72	4.346	0.0137
13	3.59	5.508	0.0129	13	3.10	4.963	0.0118				
14	3.32	5.567	0.0114	14	3.35	5.065	0.0110	14	2.96	4.581	0.0112
15	3.73	5.592	0.0111	15	3.28	5.116	0.0100	15	3.15	4.757	0.0103
16	3.44	5.646	0.0099	16	3.54	5.190	0.0094	16	3.16	4.766	0.0095
17	3.83	5.662	0.0097	17	3.44	5.234	0.0087	17	3.34	4.885	0.0089
18	3.57	5.703	0.0088	18	3.67	5.292	0.0083	18	3.33	4.912	0.0082
19	3.94	5.717	0.0086	19	3.58	5.330	0.0076	19	3.51	5.005	0.0078
20	3.66	5.754	0.0079	20	3.81	5.377	0.0073	20	3.49	5.029	0.0072
21	4.02	5.764	0.0078	21	3.70	5.409	0.0068	21	3.65	5.105	0.0069
22	3.75	5.795	0.0071	22	3.92	5.448	0.0066	22	3.61	5.131	0.0065
23	4.10	5.803	0.0071	23	3.79	5.477	0.0062	23	3.78	5.192	0.0062
24	3.82	5.829	0.0065	24	4.01	5.510	0.0060	24	3.73	5.217	0.0059
25	4.17	5.835	0.0065	25	3.89	5.534	0.0056	25	3.89	5.267	0.0056
26	3.91	5.858	0.0060	26	4.10	5.561	0.0055	26	3.84	5.290	0.0053

As the "frequency" parameter of the Gaussian series basis function, k_0 controls the "frequency" component of the basis function. If k_0 is too small, the transition zone is not steep because the high "frequency" is limited, but the value of k_0 cannot be too large, otherwise the "frequency" interval is too large and the error will increase.

From the previous description, it is known that the search for δ and k_0 is easy and it follows a certain pattern. The Gaussian series can approximate the sound source function well, but it can approximate the sound source function better if it is "further optimized" (it is worth noting that when N is already very large, the improvement that can be obtained through this optimization is very limited). Table 2.2

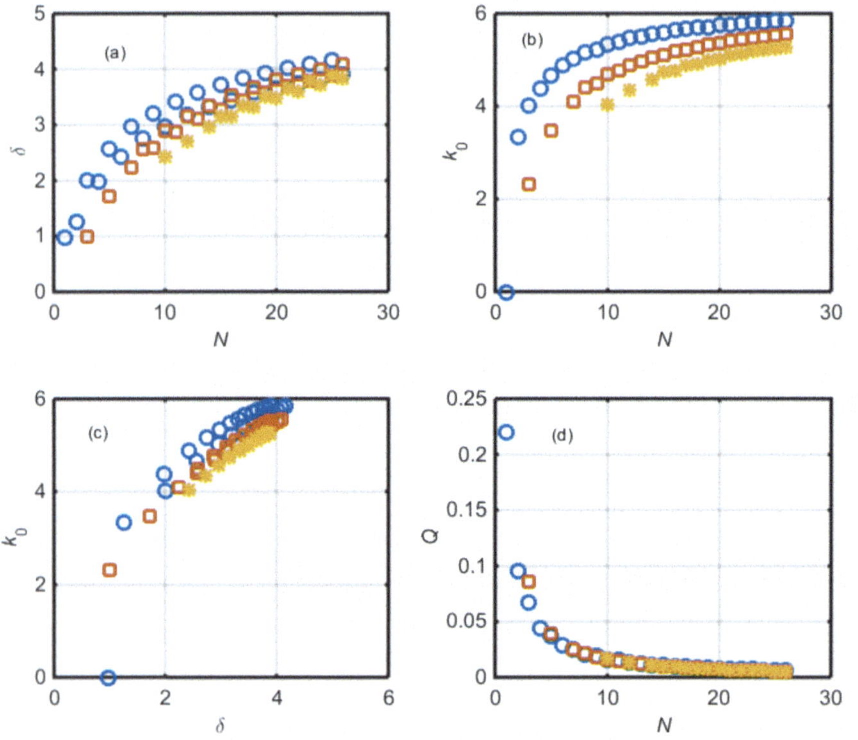

Fig. 2.1 Local minimum of $Q(\delta, k_0, N)$

gives three sets of Gaussian beam expansion coefficients obtained through "further optimization", corresponding to $N = 10$, $N = 15$ and $N = 25$. It can be observed that these coefficients appear in complex conjugate pairs, that is, $A_n = A^*_{N-n+1}$, $B_n = B^*_{N-n+1}$, $n = 1, 2, \ldots N$.

This property makes the imaginary part of the Gaussian series corresponding to these coefficients zero. It can be observed from Fig. 2.3. Figure 2.3 shows the sound source distribution function (including real and imaginary parts) calculated using the three sets of coefficients in Table 2.2. Analyzing these three sets of coefficients, it is found that the maximum values of the imaginary part of B_n are 22.9889, 37.2526, and 65.3980, respectively, and the transition area of the corresponding Gaussian series is getting steeper and steeper. This exactly proves the correctness of the explanation about the imaginary part of B_n being "frequency" in this section. It also shows that by increasing the number of N to improve the approximation effect of the sound source function, it is actually through increasing the maximum value of the imaginary part of B_n, which is similar to the concept of approximating a rectangular periodic function by taking a finite number of terms in the Fourier series. In addition, the Gibbs phenomenon can be clearly seen in Fig. 2.3, which further illustrates the relationship between the Gaussian series and the Fourier series. For comparison, the

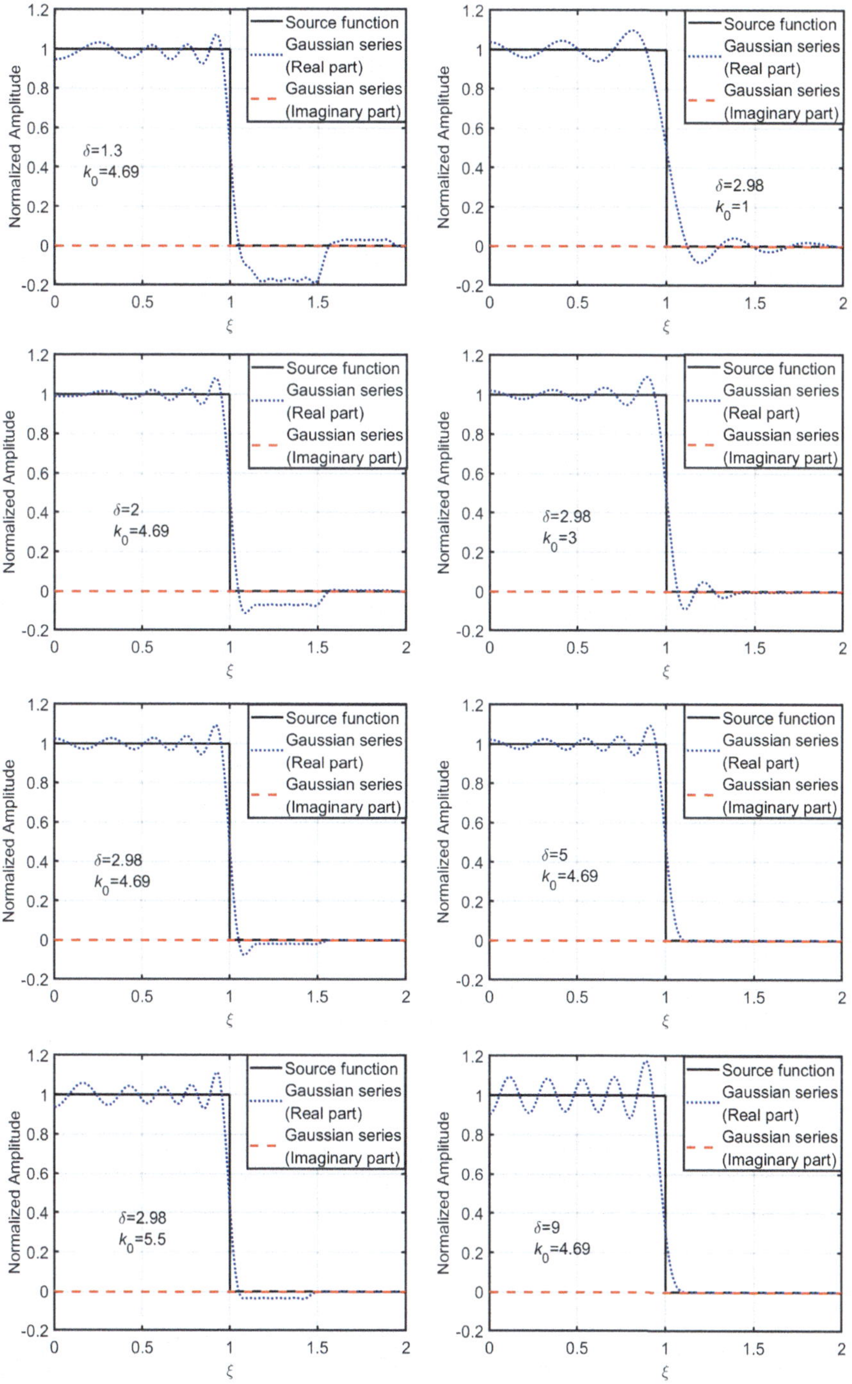

Fig. 2.2 Sound source function obtained by fitting different combinations of δ and k_0 at $N = 10$

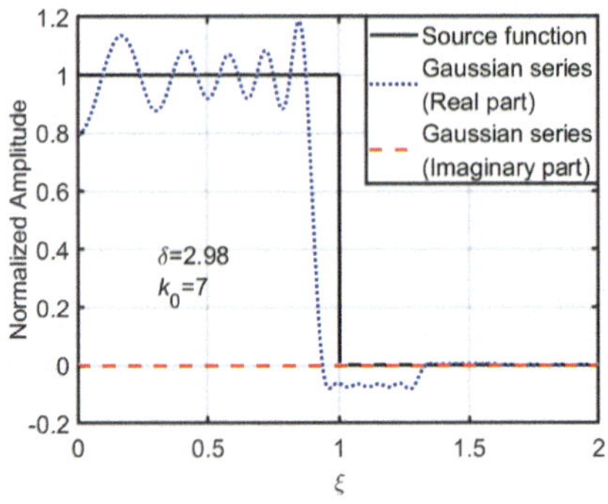

Fig. 2.2 (continued)

results of three sets of coefficients from related research are also given in Table 2.2 and Fig. 2.3. These three sets of coefficients come from Table I of references [101, 306, 319], respectively. It can be found that the coefficients obtained using the spatial domain algorithm in this section are more regular with higher accuracy.

The set of coefficients obtained with this section's algorithm, where $N = 15$, is used to calculate the diffraction sound field of a circular sound source, and the results are given in Fig. 2.4. Figures 2.4a and b show the relative axial sound pressure and the relative errors of the far-field axial sound pressure (where $z_0 = \frac{ka^2}{2}\pi$ is the Fresnel distance). Figure 2.4c–f show the off-axis sound pressure distribution at $z = 0.13z_0$, $z = 0.25z_0$, $z = 0.45z_0$, and $z = z_0$. The results of Wen and Breazeale [306] are also given for comparison in these figures. It can be seen from the figures that the Gaussian beam calculated using the coefficients of this section matches well with the Rayleigh integral (where $ka = 107.8$), except in the vicinity of the sound source (where $z = 0.13z_0$), where the result using the Gaussian beam shows a significant error in the near-axis region compared to the Rayleigh integral. Compared with the results of Wen and Breazeale, the Gaussian beam calculated using the coefficients in this section has higher accuracy in the far field (relative error is 0.2% compared to 1%). Wen and Breazeale's results [306] are more consistent with the Rayleigh integral in the near field, because they used "on-axis near field matching", introducing the axial sound field as a constraint into the calculation of the Gaussian beam expansion coefficients. If the coefficients in this section are also used for "on-axis near field matching", results with the same or better accuracy as Wen and Breazeale can be obtained in the near field.

The above examples fully demonstrate that the spatial domain algorithm in this section can handle various sound source functions. The Gaussian beam expansion coefficients obtained using the algorithm in this section can achieve good accuracy in sound field calculations. Most importantly, the algorithm in this section reveals the connection between the complex Gaussian series and the Fourier series, clarifies the meaning of the Gaussian beam expansion coefficients, and makes the Gaussian beam expansion coefficients, which originally needed to be calculated by the nonlinear optimization algorithm, become regular.

Table 2.2 Gaussian beam expansion coefficients for a uniform piston sound source obtained by the algorithm in this section

n	A_n	B_n	n	A_n	B_n
Algorithm in this section (N = 10)			Wen's result (N = 10)		
1	−0.0817−0.0223i	1.0301–22.9889i	1	11.4280+0.9517i	4.0697+0.2273i
2	−0.1751+0.2233i	1.8122–17.9499i	2	0.0600–0.0801i	1.1531–20.9330i
3	0.2465+0.7579i	2.4698–12.5585i	3	−4.2743–8.5562i	4.4608+5.1268i
4	2.5174+0.1858i	3.1737–7.2265i	4	1.6576+2.7015i	4.3521+14.9970i
5	−2.0167−7.3736i	3.8196–2.2751i	5	−5.0418+3.2488i	4.5443+10.0030i
6	−2.0167+7.3736i	3.8196+2.2751i	6	1.1227–0.6885i	3.8478+20.0780i
7	2.5174–0.1858i	3.1737+7.2265i	7	−1.0106–0.2696i	2.5280–10.3100i
8	0.2465–0.7579i	2.4698+12.5585i	8	−2.5974+3.2202i	3.3197–4.8008i
9	−0.1751–0.2233i	1.8122+17.9499i	9	−0.1484–0.3119i	1.9002–15.8200i
10	−0.0817+0.0223i	1.0301+22.9889i	10	−0.2085–0.2385i	2.6340+25.0090i
Algorithm in this section (N = 15)			Wen 的结果 (N = 15)		
1	−0.0076–0.0504i	1.0024+37.2526i	1	−2.9716+8.6187i	4.1869–5.1560i
2	0.1237–0.0970i	1.7799+32.1297i	2	−3.4811+0.9687i	3.8398–10.8004i
3	0.3759+0.0388i	2.4256+26.5911i	3	−1.3982–0.8128i	3.4355–16.3582i
4	0.5909+0.6979i	3.0891+21.0216i	4	0.0773–0.3303i	2.4618–27.7134i
5	−0.3890+2.1577i	3.6954+15.5508i	5	2.8798+1.6109i	5.4699+28.6319i
6	−5.0111+1.9202i	4.2266+10.1358i	6	0.1259–0.0957i	1.9833–33.2885i
7	−7.0574–12.0770i	4.7656+4.8823i	7	−0.2641–0.6723i	2.9335–22.0151i
8	23.7607	5.0707	8	18.0190+7.8291i	6.3036+36.7772i
9	−7.0574+12.0770i	4.7656–4.8823i	9	0.0518+0.0182i	1.3046–38.4650i
10	−5.0111–1.9202i	4.2266–10.1358i	10	−16.9438–9.9384i	6.5889+37.0680i
11	−0.3890–2.1577i	3.6954–15.5508i	11	0.3708+5.4522i	5.5518+22.4255i
12	0.5909–0.6979i	3.0891–21.0216i	12	−6.6929+4.0722i	5.4013+16.7326i
13	0.3759–0.0388i	2.4256–26.5911i	13	−9.3638–4.9998i	5.1498+11.1249i
14	0.1237+0.0970i	1.7799–32.1297i	14	1.5872–15.4212i	4.9665+5.6855i
15	−0.0076+0.0504i	1.0024–37.2526i	15	19.0024+3.6850i	4.6296+0.3055i
Algorithm in this section (N = 25)			Kim 的结果 (N = 25)		
1	0.0005–0.0436i	1.4108–65.3980i	1	−0.0519+0.0749i	1.9598–68.4910i
2	−0.1345+0.0070i	2.3570–61.2410i	2	−0.0019+0.1334i	2.2259–62.8010i
3	−0.1226+0.2101i	2.7739–55.8478i	3	0.2038+0.1560i	2.6482–51.1480i
4	0.1443+0.3864i	3.1740–50.4954i	4	0.4931–0.0546i	3.0329–39.3090i
5	0.5799+0.2649i	3.4873–44.9934i	5	−0.0140–0.0179i	0.9730–79.0310i
6	0.8938–0.2809i	3.7302–39.4377i	6	0.7515–0.7956i	3.4613–27.3790i
7	0.7220–1.1279i	3.9315–33.8400i	7	−4.6458–6.3564i	4.4953–9.5366i
8	−0.1934–1.8769i	4.0905–28.2074i	8	17.8990–9.5721i	5.3440+6.6418i

(continued)

Table 2.2 (continued)

n	A_n	B_n	n	A_n	B_n
Algorithm in this section ($N = 10$)			Wen's result ($N = 10$)		
9	−1.8527–1.9240i	4.2214–22.5548i	9	0.0205+0.4854i	4.2603+45.7700i
10	−3.7671–0.6152i	4.2827–16.8613i	10	−4.2364–3.8044i	4.9867+17.9350i
11	−5.7399+2.7953i	4.3994–11.0296i	11	0.1324–0.0406i	3.8823+60.8690i
12	−2.6550+14.2718i	4.7582–5.2846i	12	−0.0642–10.4500i	5.1546+12.1720i
13	25.2486	5.0535	13	−0.2405+0.9662i	4.9921+36.9520i
14	−2.6550–14.2718i	4.7582+5.2846i	14	−3.1798+0.0571i	4.8183+23.8730i
15	−5.7399–2.7953i	4.3994+11.0296i	15	0.2452+0.1456i	3.6069+52.5310i
16	−3.7671+0.6152i	4.2827+16.8613i	16	−1.3080+1.0953i	4.6580+30.0990i
17	−1.8527+1.9240i	4.2214+22.5548i	17	−0.0005–0.0209i	0.8307+68.9910i
18	−0.1934+1.8769i	4.0905+28.2074i	18	0.6423–0.3211i	3.2362–33.3510i
19	0.7220+1.1279i	3.9315+33.8400i	19	0.0864+0.1653i	2.4479–57.0080i
20	0.8938+0.2809i	3.7302+39.4377i	20	−0.0543+0.0113i	1.5995–73.9940i
21	0.5799–0.2649i	3.4873+44.9934i	21	14.4330+29.2290i	5.3897+1.3751i
22	0.1443–0.3864i	3.1740+50.4954i	22	0.6891–1.6732i	3.7246–21.4010i
23	−0.1226–0.2101i	2.7739+55.8478i	23	−0.1111–3.4071i	4.0538–15.4360i
24	−0.1345–0.0070i	2.3570+61.2410i	24	−21.0300+3.9134i	5.0621–3.8817i
25	0.0005+0.0436i	1.4108+65.3980i	25	0.3422+0.0904i	2.8400–45.2450i

2.2.2.2 Numerical Results in the k-Space Domain

The numerical results of the k-space domain are used to verify the extension of the algorithm in this section in the k space domain, the following uses the k space domain algorithm in this section to calculate the Gaussian beam expansion coefficients.

The first example to be processed is $\sin(k_\xi)/k_\xi$, the results are given in Fig. 2.5, the three Gaussian series curves in the figure correspond to ($N = 3$, $\delta = 2$, $k_0 = 4.021$), ($N = 5$, $\delta = 2.57$, $k_0 = 4.665$) and ($N = 7$, $\delta = 2.24$, $k_0 = 4.105$). As can be seen, as N increases, the curve of the Gaussian series can approximate $\sin(k_\xi)/k_\xi$ very well in a larger area of k_ξ.

As mentioned earlier, there is a Fourier transform relationship between the spatial domain and the k space domain, and the Gaussian beam expansion coefficients can be converted between these two domains. To illustrate that the k space domain algorithm in this section is not simply transforming the results of the spatial domain to the k space domain, here is a second example, the function being processed is $J_0(k_\xi)$. The differnce between $J_0(k_\xi)$ and $\sin(k_\xi)/k_\xi$ is that the Fourier transform of $J_0(k_\xi)$ does not yield a result, that is, the spatial domain function $J_0(k_\xi)$ does not have a corresponding function in the spatial domain. Therefore, $J_0(k_\xi)$ can only be expanded into a Gaussian series in the k-space domain. Table 2.3 gives two sets of Gaussian beam expansion coefficients for $N = 9$ and $N = 11$ and it can be evident

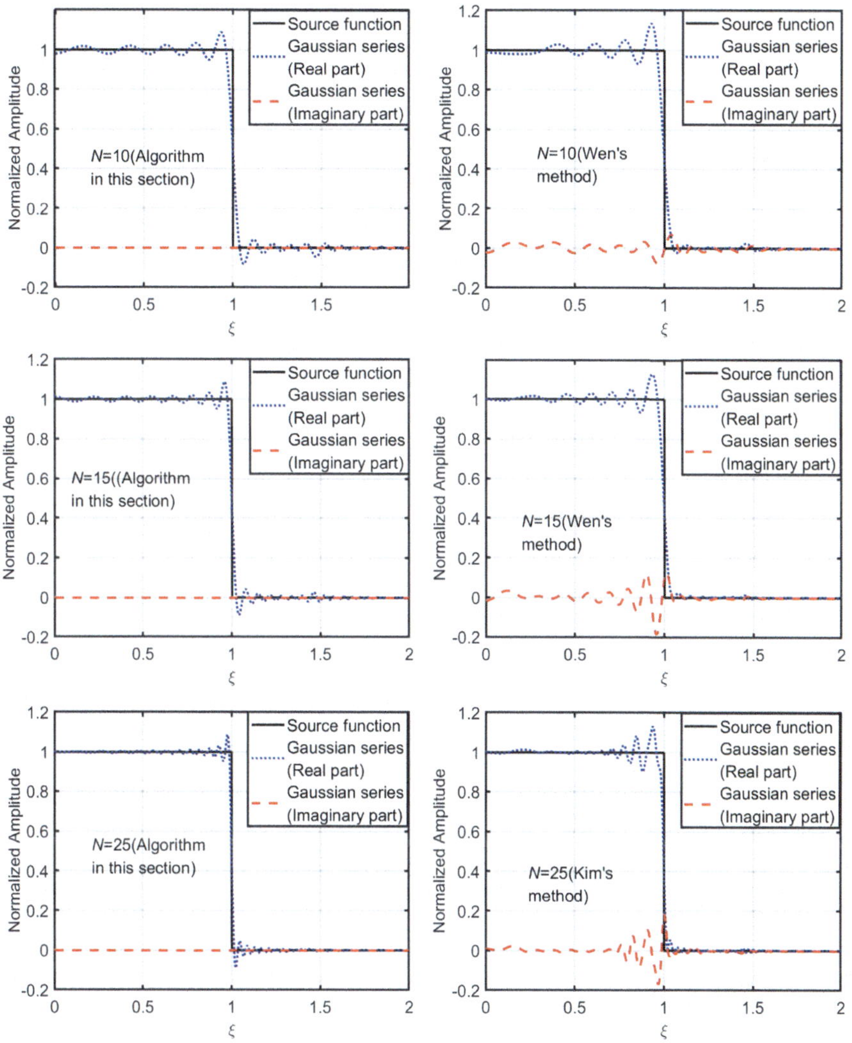

Fig. 2.3 Comparison of the sound source function obtained by the algorithm in this section with the results of related research

that the coefficients still appear in conjugate pairs. The results are given in Fig. 2.6, and it can be observed that the Gaussian series approximation to $J_0\left(k_\xi\right)$ is very good.

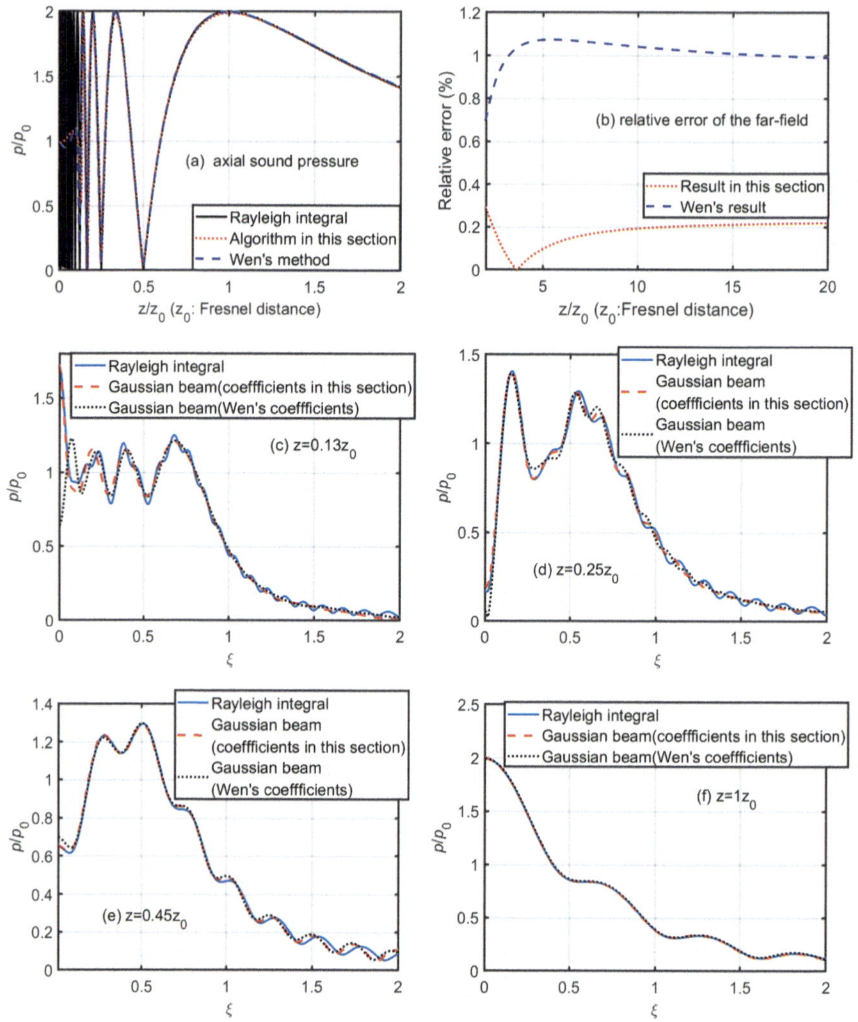

Fig. 2.4 Calculation results of the sound field of a circular sound source with uniform surface sound pressure

2.2.2.3 Numerical Results of the On-Axis Near-Field Approximation

It is known from the "on-axis near field matching" algorithm introduced earlier: the Fresnel integral is an approximation of the Rayleigh integral when $ka >> 1$. To give an intuitive impression, Fig. 2.7 shows the axial sound pressure of the circular piston sound source. It can be seen from the figure that the curve of the Rayleigh integral is different when the value of ka is different. The Fresnel integral (independent of ka) is more consistent with the Rayleigh integral curve when the value of ka is larger, but there are still obvious differences in the near field.

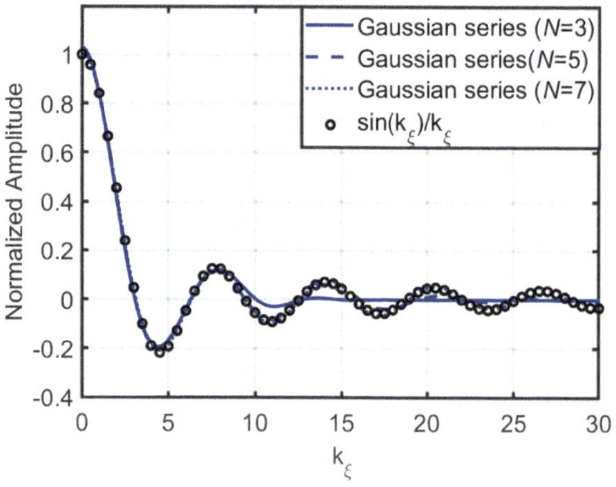

Fig. 2.5 Gaussian series fitting result of $\sin(k_\xi)/k_\xi$

Table 2.3 Gaussian beam expansion coefficients of $J_0(k_\xi)$

$N = 9$			$N = 11$		
n	A_n^k	B_n^k	n	A_n^k	B_n^k
1	0.0014–0.0605i	0.0004–0.0134i	1	−0.0289–0.0525i	0.0005–0.0104i
2	0.1734+0.0397i	0.0016–0.0174i	2	0.1663–0.1043i	0.0010–0.0127i
3	−0.0901+0.3367i	0.0042–0.0257i	3	0.2885+0.4559i	0.0027–0.0163i
4	−0.7043–0.2653i	0.0164–0.0474i	4	−0.7919+0.6490i	0.0060–0.0234i
5	2.2915	0.1474	5	−1.6103–1.6821i	0.0217–0.0381i
6	−0.7043+0.2653i	0.0164+0.0474i	6	4.9671	0.0860
7	−0.0901–0.3367i	0.0042+0.0257i	7	−1.6103+1.6821i	0.0217+0.0381i
8	0.1734–0.0397i	0.0016+0.0174i	8	−0.7919–0.6490i	0.0060+0.0234i
9	0.0014+0.0605i	0.0004+0.0134i	9	0.2885–0.4559i	0.0027+0.0163i
			10	0.1663+0.1043i	0.0010+0.0127i
			11	−0.0289+0.0525i	0.0005+0.0104i

The Gaussian beam is derived from the Fresnel integral. When the expansion coefficients of the Gaussian beam are obtained only by approximating the sound source (as given in the three sets of coefficients in Table 2.2), the Gaussian beam will only coincide with the Fresnel integral in the near field. Using the "on-axis near field matching" algorithm given in this section (with the set of coefficients where $N = 15$ in Table 2.2 as the initial value), a new set of Gaussian beam expansion coefficients (for $ka = 107.8$) were obtained. The coefficient values (which no longer have the property of conjugate pairs) are given in Table 2.4 for the 15 Gaussian beam expansion coefficients processed by axial near field approximation, and the coefficients of Wen and Breazeale (reordered) are also given for comparison. Figure 2.8 shows the sound field results of the Gaussian beam calculated using this new set of coefficients. It can

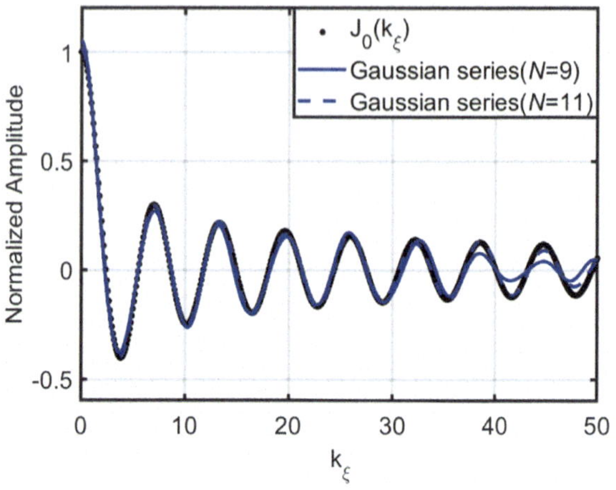

Fig. 2.6 Gaussian series fitting result of $J_0(k_\xi)$

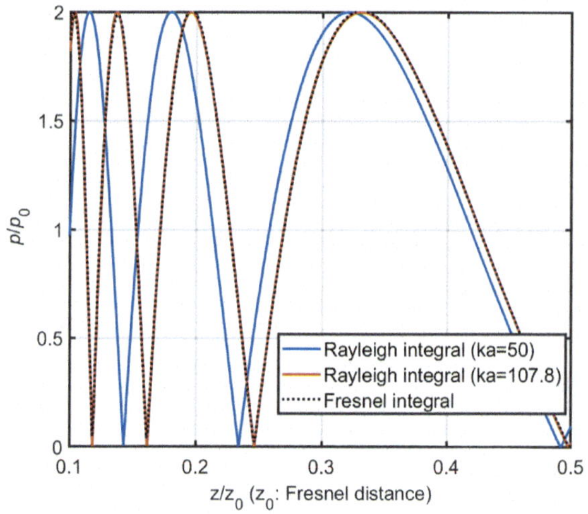

Fig. 2.7 Axial near-field sound pressure of circular piston sound source

be obvious that whether it is the axial near-field sound pressure in Fig. 2.8a or the near-field off-axis sound pressure in Fig. 2.8b, the calculation results of the Gaussian beam are in good agreement with the Rayleigh integral. Compared with the results of Wen and Breazeale, the results of this section can provide similar accuracy, and can provide high-precision simulation from $0.08z_0$ away from the surface of the sound source.

To further improve the accuracy of the near field, the "on-axis near field matching" was performed on the set of coefficients where $N = 25$ in Table 2.2. It is worth noting that a set of Gaussian beam expansion coefficients that have undergone "on-axis near field matching" are dependent on ka and are not suitable for simulating the near field of the sound field when other ka values are used. Therefore, $ka = 107.8$ and $ka = 125.66$ are chosen to implement the "on-axis near field matching". The coefficients

Table 2.4 **15** Gaussian beam expansion coefficients processed by the axial near-field approximation

Coefficients in this section			Wen's coefficients (reordered)		
n	A_n	B_n	n	A_n	B_n
1	−0.0064+0.0482i	1.0140–37.2519i	1	0.0518+0.0182i	1.3046–38.4650i
2	0.1239+0.0911i	1.7850–32.1199i	2	0.1259–0.0957i	1.9833–33.2885i
3	0.3708–0.0485i	2.4214–26.5764i	3	0.0773–0.3303i	2.4618–27.7134i
4	0.5738–0.7058i	3.0741–21.0055i	4	−0.2641–0.6723i	2.9335–22.0151i
5	−0.4127–2.1427i	3.6696–15.5388i	5	−1.3982–0.8128i	3.4355–16.3582i
6	−4.9757–1.8746i	4.1928–10.1373i	6	−3.4811+0.9687i	3.8398–10.8004i
7	−6.8750+11.9016i	4.7378–4.9044i	7	−2.9716+8.6187i	4.1869–5.1560i
8	23.2746–0.2695i	5.0592–0.0168i	8	19.0024+3.6850i	4.6296+0.3055i
9	−7.0984–11.3793i	4.7124+4.8475i	9	1.5872–15.4212i	4.9665+5.6855i
10	−4.8397+2.1589i	4.3314+10.1382i	10	−9.3638–4.9998i	5.1498+11.1249i
11	0.0675+2.1632i	3.8904+15.6115i	11	−6.6929+4.0722i	5.4013+16.7326i
12	0.8532+0.3079i	3.4317+21.2367i	12	0.3708+5.4522i	5.5518+22.4255i
13	0.1537–0.3714i	2.9885+26.7755i	13	2.8798+1.6109i	5.4699+28.6319i
14	−0.2320–0.0052i	2.7294+32.1497i	14	18.0190+7.8291i	6.3036+36.7772i
15	0.0434+0.1103i	2.2337+37.5774i	15	−16.9438–9.9384i	6.5889+37.0680i

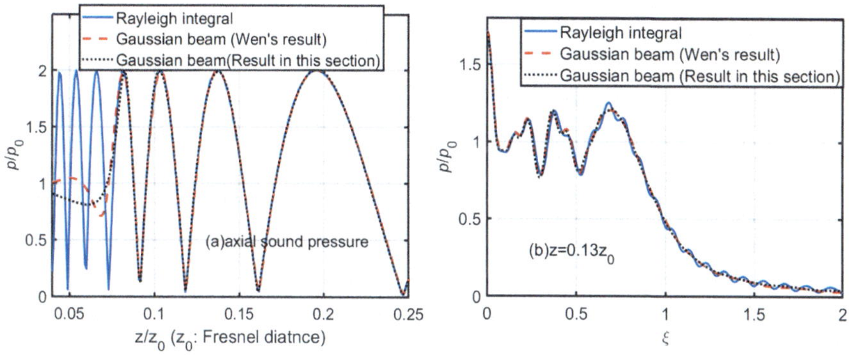

Fig. 2.8 Sound field calculated using the Gaussian beam expansion coefficients in Table 2.4

are given in Table 2.5, and the near-field acoustic fields calculated using these two sets of coefficients are provided in Fig. 2.9. It is evident that the near-field acoustic fields of the two ka values are not quite the same. The Gaussian beams using the two sets of coefficients from Table 2.5 can approximate the near-field to positions $0.05z_0$ close to the sound source surface, which is superior to the results of existing studies. In Fig. 2.9d, within the range of $1.6 < \xi < 2$, there is a clear separation between the Gaussian beams and the Rayleigh integral. This is because this area has exceeded the effective range of the parabolic approximation.

Table 2.5 Gaussian beam expansion coefficients after axial near-field approximation

Gaussian beam expansion coefficients ($ka = 107.8$)			Gaussian beam expansion coefficients ($ka = 125.66$)		
n	A_n	B_n	n	A_n	B_n
1	−0.0080–0.0359i	1.4094–65.3996i	1	−0.0041–0.0405i	1.4077–65.4021i
2	−0.1306+0.0360i	2.3567–61.2430i	2	−0.1350+0.0213i	2.3568–61.2451i
3	−0.0946+0.2481i	2.7748–55.8504i	3	−0.1124+0.2313i	2.7760–55.8529i
4	0.2119+0.4169i	3.1762–50.4979i	4	0.1745+0.4086i	3.1788–50.4999i
5	0.6870+0.2607i	3.4908–44.9955i	5	0.6340+0.2759i	3.4945–44.9965i
6	1.0237–0.3495i	3.7351–39.4390i	6	0.9696–0.2972i	3.7396–39.4388i
7	0.8405–1.2817i	3.9378–33.8400i	7	0.8091–1.1878i	3.9431–33.8380i
8	−0.1326–2.1189i	4.0983–28.2054i	8	−0.1134–1.9940i	4.1038–28.2010i
9	−1.8997–2.2356i	4.2308–22.5497i	9	−1.8062–2.1085i	4.2356–22.5420i
10	−3.9530–0.9559i	4.2936–16.8509i	10	−3.7842–0.8708i	4.2958–16.8390i
11	−6.1068+2.4052i	4.4105–11.0092i	11	−5.8669+2.4139i	4.4061–10.9926i
12	−3.5965+13.9197i	4.7589–5.2430i	12	−3.3319+13.7130i	4.7387–5.2269i
13	23.852+1.0629i	4.9979+0.0436i	13	23.6455+0.7113i	4.9653+0.0231i
14	−2.6100–12.8762i	4.6859+5.2642i	14	−2.5658–13.1102i	4.7085+5.2816i
15	−5.27–1.6802i	4.3840+10.9523i	15	−5.4481–1.7968i	4.3798+10.9416i
16	−2.5901+1.5705i	4.3278+16.8164i	16	−2.8466+1.4940i	4.3141+16.7790i
17	0.0675+1.7753i	4.2633+22.5692i	17	−0.2710+1.9944i	4.2967+22.5376i
18	0.9313+0.1270i	4.0877+28.2329i	18	1.0082+0.5740i	4.1263+28.2576i
19	−0.0894–0.4989i	3.9189+33.8332i	19	0.3238–0.5381i	3.8968+33.8624i
20	−0.2506+0.2393i	3.7421+39.4316i	20	−0.3830–0.0663i	3.7287+39.4037i
21	0.2003+0.0152i	3.4876+45.0030i	21	0.0803+0.2504i	3.5180+45.0059i
22	−0.1105–0.0343i	3.1726+50.4895i	22	0.0625–0.1274i	3.1574+50.5018i
23	0.0689+0.0332i	2.7786+55.8534i	23	−0.0668+0.0678i	2.8010+55.8438i
24	−0.0444–0.0045i	2.3600+61.2361i	24	0.0436–0.0224i	2.3558+61.2392i
25	0.0177–0.0060i	1.4092+65.4013i	25	−0.0089+0.0177i	1.4115+65.4010i

2.2.2.4 Comparison of Prony Method

The Prony method in the k space domain was proposed by Schmerr et al. [329]. The main idea of this method is to first calculate the Gaussian beam expansion coefficients in the k-space domain, and then transform them into the spatial domain. The main advantage of this algorithm is its fast computation speed, because there is no iterative process in the algorithm. However, this algorithm has stability issues.

The following will compare the algorithm in this section with the Prony method using an example from reference [329], that is, for a circular focusing transducer with a radius of 6.35 mm, working at a frequency of 5 MHz, with a focusing distance F of 121.62 mm. The source function is

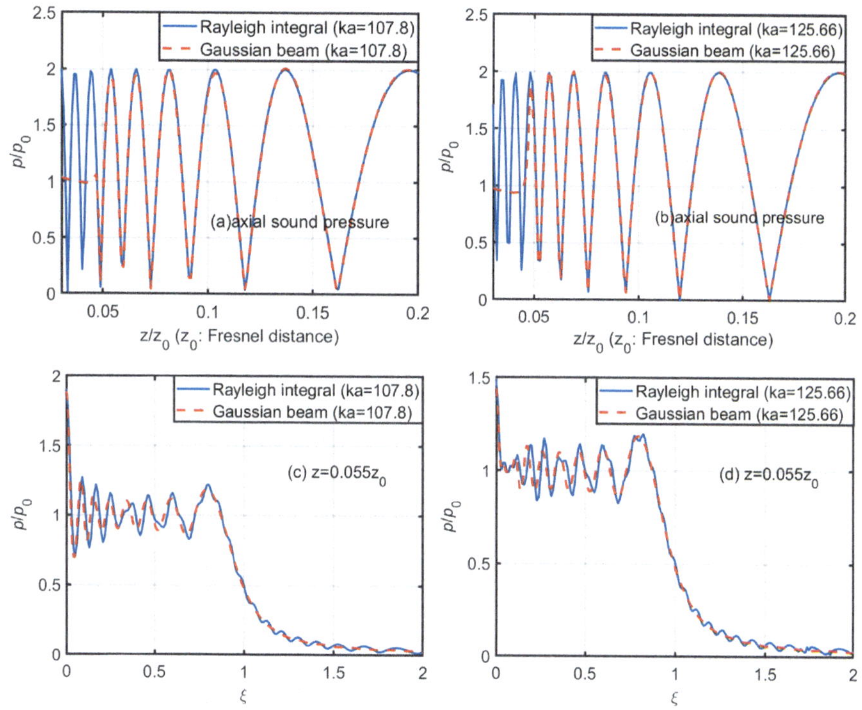

Fig. 2.9 Sound field calculated using the Gaussian beam expansion coefficients in Table 2.5

$$v(\xi) = \begin{cases} \exp\left(-i\xi^2 z_R/F\right) & |\xi| \leq 1 \\ 0 & \text{others} \end{cases} \tag{2.61}$$

This source function only differs from the source function of the previous Eq. (2.60) by a factor of $\exp\left(-i\xi^2 z_R/F\right)$. Adding iz_R/F to the corresponding Gaussian beam expansion coefficient B_n of Eq. (2.60), the Gaussian beam expansion coefficient of Eq. (2.61) can be obtained, which can be expressed as

$$B_n \to B_n + iz_R/F \tag{2.62}$$

Table 2.6 provides two sets of Gaussian beam expansion coefficients obtained using the algorithm in this section and the Prony method. The parameters used in the Prony method are $N = 9$, $\varepsilon = 0.3$, $a = 6.35$, from Fig. 3 in reference [329]. Figure 2.10 illustrates the comparison of the results of fitting the sound source function with these two sets of coefficients. It can be observed that the results obtained by the algorithm in this section are closer to the sound source function, which is because the Prony method does not aiming at minimizing Q. Figure 2.11 shows the results of sound field calculation. The Gaussian beams calculated with the two sets of coefficients match very well, with almost no difference.

Table 2.6 Comparison of the coefficients obtained by the method in this section with those obtained by the Prony method

Algorithm of this section ($N = 9$)			Prony method ($N = 9$)		
n	A_n	B_n	n	A_n	B_n
1	0.0734–0.1293i	1.2952+18.8732i	1	5.2880e−1–8.0669e+0i	4.8330–14.5173i
2	0.7756+0.4589i	2.8644+14.2064i	2	−1.3338e+2+1.3829e+2i	8.1779–10.5484i
3	−1.6242+2.3064i	3.6195+9.5665i	3	1.1737e+3–5.2632e+2i	10.1542–6.9026i
4	−5.1523–5.2004i	4.0753+4.7063i	4	−3.6322e+3+6.6278e+2i	11.2592–3.4073i
5	12.8714	4.3354	5	5.1841e+3	11.6236
6	−5.1523+5.2004i	4.0753–4.7063i	6	−3.6322e+3–6.6278e+2i	11.2592+3.4073i
7	−1.6242–2.3064i	3.6195–9.5665i	7	1.1737e+3+5.2632e+2i	10.1542+6.9026i
8	0.7756–0.4589i	2.8644–14.2064i	8	−1.3338e+2–1.3829e+2i	8.1779+10.5484i
9	0.0734+0.1293i	1.2952–18.8732i	9	5.2880e–1+8.0669e+0i	4.8330+14.5173i

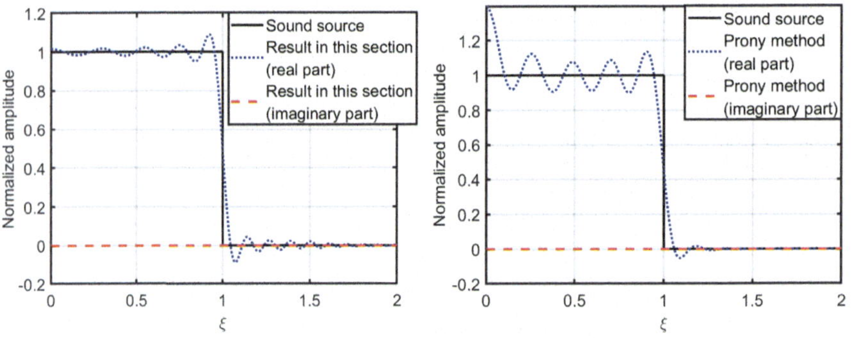

Fig. 2.10 Comparison of sound source functions calculated using the coefficients in Table 2.6

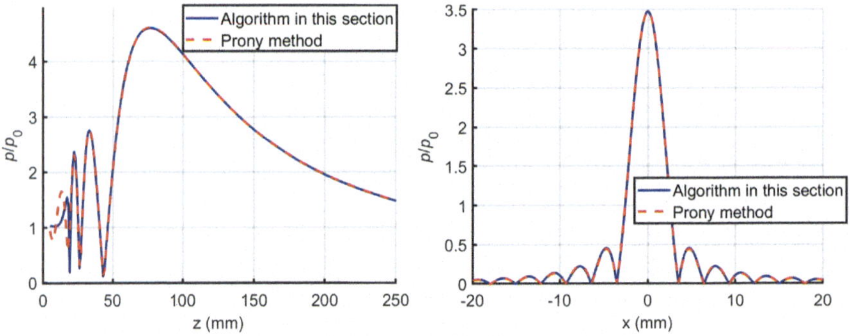

Fig. 2.11 Comparison of sound fields calculated using the coefficients in Table 2.6

From the above example, it can be observed that the Prony method can achieve good results in sound field simulation, but this algorithm does not aim at minimizing Q, so it is not superior to the algorithm in this section in approximating the sound source distribution function. The coefficients obtained by the Prony method all appear in conjugate pairs, and the imaginary part of B_n is separated by a certain "frequency" interval (i.e., k_0 is different), which is very similar to the results obtained by the algorithm in this section. However, the real part of B_n obtained by the Prony method is larger, which makes the absolute value of A_n larger. In the above example, the maximum real part of B_n in this set of coefficients is 11.6236, and the absolute value of A_n reaches the order of magnitude of 10^3. This property essentially indicates that the Prony method has instability. In the Prony method, there is a step of matrix inversion, and whether this matrix is singular directly relates to the stability of the algorithm. In the tests conducted on the Prony method, it was found that when the matrix is close to being singular, the absolute value of A_n becomes very large, and when N is large, the matrix is likely to become singular, leading to algorithm collapse. Additionally, it was found that the Prony method is sensitive to the parameter ε. A slight change in the value of ε can have unpredictable effects on the results, and possibly, this could cause the algorithm to collapse, potentially yield very good results, or lead to an indeterminate situation where the real part of B_n is less than 0 (which is absolutely not allowed). Furthermore, the Prony method does not have an "on-axis near field matching" part, so the accuracy that can be achieved when simulating the near field of the sound field is limited.

2.3 Three-Dimensional Time-Domain Finite Difference Algorithm

The Gaussian beam expansion method of the three-dimensional time-domain finite difference algorithm can be used to quickly calculate the directional beam under weak non-linear conditions, which is very useful for obtaining a large amount of intuitive quantitative results. For example, in the application of parametric arrays, the directivity of the difference frequency sound is very concerned. By using the Gaussian beam expansion method, the directivity results of the difference frequency sound under various conditions (different frequencies, different sound source sizes, different sound source shapes, etc.) can be easily obtained, giving people an intuitive feeling and thus summarizing the pattern. However, when the sound pressure intensity is high, the Gaussian beam expansion method is no longer applicable, and only pure numerical algorithms can be used, which are generally based on the KZK equation for frequency domain or time domain finite difference solution.

This section introduces a time-domain numerical algorithm to solve the three-dimensional nonlinear sound field [335, 336]. This algorithm is an extension of the two-dimensional time-domain algorithm [281]. It is utilized to predict the propagation sound fields generated by parametric array sound sources in the air, and employs an extended version of the KZK equation, considering diffraction, thermal viscous absorption, nonlinearity and relaxation effects. In the solution process, the KZK equation is first converted into the TBE (Transformed beam equation) [272, 273] equation to adapt to the far-field spherical propagation of the parametric array, then the operator splitting method is used to decompose the equation, and finally the numerical solution is obtained through the finite difference method. This algorithm based on the TBE equation is mainly for the case of non-focusing sound sources, but after modification, it can also be used to simulate the case of focusing sound sources. TBE was first introduced by Hamilton et al. [272] into the frequency domain algorithm to calculate the sound field distribution of a single-frequency finite amplitude wave in the far field. Subsequently, TBE is widely used in frequency domain algorithms, such as calculating the nonlinear sound field of dual-frequency sound sources [273] and non-axisymmetric sound sources [98], and has been extended to calculate focused sound sources (TEFB [274, 277]: Transformed equation for focused beam). Lee presented two sets of two-dimensional time-domain algorithms in his doctoral thesis [337]: one for non-focused situations, using TBE; the other for weakly focused situations. But it was not adopted because TEFB can only reduce a limited amount of computation when applied to weakly focused situations. This section mainly considers non-focused beams, so the first set of algorithms by Lee was extended to three dimensions. The harmonic imaging studied by Yang and Cleveland [286] belongs to the weakly focused situation, and their three-dimensional extension of the Texas code is based on Lee's second set of algorithms.

Compared with existing research, this section has the following differences. First, this section includes the algorithm about the relaxation effect by Cleveland et al. [5], making the algorithm more complete, capable of handling any frequency-dependent sound absorption situation. Second, in existing theoretical research, the sound source is simply set as a single circular, square or rectangular sound source, while in actual applications, arrays composed of multiple transducer units are usually used. The nonlinear sound field of the array sound source is calculated in this paper, and it is verified with the results calculated using the Gaussian beam expansion method.

2.3.1 Non-focused Beam

The original KZK equation does not include the relaxation effect, so it is only applicable to the thermal viscous absorption medium. But in actual media (such as air, seawater, etc.), there is often relaxation effect, so this section uses the extended version of the KZK equation.

$$\frac{\partial p}{\partial z} = \frac{c_0}{2}\int_{-\infty}^{t'}\left(\frac{\partial^2 p}{\partial x^2} + \frac{\partial^2 p}{\partial y^2}\right)dt'' + \frac{\delta}{2c_0^3}\frac{\partial^2 p}{\partial t'^2} + \frac{\beta}{2\rho_0 c_0^3}\frac{\partial p^2}{\partial t'}$$

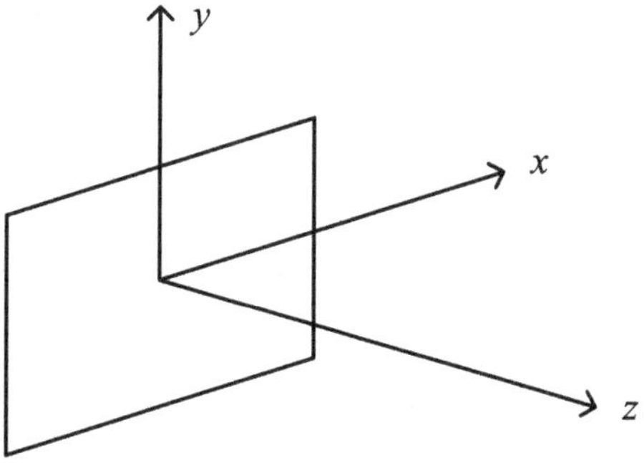

Fig. 2.12 Geometric coordinate system

$$+\sum_{\nu}\frac{c'_{\nu}}{c_0^2}\int_{-\infty}^{t'}\frac{\partial^2 p}{\partial t''^2}\exp\left(-\frac{t'-t''}{t_{\nu}}\right)dt'' \tag{2.63}$$

where p is the sound pressure, $t' = t - z/c_0$ is the delay time, c_0 is the sound speed of small signal, δ is the diffusion coefficient of the viscous medium, β is the non-linear coefficient of the medium, ρ_0 is the density of the medium, c'_{ν} and t_{ν} are the small signal speed increment and relaxation time of the relaxation process $\nu(\nu = 1, 2, \ldots)$ respectively. The right side of Eq. (2.63) represents the diffraction effect, thermoviscous absorption effect, non-linear effect, and relaxation effect in order. The coordinate system of the equation is shown in Fig. 2.12, z is the direction of sound wave propagation, x and y are the transverse directions.

For a sound wave with a frequency of f, the total sound absorption coefficient [248] is calculated as follows

$$\alpha(f) = \alpha_0(f) + \sum\nolimits_{\nu}\alpha_{\nu}(f) = \frac{\delta}{2c_0^3}(2\pi f)^2 + \sum\nolimits_{\nu}\frac{c'_{\nu}}{c_0^2}\frac{2\pi f^2 f_{\nu}}{f^2 + f_{\nu}^2} \tag{2.64}$$

where $f_{\nu} = 1/2\pi t_{\nu}$ is the relaxation frequency, α_0 is the thermal viscous absorption coefficient, α_{ν} is the absorption coefficient of the νth relaxation process.

2.3.1.1 Derivation of the TBE Equation

First the following variable transformation is required for Eq. (2.63)

$$\sigma = z/d \tag{2.65}$$

$$X = \frac{x}{a(1+\sigma)} \tag{2.66}$$

$$Y = \frac{y}{b(1+\sigma)} \tag{2.67}$$

$$\tau = \omega_0 t' - \frac{\omega_0 (x^2 + y^2)}{2c_0 d(1+\sigma)} = \omega_0 t' - (G_x X^2 + G_y Y^2)(1+\sigma) \tag{2.68}$$

$$P = \frac{p(1+\sigma)}{p_0} \tag{2.69}$$

where σ is the dimensionless axial coordinate, d is the characteristic distance in the direction of sound wave propagation (usually the Rayleigh distance), X and Y are the dimensionless transverse coordinates, a is the characteristic length of the sound source in the x axis direction (such as half the length of a rectangular sound source), b is the characteristic length of the sound source in the y axis direction (such as half the width of a rectangular sound source), τ is the dimensionless delay time, ω_0 is the angular frequency of the characteristic frequency of the sound wave f_0, $G_x = \omega_0 a^2/(2c_0 d)$, $G_y = \omega_0 b^2/(2c_0 d)$, P is the dimensionless sound pressure, p_0 is the characteristic sound pressure amplitude.

By substituting the formulas into the equation, each term in the equation can be obtained.

$$\begin{aligned}\frac{\partial p}{\partial z} =& \frac{-p_0}{d(1+\sigma)^2}P + \frac{(G_x X^2 + G_y Y^2)p_0}{d(1+\sigma)}\frac{\partial P}{\partial \tau}\\ &+ \frac{-p_0 Y}{d(1+\sigma)^2}\frac{\partial P}{\partial Y} + \frac{-p_0 X}{d(1+\sigma)^2}\frac{\partial P}{\partial X} + \frac{p_0}{d(1+\sigma)}\frac{\partial P}{\partial \sigma}\end{aligned} \tag{2.70}$$

$$\begin{aligned}\frac{c_0}{2}\int_{-\infty}^{t'}\left(\frac{\partial^2 p}{\partial x^2} + \frac{\partial^2 p}{\partial y^2}\right)dt'' =& \frac{p_0}{4d(1+\sigma)^3}\int_{-\infty}^{\tau}\left(\frac{1}{G_x}\frac{\partial^2 P}{\partial X^2} + \frac{1}{G_y}\frac{\partial^2 P}{\partial Y^2}\right)d\tau'\\ &+ \frac{-p_0}{d(1+\sigma)^2}P + \frac{(G_x X^2 + G_y Y^2)p_0}{d(1+\sigma)}\frac{\partial P}{\partial \tau}\\ &+ \frac{-p_0 Y}{d(1+\sigma)^2}\frac{\partial P}{\partial Y} + \frac{-p_0 X}{d(1+\sigma)^2}\frac{\partial P}{\partial X}\end{aligned} \tag{2.71}$$

$$\frac{\delta}{2c_0^3}\frac{\partial^2 p}{\partial t'^2} = \frac{\delta p_0 \omega_0^2}{2c_0^3(1+\sigma)}\frac{\partial^2 P}{\partial \tau^2} \tag{2.72}$$

$$\frac{\beta}{2\rho_0 c_0^3}\frac{\partial p^2}{\partial t'} = \frac{\beta \omega_0 p_0 p_0 P}{\rho_0 c_0^3 (1+\sigma)^2}\frac{\partial P}{\partial \tau} \tag{2.73}$$

$$\begin{aligned}&\sum\nolimits_\nu \frac{c'_\nu}{c_0^2}\int_{-\infty}^{t'}\frac{\partial^2 p}{\partial t''^2}\exp\left(-\frac{t'-t''}{t_\nu}\right)dt''\\ &= \sum\nolimits_\nu \frac{c'_\nu \omega_0 p_0}{c_0^2(1+\sigma)}\int_{-\infty}^{\tau}\frac{\partial^2 P}{\partial \tau'^2}\exp\left(-\frac{\tau-\tau'}{\theta_\nu}\right)d\tau'\end{aligned} \tag{2.74}$$

Substituting Eqs. (2.70) and (2.74) into Eq. (2.63), and rearranging yields

$$\frac{\partial P}{\partial \sigma} = \frac{1}{4(1+\sigma)^2} \int_{-\infty}^{\tau} \left(\frac{1}{G_x} \frac{\partial^2 P}{\partial X^2} + \frac{1}{G_y} \frac{\partial^2 P}{\partial Y^2} \right) d\tau' + A \frac{\partial^2 P}{\partial \tau^2}$$
$$+ \frac{NP}{(1+\sigma)} \frac{\partial P}{\partial \tau} + \sum_{\nu} \frac{C_\nu}{\theta_\nu} \int_{-\infty}^{\tau} \frac{\partial^2 P}{\partial \tau'^2} \exp\left(-\frac{\tau - \tau'}{\theta_\nu} \right) d\tau' \tag{2.75}$$

where $A = \alpha_0(f_0)d$, $N = \mathrm{d}/\overline{z}$, $\theta_\nu = f_0/f_\nu$, $C_\nu = d\alpha_\nu(f_0)\frac{f_0^2+f_\nu^2}{f_\nu^2}$, $\overline{z} = \frac{\rho_0 c_0^3}{\beta \omega_0 p_0}$ is the distance of shock wave generation,$\alpha_0(f_0) = \frac{\delta \omega_0^2}{2c_0^3}$ and $\alpha_\nu(f_0) = \frac{c_{\nu'}}{c_0^2} \frac{2\pi f_0^2 f_\nu}{f_0^2+f_\nu^2}$ is the thermal viscous absorption coefficient and relaxation absorption coefficient mentioned in Eq. (2.64). Equation (2.75) is a special dimensionless KZK equation, which is a type of TBE, suitable for describing the spherical propagation of non-focusing beams in the far field.

To illustrate the characteristics of TBE, Fig. 2.13 shows the correspondence between the actual coordinates and the normalized coordinates in the TBE equation. As can be seen from the figure, the trapezoidal area in the (x, y, z) coordinate system corresponds to the rectangular area in the (X, Y, σ) coordinate system, that is, the originally diffused area is mapped to a non-diffused area. The reason for doing such a mapping is that: in actual space, sound waves begin to spread in the form of spherical waves after leaving the sound source at a Rayleigh distance, and the further away from the transducer, the wider the beam spreads. If the coordinate system is transformed linearly (as in reference [286]), as σ increases, the area used to describe the beam on the X-Y plane also needs to increase accordingly, which is very unrealistic when $\sigma \gg 1$. For TBE, when $\sigma \gg 1$, $\sqrt{X^2+Y^2}$ is always less than or equal to 1, which means that the area used to describe the beam on the X-Y plane does not increase with the increase of σ. This is more realistic. For TBE, when $\sigma \gg 1$, there is a fixed proportional relation between $\sqrt{X^2+Y^2}$ and $\tan\theta = \sqrt{x^2+y^2}/z$, that is $\tan\theta = \frac{a}{r}\left(1+\sigma^{-1}\right)\sqrt{X^2+Y^2} \approx \frac{a}{r}\sqrt{X^2+Y^2}$, $\sigma \gg 1$.

The wave beam that diffuses and propagates in the (x, y, z) coordinate system is mapped to non-diffusive propagation in the (X, Y, σ) coordinate system, which facilitates the description of beam propagation in a smaller region of the $X-Y$ plane. Therefore, TBE is suitable for describing the spherical propagation of non-focused beams in the far field. The form of the TBE equation in a three-dimensional Cartesian coordinate system is given in Eq. (2.75). By choosing the parameter values in the coordinate transformation, the TBE equation used in existing research can be easily matched.

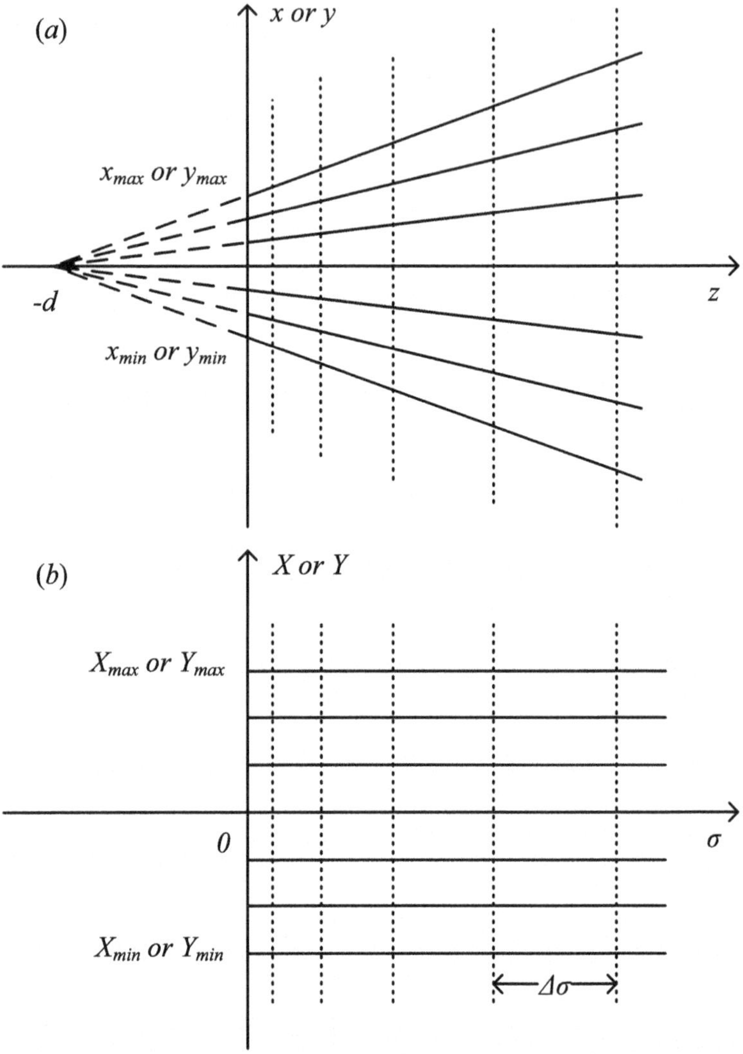

Fig. 2.13 Transformation of TBE equation from **a** actual coordinates to **b** normalized coordinates

2.3.1.2 Boundary Conditions

Some boundary conditions need to be set in the algorithm of this section. The first to be set is the sound source condition (that is, the sound pressure condition on the $z = 0$ plane), which can be represented by the formula

$$p(x, y, z = 0, t) = p_0 f(t) s(x, y) \tag{2.76}$$

where $f(t)$ is the time domain waveform, and $s(x, y)$ is the sound source distribution function. The representation of the sound source condition in the TBE equation is

$$P(X, Y, \sigma = 0, \tau) = F(\tau)S(X, Y) = f\left(\frac{\tau + G_x X^2 + G_y Y^2}{\omega_0}\right)s(aX, bY) \quad (2.77)$$

Taking the square piston sound source with uniform surface sound pressure distribution as an example, the sound source distribution function can be set as

$$S(X, Y) = \begin{cases} 1, & |X| \le 1 \text{ and } |Y| \le 1 \\ 0, & \text{others} \end{cases} \quad (2.78)$$

The next step is to set the lateral boundary conditions. In this part of the algorithm, the lateral spatial interval is set to $X_{\min} \ll X \ll X_{max}$ and $Y_{\min} \ll Y \ll Y_{max}$, and the sound pressure outside this interval is considered to be 0, and the lateral boundary conditions are expressed by the formula

$$P(X_{\min}, Y, \sigma, \tau) = 0 \quad (2.79)$$

$$P(X_{\max}, Y, \sigma, \tau) = 0 \quad (2.80)$$

$$P(X, Y_{\min}, \sigma, \tau) = 0 \quad (2.81)$$

$$P(X, Y_{\max}, \sigma, \tau) = 0 \quad (2.82)$$

The schematic diagram of the boundary on the X–Y plane is given in Fig. 2.14. It should be noted that the spatial boundary conditions can cause reflections, which will affect the calculation results. Relevant studies [272, 273, 337] have highlighted that if the boundary is established at a sufficient distance from the axis, the impact of boundary reflections on the computational outcomes in the near-axis region is minimal. Compared with the reference [286], the TBE used in this part has obvious advantages. As can be seen from Fig. 2.13, the boundaries $X_{\min}$, $X_{\max}$, $Y_{\min}$, and $Y_{\max}$ are fixed, while the actual boundaries $X_{\min}$, $X_{\max}$, $Y_{\min}$, and $Y_{\max}$ are variable and gradually increase with the increase of z (the actual boundaries $x_{\min}$, $x_{\max}$, $y_{\min}$ and $y_{\max}$ are fixed in the reference [286]), which exactly matches the diffusion characteristics of the non-focusing beam, and is beneficial to reduce the boundary reflection. In addition, once the values of ΔX and ΔY are determined, the boundaries $X_{\min}$, $X_{\max}$, $Y_{\min}$ and $Y_{\max}$ actually determine the number of grid points in the finite difference calculation, and the number of grid points is proportional to the memory required for calculation and the calculation time. The algorithm in this section can calculate with fewer grid points in the far field (compared to reference [286]), reducing the memory and calculation time required for calculation. For example, at the position of $\sigma = 9$, the number of grid points required by the algorithm in this section is

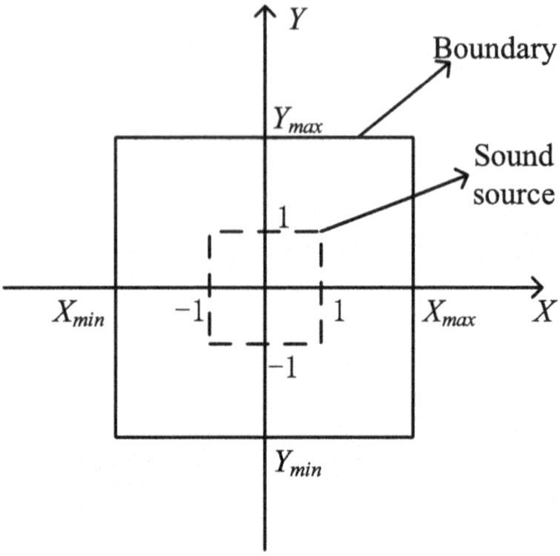

Fig. 2.14 Boundary schematic diagram

$\frac{(x_{\max}-x_{\min})(y_{\max}-y_{\min})}{a^2(1+\sigma)^2\Delta X\,\Delta Y}$, which is 1/100 of the number of grid points required by the algorithm in reference [286]. Absorbing boundaries [338–340] can effectively reduce boundary reflections, but this makes the algorithm more complex, so it is not used in this section.

The last thing to set is the time boundary condition. The time interval (also known as the time window) in the algorithm is set as $\tau_{\min} \ll \tau \ll \tau_{max}$. Outside of this interval, the sound pressure is considered to be 0, so the signal is a pulse signal, but if the time interval is set very large, the signal can also be considered continuous. The time boundary conditions are expressed in the formula as follows

$$P(X, Y, \sigma, \tau_{\min}) = 0 \tag{2.83}$$

$$P(X, Y, \sigma, \tau_{\max}) = 0 \tag{2.84}$$

Figure 2.15 depictsa schematic representation of a time window, where the pulse signal formula is

$$F(\tau) = \exp\left[-(\tau/n\pi)^{2m}\right]\sin\tau \tag{2.85}$$

where n is used to control the pulse width, m is used to control the shape of the amplitude envelope, and both quantities are positive. In Fig. 2.15, $n = 4$, $m = 2$.

In actual calculations, $X_{\min}$, $X_{\max}$, $Y_{\min}$, $Y_{\max}$, $\tau_{\min}$ and $\tau_{\max}$ values should be carefully set, as improper settings can introduce errors into the calculation results. In addition, in some special cases, there will be some special boundary conditions. For example, in the case where the sound field is axisymmetric with respect to both the X and Y axes, the following spatial boundary conditions can be given

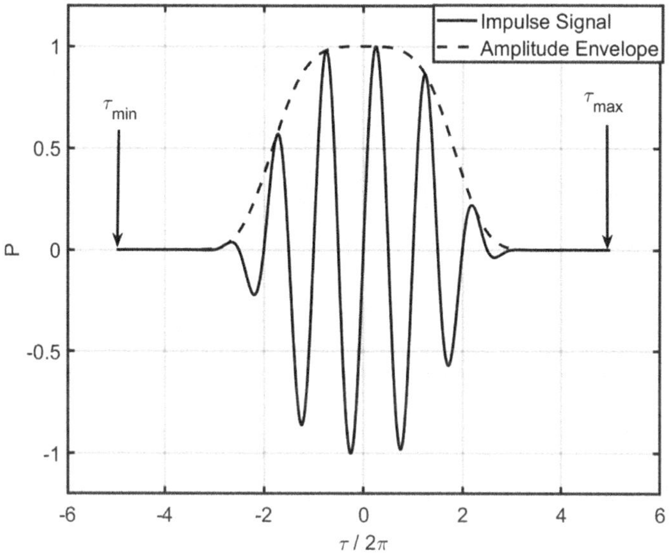

Fig. 2.15 Time window schematic diagram

$$\left.\frac{\partial P}{\partial X}\right|_{X=0} = 0 \tag{2.86}$$

$$\left.\frac{\partial P}{\partial Y}\right|_{Y=0} = 0 \tag{2.87}$$

This boundary condition can reduce the computational region of the space to $1/4$ of the original, that is, $0 \le X \le X_{\max}$, $0 \le Y \le Y_{\max}$, thereby reducing the data storage space and computational load.

The algorithm used in this section is the finite difference method, which requires the sound pressure P to be discretized both in space (X, Y, σ) and time (τ). The value of P at any point in space and time is represented as

$$P_{j,h}^{k,i} = P(j\Delta X, h\Delta Y, \sigma_k, \tau_{\min} + i\Delta\tau) \tag{2.88}$$

where ΔX and ΔY are the spatial sampling steps in the X axis and Y axis direction respectively, $\Delta\tau$ is the time sampling step, the index j, h and i are all integers, and their corresponding relationship with time boundaries and space boundaries is

$$j_{\min} \le j \le j_{\max} \Leftrightarrow X_{\min} \le X \le X_{\max} \tag{2.89}$$

$$h_{\min} \le h \le h_{\max} \Leftrightarrow Y_{\min} \le Y \le Y_{\max} \tag{2.90}$$

$$0 \le i \le i_{\max} \Leftrightarrow \tau_{\min} \le \tau \le \tau_{\max} \tag{2.91}$$

Therefore, the boundary conditions from Eqs. (2.79)–(2.84) can be expressed as

$$P^{k,i}_{j_{\min},h} = 0 \tag{2.92}$$

$$P^{k,i}_{j_{\max},h} = 0 \tag{2.93}$$

$$P^{k,i}_{j,h_{\min}} = 0 \tag{2.94}$$

$$P^{k,i}_{j,h_{\max}} = 0 \tag{2.95}$$

$$P^{k,0}_{j,h} = 0 \tag{2.96}$$

$$P^{k,i_{\max}}_{j,h} = 0 \tag{2.97}$$

The step size in the σ-axis direction varies with distance, so

$$\sigma_k = \sum_{l=0}^{k-1} (\Delta\sigma)_l \tag{2.98}$$

where $(\Delta\sigma)_k$ is the step size of the $k + 1$ step, $k = 0, 1, 2, \ldots$, and the step size of the first step is $(\Delta\sigma)_0$, $\sigma_1 = (\Delta\sigma)_0$.

2.3.1.3 Overview Description of the Algorithm

The operator splitting method is used on Eq. (2.75), resulting in

$$\frac{\partial P}{\partial \sigma} = \frac{1}{4G_x(1+\sigma)^2} \int_{-\infty}^{\tau} \frac{\partial^2 P}{\partial X^2} d\tau' \tag{2.99}$$

$$\frac{\partial P}{\partial \sigma} = \frac{1}{4G_y(1+\sigma)^2} \int_{-\infty}^{\tau} \frac{\partial^2 P}{\partial Y^2} d\tau' \tag{2.100}$$

$$\frac{\partial P}{\partial \sigma} = A \frac{\partial^2 P}{\partial \tau^2} \tag{2.101}$$

$$\left(1 + \theta_\nu \frac{\partial}{\partial \tau}\right) \frac{\partial P}{\partial \sigma} = C_\nu \frac{\partial^2 P}{\partial \tau^2}, \quad \nu = 1, 2, \ldots \tag{2.102}$$

$$\frac{\partial P}{\partial \sigma} = \frac{NP}{(1+\sigma)}\frac{\partial P}{\partial \tau} \tag{2.103}$$

where Eq. (2.102) uses the relationship

$$\phi(\tau) = \left(\frac{1}{\theta_\nu} + \frac{\partial}{\partial \tau}\right)\int_{-\infty}^{\tau} \phi(\tau')\exp\left(-\frac{\tau-\tau'}{\theta_\nu}\right)d\tau'$$

From Eqs. (2.99)–(2.103), it is obvious the TBE equation is decomposed into 5 separate equations, and each equation contains one kind of effect, from top to bottom considering the diffraction effect in the X direction, the diffraction effect in the Y direction, thermoviscous absorption effect, relaxation effect, and nonlinear effect. Figure 2.16 provides the flowchart of the algorithm. As can be seen from the figure, the algorithm process is as follows. Knowing the sound pressure on the σ plane, take the step length $\Delta\sigma$ along positive direction of σ, solve each equation from Eqs. (2.99)–(2.103) in turn, use the sound pressure value obtained from the previous equation as the initial value of the next equation, make corrections to the time domain waveform in turn by diffraction, absorption and other effects, thus obtain the sound pressure waveform on the $\sigma + \Delta\sigma$ plane. IBFD and CNFD in Fig. 2.16 are two finite difference methods, which will be mentioned in the detailed description of the algorithm later. The calculation methods for each function in Eqs. (2.99)–(2.103) are introduced separately below.

2.3.1.4 Diffraction Effect

For convenience, rewrite Eqs. (2.99) and (2.100) here

$$\frac{\partial P}{\partial \sigma} = \frac{1}{4G_x(1+\sigma)^2}\int_{-\infty}^{\tau} \frac{\partial^2 P}{\partial X^2} d\tau' \tag{2.104}$$

$$\frac{\partial P}{\partial \sigma} = \frac{1}{4G_y(1+\sigma)^2}\int_{-\infty}^{\tau} \frac{\partial^2 P}{\partial Y^2} d\tau' \tag{2.105}$$

These two equations have the same form and are treated in the same way. The partial derivatives are treated with finite differences, and the integrals are handled using the trapezoidal method. In the near field area, the truncation of the sound source distribution function (especially in the case of piston sound sources) will cause waveform oscillation. To reduce this oscillation, this section uses the implicit backward finite difference (IBFD) method in the near field area [341]. IBFD is a conditionally stable difference method, for $(\Delta\sigma)_k$ it has a first-order approximation

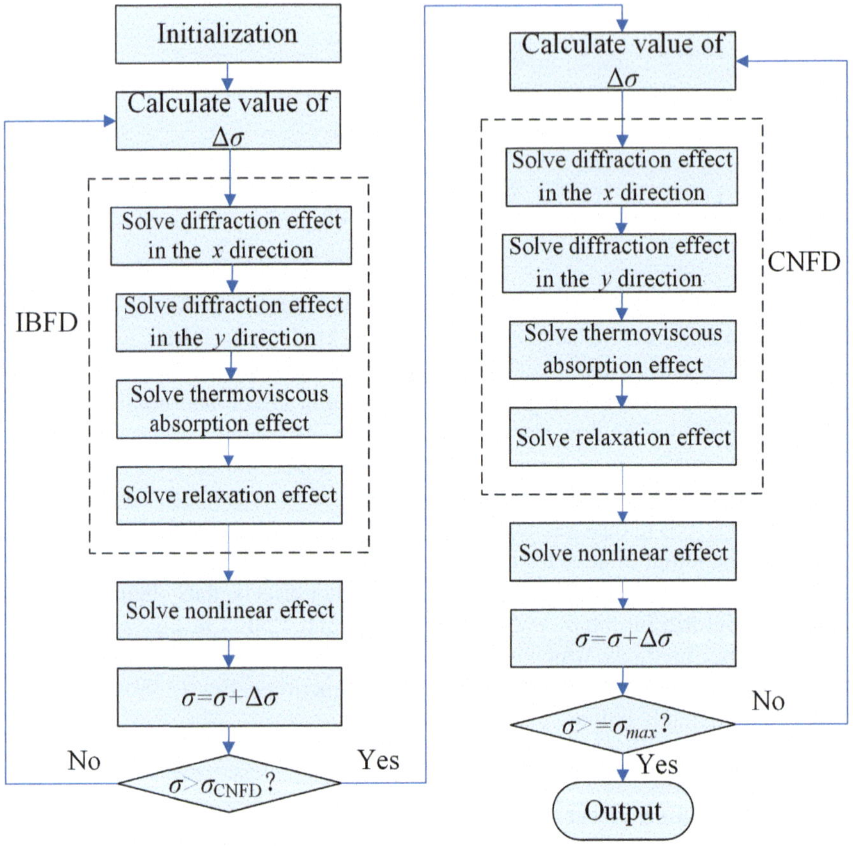

Fig. 2.16 Algorithm flowchart

accuracy, and a smaller step size must be taken to ensure the accuracy of numerical calculation results. Outside the oscillation zone, the Crank-Nicolson finite difference (CNFD) [341] method is used. CNFD is an unconditionally stable central difference method. For $(\Delta\sigma)_k$, it has a second-order approximation accuracy, the step size can be larger, and this difference method can improve computational efficiency.

(1) IBFD

First, introduce the IBFD method. The partial derivatives in Eqs. (2.104) and (2.105) are approximated as follows

$$\frac{\partial P}{\partial \sigma} \Rightarrow \frac{P_{j,h}^{k+1,i} - P_{j,h}^{k,i}}{(\Delta\sigma)_k} \tag{2.106}$$

$$\frac{\partial^2 P}{\partial X^2} \Rightarrow \frac{P_{j+1,h}^{k+1,i} - 2P_{j,h}^{k+1,i} + P_{j-1,h}^{k+1,i}}{\Delta X^2} \tag{2.107}$$

$$\frac{\partial^2 P}{\partial Y^2} \Rightarrow \frac{P_{j,h+1}^{k+1,i} - 2P_{j,h}^{k+1,i} + P_{j,h-1}^{k+1,i}}{\Delta Y^2} \tag{2.108}$$

Substitute Eqs. (2.107) and (2.108) into Eqs. (2.104) and (2.105), and handle the integral using the following trapezoidal method

$$\int_{-\infty}^{\tau} \psi(\tau')d\tau' \Rightarrow \int_{\tau_{min}}^{\tau} \psi(\tau')d\tau' \Rightarrow \Delta\tau\left[\sum_{m=1}^{i-1} \psi_m + \frac{1}{2}(\psi_0 + \psi_i)\right] \tag{2.109}$$

Thus,

$$\int_{-\infty}^{\tau} \frac{\partial^2 P}{\partial X^2}d\tau' \Rightarrow \frac{\Delta\tau}{\Delta X^2} \Rightarrow \left[\sum_{m=1}^{i-1}\left(P_{j+1,h}^{k+1,m} - 2P_{j,h}^{k+1,m} + P_{j-1,h}^{k+1,m}\right) + \frac{1}{2}\left(P_{j+1,h}^{k+1,i} - 2P_{j,h}^{k+1,i} + P_{j-1,h}^{k+1,i}\right)\right] \tag{2.110}$$

$$\int_{-\infty}^{\tau} \frac{\partial^2 P}{\partial Y^2}d\tau' \Rightarrow \frac{\Delta\tau}{\Delta Y^2}\left[\sum_{m=1}^{i-1}\left(P_{j,h+1}^{k+1,m} - 2P_{j,h}^{k+1,m} + P_{j,h-1}^{k+1,m}\right) + \frac{1}{2}\left(P_{j,h+1}^{k+1,i} - 2P_{j,h}^{k+1,i} + P_{j,h-1}^{k+1,i}\right)\right] \tag{2.111}$$

The above formula uses the time boundary conditions of replace the corresponding terms in Eqs. (2.104) and (2.105) with those from Eqs. (2.106), (2.110), and (2.111), and perform some rearrangements to yield.

For Eq. (2.104), when $j = j_{\min} + 1$

$$\left(1 + \frac{R_x}{4}\right)P_{j,h}^{k+1,i} - \frac{R_x}{8}P_{j+1,h}^{k+1,i} = \frac{R_x}{4}\sum_{m=1}^{i-1}\left(-2P_{j,h}^{k+1,m} + P_{j+1,h}^{k+1,m}\right) + P_{j,h}^{k,i} \tag{2.112}$$

When $j_{min} + 2 \leq j \leq j_{\max} - 2$

$$-\frac{R_x}{8}P_{j-1,h}^{k+1,i}+\left(1+\frac{R_x}{4}\right)P_{j,h}^{k+1,i}-\frac{R_x}{8}P_{j+1,h}^{k+1,i}$$
$$=\frac{R_x}{4}\sum_{m=1}^{i-1}\left(P_{j-1,h}^{k+1,m}-2P_{j,h}^{k+1,m}+P_{j+1,h}^{k+1,m}\right)+P_{j,h}^{k,i} \tag{2.113}$$

When $j=j_{max}-1$

$$-\frac{R_x}{8}P_{j-1,h}^{k+1,i}+\left(1+\frac{R_x}{4}\right)P_{j,h}^{k+1,i}$$
$$=\frac{R_x}{4}\sum_{m=1}^{i-1}\left(P_{j-1,h}^{k+1,m}-2P_{j,h}^{k+1,m}\right)+P_{j,h}^{k,i} \tag{2.114}$$

For Eq. (2.105), when $h=h_{\min}+1$

$$\left(1+\frac{R_y}{4}\right)P_{j,h}^{k+1,i}-\frac{R_y}{8}P_{j,h+1}^{k+1,i}$$
$$=\frac{R_y}{4}\sum_{m=1}^{i-1}\left(-2P_{j,h}^{k+1,m}+P_{j,h+1}^{k+1,m}\right)+P_{j,h}^{k,i} \tag{2.115}$$

When $h_{min}+2\le h\le h_{\max}-2$

$$-\frac{R_y}{8}P_{j,h-1}^{k+1,i}+\left(1+\frac{R_y}{4}\right)P_{j,h}^{k+1,i}-\frac{R_y}{8}P_{j,h+1}^{k+1,i}$$
$$=\frac{R_y}{4}\sum_{m=1}^{i-1}\left(P_{j,h-1}^{k+1,m}-2P_{j,h}^{k+1,m}+P_{j,h+1}^{k+1,m}\right)+P_{j,h}^{k,i} \tag{2.116}$$

When $h=h_{\max}-1$

$$-\frac{R_y}{8}P_{j,h-1}^{k+1,i}+\left(1+\frac{R_y}{4}\right)P_{j,h}^{k+1,i}$$
$$=\frac{R_y}{4}\sum_{m=1}^{i-1}\left(P_{j,h-1}^{k+1,m}-2P_{j,h}^{k+1,m}\right)+P_{j,h}^{k,i} \tag{2.117}$$

Equations (2.112) to (2.117) used the lateral boundary conditions from Eqs. (2.92) to (2.95) and

$$R_x=\frac{(\Delta\sigma)_k\Delta\tau}{G_x(1+\sigma_{k+1})^2\Delta X^2} \tag{2.118}$$

$$R_y=\frac{(\Delta\sigma)_k\Delta\tau}{G_y(1+\sigma_{k+1})^2\Delta Y^2} \tag{2.119}$$

Express Eqs. (2.112) to (2.114), Eqs. (2.115) to (2.117) in the form of three-diagonal matrices, respectively, to obtain

$$\begin{pmatrix} \eta_x & \zeta_x & & & & \\ \zeta_x & \eta_x & \zeta_x & & & \\ & \cdot & \cdot & \cdot & & \\ & & \zeta_x & \eta_x & \zeta_x & \\ & & & \cdot & \cdot & \cdot \\ & & & \zeta_x & \eta_x & \zeta_x \\ & & & & \zeta_x & \eta_x \end{pmatrix} \begin{pmatrix} P^{k+1,i}_{j_{\min}+1,h} \\ P^{k+1,i}_{j_{\min}+2,h} \\ \cdot \\ P^{k+1,i}_{j,h} \\ \cdot \\ P^{k+1,i}_{j_{\max}-2,h} \\ P^{k+1,i}_{j_{\max}-1,h} \end{pmatrix} = \begin{pmatrix} \mu_{j_{\min}+1} \\ \mu_{j_{\min}+2} \\ \cdot \\ \mu_j \\ \cdot \\ \mu_{j_{\max}-2} \\ \mu_{j_{\max}-1} \end{pmatrix} \tag{2.120}$$

$$\begin{pmatrix} \eta_y & \zeta_y & & & & \\ \zeta_y & \eta_y & \zeta_y & & & \\ & \cdot & \cdot & \cdot & & \\ & & \zeta_y & \eta_y & \zeta_y & \\ & & & \cdot & \cdot & \cdot \\ & & & \zeta_y & \eta_y & \zeta_y \\ & & & & \zeta_y & \eta_y \end{pmatrix} \begin{pmatrix} P^{k+1,i}_{j,h_{\min}+1} \\ P^{k+1,i}_{j,h_{\min}+2} \\ \cdot \\ P^{k+1,i}_{j,h} \\ \cdot \\ P^{k+1,i}_{j,h_{\max}-2} \\ P^{k+1,i}_{j,h_{\max}-1} \end{pmatrix} = \begin{pmatrix} \mu_{h_{\min}+1} \\ \mu_{h_{\min}+2} \\ \cdot \\ \mu_h \\ \cdot \\ \mu_{h_{\max}-2} \\ \mu_{h_{\max}-1} \end{pmatrix} \tag{2.121}$$

where $\eta_x = 1 + \frac{R_x}{4}$, $\zeta_x = -\frac{R_x}{8}$, $\mu_j = \frac{R_x}{4}\sum_{m=1}^{i-1}\left(P^{k+1,m}_{j-1,h} - 2P^{k+1,m}_{j,h} + P^{k+1,m}_{j+1,h}\right) + P^{k,i}_{j,h}$, $\eta_y = 1 + \frac{R_y}{4}$, $\zeta_y = -\frac{R_y}{8}$, $\mu_h = \frac{R_y}{4}\sum_{m=1}^{i-1}\left(P^{k+1,m}_{j,h-1} - 2P^{k+1,m}_{j,h} + P^{k+1,m}_{j,h+1}\right) + P^{k,i}_{j,h}$, and these are known values. The unknown values $P^{k+1,i}_{j,h}$ can be solved by using the Thomas algorithm (also known as the chasing method). However, it should be noted that, μ_j and μ_h in the calculation formula need to use the sound pressure value before time i on the σ_{k+1} plane (i.e., $P^{k+1,m}_{j,h}$, $m < i$), therefore, the sound pressure value on the σ_{k+1} plane must be solved in the time order from $i = 1$ to $i = i_{\max} - 1$.

Finally, it should be noted that within the algorithm, Eq. (2.105) is solved first, followed by the solution of Eq. (2.104). The two equations are discussed together simply because they are similar.

(B) CNFD

The CNFD method makes the following difference approximation to the partial derivatives in Eqs. (2.104) and (2.105), and yields

$$\frac{\partial P}{\partial \sigma} \Rightarrow \frac{P^{k+1,i}_{j,h} - P^{k,i}_{j,h}}{(\Delta\sigma)_k} \tag{2.122}$$

$$\frac{\partial^2 P}{\partial X^2} \Rightarrow \frac{1}{2}\left(\frac{P^{k+1,i}_{j+1,h} - 2P^{k+1,i}_{j,h} + P^{k+1,i}_{j-1,h}}{\left(\Delta X^2\right)} + \frac{P^{k,i}_{j+1,h} - 2P^{k,i}_{j,h} + P^{k,i}_{j-1,h}}{\left(\Delta X^2\right)}\right) \tag{2.123}$$

$$\frac{\partial^2 P}{\partial Y^2} \Rightarrow \frac{1}{2}\left(\frac{P_{j,h+1}^{k+1,i} - 2P_{j,h}^{k+1,i} + P_{j,h-1}^{k+1,i}}{\Delta Y^2} + \frac{P_{j,h+1}^{k,i} - 2P_{j,h}^{k,i} + P_{j,h-1}^{k,i}}{\Delta Y^2}\right) \tag{2.124}$$

Apply Eqs. (2.123), (2.124) and (2.109) to the integral in Eqs. (2.104) and (2.105), and use the boundary conditions of Eq. (2.96) to obtain

$$\int_{-\infty}^{\tau} \frac{\partial^2 P}{\partial X^2} d\tau' \Rightarrow \frac{\Delta\tau}{2\Delta X^2}\left[\sum_{m=1}^{i-1}\left(Q_{j+1,h}^{k+1,m} - 2Q_{j,h}^{k+1,m} + Q_{j-1,h}^{k+1,m}\right) + \frac{1}{2}\left(Q_{j+1,h}^{k+1,i} - 2Q_{j,h}^{k+1,i} + Q_{j-1,h}^{k+1,i}\right)\right] \tag{2.125}$$

$$\int_{-\infty}^{\tau} \frac{\partial^2 P}{\partial Y^2} d\tau' \Rightarrow \frac{\Delta\tau}{2\Delta Y^2}\left[\sum_{m=1}^{i-1}\left(Q_{j,h+1}^{k+1,m} - 2Q_{j,h}^{k+1,m} + Q_{j,h-1}^{k+1,m}\right) + \frac{1}{2}\left(Q_{j,h+1}^{k+1,i} - 2Q_{j,h}^{k+1,i} + Q_{j,h-1}^{k+1,i}\right)\right] \tag{2.126}$$

where $Q_{j,h}^{k+1,i} = P_{j,h}^{k+1,i} + P_{j,h}^{k,i}$. Replace the corresponding items in Eqs. (2.104) and (2.105) with those from Eqs. (2.122), (2.125) and (2.126), and the following relationship can be derived after rearranging.

For Eq. (2.104), when $j = j_{\min} + 1$

$$\left(1 + \frac{R_x}{8}\right)Q_{j,h}^{k+1,i} - \frac{R_x}{16}Q_{j+1,h}^{k+1,i} = \frac{R_x}{8}\sum_{m=1}^{i-1}\left(-2Q_{j,h}^{k+1,m} + Q_{j+1,h}^{k+1,m}\right) + 2P_{j,h}^{k,i} \tag{2.127}$$

When $j_{min} + 2 \le j \le j_{\max} - 2$

$$-\frac{R_x}{16}Q_{j-1,h}^{k+1,i} + \left(1 + \frac{R_x}{8}\right)Q_{j,h}^{k+1,i} - \frac{R_x}{16}Q_{j+1,h}^{k+1,i} = \frac{R_x}{8}\sum_{m=1}^{i-1}\left(Q_{j-1,h}^{k+1,m} - 2Q_{j,h}^{k+1,m} + Q_{j+1,h}^{k+1,m}\right) + 2P_{j,h}^{k,i} \tag{2.128}$$

When $j = j_{\max} - 1$

$$-\frac{R_x}{16}Q_{j-1,h}^{k+1,i} + \left(1 + \frac{R_x}{8}\right)Q_{j,h}^{k+1,i} = \frac{R_x}{8}\sum_{m=1}^{i-1}\left(Q_{j-1,h}^{k+1,m} - 2Q_{j,h}^{k+1,m}\right) + 2P_{j,h}^{k,i} \tag{2.129}$$

For Eq. (2.105), when $h = h_{\min} + 1$

$$\left(1+\frac{R_y}{8}\right)Q_{j,h}^{k+1,i} - \frac{R_y}{16}Q_{j,h+1}^{k+1,i} = \frac{R_y}{8}\sum_{m=1}^{i-1}\left(-2Q_{j,h}^{k+1,m} + Q_{j,h+1}^{k+1,m}\right) + 2P_{j,h}^{k,i} \tag{2.130}$$

When $h_{min} + 2 \le h \le h_{\max} - 2$

$$-\frac{R_y}{16}Q_{j,h-1}^{k+1,i} + \left(1+\frac{R_y}{8}\right)Q_{j,h}^{k+1,i} - \frac{R_y}{16}Q_{j,h+1}^{k+1,i} = \frac{R_y}{8}\sum_{m=1}^{i-1}\left(Q_{j-1,h}^{k+1,m} - 2Q_{j,h}^{k+1,m} + Q_{j+1,h}^{k+1,m}\right) + 2P_{j,h}^{k,i} \tag{2.131}$$

When $h = h_{max} - 1$

$$-\frac{R_y}{16}Q_{j,h-1}^{k+1,i} + \left(1+\frac{R_y}{8}\right)Q_{j,h}^{k+1,i} = \frac{R_y}{8}\sum_{m=1}^{i-1}\left(Q_{j,h-1}^{k+1,m} - 2Q_{j,h}^{k+1,m}\right) + 2P_{j,h}^{k,i} \tag{2.132}$$

The above formulas use

$$R_x = \frac{(\Delta\sigma)_k \Delta\tau}{G_x\left(1+\sigma_{k+1/2}\right)^2 \Delta X^2} \tag{2.133}$$

$$R_y = \frac{(\Delta\sigma)_k \Delta\tau}{G_y\left(1+\sigma_{k+1/2}\right)^2 \Delta Y^2} \tag{2.134}$$

$$\sigma_{k+1/2} = \sigma_k + \frac{(\Delta\sigma)_k}{2} \tag{2.135}$$

As in IBFD, write Eqs. (2.127) to (2.129), Eqs. (2.130) to (2.132) in the form of three-diagonal matrices respectively, and solve for $Q_{j,h}^{k+1,i}$ using the Thomas algorithm, then through $P_{j,h}^{k+1,i} = Q_{j,h}^{k+1,i} - P_{j,h}^{k,i}$ calculate the value of $P_{j,h}^{k+1,i}$. The solution here also needs to be solved also in time order from $i = 1$ to $i = i_{\max} - 1$.

(C) Treatment of axisymmetric boundary conditions

Among the boundary conditions introduced earlier, there is a special boundary condition, that is, when the sound field is axisymmetric in the X axis direction and Y axis direction,

$$\left.\frac{\partial P}{\partial X}\right|_{X=0} = 0 \tag{2.136}$$

$$\left.\frac{\partial P}{\partial Y}\right|_{Y=0} = 0 \tag{2.137}$$

The second-order partial derivatives under boundary conditions of Eq. (2.136) can be obtained using the following Taylor series expansion

$$P|_{X=\Delta X} = P|_{X=0} + \Delta X \left.\left(\frac{\partial P}{\partial X}\right)\right|_{X=0} + \frac{(\Delta X)^2}{2}\left.\left(\frac{\partial^2 P}{\partial^2 X}\right)\right|_{X=0} + O\big((\Delta X)^3\big)$$

$$\Downarrow$$

$$\left.\left(\frac{\partial^2 P}{\partial^2 X}\right)\right|_{X=0} = 2\frac{P|_{X=\Delta X} - P|_{X=0}}{(\Delta X)^2} + O(\Delta X) \tag{2.138}$$

Equation (2.138) can be used to derive the second order partial derivative IBFD difference approximation as

$$\left.\left(\frac{\partial^2 P}{\partial^2 X}\right)\right|_{X=0} \Rightarrow 2\frac{P_{j+1,h}^{k+1,i} - P_{j,h}^{k+1,i}}{\Delta X^2} \quad \text{at}\, j = 0 \tag{2.139}$$

Substitute Eq. (2.139) into the integral on the right side of Eq. (2.104) to yield

$$\int_{-\infty}^{\tau} \frac{\partial^2 P}{\partial X^2} d\tau' \Rightarrow \frac{\Delta\tau}{\Delta X^2}\left[\sum_{m=1}^{i-1} 2\left(P_{j+1,h}^{k+1,m} - P_{j,h}^{k+1,m}\right) + \left(P_{j+1,h}^{k+1,i} - P_{j,h}^{k+1,i}\right)\right] \text{at}\, j = 0 \tag{2.140}$$

After considering Eq. (2.140), the corresponding IBFD differential equation group of Eq. (2.104) becomes

When $j = 0$

$$\left(1 + \frac{R_x}{4}\right)P_{j,h}^{k+1,i} - \frac{R_x}{4}P_{j+1,h}^{k+1,i} = \frac{R_x}{2}\sum_{m=1}^{i-1}\left(-P_{j,h}^{k+1,m} + P_{j+1,h}^{k+1,m}\right) + P_{j,h}^{k,i} \tag{2.141}$$

When $1 \leq j \leq j_{\max} - 2$

$$-\frac{R_x}{8}P_{j-1,h}^{k+1,i} + \left(1 + \frac{R_x}{4}\right)P_{j,h}^{k+1,i} - \frac{R_x}{8}P_{j+1,h}^{k+1,i}$$

$$= \frac{R_x}{4}\sum_{m=1}^{i-1}\left(P_{j-1,h}^{k+1,m} - 2P_{j,h}^{k+1,m} + P_{j+1,h}^{k+1,m}\right) + P_{j,h}^{k,i} \tag{2.142}$$

When $j = j_{\max} - 1$

$$-\frac{R_x}{8}P_{j-1,h}^{k+1,i} + \left(1 + \frac{R_x}{4}\right)P_{j,h}^{k+1,i} = \frac{R_x}{4}\sum_{m=1}^{i-1}\left(P_{j-1,h}^{k+1,m} - 2P_{j,h}^{k+1,m}\right) + P_{j,h}^{k,i} \tag{2.143}$$

Here, R_x is the same as in Eq. (2.118). The above difference equation group is written in the form of a tridiagonal matrix system, which is

$$\begin{pmatrix} \eta_x\ 2\zeta_x & & & & \\ \zeta_x\ \ \eta_x\ \ \zeta_x & & & & \\ \cdot\ \ \cdot\ \ \cdot & & & & \\ & \zeta_x\ \eta_x\ \zeta_x & & & \\ & & \cdot\ \ \cdot\ \ \cdot & & \\ & & & \zeta_x\ \eta_x\ \zeta_x & \\ & & & & \zeta_x\ \eta_x \end{pmatrix}\begin{pmatrix} P_{0,h}^{k+1,i} \\ P_{1,h}^{k+1,i} \\ \cdot \\ P_{j,h}^{k+1,i} \\ \cdot \\ P_{j_{\max}-2,h}^{k+1,i} \\ P_{j_{\max}-1,h}^{k+1,i} \end{pmatrix} = \begin{pmatrix} \mu_0 \\ \mu_1 \\ \cdot \\ \mu_j \\ \cdot \\ \mu_{j_{\max}-2} \\ \mu_{j_{\max}-1} \end{pmatrix} \tag{2.144}$$

where $\mu_0 = \frac{R}{2}\sum_{m=1}^{i-1}\left(-P_{0,h}^{k+1,m} + P_{1,h}^{k+1,m}\right) + P_{0,h}^{k,i}$. Other parameters in Eq. (2.144) are the same as in Eq. (2.120). The solution of Eq. (2.144) is also obtained by using the Thomas algorithm.

The corresponding CNFD difference approximation of Eq. (2.138) is

$$\left(\frac{\partial^2 P}{\partial^2 X}\right)\Bigg|_{X=0} \Rightarrow \frac{\left(P_{j+1,h}^{k+1,i} - P_{j,h}^{k+1,i} + P_{j+1,h}^{k,i} - P_{j,h}^{k,i}\right)}{\Delta X^2}\ \text{at}\ j = 0 \tag{2.145}$$

Substituting Eq. (2.145) into the integral on the right side of Eq. (2.104) to obtain When $j = 0$

$$\int_{-\infty}^{\tau}\frac{\partial^2 P}{\partial X^2}d\tau' \Rightarrow \frac{\Delta\tau}{\Delta X^2}\left[\sum_{m=1}^{i-1}\left(Q_{j+1,h}^{k+1,m} - Q_{j,h}^{k+1,m}\right) + \frac{1}{2}\left(Q_{j+1,h}^{k+1,i} - Q_{j,h}^{k+1,i}\right)\right] \tag{2.146}$$

where $Q_{j,h}^{k+1,i} = P_{j,h}^{k+1,i} + P_{j,h}^{k,i}$. After considering Eq. (2.146), the corresponding CNFD differential equation group of Eq. (2.104) becomes

When $j = 0$

$$\left(1+\frac{R_x}{8}\right)Q_{j,h}^{k+1,i}-\frac{R_x}{8}Q_{j+1,h}^{k+1,i}=\frac{R_x}{4}\sum_{m=1}^{i-1}\left(-Q_{j,h}^{k+1,m}+Q_{j+1,h}^{k+1,m}\right)+2P_{j,h}^{k,i} \tag{2.147}$$

When $1\le j\le j_{\max}-2$

$$-\frac{R_x}{16}Q_{j-1,h}^{k+1,i}+\left(1+\frac{R_x}{8}\right)Q_{j,h}^{k+1,i}-\frac{R_x}{16}Q_{j+1,h}^{k+1,i}$$
$$=\frac{R_x}{8}\sum_{m=1}^{i-1}\left(Q_{j-1,h}^{k+1,m}-2Q_{j,h}^{k+1,m}+Q_{j+1,h}^{k+1,m}\right)+2P_{j,h}^{k,i} \tag{2.148}$$

When $j=j_{max}-1$

$$-\frac{R_x}{16}Q_{j-1,h}^{k+1,i}+\left(1+\frac{R_x}{8}\right)Q_{j,h}^{k+1,i}=\frac{R_x}{8}\sum_{m=1}^{i-1}\left(Q_{j-1,h}^{k+1,m}-2Q_{j,h}^{k+1,m}\right)+2P_{j,h}^{k,i} \tag{2.149}$$

Here, R_x is the same as in Eq. (2.133). The system of equations above is also solved using the Thomas algorithm.

The process of handling the boundary conditions of Eq. (2.137) is similar to that of Eq. (2.136), and the specific content is not repeated. Under this boundary condition, the IBFD difference equation group corresponding to Eq. (2.105) can be expressed as follow

When $h=0$

$$\left(1+\frac{R_y}{4}\right)P_{j,h}^{k+1,i}-\frac{R_y}{4}P_{j,h+1}^{k+1,i}$$
$$=\frac{R_y}{2}\sum_{m=1}^{i-1}\left(-P_{j,h}^{k+1,m}+P_{j,h+1}^{k+1,m}\right)+P_{j,h}^{k,i} \tag{2.150}$$

When $1\le h\le h_{\max}-2$

$$-\frac{R_y}{8}P_{j,h-1}^{k+1,i}+\left(1+\frac{R_y}{4}\right)P_{j,h}^{k+1,i}-\frac{R_y}{8}P_{j,h+1}^{k+1,i}$$
$$=\frac{R_y}{4}\sum_{m=1}^{i-1}\left(P_{j,h-1}^{k+1,m}-2P_{j,h}^{k+1,m}+P_{j,h+1}^{k+1,m}\right)+P_{j,h}^{k,i} \tag{2.151}$$

When $h=h_{\max}-1$

$$-\frac{R_y}{8}P_{j,h-1}^{k+1,i}+\left(1+\frac{R_y}{4}\right)P_{j,h}^{k+1,i}$$

$$= \frac{R_y}{4}\sum_{m=1}^{i-1}\left(P_{j,h-1}^{k+1,m} - 2P_{j,h}^{k+1,m}\right) + P_{j,h}^{k,i} \tag{2.152}$$

where R_y is the same as in Eq. (2.119). The following is the CNFD differential equation group corresponding to Eq. (2.105) under the boundary conditions of Eq. (2.137)

When $h = 0$

$$\left(1 + \frac{R_y}{8}\right)Q_{j,h}^{k+1,i} - \frac{R_y}{8}Q_{j,h+1}^{k+1,i}$$
$$= \frac{R_y}{4}\sum_{m=1}^{i-1}\left(-Q_{j,h}^{k+1,m} + Q_{j,h+1}^{k+1,m}\right) + 2P_{j,h}^{k,i} \tag{2.153}$$

when $1 \le h \le h_{\max} - 2$

$$-\frac{R_y}{16}Q_{j,h-1}^{k+1,i} + \left(1 + \frac{R_y}{8}\right)Q_{j,h}^{k+1,i} - \frac{R_y}{16}Q_{j,h+1}^{k+1,i}$$
$$= \frac{R_y}{8}\sum_{m=1}^{i-1}\left(Q_{j,h-1}^{k+1,m} - 2Q_{j,h}^{k+1,m} + Q_{j,h+1}^{k+1,m}\right) + 2P_{j,h}^{k,i} \tag{2.154}$$

When $h = h_{\max} - 1$

$$-\frac{R_y}{16}Q_{j,h-1}^{k+1,i} + \left(1 + \frac{R_y}{8}\right)Q_{j,h}^{k+1,i}$$
$$= \frac{R_y}{8}\sum_{m=1}^{i-1}\left(Q_{j,h-1}^{k+1,m} - 2Q_{j,h}^{k+1,m}\right) + 2P_{j,h}^{k,i} \tag{2.155}$$

Here, R_y is the same as in Eq. (2.134).

(D) Variable step length

From a mathematical perspective, to ensure the convergence of the diffraction algorithm, the step length $(\Delta\sigma)_k$ should be roughly proportional to $(1+\sigma_k)^2(\Delta X)^2$ and $(1+\sigma_k)^2(\Delta Y)^2$. Therefore, in the far field, the step size $(\Delta\sigma)_k$ can increase as σ_k increases. From a physical perspective, sound pressure propagates as spherical waves in the far field, and the amplitude will show a significant decrease. The change in sound pressure becomes more gently, which allows the step size $(\Delta\sigma)_k$ in far-field calculations to be increased without losing accuracy (compared to the near field).

For the focusing situation in reference [286], the step size $(\Delta\sigma)_k$ should be proportional to $(\Delta X)^2$ and $(\Delta Y)^2$, and generally only a fixed step size can be used. Therefore, the algorithm in this section is superior to the algorithm in reference [80] in terms of far-field calculations. According to the experience of predecessors [63, 74] in numerical calculations, the selection criteria for the step size $(\Delta\sigma)_k$ are as follows

$$(\Delta\sigma)_k = \begin{cases} 0.001 \times (1+\sigma_k)^2, & \sigma_k \le 0.1 \text{ (IBFD)} \\ 0.0035 \times (1+\sigma_k)^2, & \sigma_k > 0.1 \text{ (CNFD)} \end{cases} \tag{2.156}$$

2.3.1.5 Effect of Thermal Viscous Absorption

This section deals with the effect of thermal viscous absorption. For convenience, Eq. (2.101) is rewritten here

$$\frac{\partial P}{\partial \sigma} = A\frac{\partial^2 P}{\partial \tau^2} \tag{2.157}$$

(1) IBFD

The partial derivatives in Eq. (2.157) are approximated using the IBFD method as follows

$$\frac{\partial P}{\partial \sigma} \Rightarrow \frac{P_{j,h}^{k+1,i} - P_{j,h}^{k,i}}{(\Delta\sigma)_k} \tag{2.158}$$

$$\frac{\partial^2 P}{\partial \tau^2} \Rightarrow \frac{P_{j,h}^{k+1,i+1} - 2P_{j,h}^{k+1,i} + P_{j,h}^{k+1,i-1}}{\Delta\tau^2} \tag{2.159}$$

Substitute Eqs. (2.158) and (2.159) into Eq. (2.157), and then sort to obtain
When $i = 1$

$$(1+2R_\tau)P_{j,h}^{k+1,i} - R_\tau P_{j,h}^{k+1,i+1} = P_{j,h}^{k,i} \tag{2.160}$$

When $2 \le i \le i_{\max} - 2$

$$-R_\tau P_{j,h}^{k+1,i-1} + (1+2R_\tau)P_{j,h}^{k+1,i} - R_\tau P_{j,h}^{k+1,i+1} = P_{j,h}^{k,i} \tag{2.161}$$

When $i = i_{\max} - 1$

$$-R_\tau P_{j,h}^{k+1,i-1} + (1+2R_\tau)P_{j,h}^{k+1,i} = P_{j,h}^{k,i} \tag{2.162}$$

where

$$R_\tau = \frac{A(\Delta\sigma)_k}{\Delta\tau^2} \tag{2.163}$$

The solution process of the above equation group also uses the Thomas algorithm.

(B) CNFD

The CNFD method is used to approximate the partial derivatives in Eq. (2.157) as follows

$$\frac{\partial P}{\partial \sigma} \Rightarrow \frac{P_{j,h}^{k+1,i} - P_{j,h}^{k,i}}{(\Delta\sigma)_k} \tag{2.164}$$

$$\frac{\partial^2 P}{\partial \tau^2} \Rightarrow \frac{1}{2\Delta\tau^2}\left(P_{j,h}^{k+1,i+1} - 2P_{j,h}^{k+1,i} + P_{j,h}^{k+1,i-1} + P_{j,h}^{k,i+1} - 2P_{j,h}^{k,i} + P_{j,h}^{k,i-1}\right) \tag{2.165}$$

Substitute Eqs. (2.164) and (2.165) into Eq. (2.157), and then sort to obtain When $i = 1$

$$(1 + R_\tau)P_{j,h}^{k+1,i} - \frac{R_\tau}{2}P_{j,h}^{k+1,i+1} = (1 - R_\tau)P_{j,h}^{k,i} + \frac{R_\tau}{2}P_{j,h}^{k,i+1} \tag{2.166}$$

When $2 \le i \le i_{\max} - 2$

$$\begin{aligned} &-\frac{R_\tau}{2}P_{j,h}^{k+1,i-1} + (1 + R_\tau)P_{j,h}^{k+1,i} - \frac{R_\tau}{2}P_{j,h}^{k+1,i+1} \\ &= \frac{R_\tau}{2}P_{j,h}^{k,i-1} + (1 - R_\tau)P_{j,h}^{k,i} + \frac{R_\tau}{2}P_{j,h}^{k,i+1} \end{aligned} \tag{2.167}$$

When $i = i_{\max} - 1$

$$-\frac{R_\tau}{2}P_{j,h}^{k+1,i-1} + (1 + R_\tau)P_{j,h}^{k+1,i} = \frac{R_\tau}{2}P_{j,h}^{k,i-1} + (1 - R_\tau)P_{j,h}^{k,i} \tag{2.168}$$

where R_τ is the same as in Eq. (2.163). The solution process of the equation group uses the Thomas algorithm.

2.3.1.6 Relaxation Absorption Effects

For convenience, Eq. (2.102) is rewritten here

$$\left(1 + \theta_\nu \frac{\partial}{\partial \tau}\right)\frac{\partial P}{\partial \sigma} = C_\nu \frac{\partial^2 P}{\partial \tau^2}, \quad \nu = 1, 2, \ldots \tag{2.169}$$

(1) IBFD

Use the IBFD method to approximate the partial derivatives in Eq. (2.169) as follows

$$\frac{\partial P}{\partial \sigma} \Rightarrow \frac{P_{j,h}^{k+1,i} - P_{j,h}^{k,i}}{(\Delta\sigma)_k} \tag{2.170}$$

$$\frac{\partial^2 P}{\partial \tau \partial \sigma} \Rightarrow \frac{\frac{P_{j,h}^{k+1,i+1} - P_{j,h}^{k,i+1}}{(\Delta\sigma)_k} - \frac{P_{j,h}^{k+1,i-1} - P_{j,h}^{k,i-1}}{(\Delta\sigma)_k}}{2\Delta\tau} \tag{2.171}$$

$$\frac{\partial^2 P}{\partial \tau^2} \Rightarrow \frac{P_{j,h}^{k+1,i+1} - 2P_{j,h}^{k+1,i} + P_{j,h}^{k+1,i-1}}{\Delta\tau^2} \tag{2.172}$$

Substitute Eqs. (2.170) to (2.172) into Eq. (2.169), and then rearrange to obtain When $i = 1$

$$(1 + 2\lambda)P_{j,h}^{k+1,i} + (\mu - \lambda)P_{j,h}^{k+1,i+1} = P_{j,h}^{k,i} + \mu P_{j,h}^{k,i+1} \tag{2.173}$$

When $2 \le i \le i_{\max} - 2$

$$\begin{aligned}(-\mu - \lambda)P_{j,h}^{k+1,i-1} + (1 + 2\lambda)P_{j,h}^{k+1,i} + (\mu - \lambda)P_{j,h}^{k+1,i+1} \\ = -\mu P_{j,h}^{k,i-1} + P_{j,h}^{k,i} + \mu P_{j,h}^{k,i+1}\end{aligned} \tag{2.174}$$

When $i = i_{\max} - 1$

$$(-\mu - \lambda)P_{j,h}^{k+1,i-1} + (1 + 2\lambda)P_{j,h}^{k+1,i} = -\mu P_{j,h}^{k,i-1} + P_{j,h}^{k,i} \tag{2.175}$$

where

$$\mu = \frac{\theta_\nu}{2\Delta\tau} \tag{2.176}$$

$$\lambda = \frac{(\Delta\sigma)_k C_\nu}{\Delta\tau^2} \tag{2.177}$$

The solution process of the equation group uses the Thomas algorithm.

(B) CNFD

The CNFD method is used to approximate the partial derivatives in Eq. (2.169) as follows,

$$\frac{\partial P}{\partial \sigma} \Rightarrow \frac{P_{j,h}^{k+1,i} - P_{j,h}^{k,i}}{(\Delta\sigma)_k} \tag{2.178}$$

$$\frac{\partial^2 P}{\partial \tau \partial \sigma} \Rightarrow \frac{\frac{P_{j,h}^{k+1,i+1} - P_{j,h}^{k,i+1}}{(\Delta\sigma)_k} - \frac{P_{j,h}^{k+1,i-1} - P_{j,h}^{k,i-1}}{(\Delta\sigma)_k}}{2\Delta\tau} \tag{2.179}$$

$$\frac{\partial^2 P}{\partial \tau^2} \Rightarrow \frac{1}{2\Delta\tau^2}\left(P_{j,h}^{k+1,i+1} - 2P_{j,h}^{k+1,i} + P_{j,h}^{k+1,i-1} + P_{j,h}^{k,i+1} - 2P_{j,h}^{k,i} + P_{j,h}^{k,i-1}\right) \tag{2.180}$$

Substitute Eqs. (2.178) to (2.180) into Eq. (2.169) and rearrange to obtain
When $i = 1$

$$(1+\lambda)P_{j,h}^{k+1,i} + \left(\mu - \frac{\lambda}{2}\right)P_{j,h}^{k+1,i+1} = (1-\lambda)P_{j,h}^{k,i} + \left(\frac{\lambda}{2} + \mu\right)P_{j,h}^{k,i+1} \tag{2.181}$$

When $2 \le i \le i_{\max} - 2$

$$\begin{aligned}&\left(-\mu - \frac{\lambda}{2}\right)P_{j,h}^{k+1,i-1} + (1+\lambda)P_{j,h}^{k+1,i} + \left(\mu - \frac{\lambda}{2}\right)P_{j,h}^{k+1,i+1}\\ &= \left(\frac{\lambda}{2} - \mu\right)P_{j,h}^{k,i-1} + (1-\lambda)P_{j,h}^{k,i} + \left(\frac{\lambda}{2} + \mu\right)P_{j,h}^{k,i+1}\end{aligned} \tag{2.182}$$

When $i = i_{\max} - 1$

$$\left(-\mu - \frac{\lambda}{2}\right)P_{j,h}^{k+1,i-1} + (1+\lambda)P_{j,h}^{k+1,i} = \left(\frac{\lambda}{2} - \mu\right)P_{j,h}^{k,i-1} + (1-\lambda)P_{j,h}^{k,i} \tag{2.183}$$

where μ and λ are the same as in Eqs. (2.176) and (2.177). The solution process of the equation group uses the Thomas algorithm.

2.3.1.7 Nonlinear Effect

(1) Implicit solution algorithm

For convenience, the formula is rewritten here

$$\frac{\partial P}{\partial \sigma} = \frac{NP}{1+\sigma}\frac{\partial P}{\partial \tau} \tag{2.184}$$

The solution process of this equation is as follows.
Define a new variable

$$\xi(\sigma) = \ln(1+\sigma) \tag{2.185}$$

Introduce ξ into the equation to obtain

$$\frac{d\xi}{d\sigma}\frac{\partial P}{\partial \xi} = \frac{NP}{1+\sigma}\frac{\partial P}{\partial \tau} \tag{2.186}$$

Substitute

$$\frac{d\xi}{d\sigma} = \frac{1}{1+\sigma} \tag{2.187}$$

into Eq. (2.186) to obtain

$$\frac{\partial P}{\partial \xi} = NP\frac{\partial P}{\partial \tau} \tag{2.188}$$

This equation has an implicit solution

$$P(\xi, \tau) = F(\tau + NP\xi) \tag{2.189}$$

where $F(\tau)$ is an arbitrary function of time. Substitute Eq. (2.185) into Eq. (2.189) to obtain

$$P(\sigma, \tau) = F[\tau + NP\ln(1+\sigma)] \tag{2.190}$$

When the value of $P\left(\sigma_k, \tau_{j,h}^{k,i}\right)$ is known, to calculate the value of $P\left(\sigma_{k+1}, \tau_{j,h}^{k+1,i}\right)$, the following derivation can be deduced

$$\begin{aligned}
P\left(\sigma_{k+1}, \tau_{j,h}^{k+1,i}\right) &= F\left[\tau_{j,h}^{k+1,i} + NP\ln(1 + \sigma_k + (\Delta\sigma)_k)\right] \\
&= F\left[\tau_{j,h}^{k+1,i} + NP\ln\left(1 + \frac{(\Delta\sigma)_k}{1+\sigma_k}\right) + NP\ln(1+\sigma_k)\right] \\
&= P\left[\sigma_k, \tau_{j,h}^{k+1,i} + NP\ln\left(1 + \frac{(\Delta\sigma)_k}{1+\sigma_k}\right)\right] \\
&= P\left(\sigma_k, \tau_{j,h}^{k,i}\right) = P_{j,h}^{k,i}
\end{aligned} \tag{2.191}$$

Therefore, the following relationship exists

$$\tau_{j,h}^{k+1,i} = \tau_{j,h}^{k,i} - NP_{j,h}^{k,i}\ln\left(1 + \frac{(\Delta\sigma)_k}{1+\sigma_k}\right) \tag{2.192}$$

It is known from Eq. (2.88) that

$$\tau_{j,h}^{k,i} = \tau_{\min} + \mathrm{i}\,\Delta\tau \tag{2.193}$$

Therefore, when $P_{j,h}^{k,i} \neq 0$, $P\left(\sigma_{k+1}, \tau_{j,h}^{k+1,i}\right)$ is not the expected value $P_{j,h}^{k+1,i}$ at $\tau_{\min}$+ i $\Delta\tau$ time. To obtain $P_{j,h}^{k+1,i}$, the linear interpolation algorithm is required.

First consider the case where $P_{j,h}^{k,i} < 0$, and at this time $\tau_{j,h}^{k+1,i} > \tau_{\min} + \mathrm{i}\Delta\tau$, so

$$P_{j,h}^{k+1,i} = P\left(\sigma_{k+1}, \tau_{j,h}^{k+1,i-1}\right)$$

$$+\frac{P\left(\sigma_{k+1}, \tau_{j,h}^{k+1,i}\right) - P\left(\sigma_{k+1}, \tau_{j,h}^{k+1,i-1}\right)}{\tau_{j,h}^{k+1,i} - \tau_{j,h}^{k+1,i-1}}\left(\tau_{\min} + \mathrm{i}\Delta\tau - \tau_{j,h}^{k+1,i-1}\right) \tag{2.194}$$

Substitute Eqs. (2.191) to (2.193) into Eq. (2.194) to obtain

$$\begin{aligned} P_{j,h}^{k+1,i} &= P_{j,h}^{k,i-1} + \frac{P_{j,h}^{k,i} - P_{j,h}^{k,i-1}}{\Delta\tau - N\left(P_{j,h}^{k,i} - P_{j,h}^{k,i-1}\right)\ln\left(1 + \frac{(\Delta\sigma)_k}{1+\sigma_k}\right)} \\ &\left[\Delta\tau + NP_{j,h}^{k,i-1}\ln\left(1 + \frac{(\Delta\sigma)_k}{1+\sigma_k}\right)\right] \\ &= \frac{P_{j,h}^{k,i}}{1 - N\frac{\left(P_{j,h}^{k,i} - P_{j,h}^{k,i-1}\right)}{\Delta\tau}\ln\left(1 + \frac{(\Delta\sigma)_k}{1+\sigma_k}\right)} \end{aligned} \tag{2.195}$$

Now consider the case where $P_{j,h}^{k,i} > 0$, and at this time $\tau_{j,h}^{k+1,i} < \tau_{\min} + \mathrm{i}\Delta\tau$, so

$$\begin{aligned} P_{j,h}^{k+1,i} &= P(\sigma_{k+1}, \tau_{j,h}^{k+1,i}) \\ &+ \frac{P(\sigma_{k+1}, \tau_{j,h}^{k+1,i+1}) - P(\sigma_{k+1}, \tau_{j,h}^{k+1,i})}{\tau_{j,h}^{k+1,i+1} - \tau_{j,h}^{k+1,i}}\left(\tau_{\min} + i\Delta\tau - \tau_{j,h}^{k+1,i}\right) \end{aligned} \tag{2.196}$$

Substitute Eqs. (2.191) to (2.193) into Eq. (2.196) to obtain

$$\begin{aligned} P_{j,h}^{k+1,i} &= P_{j,h}^{k,i} + \frac{P_{j,h}^{k,i+1} - P_{j,h}^{k,i}}{\Delta\tau - N\left(P_{j,h}^{k,i+1} - P_{j,h}^{k,i}\right)\ln\left(1 + \frac{(\Delta\sigma)_k}{1+\sigma_k}\right)} \\ NP_{j,h}^{k,i}\ln\left(1 + \frac{(\Delta\sigma)_k}{1+\sigma_k}\right) &= \frac{P_{j,h}^{k,i}}{1 - N\frac{\left(P_{j,h}^{k,i+1} - P_{j,h}^{k,i}\right)}{\Delta\tau}\ln\left(1 + \frac{(\Delta\sigma)_k}{1+\sigma_k}\right)} \end{aligned} \tag{2.197}$$

Figure 2.17 shows a schematic diagram of the interpolation algorithm described above.

(B) Requirements for step size in calculations

As known from the above, nonlinear effects can cause distortion in the waveform of sound waves, which originates from Eq. (2.192). According to the derivation above, the change in sound pressure waveform represented by Eq. (2.184) can be understood as follows: when the waveform sampled at equal time intervals on the σ_k plane travels to the $\sigma_k + (\Delta\sigma)_k$ plane, the sound pressure value remains the same, but the position on the time axis of each sound pressure value changes (as shown in Fig. 2.18).

It is worth noting that when the step size $(\Delta\sigma)_k$ is very large, the waveform of the sound wave will show a multi-value phenomenon, as shown in Fig. 2.19. However, in

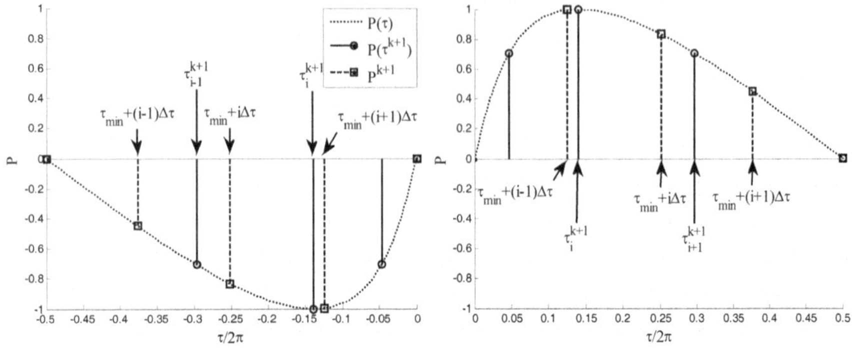

Fig. 2.17 Interpolation algorithm schematic diagram

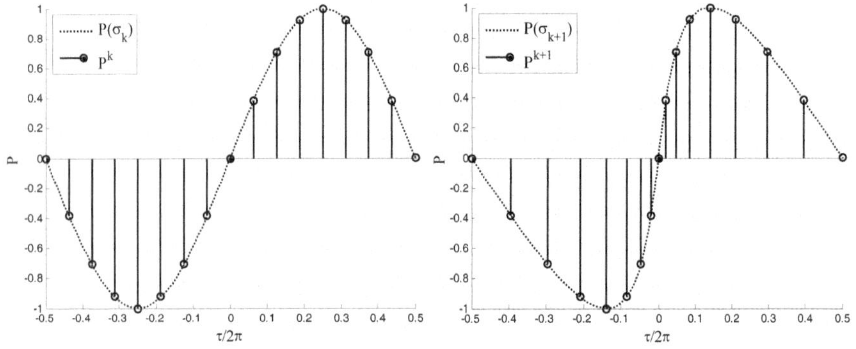

Fig. 2.18 Waveform distortion schematic diagram

actual situations, due to the effect of sound absorption, the multi-value phenomenon will not occur. To avoid the occurrence of multi-value phenomena in calculations, it is necessary to ensure that $\left(\tau_{j,h}^{k+1,i+1} - \tau_{j,h}^{k+1,i}\right) > 0$, which can be derived from Eq. (2.192).

$$\tau_{j,h}^{k+1,i+1} - \tau_{j,h}^{k+1,i} = \Delta\tau - N\ln\left(1 + \frac{(\Delta\sigma)_k}{1+\sigma_k}\right)\left(P_{j,h}^{k,i+1} - P_{j,h}^{k,i}\right) > 0 \tag{2.198}$$

Therefore

$$\frac{\Delta\tau}{N\left(P_{j,h}^{k,i+1} - P_{j,h}^{k,i}\right)} > \ln\left(1 + \frac{(\Delta\sigma)_k}{1+\sigma_k}\right) \tag{2.199}$$

Also, $\ln(1+\varepsilon) \approx \varepsilon - \frac{\varepsilon^2}{2} + \frac{\varepsilon^3}{3} + \cdots (-1)^{n+1}\frac{\varepsilon^n}{n}$. Taking its first-order approximation, the formula can be rewritten as

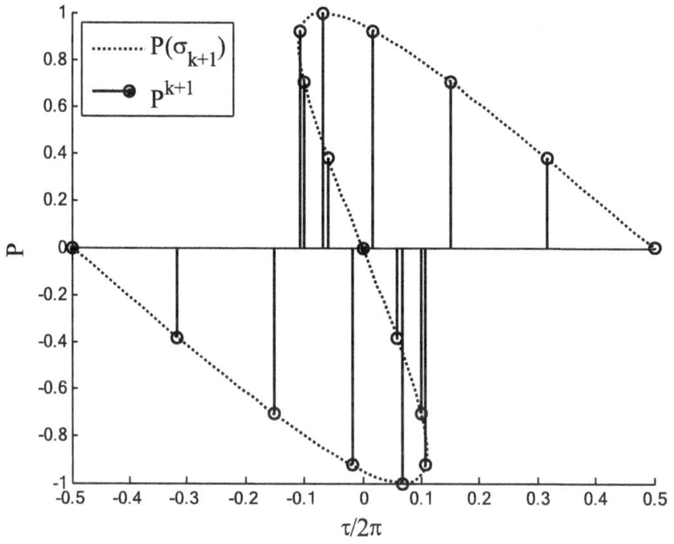

Fig. 2.19 Multi-value phenomenon schematic diagram

$$\frac{\Delta\tau}{N\left(P_{j,h}^{k,i+1} - P_{j,h}^{k,i}\right)} > \frac{(\Delta\sigma)_k}{1+\sigma_k} \tag{2.200}$$

Then the range of the step size $(\Delta\sigma)_k$ can be deduced

$$(\Delta\sigma)_k < \frac{\Delta\tau(1+\sigma_k)}{N\max\left(P_{j,h}^{k,i+1} - P_{j,h}^{k,i}\right)} \tag{2.201}$$

In reference [337], Lee provided a method for selecting $\Delta\tau$, and the size of $\Delta\tau$ is directly related to the Gol'dberg number $\Gamma = N/A = 1/\alpha_0\bar{z}$ (Γ This is related to the relative size relationship between non-linear effects and thermal viscous absorption). However, in addition to thermal viscous absorption, the algorithm in this section also includes relaxation absorption, so the method of selecting $\Delta\tau$ given by Lee cannot be directly applied to the algorithm in this section. Referencing Lee's method, and based on actual numerical simulation experience, the following method of selecting $\Delta\tau$ is given

$$\Delta\tau = \begin{cases} 2\pi/60, & \Gamma' < 1 \\ 2\pi/120, & 1 \le \Gamma' < 10 \\ 2\pi/240, & 10 \le \Gamma' \le 200 \end{cases} \tag{2.202}$$

where $\Gamma' = 1/\alpha\bar{z}$ is used to describe the relative size relationship between non-linear effects and total absorption. It should be noted that Eq. (2.202) is only a reference range and is not absolutely correct. In actual simulations, it is sometimes necessary to reduce $\Delta\tau$ to improve the accuracy of nonlinear simulation.

2.3.2 *Calculation Examples*

2.3.2.1 Circular Sound Source

Taking a circular sound source as an example, the calculation results of three-dimensional time-domain code and two-dimensional time-domain code are compared. This example is given in Figure 3 of the reference [5] by Cleveland et al. The signal is a short pulse, which includes diffraction, absorption, nonlinear effects, and a relaxation process. The initial time-domain waveform is $f(t) = \exp\left[-(\omega_0 t/3\pi)^8\right]\sin\omega_0 t$.

The sound source distribution function is

$$S(X, Y) = \begin{cases} 1, & X^2 + Y^2 \leq 1 \\ 0, & \text{others} \end{cases}$$

Other parameters are: $G = 1, A = 0.01, N = 1, C_1 = 1, \theta_1 = 1, \tau_{\min} = -18\pi$, $\tau_{\max} = 28\pi$, $\Delta\tau = 2\pi/120$, $X_{\max}=Y_{\max}=6$, $\Delta X = \Delta Y = 0.03$. For IBFD, $\Delta\sigma = 0.001 \times (1+\sigma)^2$, and for CNFD, $\Delta\sigma = 0.0035 \times (1+\sigma)^2$.

Figure 2.20 shows the calculation results. Figure 2.20a is the initial waveform, and Figs. 2.20b–d are waveforms at $\sigma = 1, \sigma = 2$ and $\sigma = 4$, respectively. From the figure, it is obvious that the calculation results of the three-dimensional time-domain code are in good agreement with the calculation results of the two-dimensional time-domain code. In addition, the figure also shows the calculation results without the relaxation process, and the waveform obtained by the three-dimensional time-domain code is also in good agreement with the waveform obtained by the two-dimensional time-domain code.

2.3.2.2 Array Sound Source

The three-dimensional time-domain algorithm is used to calculate the nonlinear sound field of an array sound source composed of multiple small transducer units Fig. 2.21 demonstrates the geometric schematic diagram of the square array sound source used in this part of the calculation. The array is composed of 4 × 4 circular transducer units, each with a radius of 2.5 cm. The excitation signal is a dual-frequency signal of 40 and 36 kHz, $p_0 = 113$ dB. The medium is air, $c_0 = 344$ m/s, $\rho_0 = 1.2$ kg/m^3, $\beta = 1.2$. The temperature is 20 °C, and the relative humidity is 50%. The surface sound pressure on each unit of the array sound source is divided into two cases, uniformly distributed (array 1)

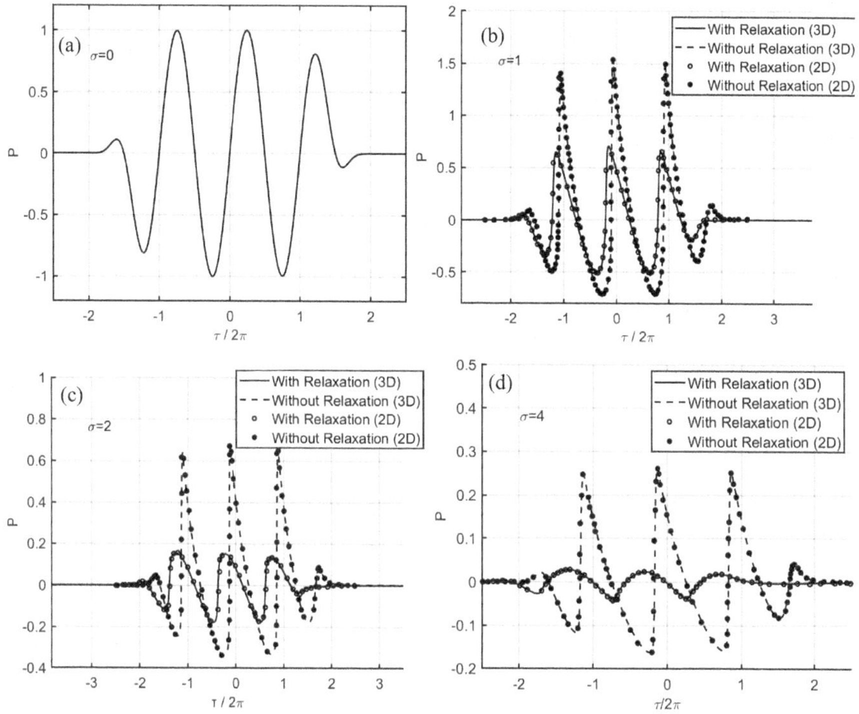

Fig. 2.20 Comparison of time-domain waveforms calculated by three-dimensional time-domain code and two-dimensional time-domain code (Calculation parameters refer to reference [5])

$$S_1(X, Y) = \begin{cases} 1, & (X - X_c)^2 + (Y - Y_c)^2 < 0.25^2 \\ 0, & \text{others} \end{cases}$$

And non-uniform distribution (Array 2)

Fig. 2.21 Geometric schematic of square array sound source

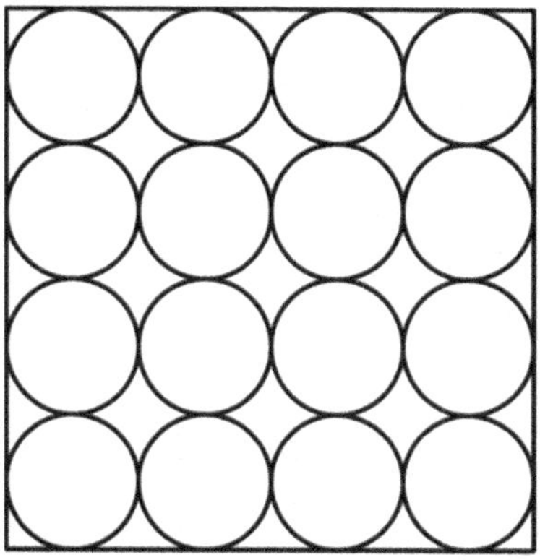

$$S_2(X,Y)=\begin{cases}1-\frac{(X-X_c)^2+(Y-Y_c)^2}{0.25^2}, & (X-X_c)^2+(Y-Y_c)^2\le 0.25^2\\ 0, & \text{others}\end{cases}$$

where (X_c, Y_c) is the center coordinate of the transducer unit.

The parameters used for calculation are: $f_0 = 36$ kHz, $r = 4.186$ m, $a = 0.1$ m, $G = 0.7854$, $N = 0.2939$, $A = 0.0998$. The initial time-domain waveform of the sound source is

$$f(t)=\exp\left[-(\omega_0 t/27\pi)^{40}\right][\sin\omega_0 t+\sin(10\omega_0 t/9)]$$

$\tau_{\min} = -32\pi, \tau_{\max} = 32\pi, \Delta\tau = 2\pi/60, X_{\max} = Y_{\max} = 6, \Delta X = \Delta Y = 0.025$. For IBFD, $\Delta\sigma = 0.001 \times (1+\sigma)^2$, and for CNFD, $\Delta\sigma = 0.003 \times (1+\sigma)^2$.

The calculation results are shown in Fig. 2.22. Among them, Fig. 2.22a is the axial sound pressure of the 40 kHz pump wave and 4 kHz difference frequency sound, and Fig. 2.22b is the off-axis sound pressure of the 36 kHz pump wave and 4 kHz difference frequency sound at 5 m. To verify the correctness of the results, the sound field results calculated using the Gaussian beam expansion method are also given for comparison. It can be observed from the figure that the calculation results of the array sound field obtained using the time domain algorithm are in good agreement with the calculation results of the array sound field obtained using the Gaussian beam expansion method. In fact, because the sound pressure level is low and it meets the quasi-linear conditions, the pump wave can be regarded as linear. The Rayleigh integral is also used to calculate the pump wave sound field of the array, and the results are also consistent with the results of the time domain algorithm in this section.

2.4 Summary

This chapter reviews the current KZK model and its solutions, and on this basis, it provides a simple and high-precision algorithm to calculate the Gaussian beam expansion coefficient and a three-dimensional time domain finite difference algorithm.

This chapter first provides a simple and high-precision algorithm to calculate the Gaussian beam expansion coefficient. Its basic principle is to process the Gaussian beam expansion coefficient A_n and B_n separately. A_n is obtained by solving a system of linear equations, B_n is determined by a simple 2-dimensional search, and then is further optimized by the steepest descent method for the Gaussian beam expansion coefficients. This makes the determination process of the Gaussian beam expansion coefficients regular and predictable. This method has some obvious advantages. (1) it can be used in the spatial domain and can be easily extended to the k space domain. The given examples prove that the algorithm in this section has a wide range of applicability. (2) it can provide high accuracy, which is proven by the numerical

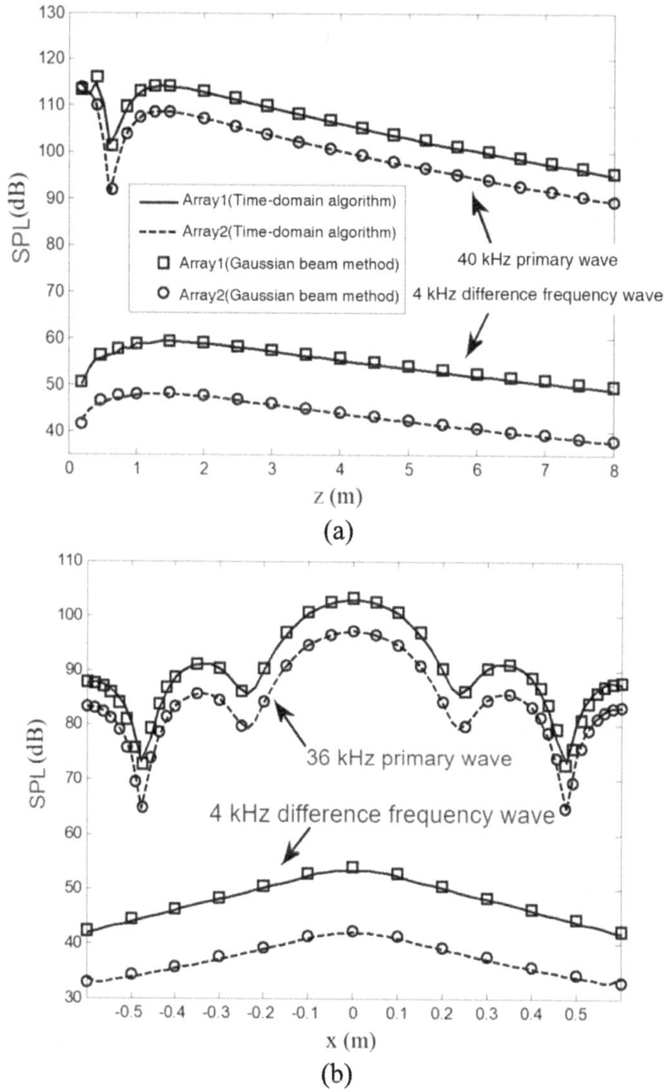

Fig. 2.22 Sound field calculation results of square array sound source at $p_0 = 113$ dB **a** axial sound pressure of 40 kHz primary wave and 4 kHz difference frequency wave **b** off-axis sound pressure of 36 kHz primary wave and 4 kHz difference frequency wave at z = 5 m

simulation results of the sound source distribution function and the sound field. (3) it is stable and highly efficient. In reference [100], Sha et al. took 4 hours on a 1.2 GHz PC to obtain a set of 10 Gaussian beam expansion coefficients. Using this algorithm, a similar set of coefficients can be obtained in just a few seconds.

This chapter then presents a three-dimensional time-domain finite difference algorithm, which is based on the extended KZK equation, including diffraction, thermal viscous absorption, nonlinearity, and relaxation. This algorithm transforms the extended KZK equation into the TBE equation to adapt to the far-field spherical wave propagation of the beam, adopts the operator splitting method to decompose various effects into five equations, and uses the IBFD and CNFD difference methods to solve each equation. To prove the effectiveness of the algorithm, the time-domain nonlinear waveform of a circular sound source was calculated, and the results are consistent with the benchmark solution of the existing related research's two-dimensional time-domain algorithm. The empirical formula of the sound absorption coefficient in the air is introduced into the time-domain algorithm to calculate the sound field of the rectangular parametric array in the air. While obtaining the correct results, it is also found that the sampling period of the time-domain waveform is an important factor affecting the calculation results of the difference frequency sound. In addition, the situation of treating the relaxation absorption coefficient as the thermal viscous absorption coefficient and inputting it into the time-domain algorithm is studied. The three-dimensional time-domain algorithm is used to calculate the parametric array sound field of a square array composed of circular transducer elements, and it was verified with the results obtained using the Gaussian beam expansion method.

Chapter 3
Preprocessing Methods of Parametric Array Loudspeakers

In 1963, Westervelt [22] established a mathematical model for the generation of a second-order sound field by two single-frequency primary wave signals based on the theory of scattering of sound by sound. In 1965, Berktay extended this theory to modulated signals. For an envelope signal modulated onto a carrier to obtain the primary wave signal $p_1 = P_0 E(t)\sin\omega_c t$, the far-field solution obtained after nonlinear action is [24]

$$p_2 = \frac{P_0^2 \beta S}{16\pi \rho_0 c_0^4 z\alpha_0} \frac{\partial^2}{\partial t^2} E^2(t) \tag{3.1}$$

where, t represents time, $E(t)$ is the envelope signal, P_0 is the initial sound pressure, ω_c is the carrier angular frequency, p_2 is the demodulated signal, β is the nonlinear coefficient of the medium, ρ_0 is the density of the medium, c_0 is the small signal sound speed, α_0 is the attenuation coefficient, S is the size of the beam cross-section of the primary wave. It is obvious that the demodulated signal is proportional to the second derivative of the square of the envelope signal.

The preprocessing method proposed by Yoneyama in 1983 based on Berktay's far-field solution [94] is double sideband amplitude modulation (DSBAM), i.e., $E(t) = 1 + m\,g(t)$, where m represents the modulation index, $g(t)$ is the modulated sound signal, and its system framework is shown in Fig. 3.1. Substituting it into Eq. (3.1), the demodulated signal can be obtained as

$$p_2 = \frac{m P_0^2 \beta S}{8\pi \rho_0 c_0^4 z\alpha_0} \frac{\partial^2}{\partial t^2} g(t) + \frac{m^2 P_0^2 \beta S}{16\pi \rho_0 c_0^4 z\alpha_0} \frac{\partial^2}{\partial t^2} g^2(t) \tag{3.2}$$

From Eq. (3.2), it becomes clear that the first term of the signal obtained by demodulation is the derivative of the modulated signal $g(t)$, its frequency component is all frequency components of $g(t)$. This mathematically explains the working principle of the parametric array loudspeaker. By modulating the audio signal onto

J. Yang and P. Ji, *Parametric Array Loudspeakers*,
https://doi.org/10.1007/978-981-96-3548-1_3

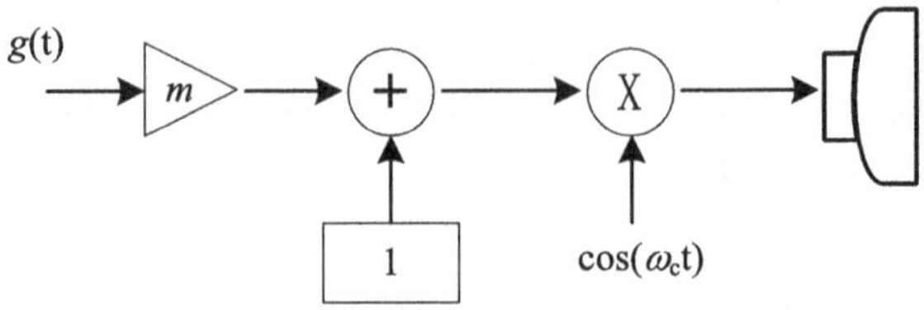

Fig. 3.1 DSBAM framework diagram

the ultrasonic carrier, the modulated ultrasonic wave can demodulate and restore the audio frequency through its nonlinear action in the air. At the same time, it can be noticed from the above formula that the second term in Eq. (3.2) is the second derivative after the square of the modulated signal $g(t)$, and this indicates that the 2nd harmonic component of $g(t)$ also appears in the demodulated signal. This mathematically points out that there is harmonic distortion in the audio frequency signal demodulated by the parametric array loudspeaker.

To eliminate the harmonic distortion in the audio frequency obtained by demodulation, scholars have carried out research on the improvement of modulation algorithms. Based on the Berktay far-field solution theory, scholars have proposed various amplitude modulation methods. Kamakura et al. [95, 130] focused on the impact of the carrier frequency and modulation method of the parametric array loudspeaker on the harmonic distortion of the demodulated audio frequency, and proposed the square root amplitude modulation (SRAM) algorithm and the single sideband amplitude modulation (SSB-AM) to reduce harmonic distortion, as shown in Figs. 3.2 and 3.3 respectively. The SRAM method is to take the square root the envelope signal before modulation, and reduce nonlinear distortion by cancelling out with the square term of demodulation, but this method requires the transducer to have a very wide bandwidth. The SSB-AM method reduces harmonic distortion by eliminating one sideband of amplitude modulation, and the bandwidth requirement for the transducer is not high, only half of the double sideband modulation technology. Single sideband modulation technology can effectively solve the harmonic distortion problem of the parametric array loudspeaker. However, these two modulation methods cannot effectively reduce intermodulation distortion. In 1998, Kite et al. [92] improved the SRAM method with the modulation signal being integrated twice before undergoing a square root operation for preprocessing, and the impact of the modulation index and transducer bandwidth on harmonic distortion and conversion efficiency has been theoretically analyzed. This technology can theoretically solve the distortion problem of parametric array loudspeakers, but it requires the ultrasonic transducer to have a large bandwidth, which is difficult to meet with current technology. At the 105th meeting of the AES (Audio Engineering Society), Pompei from MIT [342] reviewed his work, applying Kite's preprocessing method to actual systems. He stated in his article that if one wants to fully recover the envelope signal, the transducer's bandwidth must be infinitely wide. He founded Holosonic Research Labs and developed the parametric array loudspeaker product Audio Spotlight. In 2008, Tan et al. [134] proposed the MAM (Modified AM) algorithm based on orthogonal modulation, as shown in Fig. 3.4. This algorithm improves the harmonic distortion of DSBAM

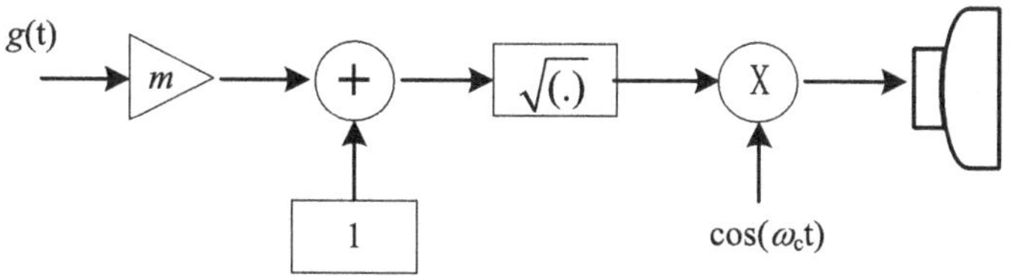

Fig. 3.2 SRAM framework diagram

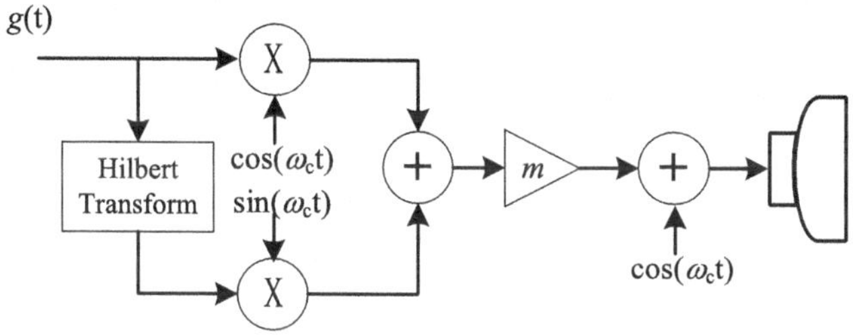

Fig. 3.3 SSBAM framework diagram

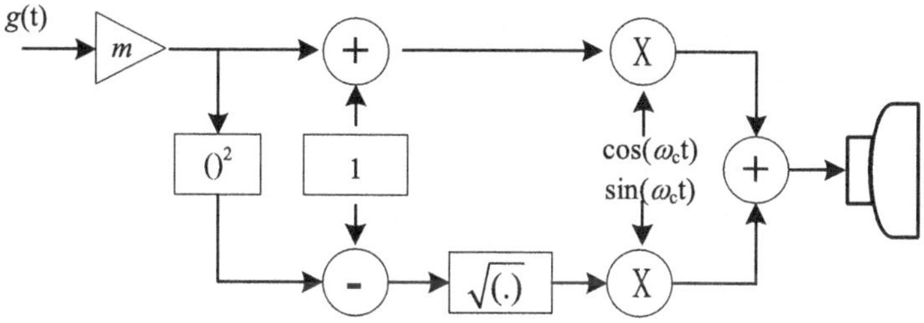

Fig. 3.4 MAM framework diagram

and SRAM, but it is more effective for single-frequency signals and not good for broadband signals. To address the shortcomings of the SSB-AM method, Croft et al. [131, 132] further developed the single-sideband modulation technology, namely the recursive single-sideband amplitude modulation (RSSB-AM) technology, using recursive technology to predict possible distortion, and reducing the error brought by SSB-AM through preprocessing technology before single-sideband modulation, to approximate the ideal SRAM result, as shown in Fig. 3.5. However, this method requires a large amount of computation and has high requirements for the speed of real-time processors. Wu et al. [133] adopted multi-rate processing technology and proposed a method for implementing RSSB-AM technology. Its feasibility was verified through experiments.

In practical applications, it has been found that the above modulation methods based on the Berktay far-field solution cannot eliminate the nonlinear distortion in the parametric array loudspeaker system. Moreover, when the sound pressure level of the

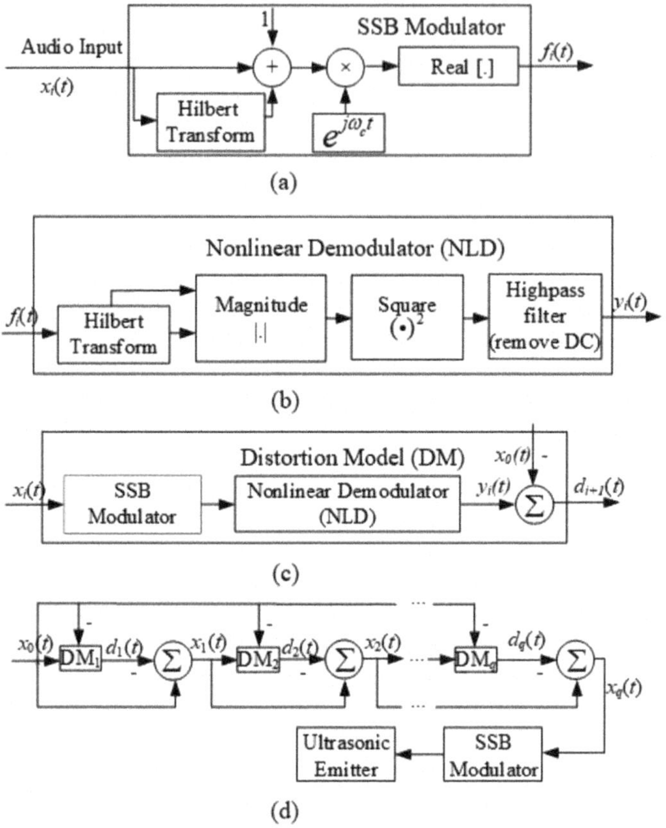

Fig. 3.5 RSSB-AM and its component block diagram **a** SSB modulator **b** NLD **c** DM and **d** RSSB-AM where $j = \sqrt{-1}$

ultrasonic primary wave signal exceeds a certain value [138], the performance of these improved modulation algorithms will significantly decrease. Therefore, it is necessary to study and introduce new algorithms to compensate for the nonlinear distortion of the parametric array loudspeaker. Currently, the Volterra Filter (VF)-based preprocessing method has been used to some extent to reduce the nonlinear distortion of the parametric array loudspeaker. VF, as a typical nonlinear filter model, is widely used to simulate various forms of nonlinear characteristics, such as nonlinear echo cancellation [343, 348], nonlinear channel identification in digital satellite communication systems [349–352], active noise control [353, 356], and image processing [357–359], etc. In 1887, the young Italian mathematician Vito Volterra extended the form of the Taylor series expansion and proposed a nonlinear filter structure. In memory of him, this form of filter is called VF [360]. In 1910, Frechet [361] conducted an in-depth theoretical study of VF and proved that the expansion terms of VF are complete. This means that any complex signal can be expanded into a series through VF. In 1930, Wiener applied VF to the nonlinear analysis of RLC circuits. Considering that the terms of VF are not orthogonal, he further developed the Wiener kernel function (Wiener G-functionals) based on VF.

Due to the complexity of the VF structure, there was less research on VF in the early stages. It was not until the 1960s and 1970s that the rapid development of computers and digital technology brought new vitality to VF research. In the 1980s, scholars discussed the uniqueness and convergence of the VF model in detail [362–365]. Current research on VF applications is divided into three directions: research on VF model establishment and parameter identification, compensation algorithms based on VF, and research on VF structure simplification. In 1969, Balakrishnan used the minimum mean-square error (MMSE) to estimate the Volterra kernel function [366]. However, this method requires the high-order statistical characteristics of the input signal to be obtained in advance, and special signals such as Gaussian white noise need to be selected. To expand the selection of input signals, Nowak et al. proposed the least squares algorithm [367]. Pearson et al. successfully estimated the kernel function of the 2nd-order VF model using this method [368]. However, this method also involves solving the inverse of high-dimensional matrices, and the identification process is time-consuming. To solve this problem, Schetzen borrowed the Wiener kernel function and proposed an identification method [369]. But the derivation of the relationship between the Wiener kernel function and the Volterra kernel function is not easy, and it is difficult to operate in actual measurements. Boyd et al. [370] provided a method for measuring the frequency domain coefficients of the 2nd-order VF model kernel function using multi-tone measurements. Then, Chua et al. extended this method to measure the parameters of the 3rd and higher order VF models [371, 372]. Subsequently, scholars discussed various ways to select the frequency of the input signal, to obtain more information with the least input signal, and reduce the measurement time [373, 376]. Currently, adaptive algorithms are most widely used in VF model identification. Scholars have successively proposed LMS (least mean square) adaptive structure [377], NLMS (normalized least mean square) adaptive structure [378], RLS (recursive least square) adaptive structure [379–382], AP (affine projection) structure [383, 384] and other adaptive iterative identification algorithms. Kajikawa compared the performance of various adaptive algorithms in VF model identification [385]. The RLS algorithm converges the fastest, but it has the highest computational complexity. NLMS converges faster than LMS, and AP algorithm converges even faster than NLMS.

Another hot topic of research is the compensation algorithm based on VF design to eliminate system nonlinear distortion. In 1980, Schetzen proposed a design method for p-order inverse filters [369]. This classic method has been used up to the present day. However, the structure of the p-order inverse filter will become extremely complex as the order increases. To address this issue, Sarti et al. proposed a recursive structure for the inverse filter [386], and Nowak introduced the gradient descent iterative algorithm into the solution of the inverse filter [387]. In response to the problem of high-order nonlinear distortion introduced by the p-order inverse filter, Lashkari proposed the concept of an exact inverse filter [388]. But this method has several problems. Firstly, the existence of solutions to nonlinear equations cannot be guaranteed. Secondly, as the order increases, the solution of nonlinear equations becomes complex. Thirdly, the choice of multiple feasible solutions should be carefully handled [389].

Because the VF structure is complex and computationally intensive [390, 391], Bernardini used multidimensional Fourier transformation to calculate multidimensional convolution in the time domain to reduce the amount of computation [392]. Kinoshita et al. proposed a method to reduce the length of VF by using downsampling [393]. Ishikawa et al. decomposed the coefficient matrix of the 2nd-order kernel function of the VF model by SVD, obtained the eigenvectors, and constructed a parallel cascade structure [394]. Frank proposed the MMD (multi-memory decomposition) structure of the filter form [395]. Kafka et al. further expanded on the basis of the MMD structure [396], replacing the linear filters with ARMA structure filters, resulting in a more universal type of nonlinear filter. Hunter et al. proposed Wiener filters and Hammerstein filters [357]. Korenberg et al. combined Wiener and Hammerstein filters to form a LNL cascade form of filter [397]. Marmarelis proposed the Volterra-Laguerre model, using Laguerre filters to represent Volterra kernel functions [398]. In recent years, Farina et al. introduced a one-dimensional Volterra filter (ODVF, One Dimensional Volterra filter) model when analyzing the distortion of nonlinear systems [399–401]. It should be emphasized that although some scholars call it a Hammerstein filter, it is different from the Hammerstein filter. The 2nd-order kernel function of the Hammerstein filter is consistent with the 1st order kernel function. While the 2nd-order kernel function of the ODVF is unrelated to the 1st order kernel function. Novak et al. implemented this model to simulate the nonlinear effects of overdrive effect pedals [402, 403]. Rebillat et al. used it to simulate the harmonic distortion of moving coil speakers and transducers [404, 405]. Tronchin [406, 407] adopted it to simulate the harmonic effects of musical instruments. Because of its simple structure and the ability to obtain its kernel function by the sweep method, ODVF has good application prospects.

VF, as the most widely used nonlinear filter model, has attracted significant attention from scholars. It has been applied in the fields of power amplifier nonlinear distortion, satellite communication systems, echo cancellation, loudspeaker nonlinear distortion compensation, active noise control, etc. However, the computational complexity of VF is high, especially when the order of system nonlinear distortion is high, the identification process of VF will be extremely time-consuming, and the resource consumption of the corresponding Volterra inverse filter is also high, making it difficult to carry out real-time filtering processing, and unable to meet the real-time requirements of applications such as loudspeaker nonlinear distortion compensation. To solve this problem, the ODVF model was introduced. The structure of this filter is simple, which can greatly reduce the computational complexity of VF.

3.1 Research on Berktay-Based Preprocessing Methods

3.1.1 *Analysis of Berktay-Based Preprocessing Methods Using KZK Model*

Some THD analysis based on the Berktay far-field solution has been conducted in reference [134], but the impact of equalization has not been studied. It should be noted that the Berktay far-field solution is an approximate solution to the weak nonlinear sound beam propagation [24]. Merklinger [97] provided a more general parametric array solution, which is $p_2 \propto P_0\partial^2\left[E(t)\tan^{-1}\left(\beta\omega_c P_0 E(t)/4\alpha\rho_0 c_0^3\right)\right]/\partial t^2$, where c_0 is the speed of small signals, ρ_0, α, β are the environmental density, the absorption coefficient of the carrier frequency [248] and the nonlinear coefficient of the medium, respectively. For $P_0 \ll 4\alpha\rho_0 c_0^3/(\beta\omega_c)$, Merklinger's solution simplifies to $p_2 \propto P_0^2\partial^2 E^2(t)/\partial t^2$, which is the same as the Berktay's solution. For $P_0 \gg 4\alpha\rho_0 c_0^3/(\beta\omega_c)$, Merklinger's solution indicates $p_2 \propto P_0\partial^2|E(t)|/\partial t^2$. Both Berktay and Merklinger's solutions ignore the absorption of difference frequency sound and higher-order effects [97, 283]. Therefore, a more accurate model is needed to explain the performance differences of various preprocessing methods in parametric array loudspeakers [94, 95, 131, 132, 134, 225]. It is well known that the KZK equation [2] can accurately describe the propagation of finite amplitude sound beams under the parabolic approximation, combining the effects of diffraction, absorption, and nonlinearity. Many scholars have verified the validity of the KZK equation [1, 283], and the Berktay's solution can be considered as an approximation of the weakly nonlinear KZK equation [283]. Although analytical solutions using the Gaussian beam expansion method in the weakly nonlinear case have been proposed in the references [101, 318, 319], the KZK equation still does not have a clear analytical solution. Given this limitation, various numerical solutions have been developed in the frequency domain, time domain, and combined time–frequency domain [1, 35, 264, 281]. The advantages of using the time-domain KZK algorithm are: ① it can calculate any time waveform; ② compared with other methods, it has high computational efficiency [27].

This section uses the time-domain KZK algorithm [281] to evaluate the THD performance of various preprocessing methods in the presence and absence of equalization, namely DSB-AM, SRAM, SSB-AM, and MAM. In the THD analysis, the recursive SSB-AM method is equivalent to the SSB-AM method, and the THD performance of these two methods is the same. Therefore, this section will only study the SSB-AM method. According to the Merklinger's general solution [97], the effect of these preprocessing methods on reducing the distortion of the parametric array loudspeaker is closely related to P_0. Therefore, to determine the effective area of P_0 for these preprocessing methods based on the KZK equation, the impact of P_0 on THD and demodulation signals was investigated in this study. In addition, two settings within the effective area of P_0 are selected to determine which preprocessing method is most applicable and most suitable for directional speech reproduction.

To measure the performance of different preprocessing methods, the performance indicator THD is used. For a single angular frequency, the definition of THD is as follows

$$\mathrm{THD} = \sqrt{\frac{T_2^2 + \cdots + T_i^2 + \cdots + T_n^2}{T_1^2 + T_2^2 + \cdots + T_i^2 + \cdots + T_n^2}} \times 100\% \tag{3.3}$$

where T_1 and T_i ($i = 2, 3, \ldots, n$) represent the fundamental angular frequency components at ω_1 and at the higher harmonic components of the angular frequency $i\omega_1$.

Here, four commonly used preprocessing methods (DSB-AM, SRAM, SSB-AM, and MAM) are investigated. Table 3.1 lists the demodulation signals of each preprocessing method. When considering a single-tone input $g(t) = \cos\omega_1 t$, the Berktay far-field solution and its corresponding THD expression are obtained. Among the four preprocessing methods listed in Table 3.1, only the demodulation signal generated by the DSB-AM method consists of the difference frequency component (expected signal) ω_1 and the second harmonic $2\omega_1$. The remaining preprocessing methods only produce difference frequency components in the demodulation signal ω_1. Therefore, only the THD value produced by the DSB-AM method is related to the modulation index, while the THD performance produced by the remaining preprocessing methods is perfect. Therefore, the THD results derived from the Berktay model are not very useful, because they cannot distinguish the performance of other preprocessing methods except for the DSB-AM method. This section will use the more complex KZK model for more detailed analysis.

In Table 3.1, the expression for the THD of the DSB-AM method predicted by the Berktay far-field solution [13] is $m/\sqrt{m^2+1} \times 100\%$, which only increases monotonically with the modulation index and is independent of different frequency components. Since the demodulation signal Berktay predicted is proportional to the second derivative of the squared envelope, there is a +12 dB/octave gain slope in all demodulation signals in Table 3.1. To balance this inherent high-pass filtering effect (+12 dB/octave, marked as a dashed line in Fig. 3.6) in the difference frequency component predicted by the Berktay model, a reverse −12 dB/octave filter (marked as a dotted line in Fig. 3.6) is introduced in the input signal. Here, the difference frequency component ranges from 0.5 to 3.5 kHz, with an interval of 500 Hz.

Table 3.1 Demodulated signals and their THD for each preprocessing method based on the Berktay far-field solution (for single-tone input)

Preprocessing method	Demodulated signal	THD
DSB-AM	$p_2 \propto \left[m\omega_1^2\cos(\omega_1 t) + m^2\omega_1^2\cos(2\omega_1 t)\right]$	$m/\sqrt{m^2+1} \times 100\%$
SRAM	$p_2 \propto \left[m\omega_1^2\cos(\omega_1 t)\right]$	0
SSB-AM	$p_2 \propto \left[m\omega_1^2\cos(\omega_1 t)\right]$	0
MAM	$p_2 \propto \left[m\omega_1^2\cos(\omega_1 t)\right]$	0

Therefore, as the difference frequency component increases, the effect of balance is equivalent to reducing the modulation index [94]. An additional modulation index $m' = 1/f_N^2$ can be introduced in these preprocessing methods to consider the impact of equalization, where f_N The difference frequency component is normalized to the lowest frequency (i.e., 500 Hz in the study). Ideally, a flat frequency response of the difference frequency component can be obtained (marked as a solid line in Fig. 3.6). For example, the envelope function of the DSB-AM method is $E(t) = 1 + mm'g(t)$, and the THD representation of the equalized DSB-AM method is

$$\frac{1}{\sqrt{\left(f_N^2/m\right)^2 + 1}} \times 100\% \tag{3.4}$$

After equalization, the THD value in Eq. (3.4) not only increases monotonically with the modulation index, but also decreases monotonically with the difference frequency component.

In this section, the demodulation process used for speech reproduction in parametric array loudspeakers is studied, and the performance of four preprocessing methods is analyzed using the THD results obtained from the KZK equation. The following parameters are used in the simulation. The radius of the ultrasonic sound source is 0.1 m, the carrier frequency is 40 kHz, and the temperature and relative humidity are set to 28 °C and 60% respectively. The observation point is on the axis, located 1.8 m from the sound source. The KZK is used to analyze the same difference frequency component within the speech bandwidth. In the preprocessing methods,

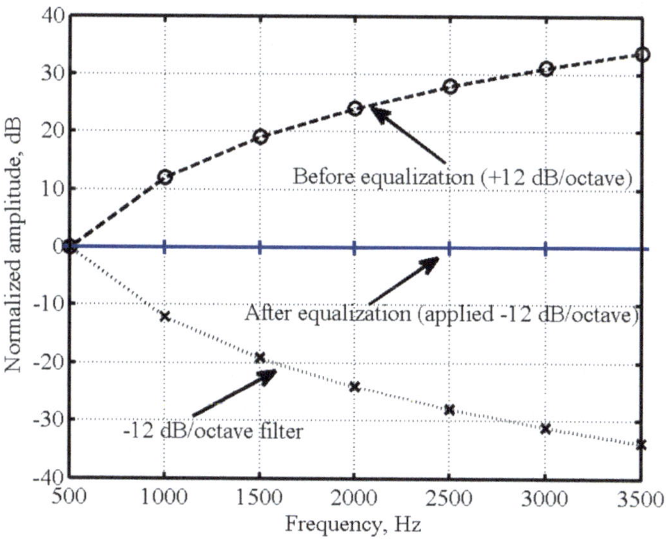

Fig. 3.6 Equalization of difference frequency components

six modulation indexes of $m = 0.5, 0.6, 0.7, 0.8, 0.9, 1.0$ are also used to cover the working range of the parametric array loudspeaker.

Assuming that the ultrasonic sound source has a relatively flat response between 24 and 56 kHz, it can be approximated by an ideal bandpass filter in the simulation. The input speech signal is band-limited by the above-mentioned bandpass filter after preprocessing, and then the filtered signal is applied to the KZK equation to simulate the nonlinear propagation process in the air.

It is generally believed that the DSB-AM method is the traditional method used in parametric array loudspeakers. Many other methods, such as SRAM, SSB-AM, and MAM, have been proposed by scholars based on the Berktay far-field solution. These Berktay-based methods attempt to reduce the distortion found in the DSB-AM method by compensating for the squared envelope predicted by the Berktay far-field solution, and assume that the second derivative effect can be compensated by the frequency response of the ultrasonic sound source. In this study using the KZK equation, the effectiveness of these methods (SRAM, SSB-AM, and MAM) is tested by benchmarking their THD values against the THD values obtained by the DSB-AM method. The SSB-AM method includes LSB-AM or USB-AM. Due to the narrow bandwidth of the speech signal, there is not much difference in attenuation and saturation effects when using these two cases [131]. Only USB-AM is used in this part of the study.

3.1.1.1 *Impact of the Initial Sound Pressure Level P_0*

Using the parameters defined in the simulation, and it is found that the constant in the Merklinger solution $4\alpha\rho_0 c_0^3/(\beta\omega_c)$ is about 130 dB [97]. Therefore, to determine the effective working region of the Berktay far-field solution, the effect of P_0 changing from 100 to 140 dB at intervals of 2 dB is studied. In addition, the difference frequency component and m are defined as 2 kHz and 0.9, respectively.

Figure 3.7 illustrates the relationship between the sound pressure level of the difference frequency component at 2 kHz and P_0 when using four preprocessing methods based on the KZK equation. It is obvious that when P_0 is greater than 132 dB, the sound pressure level of the difference frequency component becomes approximately proportional to P_0. On the contrary, when P_0 is reduced to below 132 dB, it is observed that the sound pressure level of the difference frequency component is almost proportional to P_0^2. This observation is very consistent with the Merklinger's solution, which predicts a threshold of about 130 dB. It is also observed from Fig. 3.7 that the sound pressure level of the difference frequency component obtained at 2 kHz using the DSB-AM method is the highest, followed by the SSB-AM, MAM, and SRAM methods. The reason can be explained as follows. The generation of 2 kHz can simply be regarded as the sum of the nonlinear interactions of each pair of primary waves with the 2 kHz difference frequency. For the nonlinear interaction of a pair of primary waves f_1 and f_2 (where $f_1 > f_2$), the amplitude of the difference frequency component is proportional to the amplitudes of the primary waves p_1 and p_2 is proportional to the product, i.e., $p_{f_1-f_2} \propto p_1 p_2$ [2]. At the same time,

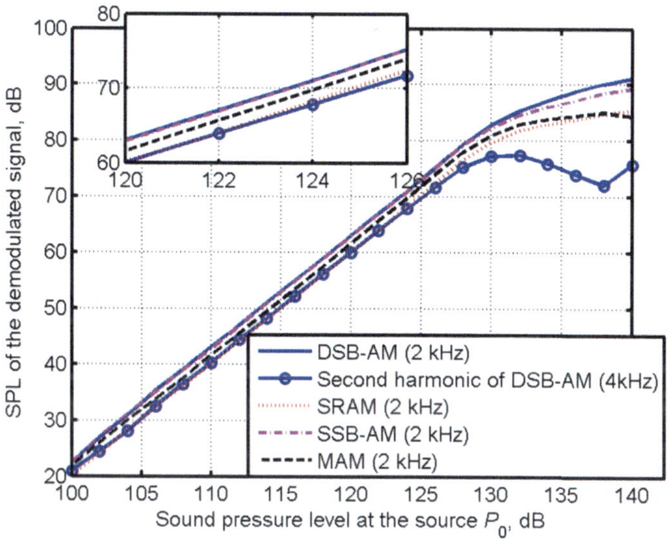

Fig. 3.7 Relationship between the sound pressure level of the difference frequency component of each preprocessing method based on the KZK equation and the second harmonic of the DSB-AM method and P_0

a pair of primary waves also satisfies the condition $\alpha_{f_2} < \alpha_{f_1}$ [248]. Table 3.2 summarizes the normalized amplitudes of the two pairs of primary waves that contribute the most to the generation of 2 kHz difference frequency components. Since there is only one sideband in the SSB-AM method, there is only one pair of primary waves, so the amplitude of the 42 kHz primary wave in the SSB-AM method is divided into two equal components for easy comparison with Table 3.2. Several issues can be seen from Table 3.2. First, the contributions of the primary wave pairs of the SRAM and MAM methods are relatively smaller than the DSB-AM method. Therefore, it is expected that the difference frequency components obtained from the DSB-AM method will be relatively higher compared to those obtained from the SRAM and MAM methods. Secondly, for the DSB-AM and SSB-AM methods, the amplitudes of the first and second primary wave pairs are the same. However, compared with the DSB-AM method, the frequency value of the second pair of primary waves in the SSB-AM method is higher. Considering the sound absorption [248], it results in higher amplitude of the difference frequency when using the DSB-AM method compared to the SSB-AM method.

Figure 3.8 illustrates the relationship between the THD performance of the four preprocessing methods derived from the KZK equation and P_0. From Fig. 3.8, it becomes clear that when P_0 is less than 132 dB, the THD values of the SRAM, SSB-AM, and MAM methods are significantly lower than the DSB-AM method. This trend is reversed when P_0 exceeds 132 dB. This is expected, as the Berktay far-field solution does not hold when P_0 is above 132 dB, under which circumstances the effectiveness of the preprocessing methods based on Berktay (SRAM, SSB-AM, and MAM) would

Table 3.2 Normalized amplitude of the primary wave pairs used to generate 2 kHz for each preprocessing method

Preprocessing method	Normalized amplitude of the first pair of primary waves		Normalized amplitude of the second pair of primary waves	
	38 kHz	40 kHz	40 kHz	42 kHz
DSB-AM	0.45	1	1	0.45
SRAM	0.2696	1	1	0.2696
SSB-AM	0.45 (42 kHz)	1	1	0.45
MAM	0.3603	1	1	0.3603

dramatically decrease. Therefore, in this section, cases where P_0 is less than 132 dB will only be considered. As previously mentioned, the main distortion of the DSB-AM method comes from the second harmonic. Interestingly, when absorption is considered in the KZK simulation [248], it is found that the sound pressure level of the second harmonic is significantly lower than the difference frequency component [94]. The additional absorption term in the KZK equation results in THD values lower than those observed from the Berktay far-field solution. Figure 3.7 shows the relationship between the sound pressure level of the second harmonic (4 kHz) of the DSB-AM method (marked with a solid line in circles) and P_0, which shows a similar trend to the 2 kHz difference frequency component, but changes more slowly. Therefore, as P_0 increases, a gradual decrease in the THD value of the DSB-AM method can be observed, as shown in Fig. 3.8.

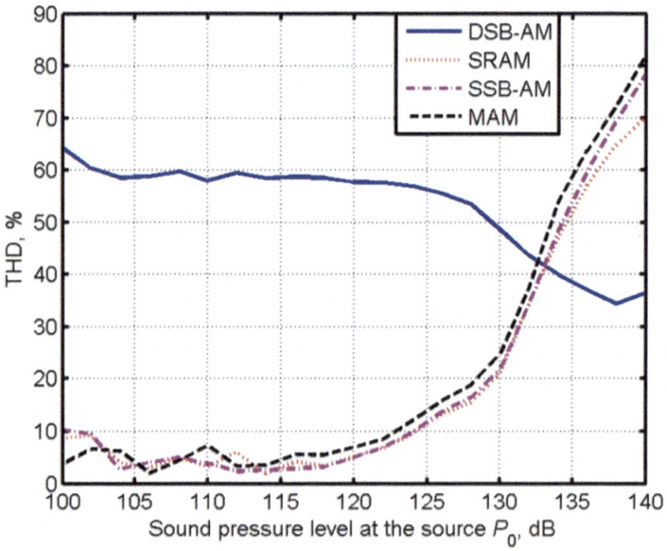

Fig. 3.8 Relationship between THD and P_0 based on the KZK equation

Further research is conducted on the relationship between the frequency response of the difference frequency component of each preprocessing method and P_0. To achieve a relatively large sound pressure level and a relatively low THD for the difference frequency component, P_0 was chosen to vary between 112 and 128 dB, with an interval of 4 dB, based on the above analysis. Under different P_0 and preprocessing methods, the normalized frequency response of the difference frequency component using the preprocessing method ($m = 0.9$) is shown in Fig. 3.9. The results based on the KZK equation show a good fit, falling between +6 and +12 dB/octave, which is also consistent with the experimental results reported in the reference [28]. When P_0 is small, the slope gain is close to +12 dB/octave, and when P_0 is high, the slope gain gradually turns to +6 dB/octave. As shown in Fig. 3.9a–d, these trends can be observed in all preprocessing methods and they can also be observed for other modulation indexes. But for simplicity, they are not shown in these figures. Therefore, the +12 dB/octave slope gain predicted by the previous Berktay far-field solution is not sufficient to describe the frequency response of the difference frequency component under all P_0 settings. Therefore, an equalizer related to P_0 must be designed to achieve a relatively flat frequency response.

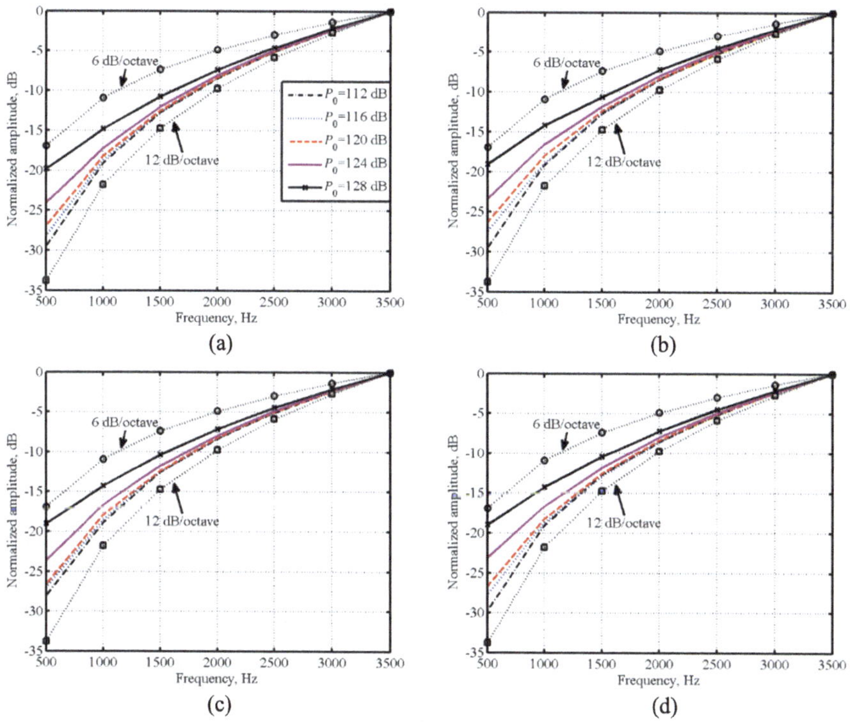

Fig. 3.9 Frequency response of difference frequency components under different P_0 **a** DSB-AM **b** SRAM **c** SSB-AM and **d** MAM (Legend in figure (a) applies to all figures)

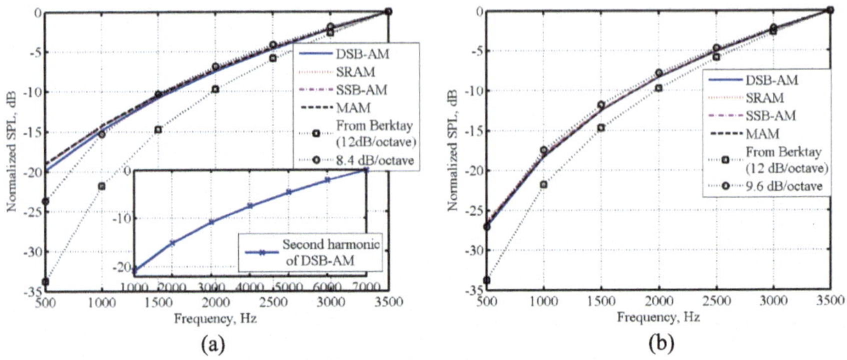

Fig. 3.10 Normalized frequency response of difference frequency components of four preprocessing methods when $m = 0.9$ **a** $P_0 = 128$ dB and **b** $P_0 = 120$ dB

Two representative P_0 settings were chosen. $P_0 = 128$ dB (close to $4\alpha\rho_0 c_0^3/(\beta\omega_c)$) and $P_0 = 120$ dB (much less than $4\alpha\rho_0 c_0^3/(\beta\omega_c)$) are selected to compare the frequency response of the difference frequency components with the same P_0, as shown in Fig. 3.10. For the same P_0, the frequency responses of these preprocessing methods are almost the same. At $P_0 = 128$ dB (Fig. 3.10a), the frequency response derived from the KZK equation is close to a +8.4 dB/octave slope. At the lower $P_0 = 120$ dB (Fig. 3.10b), the frequency response is close to a higher slope of +9.6 dB/octave. For simplicity, the inset of Fig. 3.10a only shows the normalized frequency response of the second harmonic of the DSB-AM method at $P_0 = 128$ dB. It is worth noting that the second harmonic of the DSB-AM method is similar to the frequency response between the difference frequency components using various preprocessing methods. This finding is consistent with the observations from Fig. 3.7.

3.1.1.2 THD Simulation Based on the KZK Equation

Figure 3.11 shows four subfigures, corresponding to the THD performance obtained by four preprocessing methods at $P_0 = 128$ dB without equalization. Six different modulation indexes (from $m = 0.5$ to 1, with an interval of 0.1) are verified within the speech frequency range. In these simulations, a THD of 10% is used as a performance indicator to distinguish between poor and acceptable performance. THD values of 10% and above are considered poor performance, and vice versa. The THD values of these four preprocessing methods increase monotonically with the increase of the modulation index. Although their THD values are still lower than the DSB-AM method, the THD values of these Berktay-based preprocessing methods (derived from the KZK model) are all greater than 7.5%. This is due to the relatively high P_0. For this P_0 level, the Berktay far-field solution cannot accurately predict the propagation of the sound beam. Therefore, this inaccuracy leads to higher harmonic distortion for all Berktay-based preprocessing methods at high P_0. At such a high P_0,

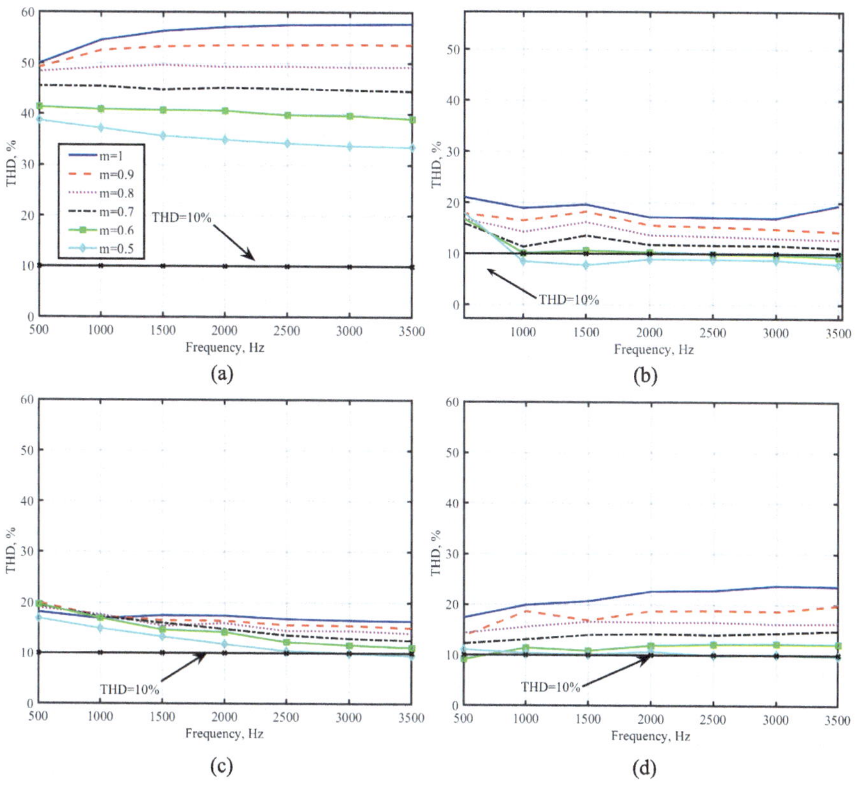

Fig. 3.11 THD values under no equalization using **a** DSB-AM **b** SRAM **c** SSB-AM and **d** MAM four preprocessing methods when $P_0 = 128$ dB (Legend in (a) applies to all figures)

it is also difficult to predict the characteristics of the distortion and THD values of these Berktay-based methods. However, for the DSB-AM method, since the significant distortion of the DSB-AM method comes from the second harmonic, the KZK and Berktay models can still achieve similar performance, that is, the THD is almost independent of the difference frequency component.

The THD performances of the preprocessing methods without equalization are summarized Table 3.3. The percentage in Table 3.3 defines the ratio of the number of combined THD values less than 10% over the total number of the combined THD values across the entire speech frequency range and accumulated over the six modulation index settings. Therefore, the higher the percentage, the less distortion obtained. It can be seen from Table 3.3 that the best preprocessing method without equalization is the SRAM method. However, in the SRAM, SSB-AM, and MAM methods without equalization, the percentage of THD values less than 10% is relatively small. Therefore, it can be concluded that these three Berktay-based methods cannot effectively reduce the harmonic distortion of $P_0 = 128$ dB without equalization.

Table 3.3 Percentage of combined THD values less than 10% in the range of m = 0.5, 0.6, 0.7, 0.8, 0.9, and 1.0 at P_0 = 128 dB

Preprocessing method	Without equalization (%)	With equalization (%)
DSB-AM	0	42.86
SRAM	21.43	85.71
SSB-AM	4.76	71.43
MAM	7.14	85.71

For simplicity, the rest of this section the THD value when $m = 0.9$ will be plotted to illustrate the performance of these four preprocessing methods. However, to compare the overall performance of these preprocessing methods, THD values for all 6 modulation indexes are still included to calculate the percentage of THD values below 10%.

In Fig. 3.10a, P0 = 128 dB is used, and the slope gain is +8.4 dB/octave. The uneven difference frequency component frequency response derived from this KZK model is not ideal for speech reproduction. Therefore, an equalization filter with a slope of −8.4 dB/octave should be applied to the input speech signal, which is similar to the equalization shown in Fig. 3.6. After equalization, when $m = 0.9$, the normalized frequency response of the difference frequency component using the preprocessing method is shown in Fig. 3.12. It can be seen from the figure that the frequency response of the difference frequency component after equalization becomes relatively flat and shows ripples less than 1.5 dB. However, it should be noted that the equalization of the input signal will reduce the electro-acoustic conversion efficiency. For the DSB-AM method, the sound pressure level of the second harmonic is proportional to m^2, and the frequency response of the second harmonic without equalization is approximately a slope of +8.4 dB/octave, as shown in the inset of Fig. 3.10a. As mentioned earlier, the effect of equalization is equivalent to reducing the modulation index [94]. Therefore, the second harmonic of the DSB-AM method is expected to attenuate about ($-2 \times 8.4 + 8.4$) dB/octave after equalization. This finding has been verified by the normalized frequency response of the second harmonic of the DSB-AM method, as shown in the inset of Fig. 3.12.

Using the KZK equation, the THD values of the four preprocessing methods with equalization when $m = 0.9$ are illustrated in Fig. 3.13. Equalizing the speech signal has the effect of reducing the modulation index, so the THD value after equalization decreases monotonically with the difference frequency component. This observation result derived from the KZK equation is also similar to Eq. (3.4). The THD performances of the preprocessing methods with equalization are also summarized in Table 3.3. The percentage of all preprocessing methods significantly increases after equalization, which means that equalization is crucial for all preprocessing methods working at P_0 = 128 dB. Interestingly, the SRAM and MAM methods provide the best THD performance after equalization. The MAM method significantly improves its THD performance after equalization. Therefore, for larger P_0 values, it is strongly recommended to equalize these Berktay-based methods.

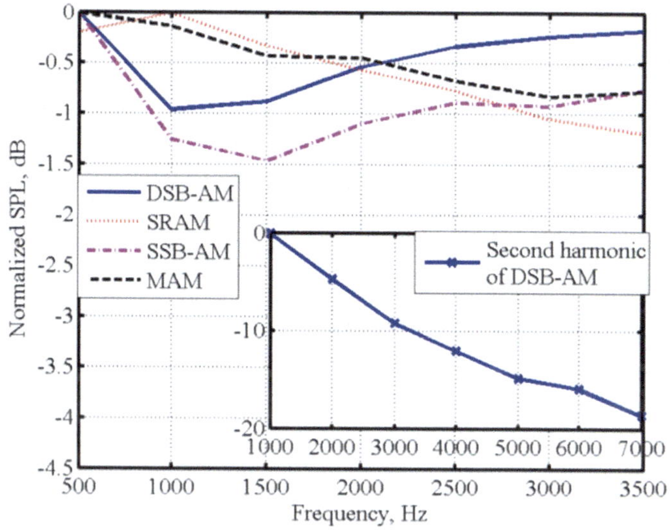

Fig. 3.12 Normalized frequency response of difference frequency components and second harmonic of DSB-AM method after equalization for each preprocessing method when $P_0 = 128$ dB and $m = 0.9$

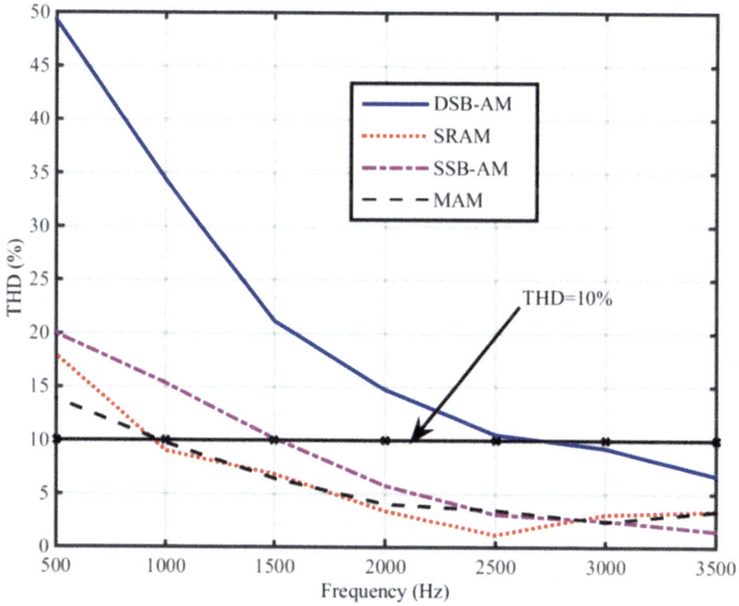

Fig. 3.13 THD values of each preprocessing method after equalization when $P_0 = 128$ dB and $m = 0.9$

Similar to the previous case, the case where P_0 is lower at 120 dB is investigated. Figure 3.14 shows the THD values obtained using four unequal preprocessing methods when $m = 0.9$. For the DSB-AM method, the distortion is almost independent of the difference frequency component due to the same reason mentioned for the case of $P_0 = 128$ dB. As can be seen from Fig. 3.14, compared with the DSB-AM method, the THD values of the SRAM, SSB-AM, and MAM methods are smaller at all frequencies. The poor frequency response of the parametric array loudspeaker at low frequencies leads to poor THD performance, where relatively high background noise at low frequencies results in a lower signal-to-noise ratio (SNR). The background noise comes from the nonlinear acoustic propagation modeling of finite amplitude waves and is also due to the finite time window length used in FFT analysis. The THD values of the SRAM, SSB-AM, and MAM methods are approximately proportional to the difference frequency component because the sound pressure level of the difference frequency component increases by a factor of +6 to +12 dB/octave [112], which is much greater than its harmonic increment. When P_0 is lower, the THD values of the SRAM, SSB-AM, and MAM methods indicate that Berktay's solution can well predict the propagation of the sound beam, so Berktay-based preprocessing methods can be used to effectively reduce distortion.

Using the KZK model, the THD values of the equalized preprocessing methods were obtained, as demonstrated in Fig. 3.15 ($m = 0.9$). The equalization filters used in these methods have a slope of −9.6 dB/octave. The THD value of the equalized

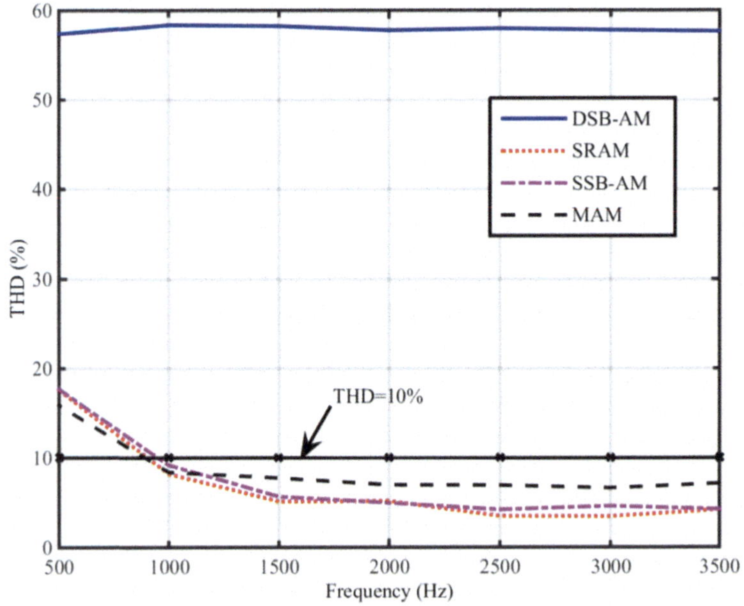

Fig. 3.14 THD values of each preprocessing method without equalization when $P_0 = 120$ dB and $m = 0.9$

DSB-AM method decreases monotonically with the difference frequency component, which is consistent with the predicted trend. There are some fluctuations at higher difference frequency components. These fluctuations are caused by background noise because under these conditions, the background noise is relatively large compared to the second harmonic and becomes the main contribution to distortion. For Berktay-based methods, most THD values become higher because the signal-to-noise ratio is lower after equalization. This is attributed to the higher reduction of the difference frequency component as compared to the noise floor and harmonics. Therefore, for lower P_0, equalization is not necessary.

When equalization is adopted and m is set to 0.9, the normalized frequency response of the difference frequency component using the preprocessing method is relatively flat, with ripples less than 1.6 dB, as shown in Fig. 3.16. Table 3.4 summarizes the THD performance of the preprocessing method for 6 different modulation indexes. From Table 3.4, it is evident that among the existing preprocessing methods, the best methods without equalization are the SRAM and SSB-AM methods, and the best method with equalization is the SSB-AM method.

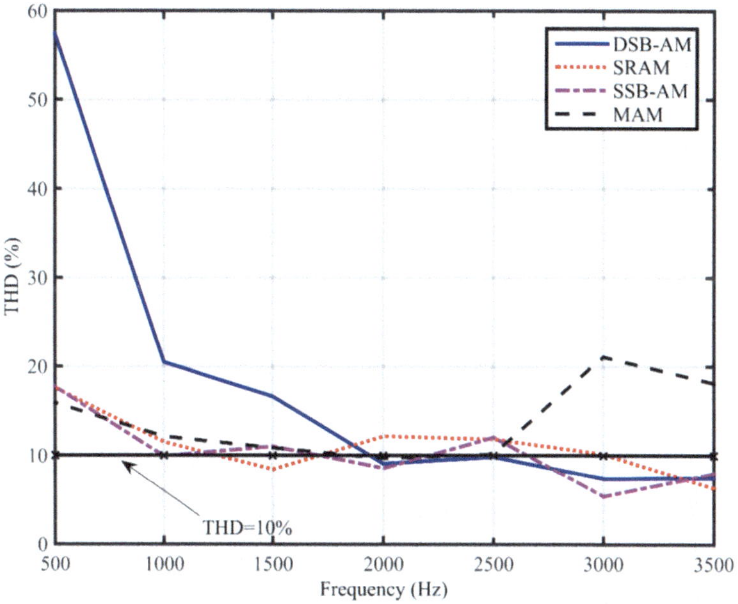

Fig. 3.15 THD values of each preprocessing method after equalization when $P_0 = 120$ dB and $m = 0.9$

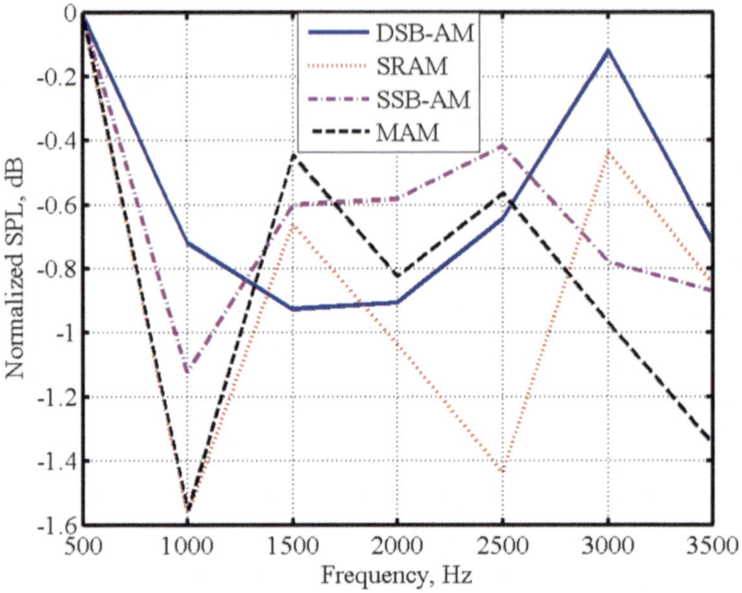

Fig. 3.16 Normalized frequency response of difference frequency components of the four preprocessing methods after equalization when $P_0 = 120$ dB and $m = 0.9$

Table 3.4 Percentage of combined total THD values below 10% within the range of $m = 0.5, 0.6, 0.7, 0.8, 0.9$, and 1.0 when $P_0 = 120$ dB

Preprocessing method	Without equalization (%)	With equalization (%)
DSB-AM	0	30.95
SRAM	85.71	26.19
SSB-AM	85.71	42.86
MAM	78.57	21.43

3.1.2 Computationally Efficient Implementation of RSSB-AM Technology for Parametric Array Loudspeakers

RSSB-AM technology has attracted wide attention in parametric array loudspeakers due to its small distortion and low transducer bandwidth requirements. In the implementation of RSSB-AM technology, two types of filters need to be implemented: Hilbert filters and equalization filters. The Hilbert filter is used to generate single-sideband signals, while the equalization filter is used to compensate for the frequency response of the ultrasonic transducer. When the conventional RSSB-AM technology is implemented, the system's working sampling frequency is usually several times the ultrasonic carrier frequency. This leads to a large computational cost for the implementation of the Hilbert filter and equalization filter. This section adopts multirate processing technology and provides an implementation method for RSSB-AM technology. Through frequency shifting technology, both the Hilbert transform and the equalization filter are performed at a lower sampling frequency to reduce the computational load.

3.1.2.1 RSSB-AM Method for Ideal Transducer

The block diagram of the conventional RSSB-AM technology is shown in Fig. 3.17, where the original audio signal is $s\ (l)$, and the sampling frequency is set to f_a. $x(n)$ is the audio signal obtained by upsampling, and the sampling frequency is f_u. $g(n)$ is the preprocessed signal, which is single-sideband modulated to obtain the input signal of the ultrasonic transducer

$$\begin{aligned} p(n) &= \left[1 + mg(n)\right]\cos\left(\frac{2\pi f_c}{f_u}n\right) + m\hat{g}(n)\sin\left(\frac{2\pi f_c}{f_u}n\right) \\ &= Re\left\{\left[1 + mg(n) - jm\hat{g}(n)\right]e^{\frac{j2\pi f_c}{f_u}n}\right\} \end{aligned} \tag{3.5}$$

where $\hat{g}(n)$ is the signal obtained after the Hilbert transform of $g(n)$, m is the modulation index, "Re" represents the real part of the complex number, f_c is the frequency of the ultrasonic carrier signal. Equation (3.5) can be converted to the frequency domain as follows

$$p(j2\pi f) = \begin{cases} \pi\delta\left[2\pi(f - f_c)\right] + mG\left[j2\pi(f - f_c)\right] f \in \left[f_c - 0.5f_a, f_c\right] \\ p(-j2\pi f)^* f \in \left[f_c, -f_c + 0.5f_a\right] \\ 0 \text{ others} \end{cases} \tag{3.6}$$

where $G(j2\pi f)$ represents the Fourier transform of the signal $g\ (n)$. To eliminate the distortion of the parametric array loudspeaker, the input signal of the ultrasonic transducer $p\ (n)$ [132] should satisfy

$$\left|\left[1 + mg(n) - jm\hat{g}(n)\right]\right|^2 = 1 + mx(n) \tag{3.7}$$

Equation (3.7) can be rewritten as

$$g(n) = \frac{x(n) - mg(n)^2 - -m\hat{g}(n)^2}{2} \tag{3.8}$$

Equation (3.8) is a nonlinear equation and it can be solved by the fixed-point iteration method

$$g_k(n) = \frac{x(n) - mg_{k-1}(n)^2 - m\hat{g}_{k-1}(n)^2}{2} \tag{3.9}$$

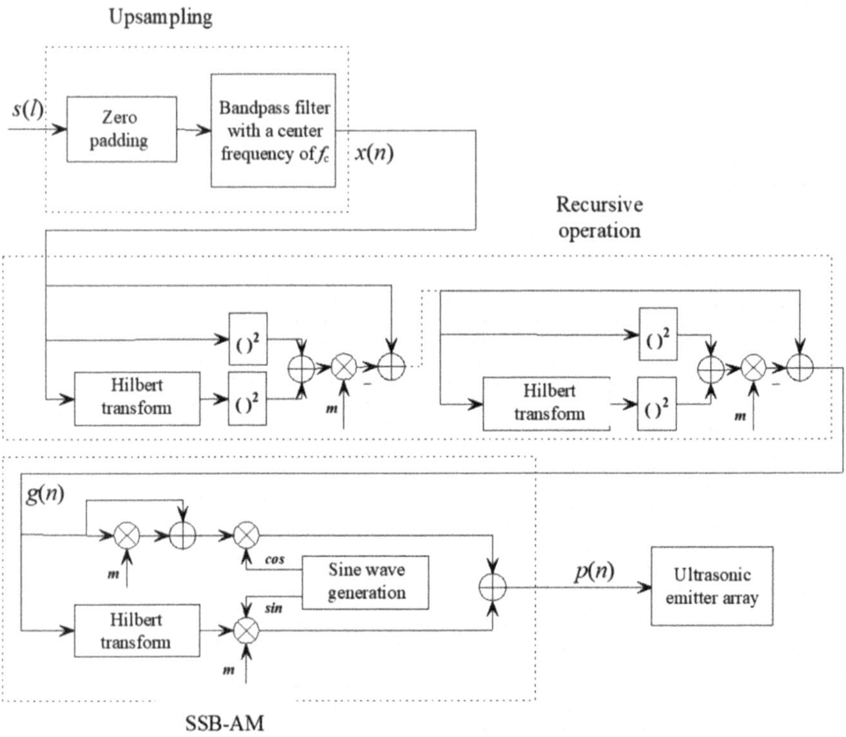

Fig. 3.17 Conventional RSSB-AM method

where k represents the number of iterations.

Hilbert transform in Eqs. (3.5) and (3.9) are usually implemented with FIR filters. The order of the Hilbert filter usually depends on the working sampling frequency. The working sampling frequency f_u in the traditional RSSB-AM algorithm, as shown in Fig. 3.17, is usually several times the ultrasonic carrier frequency, so the required order of the Hilbert filter is large.

3.1.2.2 Modified RSSB-AM Technique

The modified RSSB-AM technique is illustrated in Fig. 3.18, where the RSSB-AM is performed at a lower sampling frequency f_a. The preprocessed signal is also obtained using the fixed-point iteration method.

$$g_k(l) = \frac{s(l) - mg_{k-1}(l)^2 - m\hat{g}_{k-1}(l)^2}{2} \tag{3.10}$$

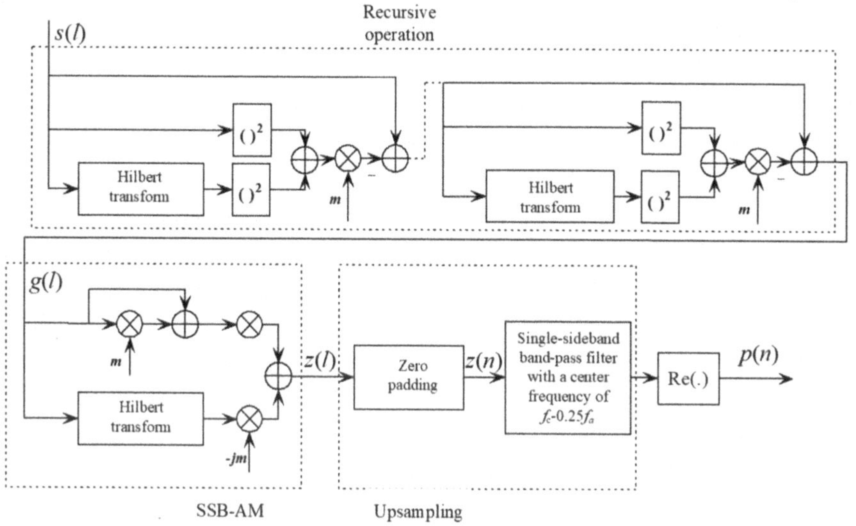

Fig. 3.18 Modified RSSB-AM technique (ideal ultrasonic array)

Unlike Eq. (3.9), the preprocessed signal sampling frequency obtained at this time is f_a. Then, single-sideband modulation is performed on the preprocessed signal to obtain

$$z(l) = \left[1 + mg(l) - jm\hat{g}(l)\right]e^{\frac{j2\pi f_c}{f_a}l} \quad (3.11)$$

Unlike Eq. (3.5), the modulated signal is a complex-domain signal. Assume that the ultrasonic carrier frequency is K times the audio sampling frequency, that is

$$f_c = Kf_a \quad (3.12)$$

Equation (3.11) can be simplified to

$$z(l) = 1 + mg(l) - jm\hat{g}(l) \quad (3.13)$$

By zero-padding $z(l)$, the sampling frequency is increased to f_u

$$z(n) = \begin{cases} z(l) & n = lM \\ 0 & \text{others} \end{cases} \quad (3.14)$$

Here it is assumed that $f_u = Mf_a$. Similarly, the frequency domain expression of Eq. (3.14) is

$$z(j2\pi f) = \begin{cases} 0 \; f \in [f_c - f_a, f_c - 0.5f_a] \\ 2\pi\delta[2\pi(f - f_c)] + 2mG[j2\pi(f - f_c)], \quad f \in [f_c - 0.5f_a, f_c] \\ 0 \; f \in [f_c, \; f_c + 0.5f_a] \end{cases} \quad (3.15)$$

Use a single-sideband bandpass filter with a passband of $[f_c - 0.5f_a, f_c]$ to filter $z(n)$. From Eq. (3.11), it is obvious that $z(n)$ is a complex domain signal, and the single-sideband bandpass filter is also complex. Take the real part of the filtered signal to get the input signal of the ultrasonic transducer. The passband is $[f_c - 0.5f_a, f_c]$ The coefficients of the single-sideband bandpass filter can be obtained through a low-pass filter with a cutoff frequency of $0.25f_a$

$$q_b(n) = q_l(n)e^{\frac{j2n\pi(f_c - 0.25f_a)}{f_u}} \quad (3.16)$$

It is found in references [409, 410] that the frequency response of the ultrasonic transducer has a significant impact on the performance of the parametric array loudspeaker. To solve this problem, an inverse filter can be added after single-sideband modulation to compensate for the frequency response of the ultrasonic transducer. Similarly, to reduce the amount of computation, the inverse filter is also implemented at a lower sampling frequency, as shown in Fig. 3.19. Assuming that the frequency response of the ultrasonic transducer in the frequency band $[f_c - 0.5f_a, f_c]$ is $H(j\omega)$, therefore, the frequency response of the inverse filter should satisfy

$$W(j2\pi f) = \begin{cases} \dfrac{1}{H[j2\pi(f + f_c)]}, \; f \in (-0.5f_a, 0) \\ W(-j2\pi f)^* \; f \in (0, 0.5f_a) \end{cases} \quad (3.17)$$

Finally, according to the above formula, the FIR structure inverse filter is designed using the filter design method [411].

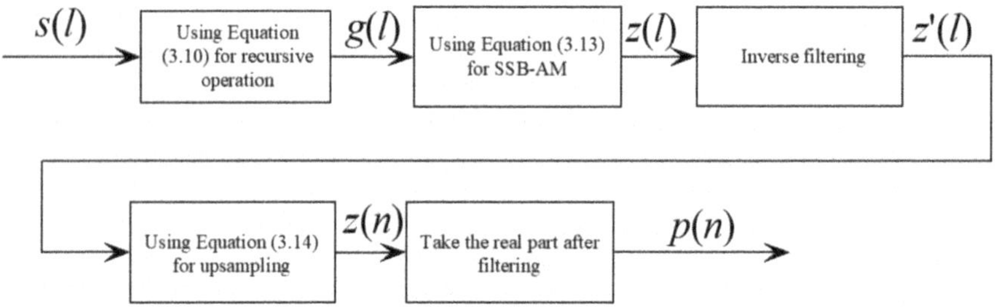

Fig. 3.19 Modified RSSB-AM method (non-ideal ultrasonic transducer array)

3.1.2.3 Computational Complexity Analysis

In this section, the computational complexity is analyzed by considering the number of multiplications required per second to evaluate the computational cost of the algorithm in this section. Equation (3.10) is used to generate the preprocessed signal and requires

$$T_r = K(5 + L)f_a \tag{3.18}$$

multiplications per second, where K represents the number of iterations, and L represents the order of the Hilbert transform filter. Equation (3.13) is used for single-sideband modulation and requires a total of

$$T_s = (2 + L)f_a \tag{3.19}$$

multiplications per second. The operation of the inverse filter requires

$$T_i = L'f_a \tag{3.20}$$

multiplications per second. L' represents the number of iterations. During the upsampling stage, the bandpass filter in single-sideband requires Nf_a complex number multiplications, that is

$$T_u = 4Nf_a \tag{3.21}$$

multiplications per second, where N is the order of the bandpass filter.

From Eqs. (3.18)–(3.21), it is evident that the overall computational complexity of the algorithm mainly depends on the order of the Hilbert transform filter, the inverse filter, and the bandpass filter. Table 3.5 shows the comparison of the order of the Hilbert transform filter required by this algorithm and the traditional algorithm under the same allowable error. The allowable error is defined as

$$E_h = 20\log10\left\{\max_{f\in\Phi}\left|H(f) - \hat{H}(f)\right|^2\right\} \tag{3.22}$$

$H(f)$ represents the required frequency response, $\hat{H}(f)$ represents the actual frequency response obtained, and Φ represents the frequency band of interest. In

Table 3.5 Filter order required under different allowable errors

E_h	L (Traditional implementation)	L (Method of this section)	N (Method of this section)
1 dB	100	24	7
0.1 dB	370	94	12
0.01 dB	3192	800	26

Table 3.5, f_a and f_u are assumed to be 40 kHz and 160 kHz, respectively. For the Hilbert transform filter, the frequency band considered is [500 19500] Hz. For the bandpass filter, the frequency band considered is [20000 40000] Hz. It becomes clear that, under the same allowable error, the order required for the bandpass filter is much less than that of the Hilbert transform filter. For the implementation method of this section, the order required for the Hilbert transform filter is only $1/M$ of the traditional method. Therefore, in the case of an ideal transducer, the computational load required for this implementation method is also only $1/M^2$ of the traditional method.

3.1.2.4 Experimental Analysis

In this section a comparison between the implementation method and the traditional implementation method is provided. In the experiment, f_a and f_u are assumed to be 40 kHz and 160 kHz, respectively, the modulation index $m = 0.9$, and the ultrasonic carrier frequency is 40 kHz. The transmitting array consists of 91 ultrasonic transducer units, and its frequency response is demonstrated in Fig. 3.20a. The allowable error for the design of the Hilbert transform filter and the inverse filter is set to 0.01 dB. For the traditional implementation method, the order required for the Hilbert transform filter and the inverse filter is 3192 and 4914, respectively. For the method of this section, the order required for the Hilbert transform filter and the inverse filter is 800 and 1014, respectively. Figure 3.20b shows the response of the inverse filter designed according to Eq. (3.17).

In the experiment, the original audio signal is two single-frequency signals as shown in Fig. 3.21a. Equation (3.10) is used to calculate the preprocessed signal. The amplitude spectrum of the preprocessed signal obtained after 5 iterations is shown in Fig. 3.21b. It is noticeable that the preprocessed signal adds 0, 2 and 3 kHz signal components to eliminate the distortion of the parametric array loudspeaker. The

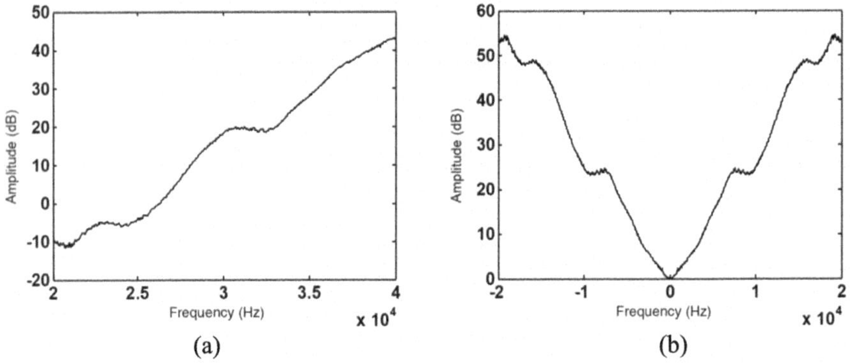

Fig. 3.20 **a** Amplitude response of the ultrasonic transducer array in the concerned frequency band **b** frequency response of the designed inverse filter

single-sideband signal generated by Eq. (3.13) is shown in Fig. 3.21c. The amplitude spectrum of the signal after inverse filtering is shown in Fig. 3.21d, where it becomes clear that the energy of the 4 kHz signal has been enhanced. The amplitude spectrum of the signal after zero-padding upsampling is shown in Fig. 3.21e, where multiple mirror components appear. Using a 128-order single-sideband bandpass filter, these mirror components are filtered out, and the resulting amplitude spectrum is shown in Fig. 3.21f. Finally, the amplitude spectrum of the signal obtained by taking the real part of the complex signal is shown in Fig. 3.21g. It is consistent with the power spectrum of the traditional method as shown in Fig. 3.21h.

The output audio test of the parametric array loudspeaker was conducted in the anechoic chamber of Institute of Acoustics, Chinese Academy of Sciences. The microphone is 6 m away from the center of the parametric array loudspeaker. The audio signals received by the modified RSSB-AM method and the traditional implementation method are shown in Fig. 3.22. It is obvious that both the modified RSSB-AM method and the traditional method can effectively reduce the distortion of the parametric array loudspeaker. The effects are similar, but the computational load required by the modified RSSB-AM method is less than that of the traditional method.

3.2 Research on Volterra-Based Preprocessing Method

This section first introduces the identification method of the VF model including the most common adaptive iterative algorithm and dual-tone measurement method. Then, it introduces the method of measuring ODVF by the sweep method. Next, it introduces the design method of the inverse filter, and gives the three common inverse filters, and conducts simulation experiments to analyze the performance of the three inverse filters. In addition, this section will also analyze the robustness of the p-order inverse filter, mainly examining the impact of the errors brought about by the identification process of the model's various order kernel functions on the inverse filter's compensation system's nonlinear distortion performance. And discuss the applicable conditions of the p-order inverse filter, that is, analyze the impact of the system's nonlinear distortion degree on the p-order inverse filter's compensation performance [412].

If a nonlinear system can be described by VF to describe its nonlinear characteristics, when inputting a signal $x(n)$, then the system's output response can be represented as

$$y(n) = \widehat{h}_0 + \sum_{i=1}^{\infty} \sum_{m_1=0}^{\infty} \sum_{m_2=0}^{\infty} \cdots \sum_{m_i=0}^{\infty} \left[\widehat{h}_i(m_1, m_2, \ldots, m_i) \cdot \prod_{j=1}^{i} x(n - m_j) \right] \tag{3.23}$$

where, $y(n)$ is the system's output signal; define $\hat{h}_i(m_1, m_2, \cdots, m_i)$ is the i-order Volterra kernel function. h_0 is the DC component, which is often omitted in some VF

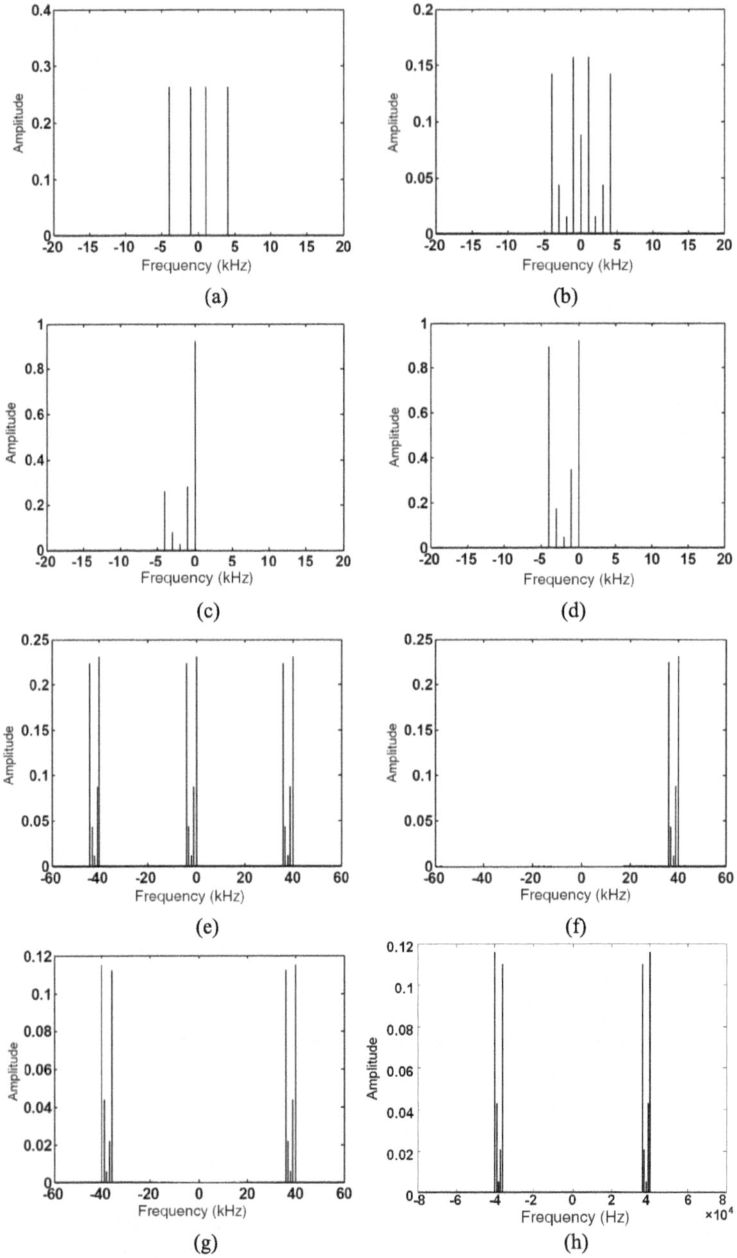

Fig. 3.21 Amplitude spectra of various related signals **a** original input signal, **b** preprocessed signal, **c** signal after single sideband, **d** signal after inverse filtering, **e** signal after zero-insertion upsampling, **f** signal after bandpass filtering, **g** transducer input signal obtained by the modified RSSB-AM method, **h** transducer input signal obtained by the traditional implementation method

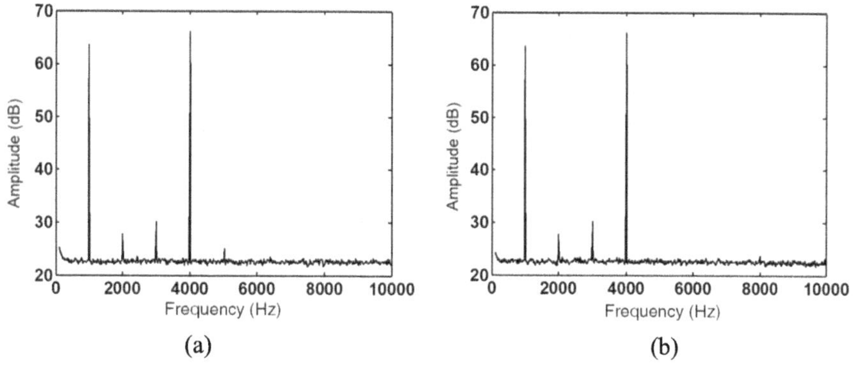

Fig. 3.22 Output audio signal of the parametric array loudspeaker **a** modified RSSB-AM method, **b** traditional implementation method

adaptive algorithm research for the convenience. The key to establishing a VF model to simulate the system's nonlinearity is to determine the Volterra kernel functions of each order. The common method is the MMSE algorithm [366], but this algorithm requires the calculation of the inverse of the matrix. For VF, because its kernel function has many coefficients, the calculation will be particularly complex. In 2000, Farina proposed an ODVF model s If a nonlinear system can be described by ODVF to describe its input–output nonlinear characteristics, when the input signal is $x(n)$, then the output response of its model is represented as

$$y(n) = \sum_{i=0}^{M_1-1} h_1(i)x(n-i) + \sum_{i=0}^{M_2-1} h_2(i)x^2(n-i) + \cdots + \sum_{i=0}^{M_p-1} h_p(i)x^p(n-i) + \cdots \tag{3.24}$$

where,$h_p(i)$ is the ODVF kernel function, $y(n)$ is the model's output, and M_p is the filter length of the p-order kernel function.

3.2.1 ODVF Model Identification

Farina [399] first gave the method of identifying model parameters using the logarithmic sweep method. Subsequently, Novak [402], Rebillat [405] and Tronchin [406] each conducted theoretical analysis of the method of measuring the ODVF model by the sweep method. However, their results are not the same, and no scholar has given a reasonable explanation about it. Moreover, Tronchin found a phase distortion problem in the process of using the logarithmic sweep to measure the parameters of the ODVF model. In response to the above problems, this section will conduct a detailed analysis of the kernel function of the VF model measured by the logarithmic sweep method, to explain why different results have appeared in existing research and why phase distortion problems occur.

For the ODVF model (as shown in Eq. (3.24)), its order is p (can be any positive integer, until it tends to infinity), when fed with the sweep signal $x(n)$, its output is

$$\begin{aligned} y(n) =& A\sum_{i=0}^{M_1-1} h_1(i)\sin[\phi(n-i)] + A^2\sum_{i=0}^{M_2-1} h_2(i)\sin^2[\phi(n-i)] \\ &+\cdots + A^p\sum_{i=0}^{M_p-1} h_p(i)\sin^p[\phi(n-i)] \end{aligned} \tag{3.25}$$

Here, it is necessary to simplify the terms $\sin^2[\phi(n-i)]$,$\sin^3[\phi(n-i)]$,...$\sin^p[\phi(n-i)]$. Take $\sin^p[\phi(n-i)]$ as an example, and first write $\sin^p[\phi(n-i)]$ in the form of a cofunction as

$$\sin^p(\phi) = \cos^p\left(\frac{\pi}{2}-\phi\right) \tag{3.26}$$

Next, expand the above formula into a complex form

$$\sin^p(\phi) = \left[\frac{e^{j(\frac{\pi}{2}-\phi)} + e^{-j(\frac{\pi}{2}-\phi)}}{2}\right]^p \tag{3.27}$$

According to the binomial theorem, the above formula can be further expanded as

$$\sin^p(\phi) = 2^{-p}\left[\sum_{r=0}^{p} C_p^r e^{j(p-2r)(\frac{\pi}{2}-\phi)}\right] \tag{3.28}$$

where the binomial coefficient is $C_p^r = \frac{p!}{r!(p-r)!}$. The above formula shows that $\sin^p(\phi)$ is formed by the superposition of $p+1$ exponential terms. Compare the *rth* term and the $(p-r)$th term to find that: $C_p^r = C_p^{p-r}$, and adding these two terms together gives

$$\begin{aligned} &C_p^r e^{j(p-2r)(\frac{\pi}{2}-\phi)} + C_p^{p-r} e^{j(p-2(p-r))(\frac{\pi}{2}-\phi)} \\ &= 2C_p^r\cos\left[(p-2r)\left(\frac{\pi}{2}-\phi\right)\right] \end{aligned} \tag{3.29}$$

From this, Eq. (3.28) can be further simplified as

$$\sin^p(\phi) = \begin{cases} 2^{-p+1}\left[\sum_{r=0}^{\frac{p-1}{2}} C_p^r \cos\left[(p-2r)\left(\frac{\pi}{2}-\phi\right)\right]\right], & p \text{ is odd} \\ 2^{-p}C_p^{p/2} + 2^{-p+1}\left[\sum_{r=0}^{\frac{p}{2}-1} C_p^r \cos\left[(p-2r)\left(\frac{\pi}{2}-\phi\right)\right]\right], & p \text{ is even} \end{cases} \tag{3.30}$$

Expand the term $\cos\left[(p-2r)\left(\frac{\pi}{2}-\phi\right)\right]$ in the above formula

$$\begin{aligned} &\cos\left[(p-2r)\left(\frac{\pi}{2}-\phi\right)\right] \\ &= \cos\left[\frac{(p-2r)\pi}{2} - \frac{\omega_1 T}{\ln(\omega_2/\omega_1)}\left[e^{\frac{(n+\gamma_{p-2r})\ln(\omega_2/\omega_1)}{N}} - 1\right]\right. \\ &\left. +(p-2r-1)\frac{\omega_1 T}{\ln(\omega_2/\omega_1)}\right] \end{aligned} \tag{3.31}$$

where $\gamma_{p-2r} = N\frac{\ln(p-2r)}{\ln(\omega_2/\omega_1)}$.

Expand Eq. (3.31) using the product-to-sum formula

$$\begin{aligned} &\cos\left[(p-2r)\left(\frac{\pi}{2}-\phi\right)\right] \\ &= \cos\left[\frac{(p-2r)\pi}{2} - \phi\left(n+\gamma_{p-2r}\right)\right]\cos\left[\frac{(p-2r-1)\omega_1 T}{\ln(\omega_2/\omega_1)}\right] \\ &- \sin\left[\frac{(p-2r)\pi}{2} - \phi\left(n+\gamma_{p-2r}\right)\right]\sin\left[\frac{(p-2r-1)\omega_1 T}{\ln(\omega_2/\omega_1)}\right] \end{aligned} \tag{3.32}$$

Substitute Eq. (3.32) into Eq. (3.30)

$$\sin^p(\phi) = \begin{cases} 2^{-p+1}\sum_{r=0}^{\frac{p-1}{2}} C_p^r(-1)^{\frac{p-1}{2}+r}\left\{\sin[\phi(n+\gamma_{p-2r})]\cos\left[\frac{(p-2r-1)\omega_1 T}{\ln(\omega_2/\omega_1)}\right]\right. \\ \left. -\cos[\phi(n+\gamma_{p-2r})]\sin\left[\frac{(p-2r-1)\omega_1 T}{\ln(\omega_2/\omega_1)}\right]\right\}, p \text{ is odd} \\ 2^{-p}C_p^{p/2} + 2^{-p+1}\sum_{r=0}^{\frac{p}{2}-1} C_p^r(-1)^{\frac{p}{2}+r}\left\{\cos[\phi(n+\gamma_{p-2r})]\cos\left[\frac{(p-2r-1)\omega_1 T}{\ln(\omega_2/\omega_1)}\right]\right. \\ \left. +\sin[\phi(n+\gamma_{p-2r})]\sin\left[\frac{(p-2r-1)\omega_1 T}{\ln(\omega_2/\omega_1)}\right]\right\}, \quad p \text{ is even} \end{cases} \tag{3.33}$$

where $\xi = \frac{\omega_1 T}{\ln(\omega_2/\omega_1)}$.

Substitute Eq. (3.33) into Eq. (3.25)

$$
\begin{aligned}
y(n) &= A\sum_{i=0}^{M_1-1} h_1(i)\sin[\phi(n-i)] \\
&+ A^2\sum_{i=0}^{M_2-1} h_2(i)\left[\frac{1}{2} - \cos\left(\frac{\omega_1 T}{\ln(\omega_2/\omega_1)}\right)\cos(\phi(n+\gamma_2-i))\right. \\
&\left. - \sin\left(\frac{\omega_1 T}{\ln(\omega_2/\omega_1)}\right)\sin(\phi(n+\gamma_2-i))\right] + \cdots \\
&+ A^{2m}\sum_{i=0}^{M_{2m}-1} h_{2m}(i)\left\{2^{-2m}C_{2m}^{m} + 2^{-2m+1}\sum_{r=0}^{m-1} C_{2m}^{r}(-1)^{m+r}\right. \\
&\left.\left\{\begin{aligned}&\cos\left[\phi(n+\gamma_{2m-2r}-i)\right]\cos\left[\frac{(2m-2r-1)\omega_1 T}{\ln(\omega_2/\omega_1)}\right] \\ &+\sin\left[\phi(n+\gamma_{2m-2r}-i)\right]\sin\left[\frac{(p-2r-1)\omega_0 T}{\ln(\omega_2/\omega_1)}\right]\end{aligned}\right\}\right\} \\
&+ A^{2m+1}\sum_{i=0}^{M_{2m+1}-1} h_{2m+1}(i)\left\{2^{-2m}\sum_{r=0}^{m} C_{2m+1}^{r}(-1)^{m+r}\right. \\
&\left.\left\{\begin{aligned}&\sin\left[\phi(n+\gamma_{2m+1-2r}-i)\right]\cos\left[\frac{(2m+1-2r-1)\omega_1 T}{\ln(\omega_2/\omega_1)}\right] \\ &-\cos\left[\phi(n+\gamma_{2m+1-2r}-i)\right]\sin\left[\frac{(2m+1-2r-1)\omega_1 T}{\ln(\omega_2/\omega_1)}\right]\end{aligned}\right\}\right\}
\end{aligned} \tag{3.34}
$$

Replace the corresponding terms in Eq. (3.34) with the sweep signal $x(n)$ to obtain

$$
\begin{aligned}
y(n) &= A\sum_{i=0}^{M_1-1} h_1(i)x(n-i) + A^2\sum_{i=0}^{M_2-1} h_2(i) \\
&\left[\frac{1}{2} + \cos(\xi)x(n+\gamma_2-i) * \delta_{hil} - \sin(\xi)x(n+\gamma_2-i)\right] + \cdots \\
&+ A^{2m}\sum_{i=0}^{M_{2m}-1} h_{2m}(i)\left\{2^{-2m}C_{2m}^{m} + 2^{-2m+1}\sum_{r=0}^{m-1} C_{2m}^{r}(-1)^{m+r}\right. \\
&\{-\cos[(2m-2r-1)\xi]x(n+\gamma_{2m-2r}-i) * \delta_{hil} \\
&\quad + \sin[(2m-2r-1)\xi]x(n+\gamma_{2m-2r}-i)\}\} \\
&+ A^{2m+1}\sum_{i=0}^{M_{2m+1}-1} h_{2m+1}(i)\left\{2^{-2m}\sum_{r=0}^{m} C_{2m}^{r}(-1)^{m+r}\{\cos[(2m-2r)\xi]\right. \\
&x(n+\gamma_{2m+1-2r}-i) + \sin[(2m-2r)\xi]x(n+\gamma_{2m+1-2r}-i) * \delta_{hil}\}\}
\end{aligned} \tag{3.35}
$$

where, δ_{hil} is a Hilbert filter, which can shift the sine function by 90° phase to become a cosine function.

Convolve both sides of Eq. (3.35) with the inverse signal of the sweep signal $\tilde{x}(n)$, to obtain the impulse response $k(n)$, and its expression is

$$\begin{aligned}
k(n) &= y(n) * \tilde{x}(n) \\
&= \sum_{i=0}^{M_1-1} h_1(i)\delta(n-i) + A\sum_{i=0}^{M_2-1} h_2(i) \\
&\left[\frac{1}{2} + \cos(\xi)\delta_{hil}(n+\gamma_2-i) - \sin(\xi)\delta(n+\gamma_2-i)\right] \\
&+\cdots \\
&+ A^{2m-1}\sum_{i=0}^{M_{2m}-1} h_{2m}(i)\left\{2^{-2m}C_{2m}^{m} + 2^{-2m+1}\sum_{r=0}^{m-1} C_{2m}^{r}(-1)^{m+r}\right. \\
&\{-\cos[(2m-2r-1)\xi]\delta_{hil}(n+\gamma_{2m-2r}-i) \\
&+\sin[(2m-2r-1)\xi]\delta(n+\gamma_{2m-2r}-i)\}\} \\
&+ A^{2m}\sum_{i=0}^{M_{2m+1}-1} h_{2m+1}(i)\left\{2^{-2m}\sum_{r=0}^{m} C_{2m}^{r}(-1)^{m+r}\{\cos[(2m-2r)\xi]\right. \\
&\delta(n+\gamma_{2m+1-2r}-i) + \sin[(2m-2r)\xi]\delta_{hil}(n+\gamma_{2m+1-2r}-i)\}\}
\end{aligned} \tag{3.36}$$

Combine like terms in Eq. (3.36), to obtain the measured impulse response $k(n)$, which is composed of a series of shifted impulse responses

$$k(n) = k_1(n) + k_2(n+\gamma_2) + \cdots k_p(n+\gamma_p) \tag{3.37}$$

where $\gamma_p = N\frac{\ln(p)}{\ln(\omega_2/\omega_1)}$ and the impulse responses $k_1(n)$, $k_2(n)$, ..., $k_p(n)$ and p-order ODVF kernel functions satisfy the following relationship

$$K = A_0 + A_1G_1 + A_2G_2 \tag{3.38}$$

where, K is a one-dimensional vector composed of impulse responses $k_1(n)$, $k_2(n)\ldots k_p(n)$ which can be written as

$$K^T = [k_1(n+\gamma_1), k_2(n+\gamma_2), \ldots, k_p(n+\gamma_p)] \tag{3.39}$$

A_0 is the DC component, which is a weighted combination of the even-order kernel functions of ODVF, and its expression is

$$A_0 = \sum_{\substack{r=1,\\ 2r\le p}}^{p} \frac{A^{2r-1}}{2^{2r}} C_{2r}^{r} \sum_{n=0}^{M_{2r}} h_{2r}(n) \tag{3.40}$$

It manifests as a noise in the final impulse response. In general, it is relatively small and can be ignored.

A_1 is a two-dimensional matrix of the same size as $p \times p$, and its matrix elements are

$$A_1(l,m)=\begin{cases}\dfrac{(-1)^{\frac{2m-l-1}{2}}A^{m-1}}{2^{m-1}}C_m^{(m-l)/2}\cdot\cos\left[(l-1)\dfrac{\omega_1 T}{\ln(\omega_2/\omega_1)}\right], \\ l \text{ is even}, m \text{ is even, and } m\geq l \\ \dfrac{(-1)^{\frac{2m-l}{2}}A^{m-1}}{2^{m-1}}C_m^{(m-l)/2}\cdot\sin\left[(l-1)\dfrac{\omega_1 T}{\ln(\omega_2/\omega_1)}\right] \\ l \text{ is odd, m is odd, and } m\geq l \\ 0 \qquad\qquad , \text{ others}\end{cases} \tag{3.41}$$

A_2 is also a two-dimensional matrix of the same size as $p \times p$, and its matrix elements are

$$A_2(l,m)=\begin{cases}\dfrac{(-1)^{\frac{2m-l-1}{2}}A^{m-1}}{2^{m-1}}C_m^{(m-l)/2}\cdot\sin\left[(l-1)\dfrac{\omega_1 T}{\ln(\omega_2/\omega_1)}\right], \\ l \text{ is even}, \ m \text{ is even, and } m\geq l \\ -\dfrac{(-1)^{\frac{2m-l}{2}}A^{m-1}}{2^{m-1}}C_m^{(m-l)/2}\cdot\cos\left[(l-1)\dfrac{\omega_1 T}{\ln(\omega_2/\omega_1)}\right], \\ l \text{ is odd}, m \text{ is odd, and } m\geq l \\ 0 \qquad\qquad , \text{ others}\end{cases} \tag{3.42}$$

G_1 is a set of one-dimensional vectors, and its size is $p \times 1$. It is composed of the shifted kernel functions of each order of ODVF, represented as

$$G_1^T = [h_1(n-\gamma_1), h_2(n-\gamma_2), \ldots, h_p(n-\gamma_p)] \tag{3.43}$$

G_2 is also a set of one-dimensional vectors, and its size is $p \times 1$. It is composed of the convolution of the kernel functions of each order of ODVF with the Hilbert filter $\delta_{hil}(n)$ and then shifted, represented as

$$\begin{aligned}G_2^T =&[h_1(n) * \delta_{hil}(n-\gamma_1), h_2(n) \\ &* \delta_{hil}(n-\gamma_2), \ldots, h_p(n) * \delta_{hil}(n-\gamma_p)]\end{aligned} \tag{3.44}$$

So far, the relationship between the kernel functions of each order of ODVF and the measured impulse response $k_i(n)$ has been derived as shown in Eq. (3.39). According to this relationship, the kernel functions of each order of ODVF can be solved from the impulse response $k_i(n)$ measured by the experiment. The following will take the 5th-order filter model as an example (since all existing work has theoretically derived the 5th-order filter model, the same 5th-order is chosen for comparison with previous work. In fact, p in Eq. (3.39) can be any positive integer), to explain in detail the steps to solve the kernel functions of each order of ODVF.

For a 5th-order ODVF model, according to the general expression given by Eqs. (3.41)–(3.44), let $p = 5$, and the parameters of the relationship equation between the kernel functions of each order of the filter and the measured impulse response are obtained as

$$A_1 = \begin{bmatrix} 1 & 0 & \frac{3}{4}A^2 & 0 & \frac{5}{8}A^4 \\ 0 & -\frac{A}{2}\sin(\xi) & 0 & -\frac{A^3}{2}\sin(\xi) & 0 \\ 0 & 0 & -\frac{A^2}{4}\cos(2\xi) & 0 & -\frac{5A^4}{16}\cos(2\xi) \\ 0 & 0 & 0 & \frac{A^3}{8}\sin(3\xi) & 0 \\ 0 & 0 & 0 & 0 & \frac{A^4}{16}\cos(4\xi) \end{bmatrix} \tag{3.45}$$

$$A_2 = \begin{bmatrix} 0 & 0 & 0 & 0 & 0 \\ 0 & \frac{A}{2}\cos(\xi) & 0 & \frac{A^3}{2}\cos(\xi) & 0 \\ 0 & 0 & -\frac{A^2}{4}\sin(2\xi) & 0 & -\frac{5A^4}{16}\sin(2\xi) \\ 0 & 0 & 0 & -\frac{A^3}{8}\cos(3\xi) & 0 \\ 0 & 0 & 0 & 0 & \frac{A^4}{16}\sin(4\xi) \end{bmatrix} \tag{3.46}$$

Obviously, solving Eq. (3.37) according to the above matrix coefficients is a huge task, because the unknowns not only include $h_i(n)$, but also $h_i(n) * \delta_{hil}(n)$. Here, four special cases are discussed.

Case One When the starting angular frequency of the sweep signal ω_1, the ending angular frequency ω_2 and the duration of the sweep signal T are chosen appropriately, so that $\xi = \frac{\pi}{2} + 2\eta\pi$, $\eta = \cdots, -1, 0, 1, 2, \cdots$, and then, Eq. (3.37) can be simplified as

$$\begin{bmatrix} 1 & 0 & \frac{3}{4}A^2 & 0 & \frac{5}{8}A^4 \\ 0 & -\frac{A}{2} & 0 & -\frac{A^3}{2} & 0 \\ 0 & 0 & \frac{A^2}{4} & 0 & \frac{5A^4}{16} \\ 0 & 0 & 0 & -\frac{A^3}{8} & 0 \\ 0 & 0 & 0 & 0 & \frac{A^4}{16} \end{bmatrix} \cdot \begin{bmatrix} h_1(n) \\ h_2(n) \\ h_3(n) \\ h_4(n) \\ h_5(n) \end{bmatrix} = \begin{bmatrix} k_1(n) \\ k_2(n) \\ k_3(n) \\ k_4(n) \\ k_5(n) \end{bmatrix} \tag{3.47}$$

Solve the above equation to get the equation to calculate $h_i(n)$ according to $k_i(n)k_i(n)$

$$
\begin{cases}
h_1 = k_1 - 3k_3 + 5k_5 \\
h_2 = -\dfrac{2}{A}k_2 + \dfrac{8}{A}k_4 \\
h_3 = \dfrac{4}{A^2}k_3 - \dfrac{20}{A^2}k_5 \\
h_4 = -\dfrac{8}{A^3}k_4 \\
h_5 = \dfrac{16}{A^4}k_5
\end{cases}
\tag{3.48}
$$

Case Two When the starting angular frequency of the sweep signal ω_1, the ending angular frequency ω_2 and the duration of the sweep signal T are chosen appropriately, so that, $\xi = -\frac{\pi}{2} + 2\eta\pi, \eta = \cdots, -1, 0, 1, 2, \cdots$, and then, Eq. (3.37) can be simplified as

$$
\begin{bmatrix}
1 & 0 & \frac{3}{4}A^2 & 0 & \frac{5}{8}A^4 \\
0 & \frac{A}{2} & 0 & \frac{A^3}{2} & 0 \\
0 & 0 & \frac{A^2}{4} & 0 & \frac{5A^4}{16} \\
0 & 0 & 0 & \frac{A^3}{8} & 0 \\
0 & 0 & 0 & 0 & \frac{A^4}{16}
\end{bmatrix}
\cdot
\begin{bmatrix}
h_1(n) \\ h_2(n) \\ h_3(n) \\ h_4(n) \\ h_5(n)
\end{bmatrix}
=
\begin{bmatrix}
k_1(n) \\ k_2(n) \\ k_3(n) \\ k_4(n) \\ k_5(n)
\end{bmatrix}
\tag{3.49}
$$

Solve the above equation to calculate $h_i(n)$ according to $k_i(n)$

$$
\begin{cases}
h_1 = k_1 - 3k_3 + 5k_5 \\
h_2 = \dfrac{2}{A}k_2 - \dfrac{8}{A}k_4 \\
h_3 = \dfrac{4}{A^2}k_3 - \dfrac{20}{A^2}k_5 \\
h_4 = \dfrac{8}{A^3}k_4 \\
h_5 = \dfrac{16}{A^4}k_5
\end{cases}
\tag{3.50}
$$

When the amplitude of the sweep signal A is 1, Eq. (3.48) is consistent with the results given by Rebillat et al. [405]. However, Rebillat et al. did not point out that only when $\xi = -\frac{\pi}{2} + 2\eta\pi, \eta = \ldots, -1, 0, 1, 2, \ldots$, the above relationship can be established.

Case Three When the starting angular frequency of the sweep signal ω_1, the ending angular frequency ω_2 and the duration of the sweep signal T are chosen appropriately, so that, $\xi = 0 + 2\eta\pi, \eta = \ldots, -1, 0, 1, 2, \ldots$, and then, Eq. (3.37) can be simplified as

$$\begin{bmatrix} 1 & 0 & \frac{3}{4}A^2 & 0 & \frac{5}{8}A^4 \\ 0 & \frac{A}{2} & 0 & \frac{A^3}{2} & 0 \\ 0 & 0 & -\frac{A^2}{4} & 0 & -\frac{5A^4}{16} \\ 0 & 0 & 0 & -\frac{A^3}{8} & 0 \\ 0 & 0 & 0 & 0 & \frac{A^4}{16} \end{bmatrix} \cdot \begin{bmatrix} h_1(n) \\ h_3(n) * \delta_{hil}(n) \\ h_2(n) * \delta_{hil}(n) \\ h_5(n) \end{bmatrix} = \begin{bmatrix} k_1(n) \\ k_2(n) \\ k_3(n) \\ k_4(n) \\ k_5(n) \end{bmatrix} \tag{3.51}$$

Solve the above equation to obtain the equation used to calculate $h_i(n)$ according to $k_i(n)$

$$\begin{cases} h_1 = k_1 + 3k_3 + 5k_5 \\ h_2 = -\frac{2}{A}k_2 * \delta_{hil} - \frac{8}{A}k_4 * \delta_{hil} \\ h_3 = -\frac{4}{A^2}k_3 - \frac{20}{A^2}k_5 \\ h_4 = \frac{8}{A^3}k_4 * \delta_{hil} \\ h_5 = \frac{16}{A^4}k_5 \end{cases} \tag{3.52}$$

When the amplitude of the sweep signal A is 1, the Eq. (3.52) is consistent with the given frequency domain expression given by Farina [399]. In Farina's article, he made an approximate derivation and obtained the above expression, but the article did not prove that only the condition when $\xi = 0 + 2\eta\pi, \quad \eta = \ldots, -1, 0, 1, 2, \ldots$ is met, there exists a relationship as shown in Eq. (3.52) between the kernel functions of ODVF and the impulse response measured.

Case Four When the starting angular frequency of the sweep signal ω_1, the ending angular frequency ω_2 and the duration of the sweep signal T are chosen appropriately, so that, $\xi = \pi + 2\eta\pi$, $\eta = \ldots, -1, 0, 1, 2, \ldots$, and then Eq. (3.37) can be simplified as follows

$$\begin{bmatrix} 1 & 0 & \frac{3}{4}A^2 & 0 & \frac{5}{8}A^4 \\ 0 & -\frac{A}{2} & 0 & -\frac{A^3}{2} & 0 \\ 0 & 0 & -\frac{A^2}{4} & 0 & -\frac{5A^4}{16} \\ 0 & 0 & 0 & \frac{A^3}{8} & 0 \\ 0 & 0 & 0 & 0 & \frac{A^4}{16} \end{bmatrix} \cdot \begin{bmatrix} h_1(n) \\ h_3(n) \\ h_2(n) * \delta_{hil}(n) \\ h_4(n) * \delta_{hil}(n) \\ h_5(n) \end{bmatrix} = \begin{bmatrix} k_1(n) \\ k_2(n) \\ k_3(n) \\ k_4(n) \\ k_5(n) \end{bmatrix} \tag{3.53}$$

Solve the above equation to obtain the equation to calculate $h_i(n)$ according to $k_i(n)$

$$\begin{cases} h_1 = k_1 + 3k_3 + 5k_5 \\ h_2 = \dfrac{2}{A}k_2 * \delta_{hil} + \dfrac{8}{A}k_4 * \delta_{hil} \\ h_3 = -\dfrac{4}{A^2}k_3 - \dfrac{20}{A^2}k_5 \\ h_4 = -\dfrac{8}{A^3}k_4 * \delta_{hil} \\ h_5 = \dfrac{16}{A^4}k_5 \end{cases} \tag{3.54}$$

This time-domain expression is actually consistent with the results given by Novak [402] and Tronchin [406]. Tronchin applied this relationship to obtain the kernel function of ODVF to simulate the harmonic distortion produced by musical instruments, and he found that the model obtained by this method has phase distortion. To correct this phase distortion, he proposed the use of fractional delay filters, but he did not point out why this phase distortion occurred.

In summary, a rigorous derivation has been conducted on the relationships of the kernel functions of each order in the swept-frequency method for measuring ODVF, establishing a unified theoretical framework as shown in Eq. (3.37) to gain a deeper understanding of the relationship between the impulse response obtained from swept-frequency measurements and the kernel functions of ODVF. From the solution steps of the 5th-order filter given above, it is obvious that when the parameters of the swept-frequency signal meet certain conditions, a concise and clear expression between the impulse response and the ODVF kernel functions can be obtained. It is also noted that different parameter choices correspond to different solution formulas. This explains why Tronchin [406] found phase distortion issues in experiments. The parameters chosen for the swept-frequency signal may not necessarily meet the conditions required for Eq. (3.54) to hold, which can introduce phase distortion in the measurements. Furthermore, comparing the four cases given above, Cases 1 and 2 are essentially similar, with the difference being that they are inversely related when solving for even-order kernel functions. Cases 3 and 4 are essentially similar, with the difference being that there is an inverse relationship when solving for even-order kernel functions. Comparing the expressions of the first two cases with the last two cases, it is known that in the latter two cases, a Hilbert filter is needed when solving for even-order kernel functions. On one hand, the Hilbert filter increases the computational complexity of system identification, and on the other hand, the Hilbert filter only satisfies the properties of the Hilbert transform under narrowband conditions, which also limits the applicability of the identified ODVF model. Therefore, in actual measurements, for convenience, the parameters of the swept-frequency signal required by Case 1 or Case 2 should be chosen.

3.2.2 *Design of Inverse Filter Based on ODVF*

3.2.2.1 Precise Inverse Filter

In the research of compensating for the nonlinear distortion of the loudspeaker system based on the establishment of a nonlinear model to analyze the nonlinear distortion characteristics of the loudspeaker system, a compensation scheme needs to be provided to eliminate or reduce the harmonic distortion of the loudspeaker system as much as possible. Specifically, it is to design a filter, process the input signal through this filter, and then feed it to the nonlinear system, so that there is no nonlinear distortion component in the system output signal or the nonlinear distortion component is as small as possible. Such a filter can be called a pre-inverse filter. Schetzen proposed a scheme to design a p-order inverse filter based on VF in 1980 [369], which continues to be used up to the present. The following will discuss how to design an inverse filter based on the established ODVF to eliminate the nonlinear distortion in the system output. Consider such an ODVF system

$$\begin{aligned} y(n) = & \sum_{i=0}^{M_1-1} h_1(i)x(n-i) + \sum_{i=0}^{M_2-1} h_2(i)x^2(n-i) + \cdots \\ & + \sum_{i=0}^{M_p-1} h_p(i)x^p(n-i) + \cdots \end{aligned} \tag{3.55}$$

where, $h_p(i)$ is the ODVF kernel function, $y(n)$ is the output of the model and M_p is the filter length of the pth-order kernel function. An infinite order system is difficult to solve for its inverse filter, so here only a pth-order system is considered.

Here, an inverse filter P needs to be designed. After the signal $x(n)$ passes through this filter, the signal $q(n)$ is obtained, which is fed into the aforementioned nonlinear system to yield

$$\begin{aligned} y(n) = & \sum_{i=0}^{M_1-1} h_1(i)q(n-i) + \sum_{i=0}^{M_2-1} h_2(i)q^2(n-i) + \cdots \\ & + \sum_{i=0}^{M_p-1} h_p(i)q^p(n-i) \end{aligned} \tag{3.56}$$

To ensure that the system output has no nonlinear distortion and only retains the linear response, it needs to satisfy

$$\sum_{i=0}^{M_1-1} h_1(i)q(n-i) + \sum_{i=0}^{M_2-1} h_2(i)q^2(n-i) + \cdots$$

$$+\sum_{i=0}^{M_p-1} h_p(i)q^p(n-i) = \sum_{i=0}^{M_1-1} h_1(i)x(n-i) \tag{3.57}$$

Assuming that the signal at past moments such as $q(n-1)$, $q(n-i)$ are known, and the current signal $q(n)$ needs to be determined. The above equation is rewritten as

$$\sum_{i=1}^{p} h_i(0)q^i(n) = A_x - B_q \tag{3.58}$$

where $A_x = \sum_{i=0}^{M_1-1} h_1(i)x(n-i)$, $B_q = \sum_{i=1}^{M_1-1} h_1(i)q(n-i) + \sum_{i=1}^{M_2-1} h_2(i)q^2(n-i) + \cdots + \sum_{i=1}^{M_p-1} h_p(i)q^p(n-i)$.

The above equation is a pth-order nonlinear equation, with the unknown being $q(n)$. At each moment, A_x can be calculated based on the current input signal $x(n)$, and B_q can also be calculated based on the historical signal $q(n-i)$, and solve the above equation to get the current signal $q(n)$, which is fed into the nonlinear system. This can ensure the complete elimination of all distortions in the nonlinear system. This scheme is also called precise inverse filter design, because it can eliminate all distortions in the nonlinear system. However, this method has some problems, such as the need to solve a nonlinear equation at each moment, and when the coefficient order p is large, the solution process of this equation is not easy. In addition, some nonlinear systems cannot guarantee that the equation will have a solution due to the coefficients of their kernel functions. These factors to some extent limit the practicality of the precise inverse filter.

3.2.2.2 *P*th-Order Inverse Filter

Revisiting Eq. (3.57), if an exact inverse filter is required, there are the aforementioned issues. In practical applications, an inverse filter that is not entirely accurate needs to be identified, yet its robustness will be enhanced. Analyzing the solution of Eq. (3.58), it is found that $q(n)$ should contain the linear component and the nonlinear component of the input signal, such as the second harmonic, the third harmonic, etc. Therefore, write $q(n)$ as

$$q(n) = q_1(n) + q_2(n) + \cdots + q_p(n) + \cdots \tag{3.59}$$

where, $q_1(n)$ only contains the linear component, $q_2(n)$ only contains the second harmonic component, $q_p(n)$ only contains the pth harmonic component. Substitute Eq. (3.59) into Eq. (3.56) to yield

$$y(n) = \sum_{k=1}^{p} h_k * \left(\sum_{i=1}^{\infty} q_i \right)^k \tag{3.60}$$

Now, the expectation is no longer to eliminate all nonlinear distortions, but to completely eliminate the 2nd to the pth harmonic distortion. That is,

$$\begin{cases} \left[\sum_{k=1}^{p} h_k * \left(\sum_{i=1}^{\infty} q_i \right)^k \right]_1 = h_1 * x \\ \left[\sum_{k=1}^{p} h_k * \left(\sum_{i=1}^{\infty} q_i \right)^k \right]_m = 0 \quad 2 \le m \le p \end{cases} \tag{3.61}$$

where $[]_m$ represents the mth harmonic distortion.

From the above formula, it can be known that

$$\left[h_k * (q_i)^k \right]_m = \begin{cases} h_k * q_i & k, i = m \\ 0 & \text{others} \end{cases} \tag{3.62}$$

Therefore

$$\left[\sum_{k=1}^{p} h_k * \left(\sum_{i=1}^{\infty} q_i \right)^k \right]_1 = h_1 * q_1 \tag{3.63}$$

Therefore, $q_1 = x$. According to the properties given by Eq. (3.62), Eq. (3.61) is written as

$$h_1 * q_m + \left[\sum_{k=2}^{m} h_k * \left(\sum_{i=1}^{m-k+1} q_i \right)^k \right]_m = 0 \quad 2 \le m \le p \tag{3.64}$$

Therefore

$$q_m = - \left[\sum_{k=2}^{m} h_k * \left(\sum_{i=1}^{m-k+1} q_i \right)^k \right]_m * h_1^{-1} \quad 2 \le m \le p \tag{3.65}$$

The above formula is the expression of the pth-order inverse filter. In this study, only the 2nd-order inverse filter and the 3rd-order inverse filter need to be considered. The structure of the two inverse filters will be discussed separately.

For the 2nd-order inverse filter, according to Eq. (3.65), it is obtained that

$$q_2 = -h_2 * q_1^2 * h_1^{-1} \tag{3.66}$$

Therefore, the framework of the 2nd-order inverse filter is shown in Fig. 3.23. If the 3rd-order inverse filter is considered, then

$$q_3 = -h_3 * q_1^3 * h_1^{-1} - 2h_2 * (q_1 q_2) * h_1^{-1} \tag{3.67}$$

The framework of the 3rd-order inverse filter is demonstrated in Fig. 3.24.

As can be seen from Figs. 3.23 and 3.24, the structure of the pth-order inverse filter designed based on ODVF is similar to the structure of the pth-order inverse filter given by Schetzen [369] based on VF. As the order p increases, it can be foreseen that the structure of the inverse filter will become extremely complex. To solve this problem, in 1992, Sarti et al. [386] proposed a recursive form of the pth-order inverse filter based on VF. It can also be applied to the inverse filter of ODVF. The framework of the 2nd-order recursive inverse filter is the same as the structure of the traditional 2nd-order inverse filter, but the structure of the 3rd-order recursive inverse filter is

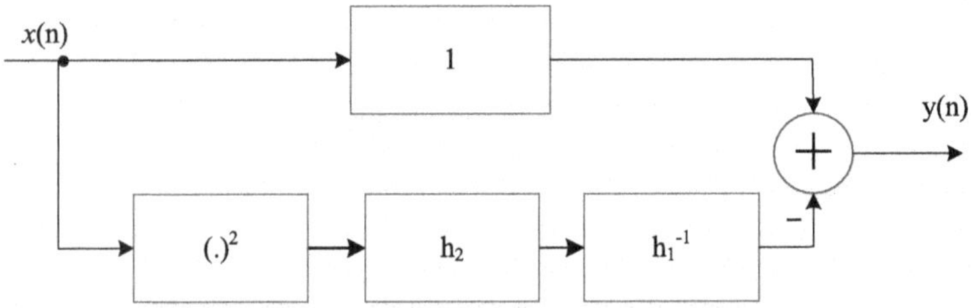

Fig. 3.23 2nd-order inverse filter framework diagram

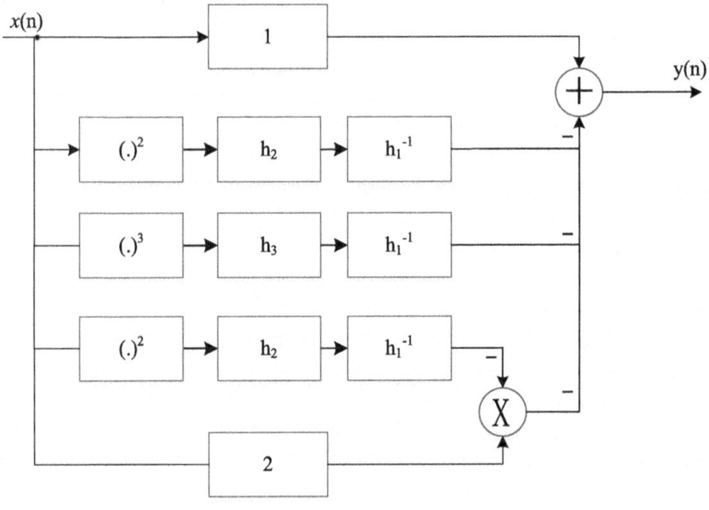

Fig. 3.24 3rd-order inverse filter framework diagram

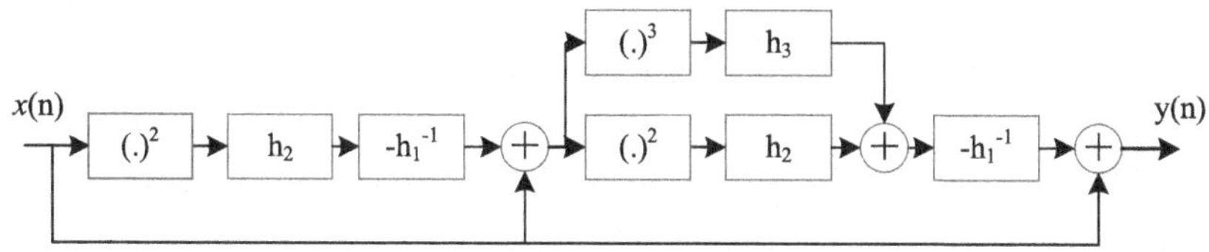

Fig. 3.25 Recursive structure of the 3rd-order inverse filter framework diagram

no longer the same as the traditional structure of the inverse filter shown in Fig. 3.24. The framework of the 3rd-order recursive inverse filter is demonstrated in Fig. 3.25.

3.2.3 Connection Between ODVF and VF

Since its proposal by Farina [399] in 2000, ODVF has been used as a simplified model of VF to this day. It is used to simulate the nonlinear characteristics of vibratory devices (such as loudspeakers) [404, 413], the nonlinear distortion of channels in echo cancellation problems [414], the nonlinear characteristics of circuit systems [415], the input–output response of electric guitar speakers [416, 417], and the distortion of devices in active noise control problems [418], etc. Different identification methods for measuring the kernel function of ODVF with the sweep frequency method are given in the literature. To distinguish the correctness of various identification methods, this section will conduct in-depth theoretical analysis on the sweep frequency method to measure the ODVF kernel function, and try to establish a unified theoretical framework to unify the existing various identification methods.

In addition, there is currently a lack of theoretical basis for the applicability of ODVF. Existing research work only introduces the ODVF model into their respective research fields and conducts experimental verification, proving its applicability through experimental results. This section attempts to explore the applicability of the ODVF model from a theoretical perspective. Considering that the traditional VF has been widely applied to the compensation of nonlinear distortion in loudspeaker systems, and there is relevant theoretical guidance for its applicability research [391, 419, 420]. This section will explore whether nonlinear systems that can be simulated by VF can also be simulated by ODVF.

This section first theoretically analyzes the basic principles of measuring the order kernel functions of ODVF by the logarithmic frequency sweep method, and derives the general expression for solving the order kernel functions of ODVF. Secondly, through simulation experiments, the correctness of the solution formula given in this section is further verified. Again, the correlation and difference between the kernel functions of VF and ODVF under the excitation of logarithmic frequency sweep signals are discussed, the errors between the two models are analyzed, some characteristic parameters affecting the size of the error are identified, and the relationship between the error and these parameters is discussed. Finally, the sufficient conditions for ODVF to replace the traditional VF model are identified.

The input–output relationship of a traditional VF satisfies

$$y(n) = \widehat{h}_0 + \sum_{i=1}^{\infty} \sum_{m_1=0}^{\infty} \sum_{m_2=0}^{\infty} \cdots \sum_{m_i=0}^{\infty} \left[\widehat{h}_i(m_1, m_2, \ldots, m_i) \cdot \prod_{j=1}^{i} x(n - m_j) \right] \tag{3.68}$$

where $y(n)$ is the output signal of the system. Define $\widehat{h}_i(m_1, m_2, \ldots, m_i)$ is the i-order Volterra kernel function. Its 1st order kernel function $\widehat{h}_1(m_1)$ is a one-dimensional vector group, the 2nd-order kernel function $\widehat{h}_2(m_1, m_2)$ is a two-dimensional matrix, correspondingly, the 3rd and higher order kernel functions are three-dimensional and higher-dimensional coefficient matrices. Farina [399] pointed out that if the non-diagonal elements of the coefficient matrix of the 2nd and higher order kernel functions of VF are considered as 0, and only its diagonal elements are retained, then

$$\widehat{h}_i(m_1, m_2, \ldots, m_i) = \begin{cases} \widehat{h}_i(m_i, m_i, \ldots, m_i), & m_1 = m_2 = \cdots = m_i \\ 0 \qquad\qquad\qquad\quad , & \text{others} \end{cases} \tag{3.69}$$

Then VF can be simplified into a simplified model, that is, ODVF. Therefore, ODVF has always been regarded as composed of the diagonal elements of VF. According to this conclusion, it can be inferred that for a system that can use VF to describe its nonlinear characteristics, if the kernel functions of VF are all concentrated on the diagonal of the coefficient matrix, and all non-diagonal elements are 0. Then the system can also use ODVF to simulate its nonlinear characteristics.

On this basis, an in-depth discussion is needed on ODVF, exploring whether it is solely composed of the diagonal elements of VF, and examining the extent of errors between ODVF and traditional VF. Section 3.2.1 discusses the measurement of ODVF by the frequency sweep method in detail, and the order kernel functions of ODVF can be obtained by measuring the frequency sweep signal. Here, the output response of VF under the same frequency sweep signal excitation will be investigated, to compare the connection and difference between VF and ODVF.

Equation (3.68) gives a general expression for an infinite order VF, which is often unachievable in practical applications, and finite order VF has a wider application. In existing research, only the performance and application of 2nd and 3rd-order VF are generally discussed. Here, based on the following considerations, the 3rd-order VF is taken as the research subject. First, the 3rd-order VF is sufficient to simulate the nonlinear characteristics of the loudspeaker system studied in this section; second, the 3rd-order VF includes linear kernel functions, even order kernel functions represented by 2nd-order kernel functions, and odd order kernel functions represented by 3rd-order kernel functions; finally, the discussion of the 3rd-order VF can be easily extended to higher orders. Equation (3.68) is written in the form of a 3rd-order VF as

$$
\begin{aligned}
y(n) = & \widehat{h}_0 + \sum_{i_1=0}^{M_1-1} \widehat{h}_1(i_1)x(n-i_1) \\
& + \sum_{i_1=0}^{M_2-1}\sum_{i_2=0}^{M_2-1} \widehat{h}_2(i_1,i_2)x(n-i_1)x(n-i_2) \\
& + \sum_{i_1=0}^{M_3-1}\sum_{i_2=0}^{M_3-1}\sum_{i_3=0}^{M_3-1} \widehat{h}_3(i_1,i_2,i_3)\cdot x(n-i_1)x(n-i_2)x(n-i_3)
\end{aligned}
\tag{3.70}
$$

where M_1, M_2, and M_3 are the memory lengths of the 1st-order Volterra kernel function $\mathbf{H}_1[\cdot]$, 2nd-order Volterra kernel function $\mathbf{H}_2[\cdot]$ and 3rd-order Volterra kernel function $\mathbf{H}_3[\cdot]$ respectively.

To facilitate analysis and discussion and obtain clear conclusions, look at VF from another angle, especially its 2nd and 3rd-order kernel functions. Taking the 2nd-order kernel function of VF as an example, $\widehat{h}_2(i_1, i_2)$ is a $M_2 \times M_2$ size matrix (as shown in Fig. 3.26), if this coefficient matrix is divided parallel to the diagonal, then the diagonal elements represented by $\widehat{h}_2(i_1, i_2)$ are a one-dimensional vector, and $\widehat{h}_2(i_1, i_1+1)$ represents a one-dimensional vector composed of elements on the sub-diagonal (or called secondary diagonal). According to this rule, rewrite the 2nd and 3rd-order kernel functions in the 3rd-order VF model in Mathews [420] into the form of diagonal vector combinations.

$$
\begin{aligned}
\widehat{y}(n) = & \widehat{h}_0(n) + \sum_{i=0}^{M_1-1} \widehat{h}_1(i)x(n-i) \\
& + \sum_{l=-M_2+1}^{M_2-1}\sum_{i=0}^{M_2-1} \widehat{h}_2(i,i+l)x(n-i)x(n-i-l) \\
& + \sum_{l_1=-M_3+1}^{M_3-1}\sum_{l_2=-M_3+1}^{M_3-1}\sum_{i=0}^{M_3-1} \widehat{h}_3(i,i+l_1,i+l_1+l_2)x(n-i) \\
& x(n-i-l_1)x(n-i-l_1-l_2)
\end{aligned}
\tag{3.71}
$$

In the above formula, it should be pointed out that for the 2nd-order kernel function $\widehat{h}_2(i, i+l)$, when l is given, i does not always take values from 0 to $M_2 - 1$, but must satisfy $i + l \geq 0$. Otherwise, if $i + l = -1$, $\widehat{h}_2(i, -1)$ is a meaningless parameter. Similarly, for the 3rd-order kernel function $\widehat{h}_3(i, i+l_1, i+l_1+l_2)$, i needs to satisfy $i \geq 0$, $i + l_1 \geq 0$ and $i + l_1 + l_2 \geq 0$.

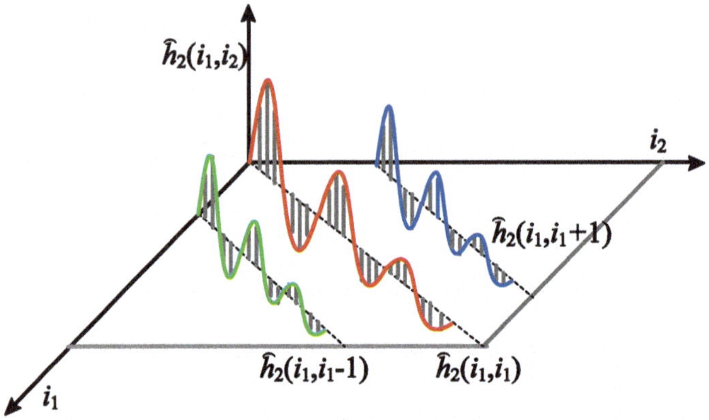

Fig. 3.26 Coefficient matrix of the 2nd-order kernel function of VF and its diagonal form

If a sweep signal $x(n)$ is fed to the 3rd-order VF model, Eq. (3.71) can be written as

$$\begin{aligned}\widehat{y}(n) &= \widehat{h}_0(n) + A\sum_{i=0}^{M_1-1}\widehat{h}_1(i)\sin[\phi(n-i)]\\ &+ A^2\sum_{l=-M_2+1}^{M_2-1}\sum_{i=0}^{M_2-1}\widehat{h}_2(i,i+l)\sin[\phi(n-i)]\sin[\phi(n-i-l)]\\ &+ A^3\sum_{l_1=-M_3+1}^{M_3-1}\sum_{l_2=-M_3+1}^{M_3-1}\sum_{i=0}^{M_3-1}\widehat{h}_3(i,i+l_1,i+l_1+l_2)\sin[\phi(n-i)]\\ &\sin[\phi(n-i-l_1)]\sin[\phi(n-i-l_1-l_2)]\end{aligned} \tag{3.72}$$

The simplification of the above formula mainly involves expanding the product terms behind the 2nd and 3rd-order kernel functions and simplifying them using the product-sum-difference formula. First consider the $\sin[\phi(n-i)]\sin[\phi(n-i-l)]$ term, expanding it gives

$$\begin{aligned}&\sin[\phi(n-i)]\sin[\phi(n-i-l)]\\ &= \cos(\xi)\sin\left[\xi e^{\frac{n-i}{N}\ln(\omega_2/\omega_1)}\right]\sin\left[\xi\left(e^{\frac{n-i-l}{N}\ln(\omega_2/\omega_1)}-1\right)\right]\\ &- \sin(\xi)\cos\left[\xi e^{\frac{n-i}{N}\ln(\omega_2/\omega_1)}\right]\sin\left[\xi\left(e^{\frac{n-i-l}{N}\ln(\omega_2/\omega_1)}-1\right)\right]\end{aligned} \tag{3.73}$$

The above formula can be obtained by using the product-sum-difference formula

$$
\begin{aligned}
&\sin[\phi(n-i)]\sin[\phi(n-i-l)]\\
&=\frac{1}{2}\cos(\xi)\Big\{\cos\Big[\xi\Big(e^{\frac{n-i-l}{N}\ln(\omega_2/\omega_1)}-e^{\frac{n-i}{N}\ln(\omega_2/\omega_1)}-1\Big)\Big]\\
&-\cos\Big[\xi\Big(e^{\frac{n-i-l}{N}\ln(\omega_2/\omega_1)}+e^{\frac{n-i}{N}\ln(\omega_2/\omega_1)}-1\Big)\Big]\Big\}\\
&-\frac{1}{2}\sin(\xi)\Big\{\sin\Big[\xi\Big(e^{\frac{n-i-l}{N}\ln(\omega_2/\omega_1)}+e^{\frac{n-i}{N}\ln(\omega_2/\omega_1)}-1\Big)\Big]\\
&+\sin\Big[\xi\Big(e^{\frac{n-i-l}{N}\ln(\omega_2/\omega_1)}-e^{\frac{n-i}{N}\ln(\omega_2/\omega_1)}-1\Big)\Big]\Big\}
\end{aligned}
\tag{3.74}
$$

Let

$$
\begin{cases}
\gamma_{21}=N\ln\Big|e^{-\frac{l}{N}\ln(\omega_2/\omega_1)}-1\Big|/\ln(\omega_2/\omega_1)\\
\gamma_{22}=N\ln\Big|e^{-\frac{l}{N}\ln(\omega_2/\omega_1)}+1\Big|/\ln(\omega_2/\omega_1)
\end{cases}
\tag{3.75}
$$

Then Eq. (3.74) can be written as

$$
\begin{aligned}
&\sin[\phi(n-i)]\sin[\phi(n-i-l)]\\
&=\frac{1}{2}\cos(\xi)\big\{\cos\big[\phi(n+\gamma_{21}-i)\big]-\cos\big[\phi(n+\gamma_{22}-i)\big]\big\}\\
&-\frac{1}{2}\sin(\xi)\big\{\sin\big[\phi(n+\gamma_{22}-i)\big]+\sin\big[\phi(n+\gamma_{21}-i)\big]\big\}
\end{aligned}
\tag{3.76}
$$

Next, for the first two sin function terms of the product term $\sin[\phi(n-i)]\sin[\phi(n-i-l_1)]\sin[\phi(n-i-l_1-l_2)]$, the product-sum-difference formula gives

$$
\begin{aligned}
&\sin[\phi(n-i)]\sin[\phi(n-i-l_1)]\sin[\phi(n-i-l_1-l_2)]\\
&=\frac{1}{2}\cos\Big[\xi\Big(e^{\frac{n-i}{N}\ln(\omega_2/\omega_1)}-e^{\frac{n-i-l_1}{N}\ln(\omega_2/\omega_1)}\Big)\Big]\sin[\phi(n-i-l_1-l_2)]\\
&-\frac{1}{2}\cos\Big[\xi\Big(e^{\frac{n-i}{N}\ln(\omega_2/\omega_1)}+e^{\frac{n-i-l_1}{N}\ln(\omega_2/\omega_1)}-2\Big)\Big]\sin[\phi(n-i-l_1-l_2)]
\end{aligned}
\tag{3.77}
$$

Expanding the cos function term in the second term of the above formula gives

$$
\begin{aligned}
&\sin[\phi(n-i)]\sin[\phi(n-i-l_1)]\sin[\phi(n-i-l_1-l_2)]\\
&=\frac{1}{2}\cos\Big[\xi\Big(e^{\frac{n-i}{N}\ln(\omega_2/\omega_1)}-e^{\frac{n-i-l_1}{N}\ln(\omega_2/\omega_1)}\Big)\Big]\sin[\phi(n-i-l_1-l_2)]\\
&-\frac{1}{2}\cos[2\xi]\cos\Big[\xi\Big(e^{\frac{n-i}{N}\ln(\omega_2/\omega_1)}+e^{\frac{n-i-l_1}{N}\ln(\omega_2/\omega_1)}\Big)\Big]\sin[\phi(n-i-l_1-l_2)]\\
&-\frac{1}{2}\sin[2\xi]\sin\Big[\xi\Big(e^{\frac{n-i}{N}\ln(\omega_2/\omega_1)}+e^{\frac{n-i-l_1}{N}\ln(\omega_2/\omega_1)}\Big)\Big]\sin[\phi(n-i-l_1-l_2)]
\end{aligned}
\tag{3.78}
$$

The trigonometric function product term in the above formula can be obtained by using the product-sum-difference formula

$$\begin{aligned}
&\sin[\phi(n-i)]\sin[\phi(n-i-l_1)]\sin[\phi(n-i-l_1-l_2)] \\
&= \frac{1}{4}\{\sin[\phi(n+\gamma_{31}-i)] + \sin[\phi(n+\gamma_{32}-i)]\} \\
&- \frac{1}{4}\cos\left[\frac{2\omega_1 T}{\ln(\omega_2/\omega_1)}\right]\{\sin[\phi(n+\gamma_{33}-i)] + \sin[\phi(n+\gamma_{34}-i)]\} \\
&- \frac{1}{4}\sin\left[\frac{2\omega_1 T}{\ln(\omega_2/\omega_1)}\right]\{\cos[\phi(n+\gamma_{33}-i)] - \cos[\phi(n+\gamma_{34}-i)]\}
\end{aligned} \tag{3.79}$$

where
$$\begin{cases}
\gamma_{31} = N\ln\left|e^{\frac{-l_1-l_2}{N}\ln(\omega_2/\omega_1)} - e^{\frac{-l_1}{N}\ln(\omega_2/\omega_1)} + 1\right| / \ln(\omega_2/\omega_1) \\
\gamma_{32} = N\ln\left|e^{\frac{-l_1-l_2}{N}\ln(\omega_2/\omega_1)} + e^{\frac{-l_1}{N}\ln(\omega_2/\omega_1)} - 1\right| / \ln(\omega_2/\omega_1) \\
\gamma_{33} = N\ln\left|e^{\frac{-l_1-l_2}{N}\ln(\omega_2/\omega_1)} - e^{\frac{-l_1}{N}\ln(\omega_2/\omega_1)} - 1\right| / \ln(\omega_2/\omega_1) \\
\gamma_{34} = N\ln\left|e^{\frac{-l_1-l_2}{N}\ln(\omega_2/\omega_1)} + e^{\frac{-l_1}{N}\ln(\omega_2/\omega_1)} + 1\right| / \ln(\omega_2/\omega_1)
\end{cases}$$

Substitute Eqs. (3.76) and (3.79) into Eq. (3.72), then the output response of the 3rd-order VF excited by the sweep signal is

$$\begin{aligned}
\widehat{y}(n) &= \widehat{h}_0 + A\sum_{i=0}^{M_1-1}\widehat{h}_1(i)x(n-i) \\
&+ \frac{A^2}{2}\sum_{l=-M_2+1}^{M_2-1}\sum_{i=0}^{M_2-1}\widehat{h}_2(i,i+l) \\
&\quad \{\cos(\xi)[-x(n+\gamma_{21}-i)*\delta_{hil}(n) + x(n+\gamma_{22}-i)*\delta_{hil}(n)] \\
&\quad - \sin(\xi)[x(n+\gamma_{21}-i) + x(n+\gamma_{22}-i)]\} \\
&+ \frac{A^3}{4}\sum_{l_2=-M_3+1}^{M_3-1}\sum_{l_1=-M_3+1}^{M_3-1}\sum_{i=0}^{M_3-1}\widehat{h}_3(i,i+l_1,i+l_1+l_2) \\
&\quad \{-\cos(2\xi)[x(n+\gamma_{33}-i) + x(n+\gamma_{34}-i)] \\
&\quad + x(n+\gamma_{31}-i) + x(n+\gamma_{32}-i) \\
&\quad - \sin(2\xi)[-x(n+\gamma_{33}-i)*\delta_{hil}(n) + x(n+\gamma_{34}-i)*\delta_{hil}(n)]\}
\end{aligned} \tag{3.80}$$

The inverse signal of the sweep signal $\tilde{x}(n)$ is convolved with the output of the VF model $\widehat{y}(n)$ to yield

$$\begin{aligned}
\widehat{z}(n) &= \widehat{y}(n) * \tilde{x}(n) \\
&= \widehat{h}_0 * \tilde{x}(n) + \widehat{h}_1(n) \\
&\quad + \frac{A}{2} \sum_{l=-M_2+1}^{M_2-1} \sum_{i=0}^{M_2-1} \widehat{h}_2(i, i+l) \\
&\quad \big\{\cos(\xi)\big[-\delta(n+\gamma_{21}-i) * \delta_{hil}(n) + \delta(n+\gamma_{22}-i) * \delta_{hil}(n)\big] \\
&\quad - \sin(\xi)\big[\delta(n+\gamma_{21}-i) + \delta(n+\gamma_{22}-i)\big]\big\} \\
&\quad + \frac{A^3}{4} \sum_{l_2=-M_3+1}^{M_3-1} \sum_{l_1=-M_3+1}^{M_3-1} \sum_{i=0}^{M_3-1} \widehat{h}_3(i, i+l_1, i+l_1+l_2) \\
&\quad \big\{-\cos(2\xi)\big[\delta(n+\gamma_{33}-i) + \delta(n+\gamma_{34}-i)\big] \\
&\quad + \delta(n+\gamma_{31}-i) + \delta(n+\gamma_{32}-i) \\
&\quad - \sin(2\xi)\big[-\delta(n+\gamma_{33}-i) * \delta_{hil}(n) + \delta(n+\gamma_{34}-i) * \delta_{hil}(n)\big]\big\}
\end{aligned} \tag{3.81}$$

In fact, $\widehat{h}_0 * \tilde{x}(n)$ in the impulse response sequence is a noise signal. It is discussed in reference [412] that in the system under study, the impact of the DC component on the impulse response is generally small, so this term can be ignored.

Consider a situation where if the starting angular frequency ω_1 of the input sweep signal $x(n)$, the cutoff angular frequency ω_2, and the signal length T are chosen appropriately to satisfy the condition that $\xi = \frac{\pi}{2} + 2\eta\pi, \eta = \ldots, -1, 0, 1, 2, \ldots$, Eq. (3.81) can be simplified to

$$\begin{aligned}
\widehat{z}(n) =& \widehat{h}_1(n) + \frac{A}{2} \sum_{l=-M_2+1}^{M_2-1} \sum_{i=0}^{M_2-1} \widehat{h}_2(i, i+l) \\
& \big\{-\big[\delta(n+\gamma_{21}-i) + \delta(n+\gamma_{22}-i)\big]\big\} \\
& + \frac{A^3}{4} \sum_{l_2=-M_3+1}^{M_3-1} \sum_{l_1=-M_3+1}^{M_3-1} \sum_{i=0}^{M_3-1} \widehat{h}_3(i, i+l_1, i+l_1+l_2) \\
& \big\{\big[\delta(n+\gamma_{33}-i) + \delta(n+\gamma_{34}-i)\big] \\
& +\delta(n+\gamma_{31}-i) + \delta(n+\gamma_{32}-i)\}
\end{aligned} \tag{3.82}$$

In general, if the length of the selected sweep signal is long enough, then the total number of sampling points N is large enough, the delay of Eqs. (3.75) and (3.78) can be expanded according to the Taylor series (and the multivariate Taylor series), only retaining its linear terms, the approximate delay calculation formula can be obtained as

$$\begin{cases} \gamma_{21} = -\infty \\ \gamma_{22} = \gamma_2 - l/2 \\ \gamma_{31} = -l_2 \\ \gamma_{32} = -2l_1 - l_2 \\ \gamma_{33} = +l_2 \\ \gamma_{34} = \gamma_3 - 2l_1/3 - l_2/3 \end{cases} \tag{3.83}$$

where, γ_2 and γ_3 are defined as in Eq. (3.37). Substitute into Eq. (3.82) to yield

$$\begin{aligned} \widehat{z}(n) =& \widehat{h}_1(n) - \frac{A}{2} \sum_{l=-M_2+1}^{M_2-1} \sum_{i=0}^{M_2-1} \widehat{h}_2(i, i+l)\delta(n+\gamma_2 - l/2 - i) \\ &+ \frac{A^2}{4} \sum_{l_2=-M_3+1}^{M_3-1} \sum_{l_1=-M_3+1}^{M_3-1} \sum_{i=0}^{M_3-1} \widehat{h}_3(i, i+l_1, i+l_1+l_2) \\ &[\delta(n+l_2-i) + \delta(n-2l_1-l_2-i) \\ &+ \delta(n-l_2-i) + \delta(n+\gamma_3-2l_1/3-l_2/3-i)] \end{aligned} \tag{3.84}$$

When analyzing Eq. (3.84), it becomes evident that the impulse response $\widehat{z}(n)$ obtained after convolution is a one-dimensional impulse response, which is the delay superposition of the various order kernel functions of VF. First look at the linear kernel function of VF, its starting point is at the origin $n = 0$; secondly, from the second term of the above formula, it is obvious that the impulse $\widehat{h}_2(i, i)$ composed of the diagonal elements of the 2nd-order kernel function of VF starts from the origin and shifts to the left by γ_2 sampling points, while the impulse $\widehat{h}_2(i, i+1)$ composed of the sub-diagonal elements starts from the origin and shifts to the left by $\gamma_2 - \frac{1}{2}$, correspondingly, the sub-diagonal impulse $\widehat{h}_2(i, i+2)$ starts from the origin and shifts to the left by $\gamma_2 - 1$. Figure 3.27 shows the phase relationship between the impulse sequence composed of the diagonal elements of the 2nd-order kernel function and the impulse response sequence composed of the non-diagonal elements. In connection with Fig. 3.26, it can be known that the diagonal impulse of the 2nd-order kernel function starts at $n = -\gamma_2$, and the sub-diagonal impulse is delayed by 0.5 sampling points, denoted as $k_2(n)$. Next, let's analyze the third term of the formula, which gives the shift and superposition changes of the 3rd-order kernel function of VF. Among them, $\delta(n + l_2 - i)$ represents the diagonal elements of the 3rd-order Volterra kernel function $\widehat{h}_3(i, i+l_1, i+l_1+l_2)$ shifted to the left from the origin by l_2 sampling points. For example, $\widehat{h}_3(i, i+1, i+1+1)$ is shifted to the left by one sampling point, and a series of diagonals of $\widehat{h}_3(i, i+2, i+2+1)$ and $\widehat{h}_3(i, i+l_1, i+l_1+1)$, etc. are all shifted to the left by one sampling point. Similarly, $\delta(n - l_2 - i)$ represents the impulse of the diagonal elements $\widehat{h}_3(i, i+l_1, i+l_1+l_2)$ shifted to the right by l_2 sampling points. $\delta(n - 2l_1 - l_2 - i)$ represents the diagonal

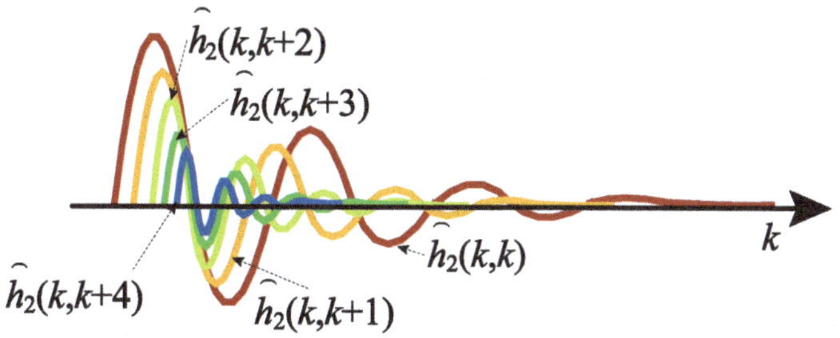

Fig. 3.27 Schematic diagram of the phase relationship between the diagonal pulse and non-diagonal pulse of the 2nd-order kernel function

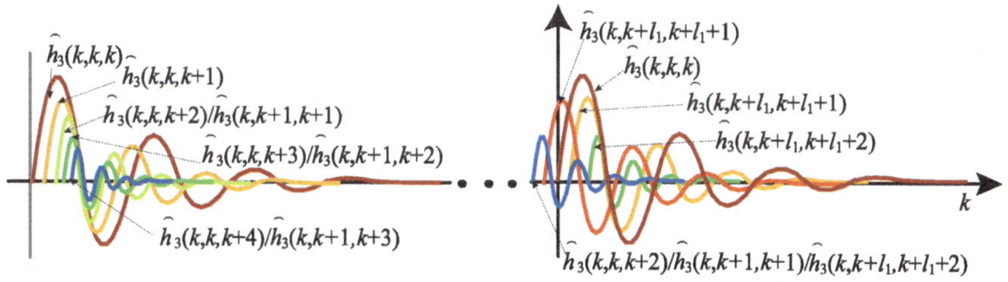

Fig. 3.28 Schematic diagram of the phase relationship between the diagonal pulse and non-diagonal pulse of the 3rd-order kernel function

elements $\widehat{h}_3(i, i + l_1, i + l_1 + l_2)$ shifted to the left by $2l_1 + l_2$ sampling points. It should be pointed out that the values of l_1 and l_2 are relatively small, so these diagonal elements are basically distributed near the origin, as shown in Fig. 3.28. Finally, let's look at $\delta(n + \gamma_3 - 2l_1/3 - l_2/3 - i)$, which indicates that the impulse of the main diagonal elements $\widehat{h}_3(i, i, i)$ is shifted to the left by γ_3 sampling points, while the sub-diagonal impulse $\widehat{h}_3(i, i, i+1)$ is shifted to the left by $\gamma_3 - 1/3$ sampling points relative to the origin, and the rest of the sub-diagonal impulses are shifted to the right by $2l_1/3 + l_2/3$ sampling points relative to $\widehat{h}_3(i, i, i)$, as shown in Fig. 3.28.

In summary, the impulse response $\widehat{z}(n)$ obtained after convolution mainly consists of three parts (as shown in Fig. 3.29). There is a string of impulse responses near the origin, mainly composed of the shift and superposition of the linear kernel function and 3rd-order kernel function of VF, denoted as $k_1(n)$. At $n = -\gamma_2$, there is an impulse response sequence, mainly composed of the shift and superposition of the diagonal and sub-diagonal impulses of the 2nd-order kernel function of VF, the starting point of this impulse sequence is $n = -\gamma_2$, denoted as $k_2(n)$. The impulse response sequence starting at $n = -\gamma_3$ on the time axis is composed of the shift and superposition of the diagonal and sub-diagonal impulses of the 3rd-order kernel function of VF, denoted as $k_3(n)$. Thus, the relationship between the impulse response sequence after convolution and the various order kernel functions of VF is obtained.

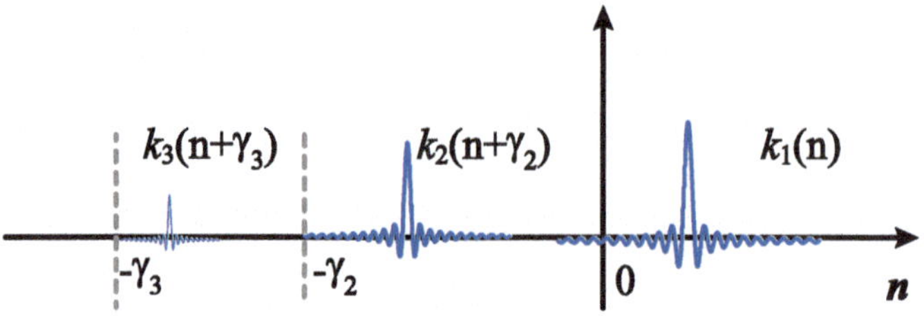

Fig. 3.29 Schematic diagram of the impulse response sequence obtained after convolution

$$\begin{cases} k_1(n) = \widehat{h}_1(n) + \dfrac{A^2}{4} \displaystyle\sum_{l_2=-M_3+1}^{M_3-1} \sum_{l_1=-M_3+1}^{M_3-1} \sum_{i=0}^{M_3-1} \widehat{h}_3(i, i+l_1, i+l_1+l_2) \\ \quad [\delta(n-l_2-i) + \delta(n-2l_1-l_2-i) + \delta(n+l_2-i)] \\ k_2(n) = -\dfrac{A}{2} \displaystyle\sum_{l=-M_2+1}^{M_2-1} \sum_{i=0}^{M_2-1} \widehat{h}_2(i, i+l)\delta(n-l/2-i) \\ k_3(n) = \dfrac{A^2}{4} \displaystyle\sum_{l_2=-M_3+1}^{M_3-1} \sum_{l_1=-M_3+1}^{M_3-1} \sum_{i=0}^{M_3-1} \widehat{h}_3(i, i+l_1, i+l_1+l_2) \\ \quad \delta(n-2l_1/3-l_2/3-i) \end{cases} \tag{3.85}$$

By measuring the output of the system's sweep signal and deconvolving to obtain the impulse response sequence, and then using Eq. (3.38) to solve for the various order kernel functions of ODVF based on the impulse response sequence. For a 3rd-order ODVF model, when the parameters such as the start and end angular frequency and time length of the sweep signal satisfy $\xi = \frac{\pi}{2} + 2\eta\pi$, $\eta = \cdots, -1, 0, 1, 2, \ldots$, Eq. (3.38) can be written as

$$\begin{cases} k_1(n) = h_1(n) + \dfrac{3A^2}{4} h_3(n) \\ k_2(n) = -\dfrac{A}{2} h_2(n) \\ k_3(n) = \dfrac{A^2}{4} h_3(n) \end{cases} \tag{3.86}$$

The impulse response after deconvolution of the sweep signal has a relationship with the various order functions of VF as shown in Eq. (3.85). It is known that the higher-order kernel functions of the ODVF can be identified through Eq. (3.86) and deconvolved impulse responses, leading to the conclusion that the higher-order kernel functions of the ODVF, obtained by the logarithmic sweep method, correspond to those of the VF in such a manner.

$$\begin{cases} h_1(n) + \frac{3}{4}A^2 h_3(n) = \overset{\frown}{h}_1(n) + \frac{A^2}{4} \\ \quad \sum_{l_2=-M_3+1}^{M_3-1} \sum_{l_1=-M_3+1}^{M_3-1} \sum_{k=0}^{M_3-1} \overset{\frown}{h}_3(k, k+l_1, k+l_1+l_2) \\ \quad [\delta(n-2l_1-l_2-k) + \delta(n+l_2-k)] \\ -\frac{A}{2}h_2(n) = -\frac{A}{2} \sum_{l=-M_2+1}^{M_2-1} \sum_{k=0}^{M_2-1} \overset{\frown}{h}_2(k, k+l)\delta(n-l/2-k) \\ \frac{A^2}{4}h_3(n) = \frac{A^2}{4} \\ \quad \sum_{l_2=-M_3+1}^{M_3-1} \sum_{l_1=-M_3+1}^{M_3-1} \sum_{k=0}^{M_3-1} \overset{\frown}{h}_3(k, k+l_1, k+l_1+l_2)\delta(n-2l_1/3-l_2/3-k) \end{cases} \tag{3.87}$$

If the amplitude of the 3rd-order kernel function of the nonlinear system being measured is much smaller than its linear kernel function, then the above formula can be further simplified to

$$\begin{cases} h_1(n) \approx \overset{\frown}{h}_1(n) \\ h_2(n) = \sum_{l=-M_2+1}^{M_2-1} \sum_{k=0}^{M_2-1} \overset{\frown}{h}_2(k, k+l)\delta(n-l/2-k) \\ h_3(n) = \sum_{l_2=-M_3+1}^{M_3-1} \sum_{l_1=-M_3+1}^{M_3-1} \sum_{k=0}^{M_3-1} \overset{\frown}{h}_3(k, k+l_1, k+l_1+l_2) \\ \delta(n-2l_1/3-l_2/3-k) \end{cases} \tag{3.88}$$

The above formula shows that the high-order kernel functions measured by the sweep method are not only the diagonal elements of the VF system's kernel functions, but also include the shift superposition of the sub-diagonal elements. The input–output relationship of ODVF is rewritten according to Eq. (3.88) as

$$\begin{aligned} y(n) = & \sum_{i=0}^{M_1-1} \overset{\frown}{h}_1(i)x(n-i) \\ & + \sum_{l=-M_2+1}^{M_2-1} \sum_{i=0}^{M_2-1} \overset{\frown}{h}_2(i, i+l)x^2(n-l/2-i) \\ & + \sum_{l_2=-M_3+1}^{M_3-1} \sum_{l_1=-M_3+1}^{M_3-1} \sum_{i=0}^{M_3-1} \overset{\frown}{h}_3(i, i+l_1, i+l_1+l_2) \end{aligned}$$

$$x^3(n - 2l_1/3 - l_2/3 - i) \tag{3.89}$$

Comparing Eqs. (3.89) and (3.71), the first term, i.e., the convolution of the linear kernel function, is the same in both formulas. The difference in the second term is reflected in the difference between $x^2(n - l/2 - i)$ and $x(n - i)x(n - i - l)$, and the difference in the third term is the difference between $x^3(n - 2l_1/3 - l_2/3 - i)$ and $x(n - i)x(n - i - l_1)x(n - i - l_1 + l_2)$.

To illustrate the size of the differences between the above relationships, let's take a simple example. Assume the input signal is $x(n) = A\sin(\omega_0 n)$, where A is the signal amplitude, ω_0 is the signal frequency. According to the mean value theorem of differentiation, it can be obtained

$$x'(n) = \frac{x(n+\triangle n) - x(n)}{\triangle n} = A\omega_0\cos(\omega_0 n) \tag{3.90}$$

The above formula shows that when the frequency of the input signal with a given amplitude ω_0 is small compared to the sampling rate $1/\Delta n$, it can be considered that $x(n+\triangle n)$ is equivalent to $x(n)$. This means that the numerical differences between $x(n-i), x(n-i-l), x(n-l/2-i)$ are small, then the difference between $x^2(n-l/2-i)$ and $x(n-i)x(n-i-l)$ is not significant. Similarly, when the frequency of the input signal is much smaller than the system sampling rate, $x^3(n - 2l_1/3 - l_2/3 - i)$ and $x(n - i)x(n - i - l_1)x(n - i - l_1 + l_2)$ will also have a weak difference. At this time, it can be considered that ODVF can approximate the VF system, and its output has an error $err(n)$, that is

$$y(n) = \widehat{y}(n) + err(n) \tag{3.91}$$

When the parameters of the sweep signal satisfy $\xi = \frac{\pi}{2} + 2\eta\pi, \eta = \ldots, -1, 0, 1, 2, \ldots$, it is discussed that the higher-order kernel functions of the ODVF, obtained through the sweep method, have a relationship with the traditional VF as described by Eq. (3.25). The following will further discuss that if the parameters of the sweep signal meet the other three conditions given in 3.2.1, the relationship between the order kernel functions of the ODVF obtained and the kernel functions of VF.

When the frequency range and time length of the sweep signal are selected to satisfy $\xi = -\frac{\pi}{2} + 2\eta\pi, \eta = \ldots, -1, 0, 1, 2, \ldots$, Eq. (3.81) can be simplified as

$$\begin{aligned}
\widehat{z}(n) = &\widehat{h}_1(n) + \frac{A}{2}\sum_{l=-M_2+1}^{M_2-1}\sum_{i=0}^{M_2-1}\widehat{h}_2(i, i+l)\delta\left(n + \gamma_2 - \frac{l}{2} - i\right) \\
&+ \frac{A^2}{4}\sum_{l_2=-M_3+1}^{M_3-1}\sum_{l_1=-M_3+1}^{M_3-1}\sum_{i=0}^{M_3-1}\widehat{h}_3(i, i+l_1, i+l_1+l_2) \\
&[\delta(n + l_2 - i) + \delta(n - 2l_1 - l_2 - i)
\end{aligned}$$

$$+\delta(n-l_2-i)+\delta\left(n+\gamma_3-\frac{2l_1}{3}-\frac{l_2}{3}-i\right)\Bigg] \tag{3.92}$$

The relationship between the kernel function of VF and the pulse response of each segment is

$$\begin{cases} k_1(n)=\widehat{h}_1(n)+\dfrac{A^2}{4}\displaystyle\sum_{l_2=-M_3+1}^{M_3-1}\sum_{l_1=-M_3+1}^{M_3-1}\sum_{i=0}^{M_3-1}\widehat{h}_3(i,i+l_1,i+l_1+l_2) \\ \quad [\delta(n-l_2-i)+\delta(n-2l_1-l_2-i)+\delta(n+l_2-i)] \\ k_2(n)=\dfrac{A}{2}\displaystyle\sum_{l=-M_2+1}^{M_2-1}\sum_{i=0}^{M_2-1}\widehat{h}_2(i,i+l)\delta(n-l/2-i) \\ k_3(n)=\dfrac{A^2}{4}\displaystyle\sum_{l_2=-M_3+1}^{M_3-1}\sum_{l_1=-M_3+1}^{M_3-1}\sum_{i=0}^{M_3-1}\widehat{h}_3(i,i+l_1,i+l_1+l_2) \\ \quad \delta(n-2l_1/3-l_2/3-i) \end{cases} \tag{3.93}$$

Under these conditions, Eq. (3.38) will be written as

$$\begin{cases} k_1(n)=h_1(n)+\dfrac{3A^2}{4}h_3(n) \\ k_2(n)=\dfrac{A}{2}h_2(n) \\ k_3(n)=\dfrac{A^2}{4}h_3(n) \end{cases} \tag{3.94}$$

In this way, the relationship between the order kernel functions of ODVF obtained by the sweep frequency method and VF can be obtained as

$$\begin{cases} h_1(n)\approx\widehat{h}_1(n) \\ h_2(n)=\displaystyle\sum_{l=-M_2+1}^{M_2-1}\sum_{k=0}^{M_2-1}\widehat{h}_2(k,k+l)\delta\left(n-\frac{l}{2}-k\right) \\ h_3(n)=\displaystyle\sum_{l_2=-M_3+1}^{M_3-1}\sum_{l_1=-M_3+1}^{M_3-1}\sum_{k=0}^{M_3-1}\widehat{h}_3(k,k+l_1,k+l_1+l_2) \\ \quad \delta\left(n-\dfrac{2}{3}l_1-\dfrac{1}{3}l_2-k\right) \end{cases} \tag{3.95}$$

The above formula is completely consistent with Eq. (3.88).

Based on the same analysis steps, the remaining two situations were analyzed. Due to space limitations, the process is not listed here. The conclusion is that when the parameters meet $\xi=0+2\eta\pi$, $\eta=\cdots,-1,0,1,2,\ldots$ and $\xi=\pi+2\eta\pi$, $\eta=$

$\cdots, -1, 0, 1, 2, \ldots$, the relationship between the order kernel functions of ODVF obtained by sweep frequency and the kernel functions of VF is always as shown in Eq. (3.88). That is, in various value situations (here refers to four specific conditions, and only under these specific conditions can the pulse response obtained by sweep convolution be used to solve the order kernel functions of ODVF), the ODVF model established by the sweep signal not only contains the effect of the diagonal elements of VF, but also the content of the sub-diagonal elements. However, it should also be noted that there is a difference between the ODVF model and the VF model, but as discussed above, when the frequency of the system input signal is much smaller than the sampling rate, the difference between the two models is relatively small.

To discern the correctness of various identification methods of the existing logarithmic sweep frequency measurement ODVF model, and to explain the phase distortion problem that appears in the model identification process in the literature, this part has carried out theoretical exploration on the sweep frequency measurement ODVF model and established a unified theoretical framework. This theoretical framework unifies the existing various solution schemes. Their formulas are just special cases that appear when the starting angular frequency, ending angular frequency, and length of the sweep signal meet different conditions. A general expression is given, which can solve the kernel functions of any order. Taking the identification of the 5th-order ODVF as an example, it further explains how to solve the kernel functions of each order of the filter. A special parameter selection scheme is given, and four special solutions are obtained, which are helpful for quickly measuring the parameters of the ODVF model in engineering.

3.2.4 Nonlinear Distortion Compensation of the Parametric Array Loudspeaker Based on OVDF

Related research proposed various improved amplitude modulation methods for the distortion problem of the parametric array loudspeaker, trying to eliminate the harmonic distortion in the audio frequency. However, when the sound pressure level of the source wave exceeds a threshold, the performance of these algorithms will decline. Therefore, some scholars have introduced VF to suppress the harmonic distortion of the parametric array loudspeaker. Considering the complexity of the Volterra inverse filter, this part will introduce the ODVF inverse filter to compensate for the nonlinear distortion in the audio frequency of the parametric array loudspeaker, mainly analyzing the harmonic distortion. references [146, 412] also includes the analysis of intermodulation distortion, which will not be expanded in this part.

3.2.4.1 Setup of the Experiment

The experiment was conducted in the anechoic chamber of Institute of Acoustics, Chinese Academy of Sciences (as shown in Fig. 3.30). In the experiment, the test signal (Gaussian white noise signal or sweep signal) was first generated by the data acquisition system NI-PXIe1075, and after amplitude modulation, it was amplified by the power amplifier circuit (the power amplifier bandwidth is 3~100 kHz, the amplitude response is flat and consistent within the entire passband, and the fluctuation is within ± 1 dB) and then fed to the ultrasonic transducer array. The audio frequency signal obtained after self-demodulation in the air was synchronously collected by the B&K 4189 microphone at 2.3 m. The microphone and the transducer array are at the same height, 1.5 m from the ground. During the experiment, the ultrasonic carrier acts on the measurement transducer, bringing spurious sound. To eliminate the influence of spurious sound on the measurement results, an acoustic filter [191] was placed in front of the microphone in the experiment, which can effectively attenuate the ultrasonic carrier signal. To eliminate the influence of the acoustic filter on the measurement results, the frequency response of the acoustic filter is compensated before the model identification. The temperature in the anechoic chamber is 21°C, and the relative humidity is 20%. The carrier frequency of the parametric array loudspeaker is 40 kHz. A total of 147 piezoelectric transducers are closely arranged to form an approximately circular transducer array (as shown in Fig. 3.30b). The equivalent radius of the array is 0.1 m. At 2.3 m, the sound pressure level of the ultrasonic primary wave measured by the B&K 4138 microphone is 132 dB. The 4138 microphone is specifically used for measuring high-intensity high-frequency sound pressure signals, with a measurement frequency range of 6.5 Hz to 140 kHz and a sound pressure range of 52.2 to 168 dB.

(a) (b)

Fig. 3.30 Experimental scene **a** diagram of the placement of the parametric array loudspeaker and ultrasonic transducer **b** ultrasonic transducer array

3.2.4.2 Model Identification and Performance Evaluation

In practical applications, DSBAM is the simplest amplitude modulation algorithm, and SSB-AM has the smallest harmonic distortion among various modulation algorithms [138], and as can be seen from the previous simulation experiments, its performance is basically equivalent to that of SRAM and MAM modulation algorithms, so in the experiment, only the parametric array loudspeaker preprocessed by these two preprocessing methods is taken as a typical representative to study the linear response and nonlinear characteristics of the ODVF simulated output of the parametric array loudspeaker. The order kernel functions of ODVF under the two amplitude modulation conditions measured by the sweep frequency method are shown in Fig. 3.31. From the figure, it is obvious that the amplitudes of the linear kernel functions in the DSBAM and SSB-AM cases are basically equivalent, while the 2nd-order kernel function (Fig. 3.31b) and 3rd-order kernel function (Fig. 3.31c) of SSB-AM are smaller than the corresponding kernel function amplitudes of DSBAM. This is because the 2nd and 3rd harmonic distortions of SSB-AM are smaller than those of DSBAM. It can also be observed from the figure that as the order of the kernel function increases, the amplitude of the impulse response also decreases, i.e., the amplitude of the 2nd-order kernel function is smaller than that of the linear kernel function, and the amplitude of the 3rd-order kernel function is smaller than that of the 2nd-order kernel function.

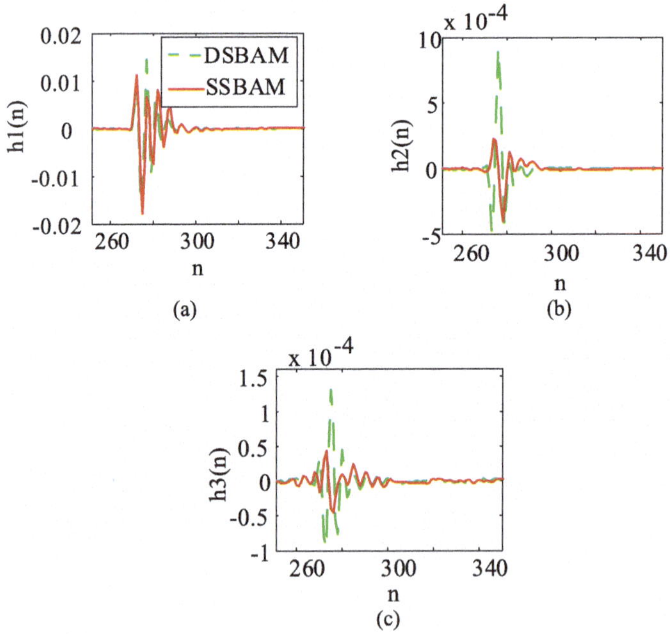

Fig. 3.31 ODVF order kernel functions under DSBAM and SSBAM preprocessing conditions **a** linear kernel function **b** 2nd-order kernel function **c** 3rd-order kernel function

To better compare the output characteristics of the ODVF simulated parametric array loudspeaker, VF is introduced for comparison. The kernel function of VF can be obtained by the NLMS adaptive iterative algorithm. Figure 3.32 shows two sections of the linear kernel function, 2nd-order kernel function matrix, and 3rd-order kernel function matrix of the VF of the parametric array loudspeaker preprocessed by DSBAM (hereinafter referred to as the DSBAM system). Figure 3.32a shows the phase frequency and amplitude frequency response curves of the linear kernel function, as well as the phase frequency and amplitude frequency response of ODVF. From the figure, it is evident that the amplitude and phase responses of the linear kernel functions of the two models basically coincide, but there is a slight difference in phase. In the previous simulation experiments, it was also found that for the DSBAM system, the difference in the linear kernel functions of the two models mainly comes from the influence of the system's 3rd harmonic distortion on the identification of the linear kernel function of ODVF. The experimental results here once again verify this point. From the coefficient distribution of the 2nd and 3rd-order kernel functions in the figure, it can also be found that the non-zero elements in the coefficient matrix are basically concentrated near the diagonal, which means that the difference between ODVF and VF models is not too significant.

Figure 3.33 illustrates two sections of the linear kernel function, 2nd-order kernel function matrix, and 3rd-order kernel function matrix of the VF of the parametric array loudspeaker preprocessed by SSB-AM (hereinafter referred to as the SSB-AM system). Figure 3.33a shows the amplitude frequency and phase frequency response curves of the linear kernel function of ODVF and VF. From the figure, it can be observed that the linear responses of the two models almost completely coincide. Especially the coincidence degree of the phase frequency response is higher than that of the DSBAM system (Fig. 3.32a). This is because the 3rd harmonic distortion of the SSB-AM system is smaller, and the influence on the identification of the linear kernel function of ODVF is smaller.

In order to quantitatively analyze the accuracy of the established ODVF model, a single frequency signal is fed to the parametric array loudspeaker system here, and the frequency of the system output signal is [0.5, 1.0 ... 3.5 kHz]. The signal is fed into both the ODVF model and the VF model to obtain the model outputs. Here, the relative error is introduced as an indicator to measure the difference between the output of each order kernel function of the model and the harmonic signal of the system output. The relative error is defined as

$$E_m = \frac{|H_m[x] - Y_m|}{Y_m} \times 100\%, \qquad m = 1, 2, 3 \tag{3.96}$$

where, m represents the order of each harmonic, Y_m represents the mth harmonic distortion of the output sound frequency of the parametric array loudspeaker system, $H_m[x]$ represents the mth harmonic distortion of the model output (including both VF model and ODVF model).

First, for the DSBAM system, measure its system output Y_m, then calculate the outputs of the two models $H_m[x]$, and calculate the relative error E_m of each harmonic

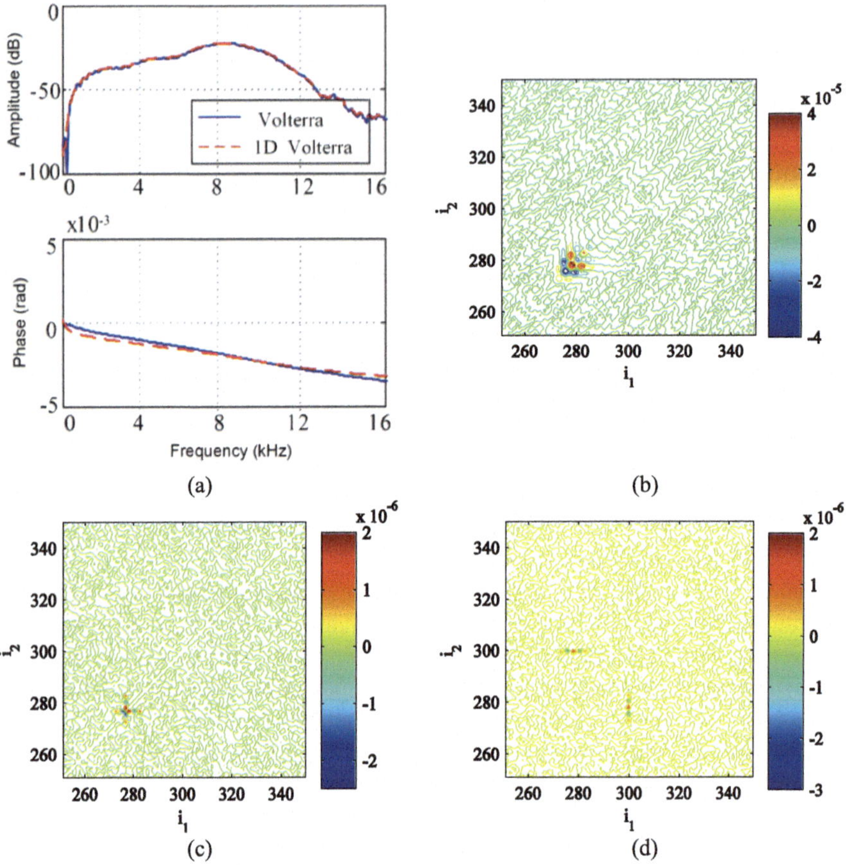

Fig. 3.32 VF order kernel functions under DSBAM preprocessing conditions. **a** Linear kernel function and amplitude and phase frequency response, **b** 2nd-order kernel function, **c** cross-section of the 3rd-order kernel function $h_3(i_1, i_2, 280)$, **d** cross-section of the 3rd-order kernel function $h_3(i_1, i_2, 300)$

component according to Eq. (3.96). Figure 3.34a demonstrates the relative error between the VF model and ODVF model of the DSBAM system and the actual system. From the figure, it is evident that the E_1 of the two models basically coincide, which is because the impulse responses of the linear kernel functions of the two models are basically the same, and the E_2 of the ODVF model is larger than that of the VF model, which comes from the error caused by the simplification of the ODVF model. Similarly, the E_3 of the ODVF model is larger than that of the VF model. From the figure, it can also be seen that E_2 is larger than E_1, and E_3 is larger than E_2. This is because the harmonic distortion of the parametric array loudspeaker system studied in this section decreases with the increase of the order, and the decrease of the harmonic distortion amplitude means that the signal-to-noise ratio (SNR, signal noise ratio) of the corresponding kernel function coefficient decreases, leading to

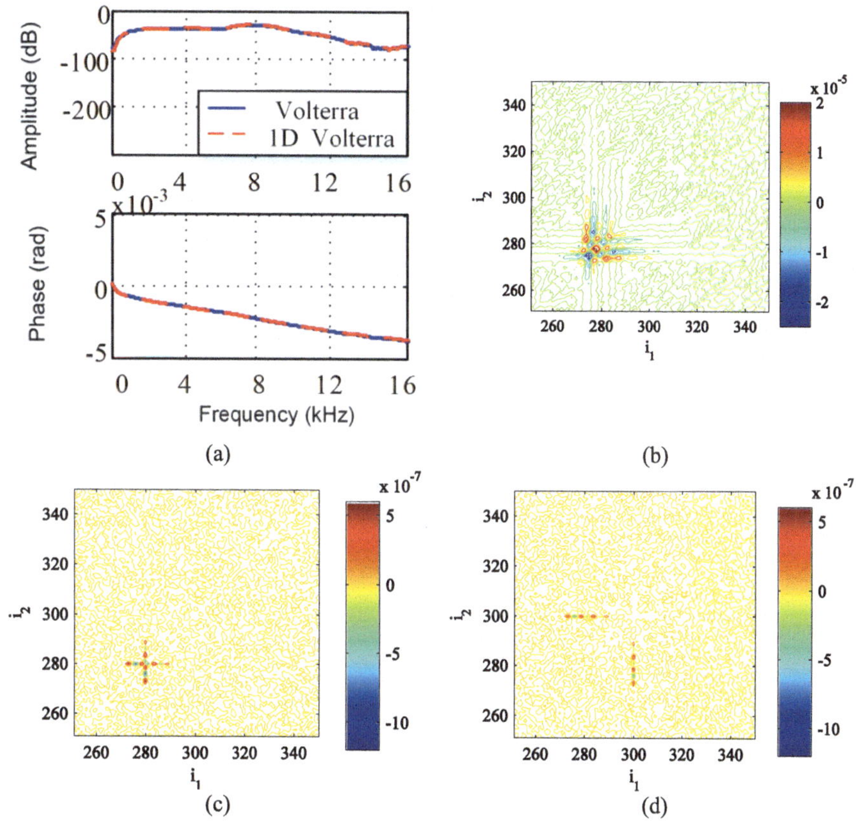

Fig. 3.33 VF order kernel functions under SSBAM preprocessing conditions **a** amplitude and phase frequency response of the linear kernel function, **b** 2nd-order kernel function, **c** cross-section of the 3rd-order kernel function $h_3(i_1, i_2, 280)$, **d** cross-section of the 3rd-order kernel function $h_3(i_1, i_2, 300)$

a larger estimation error in the model identification [403]. Figure 3.34b shows the relative error of the SSB-AM system model, and the error trend is the same as that of the DSBAM system. From Fig. 3.34, it can be concluded that ODVF can simulate the linear response and nonlinear response of the DSBAM and SSB-AM system outputs, and it is also found that there is an error in the ODVF model compared with the VF model. In subsequent work, the impact of the presence of errors on the compensation of system nonlinearities will be further compared.

3.2.4.3 Effect of 2nd-Order Inverse Filtering Compensation

It has been analyzed above that ODVF can effectively simulate the linear response and harmonic distortion of the parametric array loudspeaker, but the purpose of this

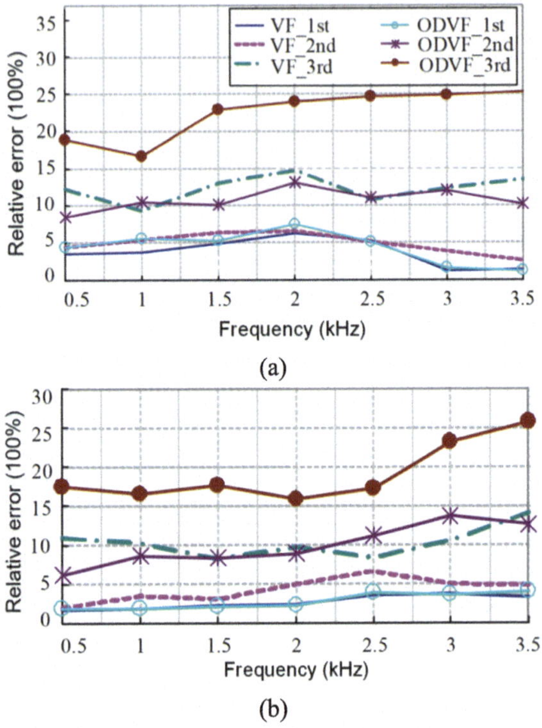

Fig. 3.34 Relative error of each order kernel function **a** DSBAM, **b** SSBAM

part is not only this, but also hopes to reduce or even eliminate the harmonic distortion in the system. That is to say, there is a need to find a filter that, when the signal passes through it, can be fed to the parametric array loudspeaker system to eliminate the harmonic distortion in the system's output. This kind of filter is also called an inverse filter, and the design of the inverse filter is detailed in Sect. 3.2.2. First, choose a 2nd-order inverse filter, and its framework is illustrated in Fig. 3.35, where $H_2[.]$ is the 2nd-order kernel function operation of the VF or ODVF model, $H_1^{-1}[.]$ is the inverse of the model's linear response. Because in the actual system, it cannot guarantee the causality of the inverse of the linear response, a delay τ is introduced into the system. It should be noted that the structure of the inverse filter designed based on the ODVF model and the VF model is consistent, and only the kernel function is different. For convenience, the inverse filters designed based on these two models are called ODVF inverse filter and Volterra inverse filter, respectively.

In the experiment, the DSBAM system and the SSB-AM system were measured separately, and the changes in the harmonic distortion of the system before and after the inverse filter compensation were compared. Figure 3.36 shows the system response of the DSBAM system before and after inverse filter compensation. Figure 3.36a is the linear response of the system. VF represents the Volterra inverse filter after compensation, and ODVF represents the ODVF inverse filter. From the figure, it can be observed that the linear response of the system remains basically

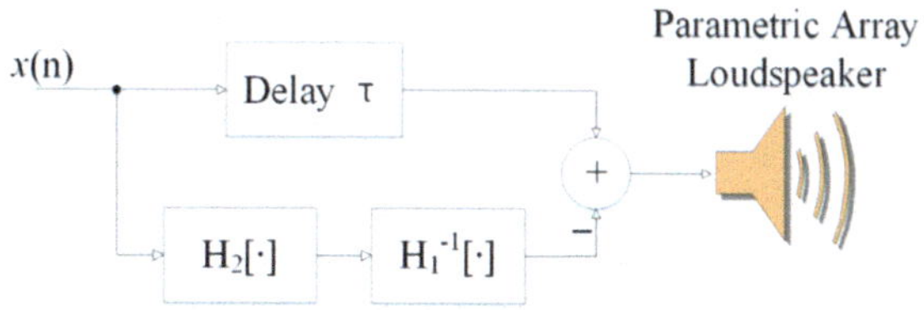

Fig. 3.35 2nd-order inverse filter framework diagram in the actual system

unchanged before and after compensation, which is also the purpose of the inverse filter designed in this part. Figure 3.36b is the 2nd harmonic response of the system. From the figure, it becomes clear that after the compensation of the ODVF inverse filter, the 2nd harmonic distortion of the system is reduced by more than 10 dB in the measured frequency band. It is also noted that the Volterra inverse filter brings more attenuation to the 2nd harmonic distortion of the system, which is mainly due to its higher model accuracy. Figure 3.36c is the 3rd harmonic response of the system. From the figure, it is obvious that the 3rd harmonic distortion of the system has not decreased but increased after the compensation of the ODVF inverse filter and the Volterra inverse filter. In fact, in the design of the 2nd-order inverse filter, only the reduction of the 2nd harmonic distortion of the system is considered, and the changes in the 3rd or higher harmonic distortion are not considered. However, in the DSBAM system, the 2nd-order inverse filter introduces new harmonic distortion. The following explains the mechanism of the generation of new harmonics. A single frequency signal $\sin(\omega t)$ is fed into the inverse filter. Due to the 2nd-order kernel function operation in the inverse filter $H_2[\cdot]$, the output of the inverse filter contains the fundamental frequency component of the signal $\sin(\omega t)$ and the 2nd harmonic component $\sin(2\omega t)$. This mixed frequency signal is fed into the nonlinear system to be compensated, and the cross product in the nonlinear action of the system $\sin(\omega t) \times \sin(2\omega t)$ generates 3rd-harmonic distortion. In fact, the nonlinear action of the system will also generate new fundamental frequency components, but compared to the original fundamental frequency signal of the system, the newly generated fundamental frequency signal is weaker and can be considered to have no effect on the linear response of the original system. From the above analysis, it is found that for the DSBAM system, after compensation by the inverse filter, the linear response of the system remains unchanged, the 2nd harmonic distortion is reduced, and the 3rd harmonic distortion is increased. To comprehensively evaluate the changes in various harmonics, THD is used to evaluate the effect of the inverse filter in suppressing the harmonic distortion of the system. Only the 4th harmonic distortion needs to be calculated, because the 5th and higher harmonic distortion components are small and will hardly affect the THD. Figure 3.36d shows the curve of THD changing with frequency before and after the DSBAM system is compensated by the inverse filter. From the figure, it becomes clear that after the ODVF inverse filter compensation, the system's THD has significantly decreased, and after the Volterra inverse filter compensation, the system's THD is basically the same as after the ODVF inverse filter compensation. At some frequencies, the THD after the Volterra inverse filter compensation is lower, while at some frequencies such as 2.5 kHz, the THD of the

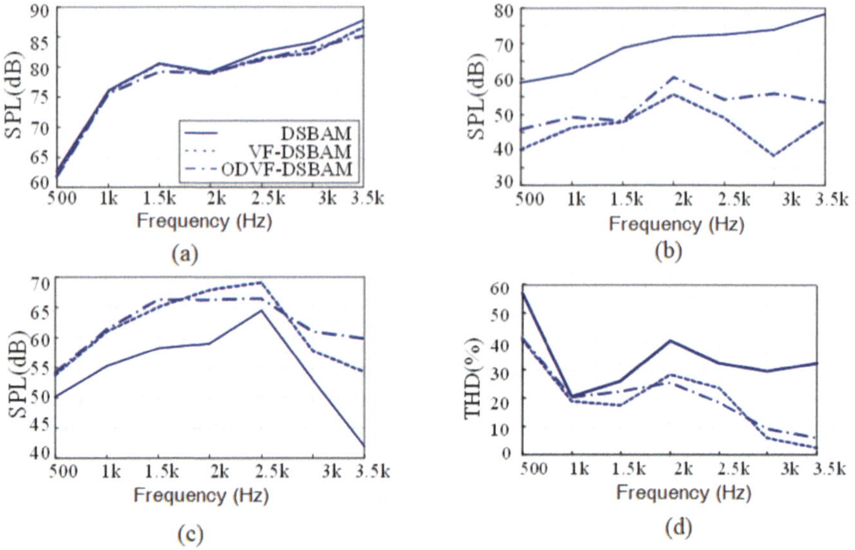

Fig. 3.36 System response before and after compensation under DSBAM preprocessing conditions **a** linear response, **b** 2nd harmonic distortion, **c** 3rd harmonic distortion, **d** THD

Volterra inverse filter is higher. This is because the 3rd harmonic distortion introduced by the Volterra inverse filter at 2.5 kHz is larger. Figure 3.37 shows the system response before and after the SSB-AM system is compensated by the inverse filter. From Fig. 3.37a, the inverse filter does not affect the linear response of the system. From Fig. 3.37b, it is obvious that the ODVF inverse filter can effectively suppress the 2nd harmonic distortion of the system, but its effect is slightly inferior to the Volterra inverse filter. From Fig. 3.37c, the inverse filter does not bring much impact to the 3rd harmonic of the system. Compared with the DSBAM system, the inverse filter raises the 3rd harmonic of the system. This difference is mainly attributed to the fact that the 2nd harmonic distortion of the SSB-AM system is smaller than that of the DSBAM system, so the 3rd harmonic introduced by the inverse filter is also smaller. Figure 3.37d shows the THD size before and after system compensation. From the figure, it can be observed that after the ODVF inverse filter and Volterra inverse filter compensation, the system's THD is basically the same. To quantitatively analyze the performance of the inverse filter compensation, Table 3.6 lists the average harmonic distortion and THD attenuation brought by the 2nd-order inverse filter to the system at each frequency point. From the table, it is obvious that the average attenuation of THD brought by the ODVF inverse filter and the Volterra inverse filter to the system is not much different. It can also be found that for the SSB-AM system, the attenuation brought by the 2nd-order inverse filter is much greater than that of the DSBAM system. This conclusion means that the effect of the 2nd-order inverse filter in suppressing THD in the SSB-AM system is more obvious than in the DSBAM system.

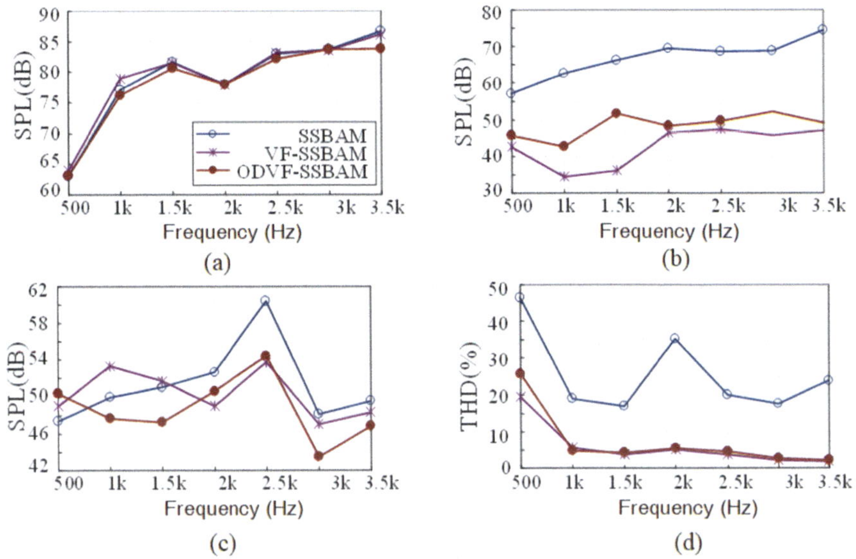

Fig. 3.37 System response before and after compensation under SSBAM preprocessing conditions **a** linear response, **b** 2nd harmonic distortion, **c** 3rd harmonic distortion, **d** THD

Table 3.6 Harmonic distortion attenuation under 2nd-order inverse filter

Modulation method	Compensation method	2nd harmonic distortion attenuation (dB)	3rd harmonic distortion attenuation (dB)	THD attenuation (%)
DSBAM	Based on VF	22.93	−6.73	14.37
	Based on ODVF	16.93	−7.65	13.58
SSB-AM	Based on VF	23.80	0.99	19.62
	Based on ODVF	18.21	2.65	18.44

3.2.4.4 Effect of 3rd-Order Inverse Filtering Compensation

It was studied experimentally in the previous section, where it was found that a 2nd-order inverse filter can effectively suppress the 2nd harmonic distortion of the parametric array loudspeaker system. However, it was noted that in the DSBAM system, the 2nd-order inverse filter led to an increase in the system's 3rd harmonic distortion. To further eliminate the system's 3rd harmonic distortion, a 3rd-order inverse filter was designed, as shown in Fig. 3.38, with the aim of eliminating the system's 2nd harmonic distortion while also eliminating the system's 3rd harmonic distortion. Compared to the 2nd-order inverse filter, the 3rd-order inverse filter takes into account the system's original 3rd-order nonlinear term, and also considers the new 3rd harmonic distortion term introduced by the 2nd-order inverse filtering operation (as shown in the dashed box in Fig. 3.38). In fact, it can be imagined that the 3rd-order

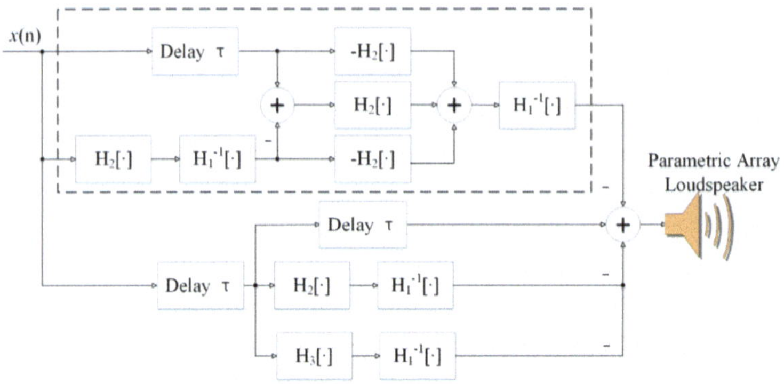

Fig. 3.38 3rd order inverse filter framework diagram

inverse filter will introduce new 4th and higher order harmonic distortions. However, in the experiment, it was found that for the parametric array loudspeaker system under study, as the order of the harmonics increases, the amplitude of the harmonic components gradually decreases. The newly introduced 4th harmonic distortion has little effect on the original system's 4th harmonic distortion.

In this section, experiments were conducted to measure the DSBAM system and the SSB-AM system, comparing the changes in each harmonic distortion before and after the inverse filter compensation. Figure 3.39 shows the system response before and after the DSBAM system is compensated by the inverse filter. Figure 3.40 demonstrates the system response before and after the SSB-AM system is compensated by the inverse filter. From the figures, it is obvious that the linear response of the system remains basically unchanged before and after the inverse filter compensation. In addition, Figs. 3.39b and 3.40b indicate that the 3rd-order inverse filter can effectively reduce the system's 2nd harmonic response, and Figs. 3.39c and 3.40c indicate that the 3rd-order inverse filter can effectively reduce the system's 3rd harmonic response. Moreover, the compensation effect of the Volterra inverse filter is better than that of the ODVF inverse filter. As can be seen from Figs. 3.39d and 3.40d, after the 3rd-order ODVF inverse filter and Volterra inverse filter compensation, the system's THD has significantly reduced, and the two types of inverse filters have little difference in reducing the system's THD.

To quantitatively analyze the compensation effect of the 3rd-order inverse filter, the average attenuation brought by the harmonic distortion of the inverse filter to the system is listed in Table 3.7. Comparing with Table 3.6, it is found that the attenuation of the 2nd harmonic distortion by the 3rd-order inverse filter is slightly smaller than that by the 2nd-order inverse filter. This may be because the 3rd-order inverse filter generates a fundamental frequency component, and the newly generated fundamental frequency component brings new 2nd harmonic distortion through the nonlinear action of the system. However, it can also be seen from the table that the difference caused by this effect is only within 2 dB. Comparing the two tables, it can be found that for the DSBAM system, the attenuation of THD by the 3rd-order

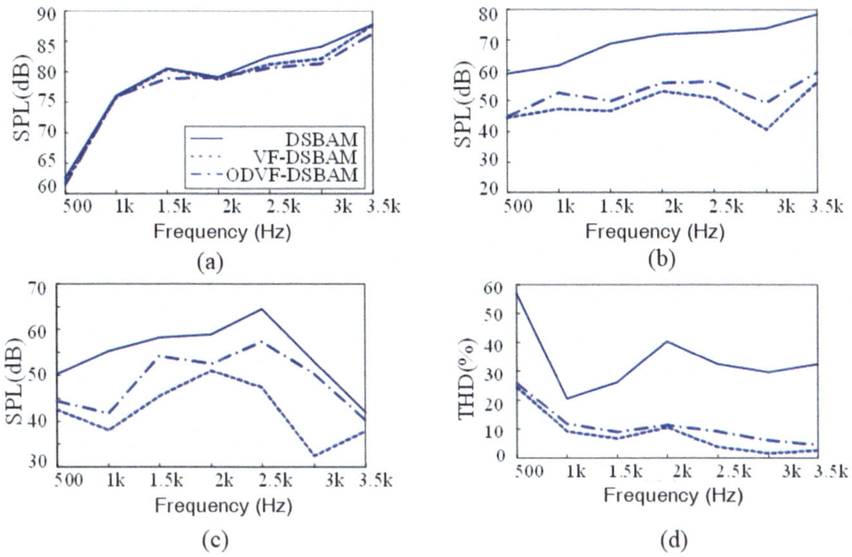

Fig. 3.39 System response before and after compensation under DSBAM preprocessing conditions **a** linear response, **b** second harmonic distortion, **c** third harmonic distortion, **d** THD

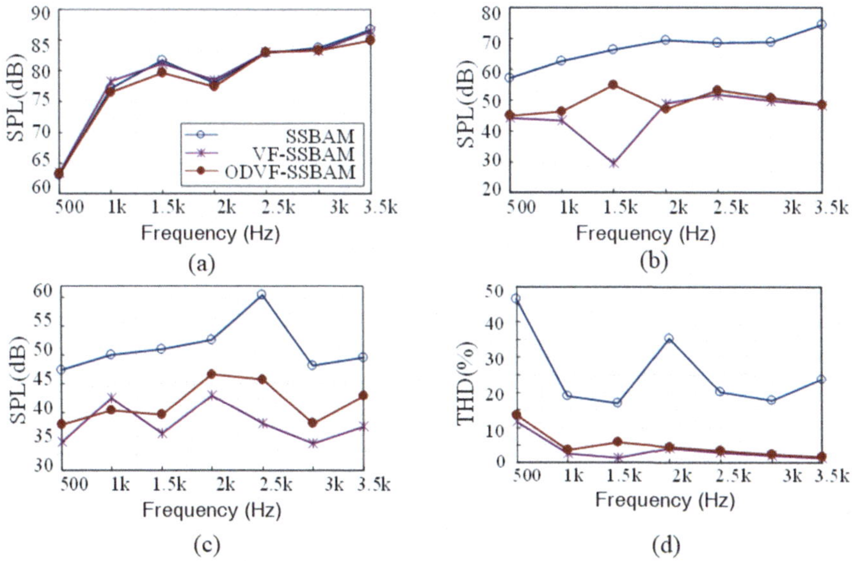

Fig. 3.40 System response before and after compensation under SSBAM preprocessing **a** linear response, **b** second harmonic distortion, **c** third harmonic distortion, **d** THD

Table 3.7 Harmonic distortion attenuation under 3rd-order inverse filter

Modulation method	Compensation method	2nd harmonic distortion attenuation (dB)	3rd harmonic distortion attenuation (dB)	THD attenuation (%)
DSBAM	VF-based	20.91	12.52	25.67
	ODVF-based	16.79	5.91	22.94
SSB-AM	VF-based	21.54	13.09	21.88
	ODVF-based	17.27	9.64	20.58

inverse filter is almost twice that of the 2nd-order inverse filter, because it eliminates the 2nd harmonic while suppressing the 3rd harmonic distortion. For the SSB-AM system, the attenuation of THD by the 3rd-order inverse filter is almost the same as that by the 2nd-order inverse filter. This is because the THD of the SSB-AM system is already very small after the 2nd-order inverse filter compensation, and the 3rd-order inverse filter of the system is also very small, contributing weakly to the calculation of THD. Therefore, although the 3rd-order inverse filter suppresses the 3rd harmonic by about 10 dB, the change in THD is very small. Based on this, it can be considered that the THD compensation effect of the 3rd-order inverse filter on the DSBAM system is more obvious than that on the SSB-AM system.

3.2.4.5 Computational Complexity Analysis

It can be seen from the experimental results of the previous section, both ODVF and VF can effectively simulate the harmonic distortion and intermodulation distortion of the parametric array loudspeaker. However, in actual use, the complexity of VF is a problem that cannot be ignored. First, the parameter identification of the VF model is time-consuming and laborious. The common NLMS algorithm requires nonlinear filtering calculations for each iteration, resulting in a large computational load, and its accuracy is not high when the signal-to-noise ratio is low. The parameter identification of ODVF can be completed within a few minutes by the sweep method. Secondly, the computational complexity of the Volterra inverse filter and the ODVF inverse filter is examined to explore the possibility of real-time processing on the hardware platform in the future. The resources consumed by the hardware implementation of DSBAM and SSB-AM will also be considered to find the optimal preprocessing scheme for the parametric array loudspeaker.

For the DSBAM system, only 2 multipliers are needed; for SSB-AM, a Hilbert filter is introduced, which increases the consumption of computing resources. Assume the length of the Hilbert filter is M_{hil}, and the total number of multipliers required by the entire SSB-AM system is $M_{hil} + 3$. For the Volterra inverse filter, assuming the lengths of its linear kernel, 2nd-order kernel, and 3rd-order kernel filters are M_1, M_2^2 and M_3^3. For the 2nd-order Volterra inverse filter, the number of multipliers required is $M_1 + 2M_2^2$; for the 3rd-order Volterra inverse filter, the number

Table 3.8 Computational complexity and harmonic distortion of various preprocessing methods

Preprocessing method	Number of multipliers	Average THD (%)
DSBAM	2	34.11
VF-DSBAM (2nd-order)	205,122	19.74
VF-DSBAM (3rd-order)	99,329,282	8.43
ODVF-DSBAM (2nd-order)	962	20.53
ODVF-DSBAM (3rd-order)	5,442	11.16
SSB-AM	103	25.62
VF-SSB-AM (2nd-order)	205,223	5.99
VF-SSB-AM (3rd-order)	99,329,383	3.73
ODVF-SSB-AM (2nd-order)	1,063	7.17
ODVF-SSB-AM (3rd-order)	5,543	5.03

of multipliers required is $4M_1 + 10M_2^2 + 3M_3^3$. Correspondingly, suppose the filter lengths of the linear kernel, 2nd-order kernel, and 3rd-order kernel of the ODVF model are M_1, M_2 and M_3. Then the number of multipliers required for the 2nd-order ODVF inverse filter is $M_1 + 2M_2$, and the number of multipliers required for the 3rd-order ODVF inverse filter is $4M_1 + 10M_2 + 3M_3$.

In actual measurement experiments, the length of the filter is chosen to be $M_1 = M_2 = M_3 = 320$, and the length of the Hilbert filter is $M_{hil} = 100$. Table 3.8 lists the combinations of various inverse filters and amplitude modulation algorithms and the computational resources consumed by the processing methods. The table also lists the average THD of the system output corresponding to various preprocessing methods as a performance reference. It can be seen from the table that the reduction of the THD of the system output in various preprocessing methods is often at the cost of an increase in computational resources. The table also shows that in the DSBAM and SSB-AM systems, the system THD after preprocessing with the ODVF inverse filter and the Volterra inverse filter can be compared, but the resources consumed by the ODVF inverse filter are much less than those of the Volterra inverse filter. Finally, considering the consumption of computational resources and the performance of system harmonic distortion, the preprocessing method of cascading the 2nd or 3rd-order ODVF inverse filter with SSB-AM has the smallest THD and consumes less resources among all the schemes discussed above.

3.3 Summary

This chapter summarizes the preprocessing methods currently used in parametric array loudspeakers.

On the one hand, this chapter uses the KZK equation to conduct a comparative theoretical analysis with and without equalization of the four commonly used preprocessing methods for reproducing directional speech signals in parametric array loudspeakers, focusing on comparing the simulation results obtained from the demodulated signal between the KZK equation and the Berktay far-field solution. After determining the relationship between THD and P_0 using the KZK equation, in the study, P_0 is limited to below 132 dB. The gain slope of the difference frequency component frequency response obtained from the KZK equation is less than the gain slope predicted by the Berktay far-field solution, between +6 and +12 dB/octave. According to the gain slope obtained from the KZK equation, an equalization filter for the preprocessing method is designed, which is then applied to the input speech signal under the conditions of $P_0 = 128$ dB and 120 dB. The equalized frequency response of the difference frequency component is relatively flat, showing ripples less than 1.5 dB and 1.6 dB, respectively. By using 10% THD as a performance indicator, under the conditions of $P_0 = 128$ dB and $P_0 = 120$ dB, the best preprocessing methods are SRAM and SSB-AM, respectively. From the analysis, the most suitable and appropriate preprocessing methods for reproducing speech are SRAM and MAM for the case of $P_0 = 128$ dB with equalization. On this basis, a RSSB-AM implementation method for parametric array loudspeakers is introduced. Recursive single-sideband modulation and ultrasonic transducers are performed at a lower sampling rate, effectively reducing the required order of Hilbert transform filters and inverse filters. The processed signal is then upsampled to a higher sampling rate by zero-padding. To remove mirror frequency components, a complex domain bandpass filter is used. Finally, the real part of the signal is extracted to obtain the input signal of the ultrasonic transducer. Experiments show that this method can ensure the same performance as traditional methods while reducing the computational load required for system implementation.

On the other hand, this chapter introduces the ODVF model into the preprocessing research of parametric array loudspeakers, and conducts corresponding theoretical and experimental research. Using the KZK equation as a mathematical model, simulation experiments have verified that ODVF can simulate the linear and nonlinear responses of the audio frequency of the parametric array loudspeaker system preprocessed by DSBAM and SSB-AM. Subsequently, the ODVF model was established through experimental means, and 2nd-order and 3rd-order inverse filters were designed respectively. A Volterra inverse filter was introduced for comparison. The Volterra inverse filter was also introduced for comparison. The experimental results show that the ODVF inverse filter can effectively distort the harmonic of the parametric array loudspeaker system. The system's THD after compensation with an ODVF inverse filter is essentially equivalent to that after compensation with a Volterra inverse filter, while the ODVF inverse filter consumes fewer resources.

Chapter 4
Measurement of Parametric Array Loudspeakers

4.1 Overview of Parametric Array Loudspeaker Measurement

The difference frequency sound of parametric array loudspeakers is generated by the nonlinear self-demodulation of finite amplitude ultrasound in the air, which means that when measuring with a microphone in the near field, the microphone will capture the high sound pressure level of the original ultrasound signal while measuring the difference frequency sound. In practical engineering applications, accurate measurement of the sound field of parametric array loudspeakers is very important for the design and development of the system. The additional difference frequency electrical signal caused by the nonlinear distortion of the microphone inevitably appears in the measurement results, this part of the signal has the same frequency as the real difference frequency sound signal, but does not exist in the real acoustic field of the air, and it is called spurious sound. The generation schematic diagram is shown in Fig. 4.1. In the measurement process of parametric array loudspeakers, spurious sound is a key factor, directly related to the accuracy of the measurement results.

Because the parametric array loudspeaker is different from the ordinary loudspeaker, the audio frequency wave it reproduces coexists with an ultrasound pump wave, which places higher demands on the accurate measurement of parametric array loudspeakers. When two ultrasound waves with frequencies f_a and f_b (where $f_a > f_b$) are emitted after the ultrasonic transducer, due to the nonlinear effect in the air, multiple different frequency sound waves can be generated, namely the difference $(f_a - f_b)$ and the sum $(f_a + f_b)$ of the two frequencies, as well as higher order harmonics. Because the sound wave attenuation coefficient is directly proportional to the square of the frequency, the high frequency components decay faster during propagation, so after a certain distance of propagation, only the difference frequency sound wave exists, forming the sound frequency. However, in the near field area of parametric array loudspeakers, there are both ultrasound and demodulated sound frequency, and because the sound pressure level of ultrasound f_a and f_b is high,

J. Yang and P. Ji, *Parametric Array Loudspeakers*,
https://doi.org/10.1007/978-981-96-3548-1_4

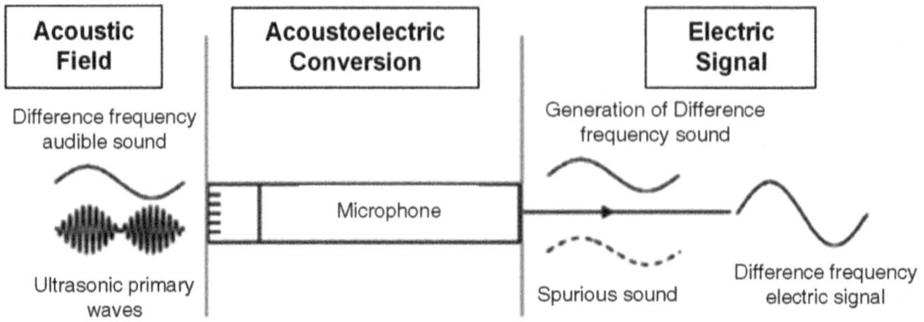

Fig. 4.1 Schematic diagram of spurious sound generation

especially in the near field, it is generally above 120 dB, so spurious sound will be generated when measuring directly with a microphone in this area. Spurious sound mainly comes from two aspects, one is due to the nonlinear effect of the microphone preamplifier circuit, and the other is due to the radiation sound pressure on the surface of the transducer. The frequency of the spurious sound formed by these two is the difference ($f_a - f_b$) of the two ultrasound frequencies f_a and f_b, and the amplitude of the spurious sound is directly proportional to the product of the amplitudes of the ultrasound f_a and f_b [112, 172]. It has been proven that the spurious sound generated by sound radiation pressure is not the main contribution in the measurement system [112, 172]. Spurious sound severely interferes with the measurement of actual sound frequencies, especially in the near field region, where spurious sound is much larger than the sound frequency generated by the parametric array loudspeaker [192], so it is necessary to design and use acoustic filters or other methods to remove the influence of spurious sound, such as designing and using acoustic filters, to obtain the real demodulated sound frequency.

Since the acoustic filter is crucial for the accurate measurement of the parametric array loudspeaker, the near-field measurement of the difference frequency sound signal of the parametric array loudspeaker is a complex issue that has received much attention in recent years. In the near field, due to the high sound pressure level of the ultrasonic primary wave, the spurious sound is much larger than the difference frequency sound signal generated by the parametric array loudspeaker, and a filter must be adopted to obtain the real sound frequency signal. In the far field with the attenuation of the ultrasonic primary wave, the demodulated sound frequency signal can be obtained directly by using a microphone. In actual measurements, there are mainly two methods to avoid the influence of spurious sound. One is to increase the measurement distance, because the absorption coefficient of air for ultrasound is much higher than that of sound frequency, so after the amplitude of ultrasound drops to a certain range, the influence of spurious sound can basically be ignored. The second is to use acoustic filters to filter out ultrasound, which can be achieved by using specific acoustic structures or materials to reduce the amplitude of ultrasound in the measurement system, thereby reducing the influence of spurious sound. In 1975, Bennett and Blackstock [93] first used a hemispherical acoustic filter made of 0.06

mm thick plastic glass paper in the experiment of parametric array loudspeaker in the air, and placed at the front end of the measurement microphone, which attenuated the difference frequency sound (5 kHz) by 3.5 dB, but attenuated the ultrasonic primary wave (18.6 and 23.5 kHz) by 20 dB, achieving low-pass characteristics, thus effectively reducing spurious sound. Moffett et al. [187] studied the nonlinear problem of measuring hydrophones in water in 1982. Garrett et al. [33] also found the influence of internal nonlinearity in the water receiving system, which is proportional to the product of the pump wave at the measurement point, and the experimental results of axial propagation and beam pattern support this view. Lucas et al. [421] used a piece of SOAB rubber (1.3 cm thick) as an acoustic filter when studying the focused parametric array in water, which attenuated the primary wave signal by 16 dB, but also attenuated the difference frequency signal by 2.5 dB. Havelock and Brammer [99] installed 4 layers of 3.18 mm felt in the experiment, which reduced the sound pressure level at 30 kHz frequency by more than 20 dB, and the sound frequency attenuated by 3–5 dB. Toda [172] proposed a complex acoustic filter with 4 layers of polymer, found that the attenuation in the range of 30–40 kHz was 30–40 dB, and there was an unwanted $\pm$ 5dB deviation in the range of 10 Hz–10 kHz. However, this acoustic filter structure is too complex and not conducive to practical application. Humphrey et al. [61] used a 30 mm thick polyurethane panel as an acoustic filter when evaluating the underwater acoustic performance of the parametric array panel. During the measurement using CMUT transducers as the emission source for parametric array loudspeakers by Wygant et al. [112], it was found that the experimental results using a layer of Ceylon synthetic fiber as an acoustic filter matched the simulation results very well, and a method for calculating the amplitude of spurious sound was provided. Kamakura et al. [188, 189] used two groups of pump waves of opposite phase as the driving signal for parametric array loudspeakers to reduce the sound pressure level of the pump wave, but the impact on the generated difference frequency sound is small. This method can also be used to reduce spurious sound. This anti-phase method has been applied in the measurement of sound pressure on the axis [190] and material performance parameters [422]. In the general measurement process, the impact of spurious sound can be overcome to a certain extent by utilizing the physical characteristics of the microphone, as shown in Figs. 4.2 and 4.3. The signal sensitivity response schematic diagram measured by the 1/4-inch microphone when the sound wave is incident on the transducer from 0° and 90°. It is obvious that when the sound wave is incident from 0°, a relatively flat sensitivity response curve can be maintained in the full frequency band. However, under the condition of 90° incidence, the sensitivity of the ultrasonic frequency band gradually decreases with the increase of frequency. At a frequency of 50 kHz, the sound pressure level result obtained by the GRAS 46BE 1/4-inch transducer measuring the sound wave at 90° incidence is 15 dB less than the amplitude value of the 0° incidence result. This plays a very important role in parametric array measurement and can reduce the spurious sound caused by the circuit part caused by ultrasound to a certain extent. Ju and Kim [423] conducted a qualitative analysis of spurious sound by utilizing the different response sensitivities of microphone at different incidence angles in the high frequency band. The following part of this

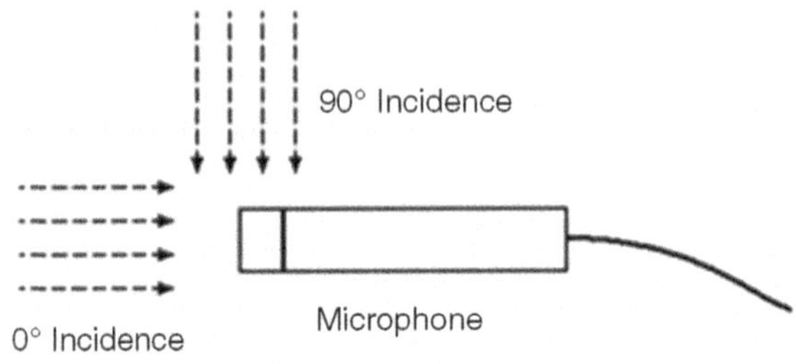

Fig. 4.2 Explanation of incident angle

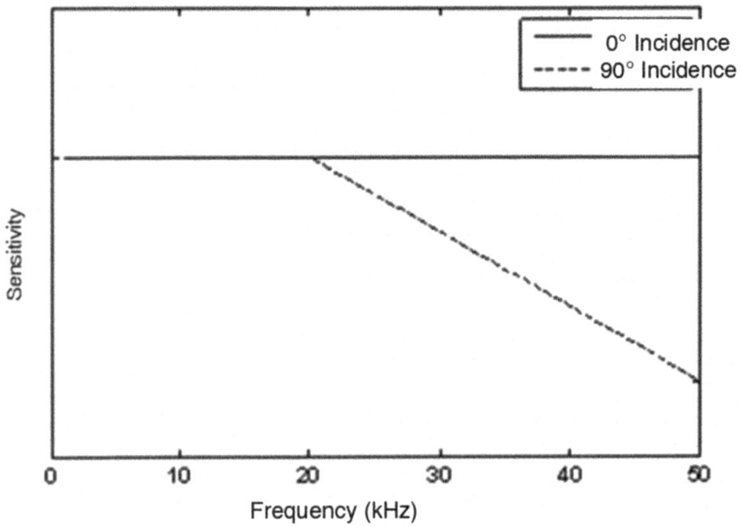

Fig. 4.3 Schematic diagram of frequency response under different incident angles of the microphone

chapter provides a detailed explanation of the two current methods for removing spurious sound and analyzes the four main parameters that affect the measurement results Fig. 4.3.

4.2 Cylindrical Ultrasonic Filter

This section designs a low-pass filter for the parametric array loudspeaker measurement system through experimental methods [191]. The filter structure is simple, with an average attenuation of ultrasonic waves exceeding 15 dB, an impact on audio frequencies of less than 5 dB, and its effectiveness has been confirmed through actual measurements of the parametric array loudspeaker.

4.2.1 *Structure and Performance of the Filter*

A new type of filter structure is demonstrated in Fig. 4.4. A circular aluminum plate with a thickness of 0.6 mm is placed at the front end of the microphone, which can cut off ultrasonic waves and attenuate them. Because the wavelength of the audio frequency is larger, it can bypass the aluminum plate to reach the microphone, so the filter's attenuation of the audio frequency is smaller. The filter is 20 mm long, and the end center has a through hole with a diameter of 7 mm for installing the microphone. To verify the performance of the filter, an experiment was conducted on a filter with a diameter of 16 mm in an anechoic chamber. The sound source is an ultrasonic transducer array and a common loudspeaker, used to transmit ultrasonic signals and audio frequency signals respectively. A 1/4-inch GRAS 46BE microphone was placed 3 m away from the sound source axis. The specific experimental steps are to first directly measure the sound pressure signal of the sound source with the microphone, and then measure the sound pressure signal of the sound source again after adding the filter. The difference between the two sound pressure signals is the transmission loss (TL) of the filter.

The TL measurement results of the cylindrical filter in the 30–50 kHz ultrasonic frequency band are shown in Fig. 4.5a. As shown in this figure, for a filter with a diameter of 16 mm, its resonance frequency is around 21.5 kHz, which is not within the frequency range of 30–50 kHz ultrasound. Therefore, to avoid resonance, the diameter of the aluminum plate should be greater than 12 mm, corresponding to the wavelength of 30 kHz ultrasound. In further experiments, a filter with a diameter of 16 mm was selected for testing (as shown in Fig. 4.4), and its TL in the audio frequency band was measured by comparison method as shown in Fig. 4.5b. As can be observed from Fig. 4.5a, the filter has an average attenuation of more than 15 dB for 30–50 kHz ultrasound, and the TL can reach 35 dB near the carrier frequency of 40 kHz. In the audio frequency band, the TL curve in Fig. 4.5b is almost flat.

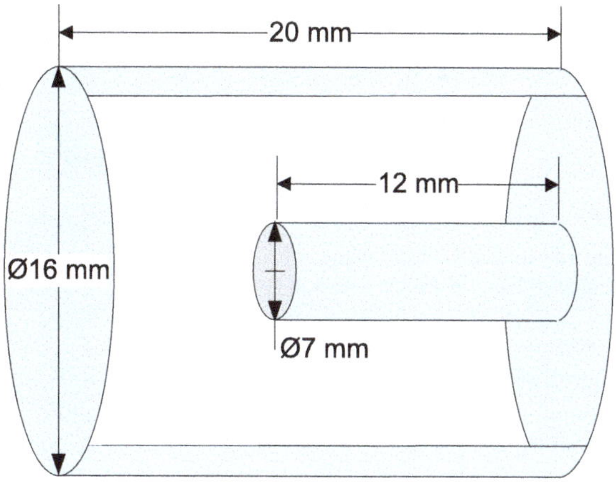

Fig. 4.4 Schematic diagram of the filter structure

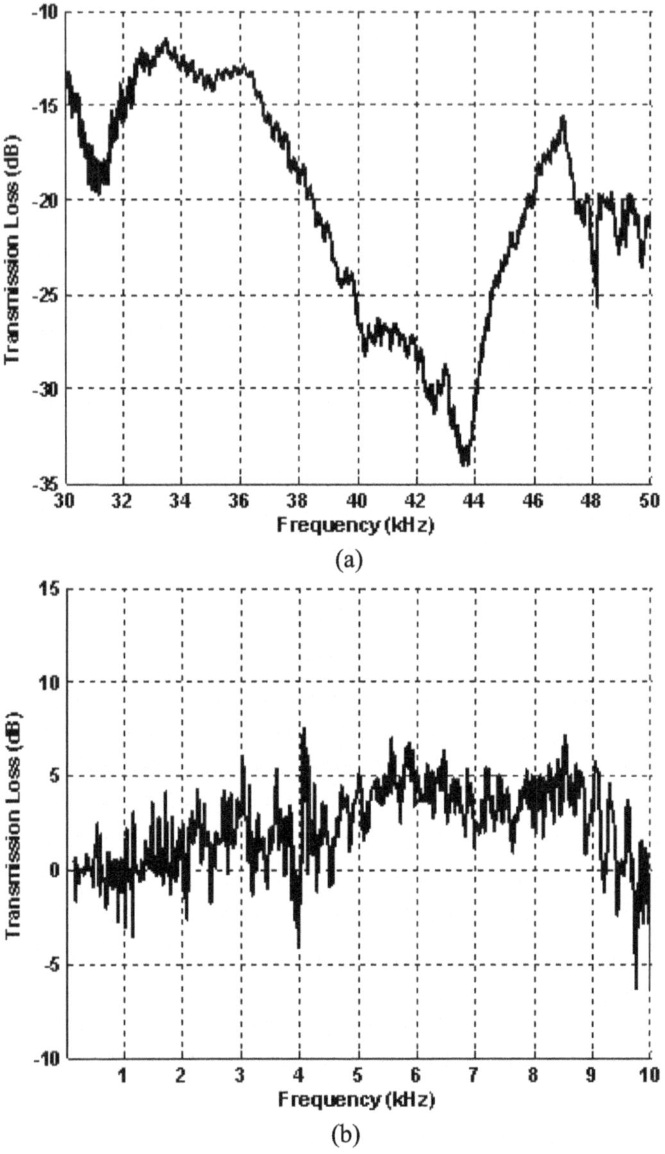

Fig. 4.5 **a** Measurement results of sound insulation in the ultrasonic frequency band and **b** audio frequency band of a filter with a diameter of 16 mm

Therefore, it is evident that this filter has obvious low-pass characteristics, greatly attenuating ultrasound, but has little effect on audio frequency.

4.2.2 Experimental Verification of the Filter

To verify the effectiveness of the filter, actual measurements were carried out on the parametric array loudspeaker in the anechoic chamber using this filter. The parametric array loudspeaker is a regular hexagon composed of 91 ultrasonic transducer units with a diameter of 16 mm in parallel, with a side length of 10 cm. The preprocessing algorithm of double-sideband modulation is adopted, and the ultrasonic carrier frequency is 40 kHz. The same microphone as in the previous experiment was placed at 3 m from the axis of the parametric array loudspeaker. The frequency response of the parametric array loudspeaker is shown in Fig. 4.6. It becomes clear that, without the filter, the amplitude of the spurious sound is independent of the frequency of the difference sound. The demodulated audio frequency amplitude produced by the parametric array loudspeaker is proportional to f_d^n [112], where f_d is the difference audio frequency. The exponent n satisfies $1 \le n \le 2$, which is determined by the ratio of the Rayleigh distance to the length of the parametric array. In the Westervelt model [21, 22], since the Rayleigh distance is much greater than the length of the parametric array, $n \approx 2$. In the Berktay model [243], since the Rayleigh distance is much less than the length of the parametric array, $n \approx 1$. As can be seen from Fig. 4.6, for the parametric array loudspeaker used in the experiment, $n \approx 1.5$ can approximate the relationship between the sound frequency amplitude and the difference frequency sound frequency well.

To further distinguish between spurious sound and the sound frequency signal generated by the parametric array loudspeaker, this section measures the axial sound pressure of the parametric array loudspeaker, and the measurement results are shown

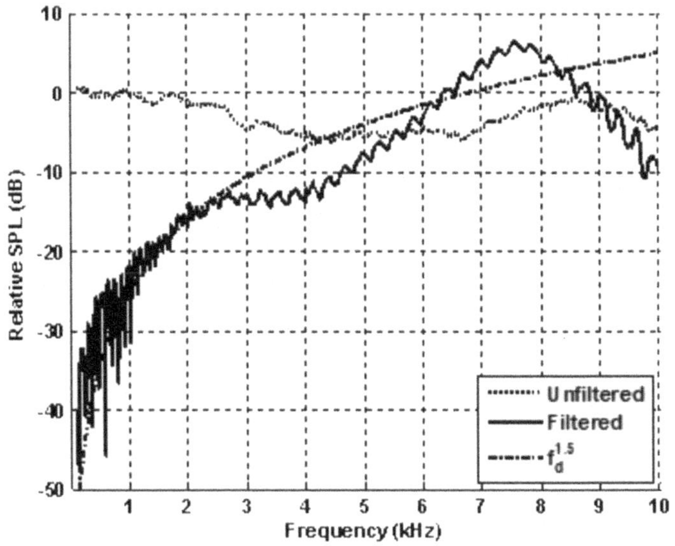

Fig. 4.6 Frequency response of the parametric array loudspeaker

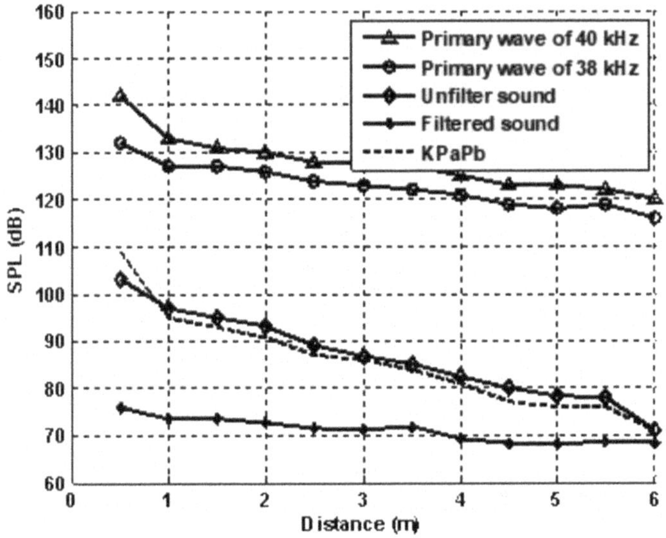

Fig. 4.7 Axial sound pressure of the parametric array loudspeaker

in Fig. 4.7. The driving signal frequencies of the parametric array loudspeaker are 40 and 38 kHz. Spurious sound can be represented as $p_{sn} = Kp_a p_b$, where p_a and p_b are the ultrasonic signal amplitudes at the transducer. Through curve fitting, the constant $K \approx 2.8 \times 10^{-4}\text{Pa}^{-1}$ can be obtained. Without a filter, the relationship between spurious sound and distance is that the sound pressure decreases by 12 dB for each doubling of distance. However, with a filter, the relationship between the sound pressure of the difference frequency sound and distance is a decrease of 5 dB. Therefore, it can be considered that the influence of spurious sound on the measurement of the demodulated signal has been basically eliminated. It can also be seen from the figure that in the near field, the sound pressure level of the ultrasonic pump wave is all above 120 dB, and the spurious sound is much larger than the sound frequency produced by the parametric array loudspeaker. However, in the far field, as the ultrasonic pump wave attenuates, the spurious sound gradually decreases, so when measuring in the far field, the sound frequency signal produced by the parametric array loudspeaker can be obtained without adding a filter.

4.3 Design of an Acoustic Filter Based on Phononic Crystal

In the past few decades, phononic crystals (PC) have aroused great interest, mainly because they can produce phononic band gaps, in which elastic waves will not propagate in the structure. The existence of band gaps provides potential applications for these artificial materials, such as filtering elastic waves. There are two mechanisms related to PC, namely the Bragg scattering mechanism [424, 425] and the local resonance mechanism [426, 427]. It is well known that compared with conventional PCs based on the Bragg scattering mechanism, PCs based on the local resonance mechanism are more suitable for lower frequencies. In the study of parametric array loudspeakers, only how to suppress the pump wave within the ultrasonic range is concerned, therefore, only need to study how to apply the PC based on the Bragg scattering mechanism to the measurement in the parametric array loudspeaker. The study of the Bragg scattering mechanism [424, 425] shows that the center frequency of the gap is given by the Bragg condition

$$f = nv/2a(n = 1, 2, 3, \ldots) \tag{4.1}$$

where v is the elastic wave speed of the base material, and a is the distance between the scatterers, that is, the lattice parameter. One advantage of PC is that by changing a, the peak of attenuation can be achieved within the selected frequency range.

In this section, an acoustic filter using PC is presented, which eliminates the spurious sound in the measurement of the parametric array loudspeaker by suppressing the amplitude of the pump wave without significantly affecting the amplitude of the sound frequency [192]. The performance of this acoustic filter is verified through simulation and experiment. To further verify the measurement results after applying the given filter, a numerical simulation of the Gaussian beam expansion method based on the parametric array loudspeaker was carried out. The performance of this filter in the directional measurement of the parametric array loudspeaker was also studied.

4.3.1 Theoretical Modeling and Numerical Simulation

In commercial Comsol Multiphysics simulation software, the finite element method is adopted to calculate the transmission loss TL with or without PC. Since a large number of grids (>10^9) to solve the wave equation in three-dimensional (3D) regions, so the 3D model is unrealistic. Since solving the wave equation in three-dimensional (3D) regions requires a large number of grids (>10^9), the 3D model is unrealistic. Therefore, in this study, to evaluate the performance of the PC-based filter, quasi-one-dimensional (quasi-1D) and two-dimensional (2D) numerical simulations of infinite length in the y direction and along the x direction were carried out.

The classical structure of a 2D PC composed of two types of elastic materials with different mechanical properties is shown in Fig. 4.8a, where d is the diameter of the cylinder. The front view of the filter model with a two-dimensional PC system is shown in Fig. 4.8b, which includes cylinders embedded in the air. The center frequency of the first Bragg bandgap is

$$f_c = c_0/2a \tag{4.2}$$

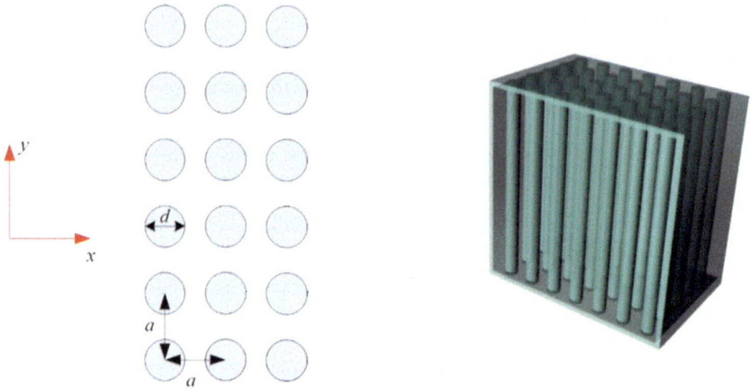

(a) (b)

Fig. 4.8 Two-dimensional PC's **a** classic structure diagram and **b** front view of the filter model

where c_0 is the small amplitude speed of sound in the air. Since the expected center frequency of the Bragg bandgap is about 40 kHz, which is the carrier frequency in parametric array loudspeakers, the lattice parameter a is defined as 4 mm.

The quasi-1D simulation model in Comsol is shown in Fig. 4.9, where the number of lattices is N, and periodic conditions are used to reduce computation time. The material of the cylinder is an aluminum alloy with a density of 2780 kg/m^3, a Young's modulus of 73.1 GPa, and a Poisson's ratio of 0.33. The ambient fluid is air with a density of 1.2 kg/m^3 and a speed of sound of 343 m/s. The boundary condition of normal acceleration of 1 m/s^2 at the left boundary is used as the radiation condition. Plane wave radiation conditions are applied at both left and right boundaries to generate plane waves. For the rigid cylinders in the PC, strict boundary conditions have been applied, i.e., the normal component of the air particle velocity in the cylinder wall is zero due to the high acoustic impedance mismatch at the air–solid boundary. The frequency range is from 0.1 to 60 kHz, with an interval of 0.1 kHz. Therefore, the minimum wavelength is 0.57 cm, and the maximum element size of the mesh is set to 1/6 of the minimum wavelength, i.e., 0.09 cm.

When $N = 5$, the influence of the cylinder size on TL is shown in Fig. 4.10a, where the diameter of the cylinder ranges from 1.6 to 3.2 mm, with an interval of 0.4 mm. For all cylinder diameters d, the Bragg bandgap is clearly observed. As the wavelength of the incident sound wave is comparable to the lattice parameter, the TL

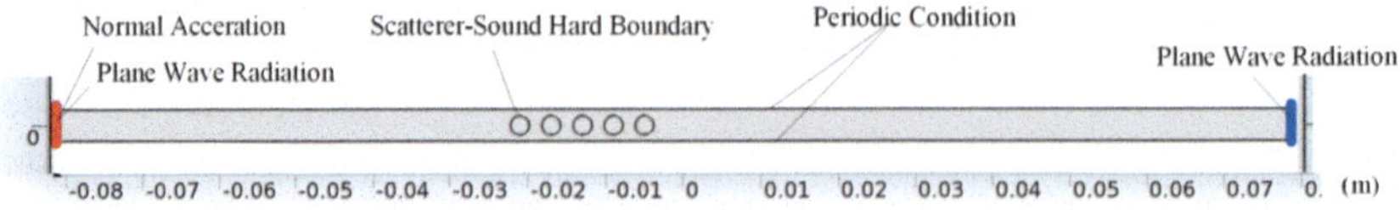

Fig. 4.9 Quasi-1D model of PC-based filter in Comsol

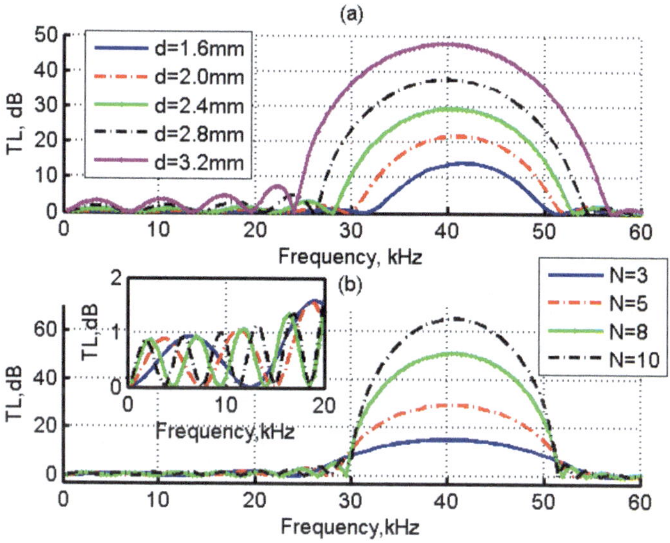

Fig. 4.10 Effect of parameter changes on TL **a** different d and **b** different *N*

designed around the center frequency of 40 kHz clearly shows the formation of the bandgap, which means that high attenuation occurs on this selective frequency band due to multiple scattering phenomena. The TL before the bandgap clearly indicates that at low frequencies, the PC system behaves as a homogeneous material, and the propagation of sound waves is hardly affected by the periodic structure. This is because the lattice parameter is much smaller than the relevant wavelength. After the bandgap, the wavelength of the sound wave is smaller than the lattice parameter of the PC system, so the TL of the sound frequency band becomes smaller.

From Fig. 4.10a, it can be observed that as the diameter of the cylinder increases, both the TL of the entire frequency band and the bandwidth of the Bragg bandgap increase. When the diameter of the cylinder increases, the center frequency of the Bragg band will slightly move to a lower frequency. Since the purpose of using a acoustic filter in a parametric array loudspeaker is to increase the TL in the ultrasonic frequency and reduce the TL in the sound frequency, a balance should be struck between the TL in these two areas. From Fig. 4.10a, it is obvious that the Bragg bandgap of $d = 2.4$ mm greater than 10 dB is between 30 and 51 kHz, and the peak at 40 kHz is almost 30 dB. At the same time, the maximum TL in the sound frequency range of $d = 2.4$ mm is less than 2 dB. Therefore, $d = 2.4$ mm is chosen.

Figure 4.10b shows the simulation results of the effect of the number of lattice *N* on TL, where $d = 2.4$ mm. For all lattice *N*, the Bragg bandgap is still easily observed, and the trend is the same as the above study. When *N* increases, the transition area between 0 and 10 dB becomes steeper, and the bandgap becomes smaller, but the bandwidth of TL greater than 10 dB is about 21 kHz and is almost unchanged. Before and after the Bragg bandgap, more ripples appear as *N* increases. The TL in the Bragg bandgap greatly increases as N increases, and its maximum value is almost proportional to N. The position of the center bandgap is almost unchanged at around 40 kHz. As illustrated in the inset of Fig. 4.10b, the TL in the audio frequency range remains relatively unchanged, with a maximum value of about 1.6 dB, and almost all have the same envelope. After considering the actual measurement requirements of the parametric array loudspeaker, N is chosen to be 5.

Therefore, from the quasi-1D simulation results of Comsol, the optimal parameters of the PC that meet most of the measurement requirements related to parametric array loudspeaker are $a = 4$ mm, $d = 2.4$ mm, and $N = 5$. It is noted that the maximum attenuation of the pump wave can be achieved by adjusting the structural parameters.

The development of 2D numerical calculations can further study the performance of PC-based acoustic filters. Figure 4.11 shows its model in Comsol, which consists of 5×7 cylinders, and its boundary conditions are almost the same as those of the quasi-cylindrical 1D simulation. The main difference between these two models is that, compared with the quasi-1D model, the filter in the 2D model consists of a finite number of scatterers, while the quasi-1D model uses periodic conditions to simulate an infinite number of cylinders in the y direction. To simulate the actual situation, all cylinders need to be installed on support, and two identical rectangular frames are placed on the bottom and top, with thicknesses and widths of 1 mm and 2 cm respectively, using the same material as the cylinders.

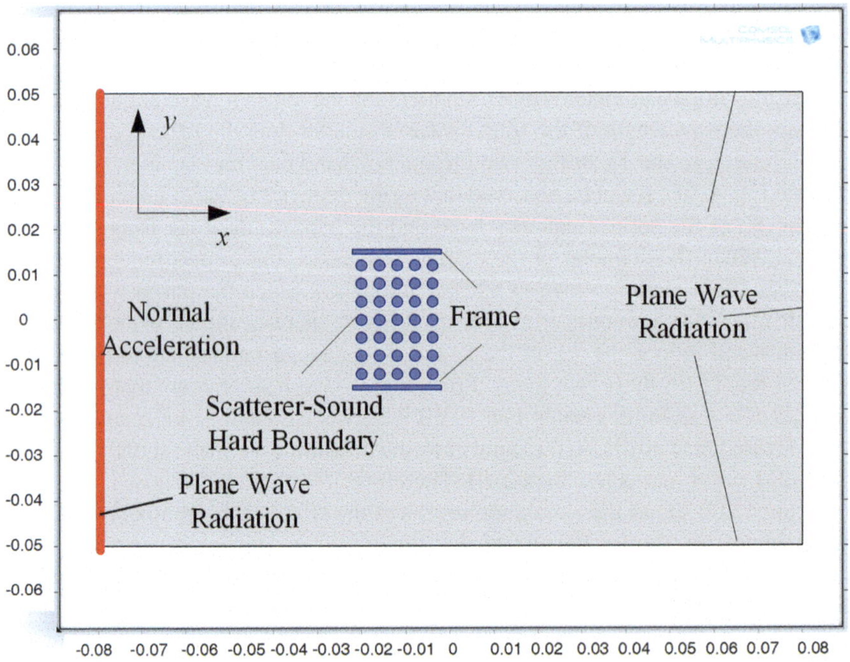

Fig. 4.11 Two-dimensional filter model in Comsol

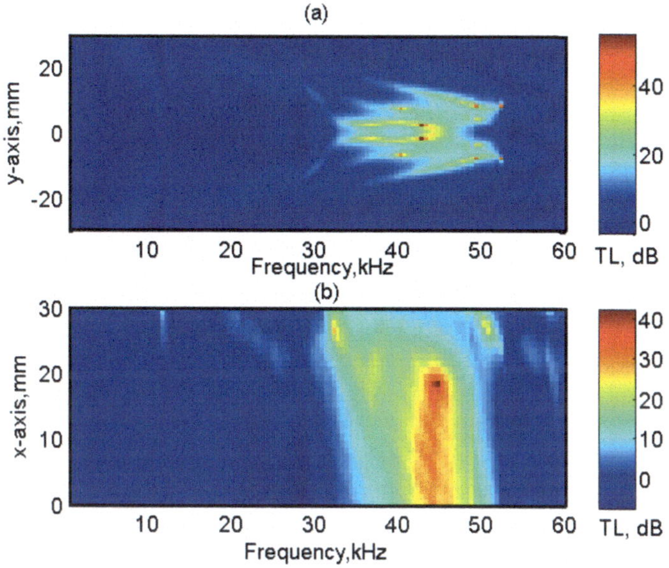

Fig. 4.12 Simulation results of the TL distribution and frequency and position relationship using the 2D model in Comsol **a** at $x = 1$ cm **b** at $y = 0$ cm

Figure 4.12 shows the changes in TL distribution on the x and y axes with frequency and position. Figure 4.12a is the TL at $x = 1$ cm, where y ranges from -3 to 3 cm, and Fig. 4.12b is the TL at $y = 0$ cm, where x ranges from 0 to 3 cm, with an interval of 1 mm. In Fig. 4.12, a trend similar to the quasi-1D model can be found, and a clear Bragg bandgap can be observed, the reason for which is the same as explained in the quasi-1D model. Due to the limited number of scatterers in the 2D model, the Bragg bandgap can only be observed in the area near the filter. When the observation point is far from the filter, there is almost no attenuation, which is particularly evident on the y-axis, as shown in Fig. 4.12a.

The comparison of TL between the above quasi-1D model and 2D model is shown in Fig. 4.13, where the TL at $x = 1$ cm and $y = 0$ cm is chosen as the representative of the 2D model. It is noted that they have almost the same trend, that is, the TL in the audio frequency is relatively small, while the TL in the ultrasonic frequency near the center of the bandgap is larger. In most frequencies, the TL of the 2D model is smaller than that of the quasi-1D model. The maximum difference between the two models is about 2 dB in the audio frequency range and about 10 dB in the ultrasonic range. Therefore, the TL in both the quasi-1D model and the 2D model can meet the measurement requirements of the parametric array loudspeaker, and can reduce or even eliminate the spurious sound in the receiving system.

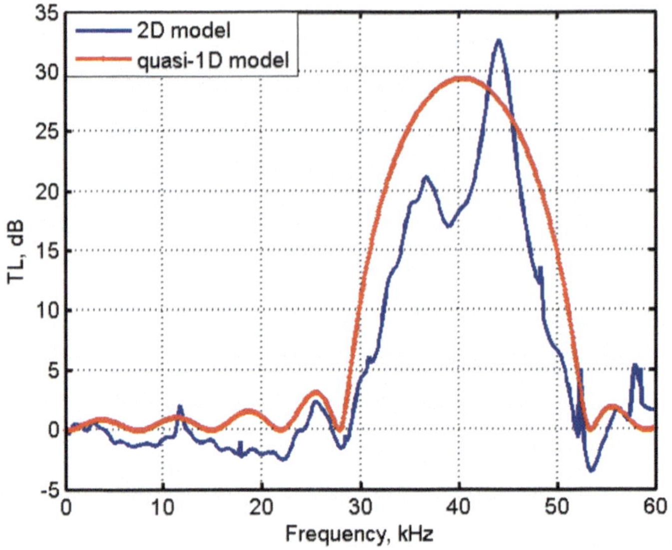

Fig. 4.13 Comparison of TL of quasi-one-dimensional and two-dimensional models of filters based on phononic crystals

4.3.2 *Experimental Verification*

According to the above simulation results, as shown in Fig. 4.14, a prototyped PC acoustic filter consisting of 5 × 7 cylinders was used, where $a = 4$ mm, $d = 2.4$ mm. The material of the cylinder is aluminum alloy, the same as in the simulation. The total size of this filter is 29.2 mm × 30.2 mm × 20 mm, which is relatively small and can be easily installed in front of the microphone.

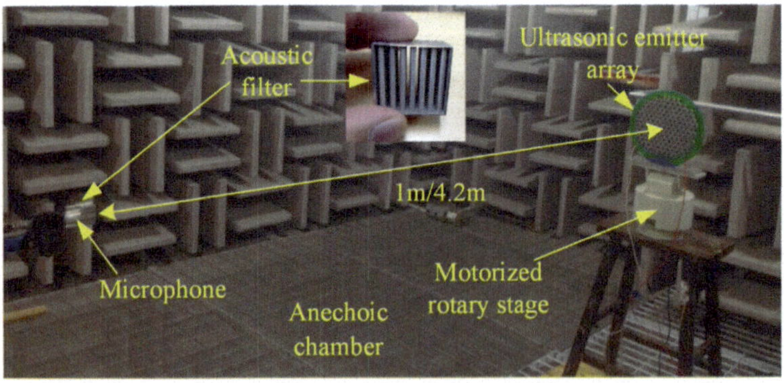

Fig. 4.14 Experimental setup

To test the performance of this filter, preliminary experiments were conducted in the anechoic chamber of Institute of Acoustics, Chinese Academy of Sciences. The temperature was 29.6 °C, and the relative humidity was 64%. A hexagonal ultrasonic source array was used in the experiment, which consisted of 147 commercial ceramic transducers (ZT40-16 type, Nicera Sensor Co., Ltd., Shanghai, China). The pump wave level at 40 kHz measured by the GRAS 46BE 1/4-inch microphone 1 m away from the ultrasonic source array was about 136 dB (reference value 20 μPa).

A B&K 4189 type pre-polarized free-field 1/2-inch microphone (Denmark, Brüel & Kjær) was used to measure the audio frequency and pump wave at two positions, namely 1 and 4.2 m, the latter being the maximum observation distance, which is limited by the size of the anechoic chamber. B&K 4189 is connected to a B&K analyzer platform with steady-state response analysis function (Brüel & Kjær in Denmark). The whole process mainly includes three steps, each step includes two sub-steps, with and without the installation of a filter in front of B&K 4189. The first and second steps are to measure the TL of the filter in the audio frequency range and ultrasonic range, respectively, and the third step is to study its performance in the parametric array loudspeaker. A 4-inch loudspeaker (HiVi H4) is used as the sound source to measure the TL in the audio frequency range, where the test frequency range is 0.1 to 19.9 kHz, with an interval of 0.1 kHz. By calculating the difference between the frequency responses of the signal source measured by B&K 4189 with and without the acoustic filter, the TL in the audio frequency range can be obtained. The same process is used to measure the TL in the ultrasonic range, where the aforementioned ultrasonic sound source array is used as the source, and the test frequency range is 20–60 kHz, with an interval of 0.1 kHz. A Class D power amplifier with extremely low phase shift and output noise can amplify up to 100 kHz, which is used to drive the ultrasonic sound source array. It should be noted that the nominal sensitivity of B&K 4189 is only applicable to the sound frequency range. If the same sensitivity is applied to the ultrasonic range, it will become inaccurate. Fortunately, TL is a relative value and will not be affected by the inaccuracy of sensitivity in the ultrasonic range. That is to say, B&K 4189 can be used to obtain the TL in the audio frequency range and ultrasonic range. This section also studies the characteristics of using an acoustic filter in a parametric array loudspeaker, and its experimental setup is illustrated in Fig. 4.14, where the test frequency range is 0.5 to 15 kHz, with an interval of 0.1 kHz. In this study, for simplicity, a lower sideband modulation (LSB-AM) with a carrier scheme was used in the parametric array loudspeaker, which was implemented in ADI's ADSP-21469.

The smooth TL of the filter at these two axial positions is shown in Fig. 4.15, where the Bragg bandgap can be easily observed. Their TL trends are almost the same, i.e., almost the same in the audio frequency range, and slightly different in the ultrasonic range. High attenuation (>10 dB) can be observed between 31 and 51 kHz, with the maximum sound pressure attenuation level being 27 dB. The TL in the audio frequency range fluctuates slightly. For these two positions, the attenuation of the entire sound frequency is less than 2 dB, which is better than any known acoustic filter used in the parametric array loudspeaker obtained from literature retrieval. For comparison, Fig. 4.15 also presents the 2D simulation results for the filter positioned

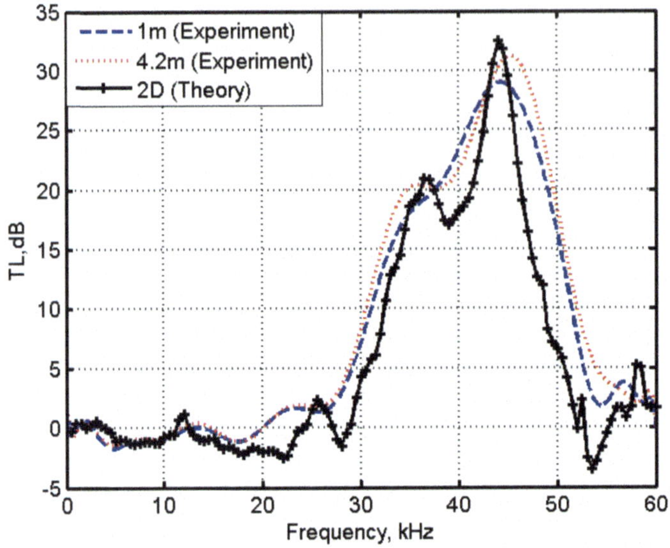

Fig. 4.15 Comparison of TL of filters in two different axial positions between simulation results and experimental results

at $x = 1$ cm and $y = 0$ cm. A good consistency can be found between the experimental results and the 2D simulation results, especially in the case of audio frequency range. Although generally consistent, small differences can still be observed and may be attributed to the following factors. First, the length of the filter is not sufficient to consider it as a 2D problem, and the theoretical study involves a 2D problem. However, the experimental results can only be considered as representing a quasi-2D problem. The finite length of the scatterer will impact the performance of the given filter, and their edges may function as sound radiators [428]. Secondly, the incident wave is not strictly a plane wave, as it will also contain an oblique component. Finally, defects in the experimental system may also contribute to some differences.

The difference signal of the parametric array loudspeaker measured at these two axial positions with or without the filter is illustrated in Fig. 4.16. To verify the results, comparative measurements were also made on the acoustic filters in the previous section. It should be noted that when using the filter, the frequency response of the following experimental audio results of the two acoustic filters has been compensated, which is called the system calibration step.

From the above analysis, it can be observed that the measured signal without any acoustic filter is a mixed signal, which includes both the difference sound from the parametric array effect and the spurious sound from the nonlinearity of the receiving system. From Fig. 4.16a, it is evident that the spurious sound is very serious at 1 m, that is, the nonlinearity of the receiving system determines the measurement results near the sound source. Note that for higher frequencies, the spurious sound becomes smaller, which is mainly due to the high quality factor characteristics of

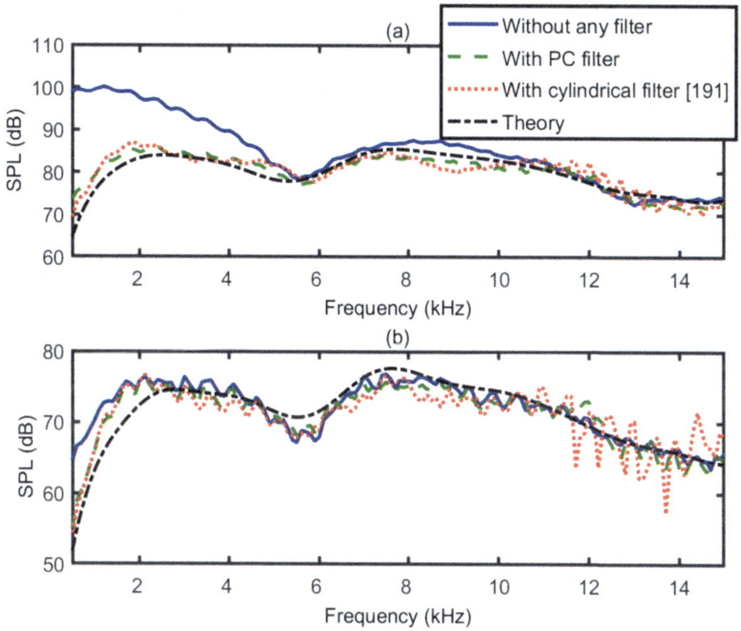

Fig. 4.16 Measurement of difference frequency signal of parametric array loudspeaker at **a** 1 m and **b** 4.2 m (Legend in (**a**) applies to all figures)

the ultrasonic source array. The frequency response of the ultrasonic source array is measured by the GRAS 46BE 1/4-inch microphone at 1 m from the sound source, as shown in Fig. 4.17. The response peak is at about 40 kHz, with a -6 dB bandwidth of only 4 kHz. Since the bandpass filter response of the ultrasonic source array will reduce the level of the sideband frequency (i.e., the lower pump wave), it will partially alleviate the nonlinearity of the microphone at higher frequencies, thereby reducing the spurious sound. Therefore, before using the filter at 1 m, the frequency response of the difference signal can be roughly divided into two areas. When <5 kHz, the spurious sound will dominate the effect of the parametric array. Outside this area, the spurious sound can be almost ignored, and the frequency response with or without the filter is almost the same. In this case, the measurement does not need a filter. As shown in Fig. 4.16b, similar results can be observed at 4.2 m, but the area where the spurious sound dominates is less than 2 kHz, much less than the distance of 1 m, and the reason is explained as follows. It is well known that spurious sound is proportional to the product of the pump wave. Due to absorption, diffraction and nonlinear effects on ultrasound, which can be accurately described by the KZK nonlinear parabolic equation [2], the pump wave will decrease significantly with the increase of propagation distance, and so will the spurious sound. It has been proven that due to the end-fire effect, the difference signal of the parametric array loudspeaker with circular and rectangular apertures first increases with the increase of distance, reaches a peak, and then gradually decreases [98, 275]. That means the

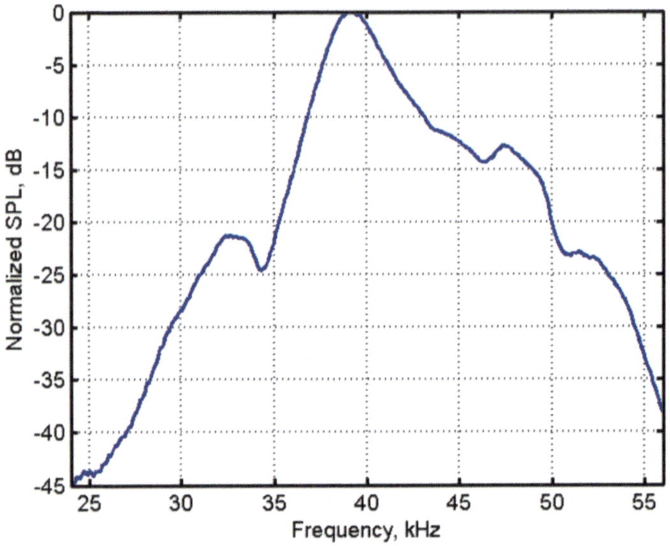

Fig. 4.17 Normalized frequency response of the ultrasonic sound source array measured at 1 m

difference-frequency signal has a slower change than the spurious sound with the propagation distance. It can be predicted the spurious sound will have a negligible effect on the mixed signal when the distance is larger, and then there is no need to adopt any acoustic filter at a larger distance [112].

To further verify the results, numerical simulations based on the Gaussian beam expansion method were carried out for the parametric array loudspeaker, considering the impact of the frequency response of the ultrasonic source array. The simulation results at 1 and 4.2 m are plotted in Fig. 4.16. In the case of weak nonlinearity, the Gaussian beam expansion method has been used to calculate the sound field of the sound beam, which can greatly reduce the calculation time, especially for the calculation of the second-order sound field. In this study, the source is roughly modeled as a circular plane piston with an effective radius of 0.08 m. For these two distances, the experimental curves using the filter almost match well with the simulation results. Apart from experimental measurement errors, the main reason for the difference between the simulation results and the experimental results may be the uneven velocity distribution on the source surface in the experiment. The results of the given filter are very consistent with the results of the filter in Sect. 4.2, and at some higher frequencies, this filter can even outperform other filters. This means that the given PC-based filter has better performance, with extremely low attenuation throughout the audio range, acceptable attenuation in the ultrasonic range, and controllable parameters to achieve the desired characteristics.

Figure 4.18 shows the normalized frequency response of the parametric array loudspeaker measured at two distances with and without the acoustic filter, where the effect of the ultrasonic source array has been eliminated. As mentioned above, the difference frequency sound produced by the parametric array loudspeaker is proportional to f_d^n where $1 \leq n \leq 2$ [112, 191]. The experimental results show that $n \approx 1.6$, which was a good approximation of the difference-frequency dependence. Figure 4.18 also plots the normalized numerical results based on the Gaussian beam expansion method at 1 and 4.2 m, which almost fit the curve of $n \approx 1.6$,, the difference being slightly at lower frequencies (<4 kHz). Therefore, from another perspective, f_d^n can be used to roughly predict the frequency response of the difference frequency signal from the parametric array loudspeaker.

For parametric array loudspeakers, it is crucial to accurately measure their directivity without any spurious sound. Therefore, it is also necessary to determine the performance of the given filter during directivity measurements. Since the directivity of the pump wave is strong, and there is a relationship between the spurious sound and the pump wave, this section will only study the characteristics of the filter at small angles ($\leq 15°$). For larger angles, due to the relatively small pump wave, there will be no spurious sound. The ultrasonic source array is installed on a computer-controlled electric rotating platform with an accuracy of 0.0125°. The experimental steps at each angle are similar to the above experiment with a rotation angle of zero. The B&K 4189 is placed at 2 m from the source. Figure 4.19 shows the smooth TL with six rotation angles (i.e., 0°, 2.5°, 5°, 7.5°, 10°, and 15°). It can be observed that all TLs at these angles have almost the same trend, and the bandgap phenomenon can be easily observed.

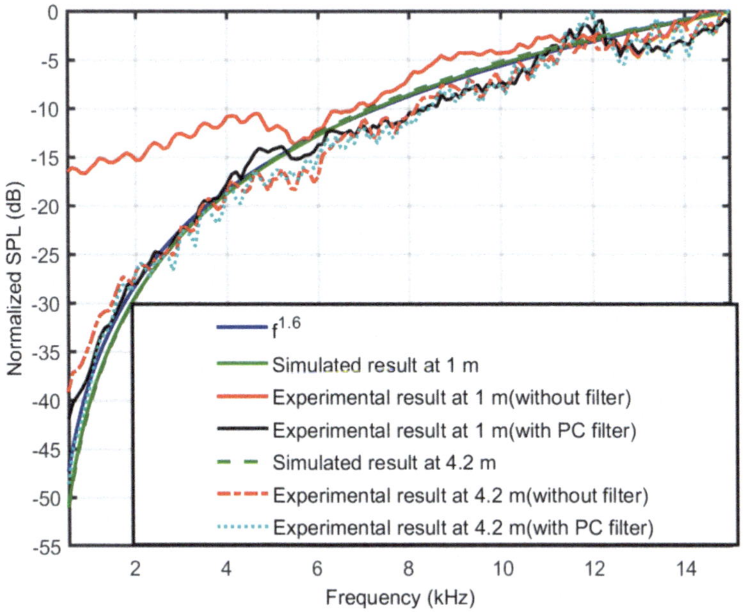

Fig. 4.18 Normalized frequency response of the difference frequency signal of the parametric array loudspeaker at 1 and 4.2 m

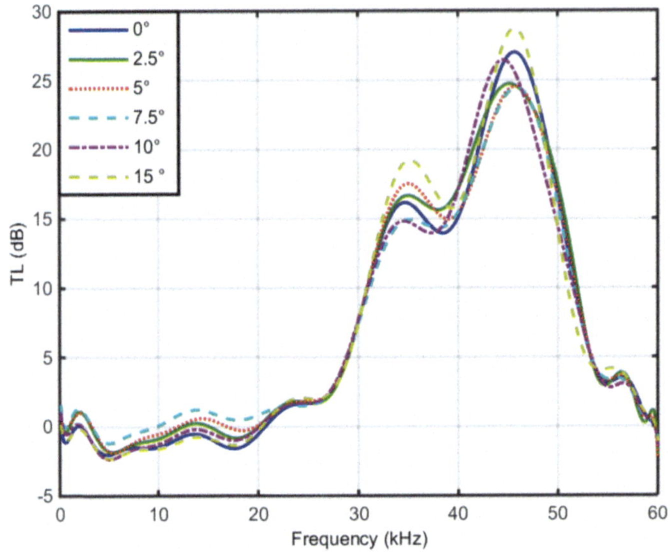

Fig. 4.19 Measured TLs of the filter at different angles

To evaluate the performance of the given filter in the directivity measurement of the parametric array loudspeaker, another experiment was conducted. Figure 4.20 shows the frequency response of the difference signal for the above six different angles. It is easy to observe that as the rotation angle increases, the difference between the mixed signal and the difference signal becomes smaller, and when the rotation angle is greater than 7.5°, there is almost no difference between the signals measured with and without the acoustic filter. This means that at such a large angle, due to the decrease in the pump wave signal, there is almost no expected spurious sound. This is because the product of the pump waves becomes smaller, and in this case, there is no need to use any acoustic filter. In this study of the ultrasonic source array, the acoustic filter is still effective in directivity measurements, and when the rotation angle is greater than 7.5°, there is no need for an acoustic filter.

From the above analysis and results, three main advantages of this filter can be clearly observed and summarized as follows. The first one is that its performance can be roughly predicted by the theory. The parameters of this structure can theoretically be predicted and adjusted according to the audio frequency and the required characteristics of the pump wave, which means that filters with given characteristics can be easily customized. The second one is its small size and easy prototype design and installation in front of the microphone. The characteristics of the given filter have been proven to be sufficient to measure parametric array loudspeakers of such a small size. The last is that a smaller attenuation for the interested audible frequency. There is even no need to do the system calibration step for some scenarios if the audible frequency response of the filter cannot be obtained without causing severe mistake. For example, the measured frequency responses of the parametric

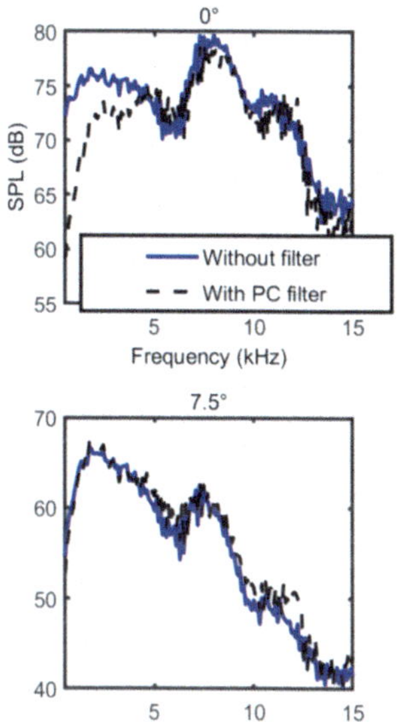

Fig. 4.20 Difference frequency sound response of the parametric array loudspeaker measured at different angles

array loudspeaker at 1 m with/without system calibration step, i.e., equalization (EQ) of above two filters are demonstrated in Fig. 4.21. It is easy to observe that the two results (with/without equalization) of the acoustic filter based on the phononic crystal almost match the same theoretical results. This is not the case with the cylindrical ultrasonic filter. Although after the system calibration step, the results of the cylindrical ultrasonic filter are almost the same as those of the filter based on the phononic crystal, it is difficult to tell its trend when the system calibration step is not adopted in the frequency region higher than 4 kHz.

4.3.3 *Case Study*

The preliminary experiment was conducted in the anechoic chamber of Institute of Acoustics, Chinese Academy of Sciences, to study the performance of different sound source apertures and modulation schemes after using an acoustic filter based on phononic crystals, thereby proving its effectiveness. The temperature was 26 °C, and the relative humidity was 69%. The experimental setup is shown in Fig. 4.14. The test frequency range was 0.5–9 kHz, with an interval of 0.1 kHz. Figure 4.14 shows the prototype acoustic filter. In the experiment, three different apertures were measured at 2 *m*: a circular aperture with an effective radius of 0.08 m (A1), a circular aperture with an effective radius of 0.04 m (A2) as shown in Fig. 4.22, and a rectangular aperture of approximately 0.26 m × 0.14 m (A3). These three arrays

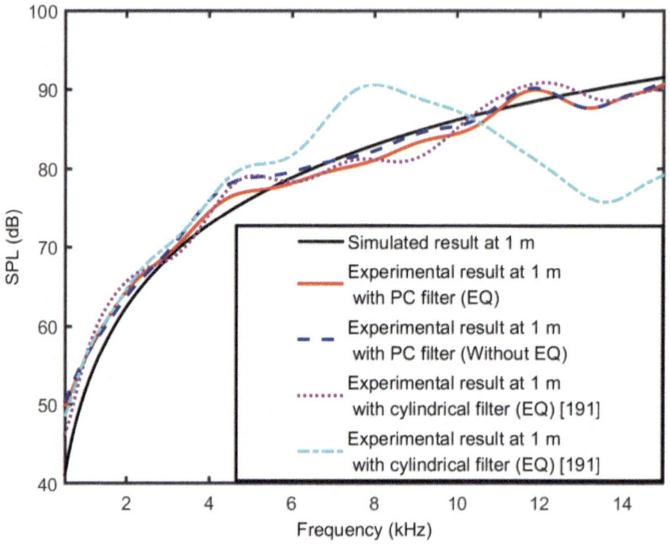

Fig. 4.21 Frequency response of the parametric array loudspeaker with/without filter at 1 m for two types of filters

were composed of 147, 60, and 187 commercial ceramic transducers, respectively. The type of ultrasonic sound source used for A1 and A3 was ZT40-16 from Nicera Sensor Co., Ltd., Shanghai, China, and the model of the ultrasonic sound source used for A2 was MA40S4S from Murata Manufacturing Co., Ltd., Japan. Figures 4.23, 4.24 and 4.25 show the measured difference frequency sound of parametric array loudspeakers with A1–A3. To further verify the results, numerical simulations of the Gaussian beam expansion method based on parametric array loudspeakers were also conducted. The demodulated sound produced by the parametric array loudspeaker matches well with the simulation results, proving the effectiveness of the acoustic filter. Note that without any acoustic filter, it is impossible to observe the parametric array effect at lower frequencies. At this distance, the results of the rectangular aperture are almost similar to those of the circular aperture, which is consistent with the findings in reference [98]. The frequency response of these three ultrasonic sound source arrays is shown in Fig. 4.26 to help explain the differences in the demodulated difference frequency sound. It is observed that the audible sound is almost the same for each aperture in the lower frequency with both two sidebands, but it is larger with upper sideband than that with lower sideband for these three apertures in the higher frequency. Croft and Norris has suggested that LSB has several advantages over USB considering the saturation levels and attenuation level [131]. In the current situation, the level of the pump wave is far less than the saturation level, and the difference in attenuation level between LSB-AM and USB-AM can also be compensated by the frequency response of the ultrasonic sound source. For example, when the difference frequency sound is 5 kHz for A1, the amplitude difference between the pump waves

Fig. 4.22 The 3 sound sources used

at 45 and 35 kHz in Fig. 4.26 is about 8.6 dB. The sound absorption between these two pump waves is about −1 dB. In this case, the sum of the pump waves of USB-AM is about 7.6 dB larger than that of LSB-AM, which is almost the same as the difference (7dB) between USB-AM and LSB-AM at 5 kHz in Fig. 4.23. Since the amplitude of the difference frequency component is almost proportional to the product of the amplitudes of the pump waves [2], the 5 kHz difference frequency sound obtained through USB-AM is larger than that obtained through LSB-AM with A1. Therefore, USB-AM was chosen as the modulation scheme in this study.

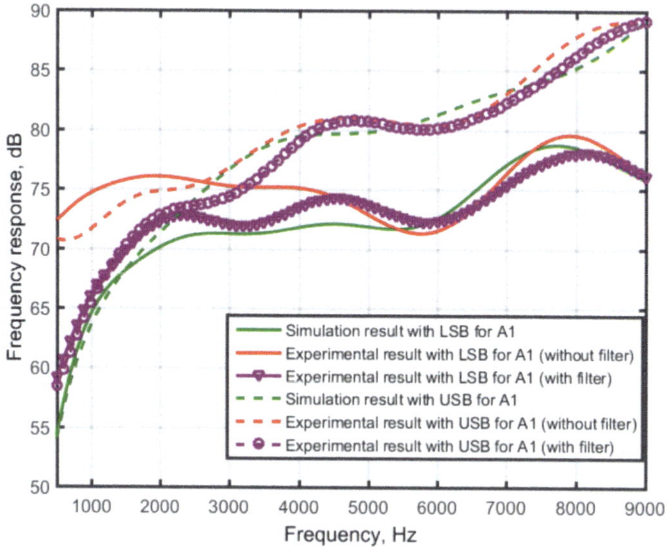

Fig. 4.23 Difference frequency sound of the parametric array loudspeaker when using A1

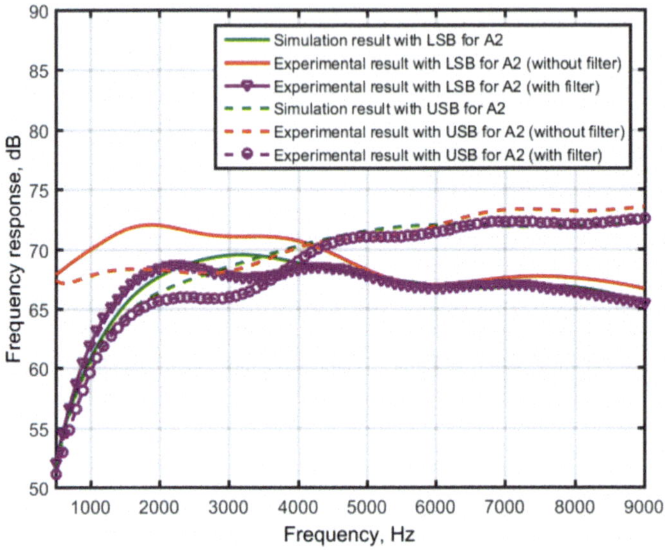

Fig. 4.24 Difference frequency sound of the parametric array loudspeaker when using A2

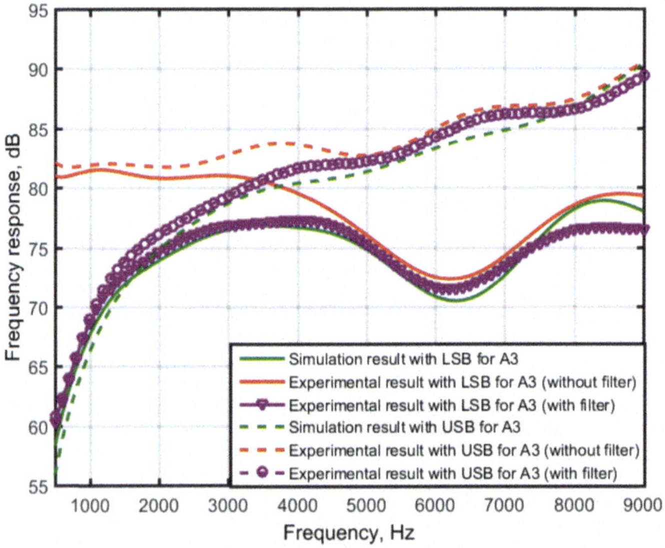

Fig. 4.25 Difference frequency sound of the parametric array loudspeaker when using A3

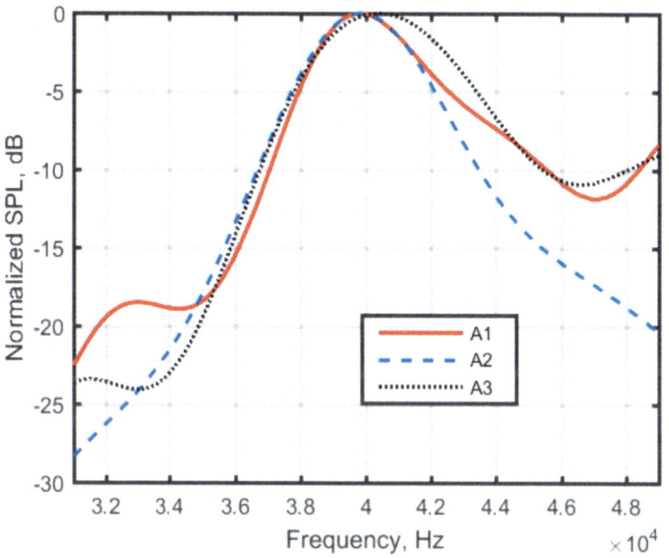

Fig. 4.26 Normalized frequency amplitude response of the three transducer arrays

The normalized frequency response of the three transducer arrays is shown in Figs. 4.27, 4.28 and 4.29, where the effect of the ultrasonic sound source array is eliminated. The demodulated difference frequency sound produced by the parametric array loudspeaker is proportional to f_d^n, where $1 \leq n \leq 2$ [112, 192]. In Figs. 4.27, 4.28 and 4.29, the experimental results roughly show that each n in the three arrays is 1.6, 1.55, and 1.65, respectively. In the current situation, the absorption lengths of these three arrays are the same because they have the same carrier frequency [2], n is determined by the Rayleigh distance. The n obtained from the experimental results is consistent with the trend of the Rayleigh distance of these three arrays. It should also be noted that for these three arrays, USB-AM and LSB-AM almost have the normalized frequency response of the audio frequency, for the same reason as mentioned above. Therefore, by using f_d^n, the frequency response of the demodulated difference sound from the parametric array loudspeaker can be roughly predicted.

4.4 Experimental Study on the Influence of Parameters on Spurious Sound in Parametric Array Loudspeakers

The signal measured at the observation point is a mixed signal, including the demodulated difference sound p_{ds} and spurious sound p_{ss}. Due to their unknown phase relationship, it is difficult to separate the demodulated difference sound from the mixed signal. In most measurements of parametric array loudspeakers, an acoustic filter with the following three characteristics can be used to eliminate spurious sound:

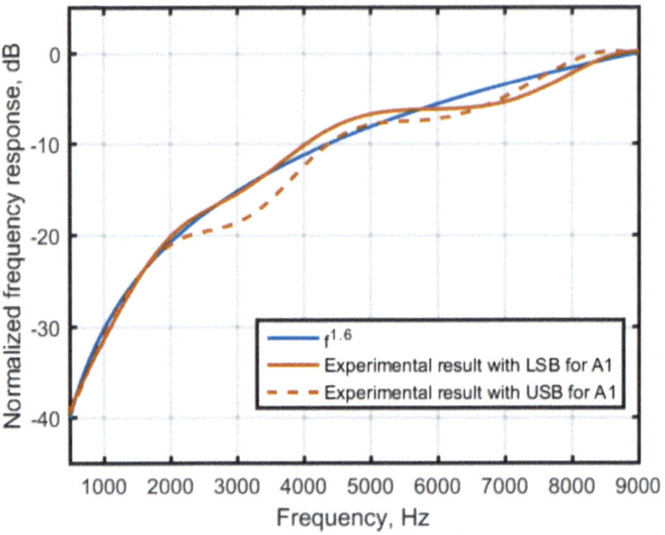

Fig. 4.27 Normalized frequency response of the difference frequency sound of the parametric array loudspeaker when using A1

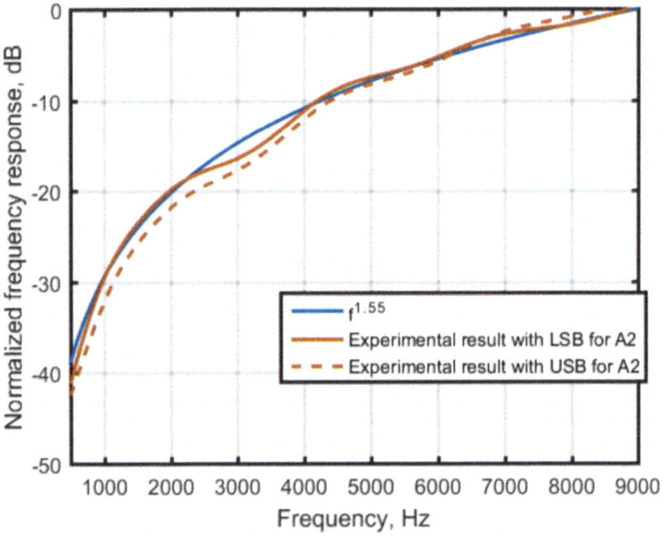

Fig. 4.28 Normalized frequency response of the difference frequency sound of the parametric array loudspeaker when using A2

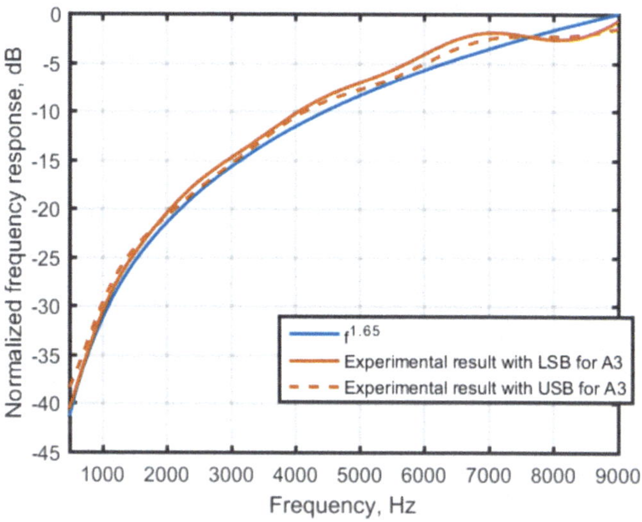

Fig. 4.29 Normalized frequency response of the difference frequency sound of the parametric array loudspeaker when using A3

attenuation below the carrier frequency exceeds 20 dB, a bandwidth around 20 kHz with about 10 dB attenuation, and a small attenuation on the difference-frequency sound (<2 dB). Several methods using different types of acoustic filters have been proposed in the measurements of the parametric array [61, 93, 95, 99, 112, 172, 191, 421]. However, these acoustic filters cannot meet all the above three conditions. Two other methods without any acoustic filters also have their own limitations [190, 423]. Currently, an acoustic filter using phononic crystals can meet the above requirements of the acoustic filter [192], and its effectiveness has also been proven by comparing the performance of parametric array loudspeakers with different sizes and modulation schemes [193] In this part of the research, with the help of the acoustic filter, a more in-depth study was conducted on the spurious sound in the measurement of parametric array loudspeakers.

The key interest in measuring parametric array loudspeakers is whether it is possible to largely eliminate or mitigate the impact of spurious sound, which is an serious issue in the near field. According to related research [112], spurious sound p_{ss} can be calculated through $Kp_{1m}p_{2m}$, where K is determined by curve fitting based on the measurement results, p_{1m} and p_{2m} are the amplitudes of the pump waves at the microphone position. It should be noted that only when $p_{ss} << p_{ds}$, the mixed signal can be assumed to be the demodulated difference frequency sound, and the spurious sound can be ignored. Therefore, when measuring parametric array loudspeakers, considering only the amplitude of the spurious sound is incomplete, p_{ds} must be considered. To determine when the spurious sound can be ignored, it is important to use the ratio R between the spurious sound p_{ss} and the demodulated difference frequency sound p_{ds}. For simplicity, a logarithmic scale is used, i.e.,

$20\log_{10}R = 20\log_{10}p_{ss} - 20\log_{10}p_{ds}$ (unit: dB). This means that the ratio R can be easily converted into the difference in sound pressure level (unit: dB, reference value: 2×10^{-5} Pa) between the spurious sound and the demodulated difference frequency sound. There are two advantages to using the ratio of the above sound pressure level differences. One is to determine the threshold to decide whether the spurious sound can be ignored, and the other is to eliminate the effect of K when comparing the above sound pressure level differences for different radii or different frequencies (assuming the same K is used at different frequencies). It is well known that p_{ss} and p_{ds} are mainly related to the amplitude of the pump wave at the source, the radius of the sound source, the frequency value, and the observation distance [2]. Therefore, when studying spurious sound, the above four parameters must be considered. This section thoroughly analyzes the impact of the above four parameters on spurious sound through experiments. To further verify the measurement results, numerical simulations were also carried out on the Gaussian beam expansion method based on the parametric array loudspeaker [101, 319].

As mentioned above, the amplitude of the spurious sound based on the Gaussian beam expansion method can be expressed as

$$
\begin{aligned}
p_{ss} &= Kp_{1m}p_{2m} = K|q_a(x,y,z)||q_b(x,y,z)| \\
&= Kp_{0a}p_{0b}\exp(-\alpha_a z - \alpha_b z)\left|\sum_{n=1}^{N_G} A_{n,xy}\frac{\exp\left(-\frac{B_{n,x}x^2/l_x^2}{1+iB_{n,x}z/z_{a,x}} - \frac{B_{n,y}y^2/l_y^2}{1+iB_{n,y}z/z_{a,y}}\right)}{\sqrt{1+iB_{n,x}z/z_{a,x}}\sqrt{1+iB_{n,y}z/z_{a,y}}}\right| \\
&\left|\sum_{n=1}^{N_G} A_{n,xy}\frac{\exp\left(-\frac{B_{n,x}x^2/l_x^2}{1+iB_{n,x}z/z_{b,x}} - \frac{B_{n,y}y^2/l_y^2}{1+iB_{n,y}z/z_{b,y}}\right)}{\sqrt{1+iB_{n,x}z/z_{b,x}}\sqrt{1+iB_{n,y}z/z_{b,y}}}\right|
\end{aligned}
\tag{4.3}
$$

Experiments were conducted in the anechoic chamber of Institute of Acoustics, Chinese Academy of Sciences, to test the influence of the aforementioned four parameters on the sound pressure level difference between spurious sound and the demodulated difference frequency sound. For simplicity, only one distance value was used when studying the influence of the amplitude of the pump wave, the radius of the sound source array, and the frequency value. The basic principle for determining this measurement distance is that it should not be within the fluctuation range of the sound source array, and the sound pressure level difference between the spurious sound and the demodulated difference frequency sound should be large. For a sound source array with a large radius, when the observation distance is small, it may still be in the fluctuation area, which means it cannot represent the relationship between the sound pressure level difference and the radius of the sound source or the frequency value of the demodulated difference frequency sound. For a sound source array with a small radius, theoretically, when the observation distance is large, the sound pressure level difference between the spurious sound and the demodulated difference frequency sound will be smaller. Therefore, due to the unknown phase relationship between these two sounds, it is difficult to separate the demodulated difference frequency sound from the spurious sound. In fact, when the distance is large, there is no need to consider spurious sound [112, 172]. After careful consideration and pilot testing, the measurement distance was chosen to be 3 *m*.

The experimental setup is shown in Fig. 4.30. The audio signal is generated by the B&K Pulse system (Brüel & Kjær, Denmark) and serves as the input to the Analog Devices ADSP-21469 signal processor, which performs single-sideband signal processing. For simplicity, a carrier scheme was used for the wideband modulation (LSB-AM) process [138]. The modulated signal is then amplified and fed into the ultrasonic sound source array. The B&K 4189 microphone (Brüel & Kjær, Denmark) and the GRAS 46BE 1/4-inch microphone (GRAS, Denmark) was used to measure the demodulated difference frequency sound and the pump wave at 3 m from the ultrasonic sound source array. The test frequency range is 0.5–10 kHz, with an interval of 0.1 kHz. In this study, the three sources are roughly modeled as circular plane pistons, with effective radii of 0.04 m (A1), 0.1 m (A2), and 0.08 m (A3), as shown in Fig. 4.30. These three sound source arrays are composed of 60, 429, and 147 commercial ceramic transducers, respectively. The type of ultrasonic sound source used for A1 and A2 is MA40S4S from Murata Manufacturing Co., Ltd., Japan, and the type used for A3 is ZT40-16 from Nicera Co., Ltd., Shanghai, China. A3 is only used to study the effect of observation distance. The measurement conditions are as follows: temperature is 13 °C, relative humidity is 48%. Before each experiment, the microphones were calibrated on-site using the B&K 4231 calibrator (Brüel & Kjær, Denmark).

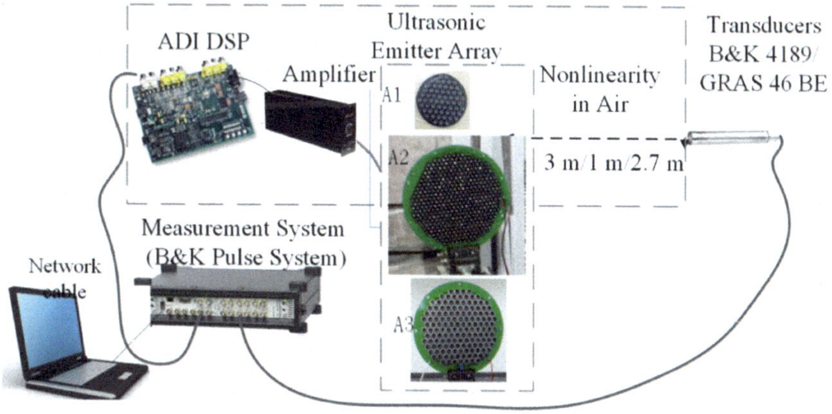

Fig. 4.30 Experimental setup

4.4.1 Effect of the Levels of the Primary Waves

It can be known from Eqs. (2.28) and (4.3) that the amplitudes of the spurious sound and the demodulated difference frequency sound are both proportional to the product of the pump wave at the sound source, i.e., it can be easily inferred that by reducing the amplitude of the pump wave at the sound source, the spurious sound can be reduced. However, this will also cause the demodulated difference frequency sound to decrease at the same rate. Therefore, when the amplitude of the pump wave at the source changes, the trend of the mixed signal at the measurement position will remain unchanged. This means that it is impossible to alleviate the nonlinear effects of the receiving system by reducing the amplitude of the pump wave at the source. There is currently no corresponding experimental verification in the literature. Therefore, to determine whether it is possible to experimentally reduce the spurious sound by reducing the amplitude of the pump wave, it is planned to conduct experiments by changing the amplitude of the audio signal drive level in the B&K Pulse software. This means that the amplitude of an pump wave will be changed according to the modulation scheme used in this study, while keeping the amplitude of the 40 kHz carrier unchanged, which will ultimately cause a change in the product of the pump waves.

Figure 4.31 shows the frequency response of a parametric array loudspeaker with and without an acoustic filter, measured at 3 m using an A2 sound source array, using four different drive levels, namely 1, 0.5, 0.25, and 0.125 V. The frequency response of the acoustic filter to the following experimental sound frequency results has been compensated. From the above analysis, it can be observed that the measured signal without any acoustic filter is a mixed signal, which includes both the demodulated difference sound from the parametric array effect and the spurious sound from the nonlinearity of the receiving system. From the related research [25] it can be determined that after using the acoustic filter based on phononic crystals, the spurious sound is removed, leaving only the demodulated difference sound. It is clear to observe from Fig. 4.31 that all the frequency responses are almost proportional to each other for both with/without the filter. For each drive level without any filter, the average sound pressure levels of the frequency response are 63.2 dB, 57.2 dB, 51.2 dB, and 45.2 dB, respectively, while the sound pressure levels of the frequency response with a filter are 59.5 dB, 53.5 dB, 47.5 dB, and 41.4 dB respectively. For the two sets of measurement data with and without filters, there is a decrease of about 6 dB between any two adjacent levels, which is equal to the ratio between the input drive voltages, meaning that the difference between the frequency responses of any two demodulated difference sounds and mixed signals is almost equal to the ratio of input levels. According to the relationship between the mixed signal, the demodulated difference sound, and the spurious sound, it can be concluded that the spurious sound is also proportional to the input level. In short, if the pump wave amplitude at the source is reduced, the demodulated difference sound and spurious sound will decrease at the same rate. When the pump wave amplitude at the source changes, the overall trend of the mixed signal remains unchanged. The simulation results are also shown in Fig. 4.31, where the sound source is assumed to be a flat circular piston with a radius of 0.1 m, and 10 Gaussian coefficients are used in reference [30]. From Fig. 8 in reference [138], it can be known that when P_0 is small, the slope gain is close to 12 dB/octave, and for higher P_0, the slope gain gradually moves to 6 dB/octave, where P_0 is the initial sound pressure level of the carrier. When the carrier $P_0 \leq 120$dB, there is a slight shift towards 6 dB/octave. As mentioned above, in practical applications, due to safety issues, the amplitude of the pump wave is controlled [125]. For these four drive voltages, using four P_0 acoustic filters, there is good consistency between the simulation results and the measured frequency response: 120.4 dB, 117.4 dB, 114.4 dB, and 111.4 dB respectively. Figure 4.32 shows the normalized frequency response of the parametric array loudspeaker with and without an acoustic filter, and the simulation results are also plotted. For clarity, only the response between 500 and 5000 Hz is given. Only one normalized frequency response of the simulation result is given in Fig. 4.32, since it is independent of the pump wave amplitude. The simulation results of the acoustic filter are in good agreement with the experimental results. As can be seen from Fig. 4.32, for different amplitudes of the pump wave with and without an acoustic filter, the changes in the normalized frequency response of the parametric array loudspeaker are small. It means that the shift towards 6 dB/octave with P_0 is smaller than the experimental error. From another perspective, when P_0 is small, it is reasonable to ignore the change of P_0 and obtain the analytical solution of the parametric array loudspeaker based on the Gaussian beam expansion method. Therefore, it is impossible to eliminate or even mitigate the impact of spurious sound by reducing the amplitude of the pump wave at the sound source.

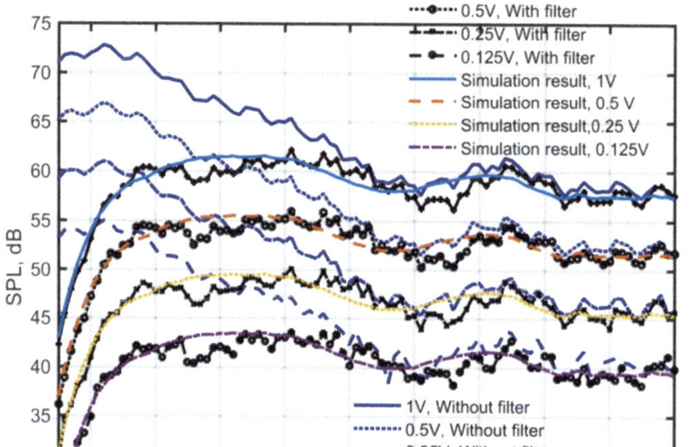

Fig. 4.31 Frequency response of the parametric array loudspeaker measured at 3 m with different inputs using A2

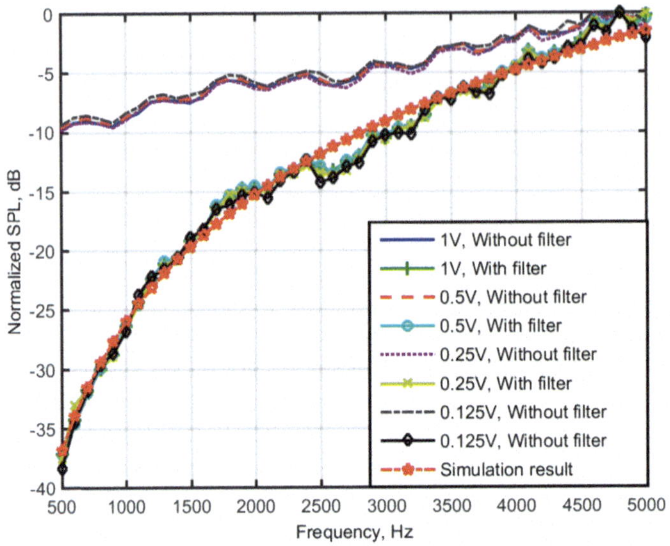

Fig. 4.32 Normalized measurement frequency response of the parametric array loudspeaker at 3 m with different inputs using A2

4.4.2 Effect of the Radius of the Emitter Array

The frequency responses of the parametric array loudspeaker measured at 3 m for both A1 and A2 are illustrated in Fig. 4.33 with/with the acoustic filter. To further verify the experimental results, a numerical simulation of the parametric array loudspeaker was performed based on the Gaussian beam expansion method. Considering the influence of the frequency response of each ultrasonic sound source array, and the results of the two sound source arrays are plotted in Fig. 4.33. It can be observed that the experimental curves of the filter and the simulation results of the two sound source arrays all show good consistency. The uneven velocity distribution on the source surface in the ultrasonic sound source array and the individual frequency response of each piezoelectric transducer (PZT) [429, 430] are the main reasons for the measurement data error and the difference between the simulation data and the experimental data.

For clarity, a frequency value of 0.5 kHz was chosen here to illustrate the influence of the radius. At this frequency value, the demodulated difference frequency sound p_{ds} is much smaller than the mixed signal, which means that p_{ds} can be ignored, and the mixed signal is mainly spurious sound p_{ds}. Therefore, for these two arrays, p_{ds} and p_{ss} at 0.5 kHz can be easily extracted from Fig. 4.33, and listed them in the Table 4.1. The simulation and experimental results prove that as the radius of the array increases as expected, the sound pressure level difference between the spurious sound and the demodulated difference frequency sound increases. The radii of the arrays are 0.04 and 0.1 m, and the calculated K values are 1.29×10^{-4} Pa^{-1} and 1.56×10^{-4} Pa^{-1}.

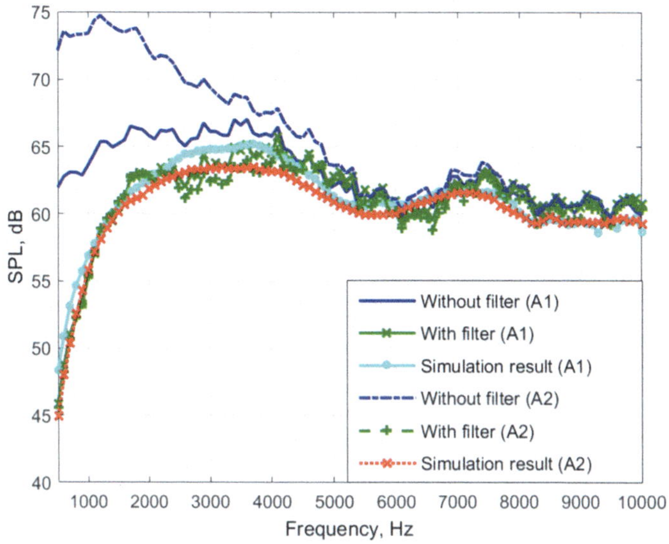

Fig. 4.33 Frequency response of the parametric array loudspeakers A1 and A2 at 3 m

The difference between these two K values is less than 2 dB, which is mainly caused by measurement errors. Assuming the K in the simulation is 1.4×10^{-4} Pa^{-1}, which is the average of the two arrays mentioned above. In the simulation, $p_{1m}p_{2m}$ value is the same as in the experiment. There is some discrepancy in the value of $20\log_{10}R$, mainly due to different K values. From Table 4.1, it can be clearly seen that there is a good match between the simulation and experimental results. The differences of $20\log_{10}R$ between these two radii by simulation and by experiment are 11.9 dB and 10.6 dB, respectively. The experimental value is 1.3 dB lower than the simulation result, which is also caused by experimental error. The sound pressure level differences between the spurious sound and the demodulated difference frequency sound in the observation region for both simulation and experimental results are proportional to $a^{1.5}$ and $a^{1.35}$ respectively.

Table 4.1 Comparison of simulation and experimental results

Radius of the sound source, m	0.04	0.1
p_{ds} (Theoretical), dB	48.3	44.9
p_{ss} (Theoretical), dB	62.7	71.3
$20\log_{10}R$ (Theoretical), dB	14.4	26.3
p_{ds} (Experimental), dB	47	46.6
p_{ss} (Experimental), dB	62	72.2
$p_{1m}p_{2m}$ (Experimental), dB	230.8	239.3
$20\log_{10}R$ (Experimental), dB	15	25.6

To further study the relationship between K and frequency values, Ks from 0.5 to 3 kHz calculated by $p_{ss}/(p_{1m}p_{2m})$ are obtained from the measurements for both these two emitter arrays, as illustrated in the left side of Fig. 4.34. The demodulated difference frequency sound is also plotted on the right side of Fig. 4.34 for a rough calculation of the threshold. Frequencies above 3 kHz are not considered here because the difference between the mixed signal and the demodulated difference frequency sound in this frequency range is very small, especially for the sound source array with a radius of 0.04 m. In this case, due to the unknown phase relationship, the spurious sound cannot be determined.

When the mixed signal is much larger than the demodulated difference frequency sound, the spurious sound dominates, and the demodulated difference frequency sound can be ignored. This is correct, especially in the near field with lower frequency values [172]. In this case, the mixed signal can be roughly regarded as spurious sound. Through this approximation, K can be obtained. It is well known that for different sound sources with different radii, K should be the same at the same frequency value. When, $20\log_{10}(p_{ss}/p_{ds}) = Y_1 - Y_2 \geq M$, that is, when $Y_1 \geq M + Y_2$, Y_2 can be ignored, where $Y_1 = 20\log_{10}p_{ss}$, $Y_2 = 20\log_{10}p_{ds}$, and M is the threshold. This means that the difference frequency sound of demodulation can be ignored in the mixed signal. Due to their symmetry and their similar contribution to the mixed signal, when $20\log_{10}(p_{ds}/p_{ss}) = Y_2 - Y_1 \geq M$, that is, $Y_1 \leq Y_2 - M$, the spurious sound can be ignored. In reference [112], the spurious sound can be ignored at 3 m, where the sound pressure levels of the demodulated difference frequency sound and the spurious sound are about 50 dB and 35 dB, respectively. In this case, the difference in sound pressure level between the spurious sound and the demodulated difference

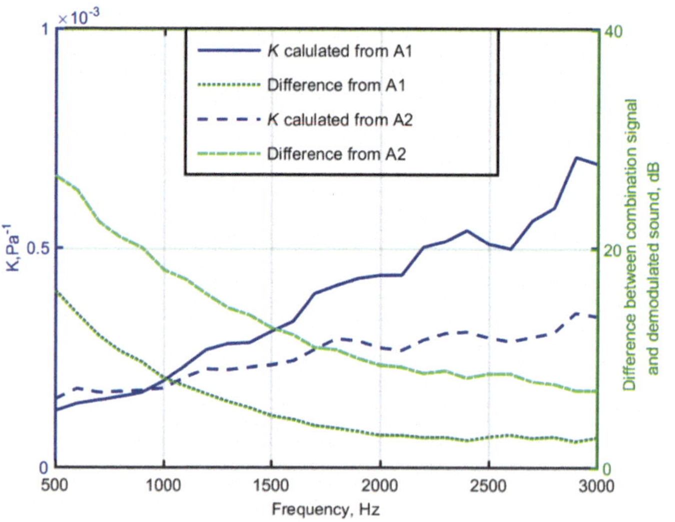

Fig. 4.34 Relationship between K and frequency values (left), difference between mixed signal and demodulated difference sound (right)

frequency sound is about -15 dB, where $K = 2 \times 10^{-4}$ Pa^{-1}. Due to their symmetry, when the above sound pressure level difference is 15 dB, the demodulated difference frequency sound can be ignored. For two different sound source arrays, at frequencies below 1.1 kHz, K is almost the same, and after this region, the difference in K becomes larger. When the frequency is 1.1 kHz, K is calculated to be around 2×10^{-4} Pa^{-1}. For a sound source array with a radius of 0.1 m, the difference between the mixed signal and the demodulated difference sound is about 17 dB, which is greater than the difference of the same K value in reference [112]. The difference of 0.04 m radius in Fig. 4.34 is about 7 dB at 1.1 kHz, where almost the same K is obtained as that for the radius of 0.1 m. This means that a threshold of about 7 dB is sufficient to ignore the demodulated difference sound in the study. For the entire frequency range when the radius is 0.1 m as shown in Fig. 4.34, the difference between the mixed signal and the demodulated difference sound is greater than about 7 dB. It signifies that Ks are valid for the radius of 0.1 m with the frequency value lower than 3 kHz. It is also noted that K increases gradually with the frequency value. However, it is noted that Ks larger than 1.1 kHz with the radius of 0.04 m need to be ignored in Fig. 4.34 since the above SPL difference is smaller than 7 dB, which means in that case the demodulated difference frequency sound cannot be directly separated from the mixed signal. In a word, in this study, when the SPL difference between the spurious sound and the demodulated difference frequency sound is smaller than −7 dB, the spurious sound can be ignored.

4.4.3 Effect of Frequency Values

The normalized frequency responses of parametric array loudspeaker measured both with/without our acoustic filter for the emitter array with A2 are illustrated in Fig. 4.35 by removing the effect of the ultrasonic emitter array as illustrated in the inset. The sound frequency produced by the parametric array loudspeaker is proportional to $f_d^{1.7}$ [112], which is a good approximation of the demodulated difference frequency dependence. The normalized numerical result based on the Gaussian-beam expansion technique is also plotted in Fig. 4.35 and almost fits with the curve of $f_d^{1.7}$, which proves the validity of the experimental data.

Based on the above analysis, when the spurious sound is about 7 dB higher than the demodulated difference sound, the demodulated difference sound can be ignored and the mixed signal can be regarded as the spurious sound. Therefore, in this case, the frequency responses of 0.5, 1, 2, and 3 kHz are selected to study the relationship between the sound pressure level difference and the frequency value. Table 4.2 shows the performance under these four frequency values.

There is a substantial increase in the demodulated difference frequency sound with the frequency value, which is much larger than that for the spurious sound. From Table 4.2, it is noted that the SPL difference between the spurious sound and the demodulated difference-frequency sound becomes smaller as the frequency value increases as expected, which is consistent with the simulation results. It is expected

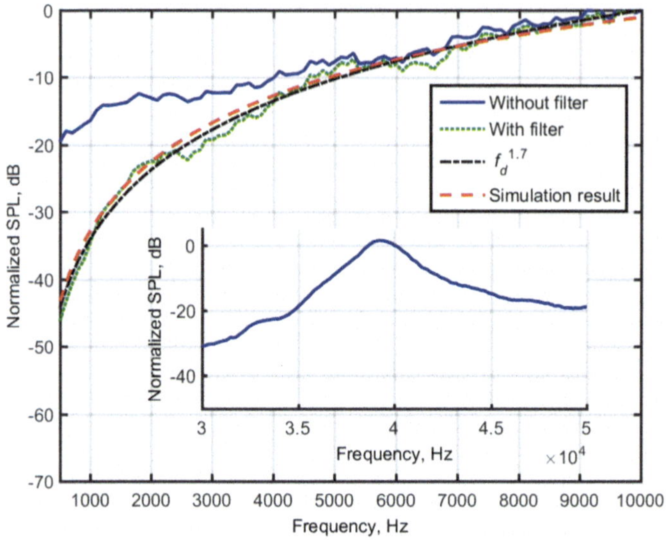

Fig. 4.35 Normalized frequency response of the parametric array loudspeaker at 3 m when using A2

Table 4.2 Comparison of performance at specified frequency values

f_d, kHz	0.5	1	2	3
Normalized p_{ss}	−19.6	−16.4	−13.1	−12.3
Normalized p_{ds}	−46.2	−34.4	−22.4	−19.3
$p_{ds} - p_{ds,\text{ at 0.5 kHz}}$	0	11.8	22.8	26.9
$20\log_{10}R$	26.6	18.0	9.4	7.0
$20\log_{10}R - 20\log_{10}R_{\text{at 0.5 kHz}}$, dB	0	−8.6	−17.3	−19.6
Normalized$(20\log_{10}R - 20\log_{10}R_{\text{at 0.5 kHz}})$, dB	0	−9.9	−22.1	−26.4
$20\log_{10}\left[f_d^n/(0.5\,kHz)^n\right]$ $(n\approx 1.7)$	0	10.2	20.5	26.5

that for a larger frequency value, the spurious sound can be ignored even at a relatively closer position to the source. The spurious sound increases with the frequency value, because K gradually increases with the frequency value from the study in Fig. 4.34. For example, the K difference between 0.5 and 1 kHz for a radius of 0.1 m in Fig. 4.34 is about 3.7 dB. If the K at 1 kHz is normalized with the K at 0.5 kHz, the normalized spurious sound at 1 kHz will be changed to about −20.1 dB, which is almost the same as the spurious sound at 0.5 kHz. Therefore, it can be observed that when all Ks at different frequencies in Fig. 4.34 are normalized with a frequency of 0.5 kHz, a good match can also be achieved between the simulation and experimental results, as shown in Table 4.2, where the effect of K has been removed. From Table 4.2, it is evident that the above sound pressure level difference fits the relationship with the frequency value as $f_d^{-1.7}$.

4.4.4 Effect of Observation Distance

To verify the effect of observation distance on the sound pressure level difference between spurious sound and demodulated difference frequency sound, another experiment with and without acoustic filter was conducted on A3, and the results of which are shown in Fig. 4.36. In this case, the critical distance of A3 is about 0.74 m. Therefore, two distances were chosen: 1 and 2.7 m, both of which are greater than the critical distance. It is easy to observe that for these two distances, the trend of frequency response with the acoustic filter is almost the same, that is, they are almost parallel to each other. It should be noted that only the lower frequency of 0.5 kHz was chosen. Because in previous studies, the threshold for ignoring the demodulated difference frequency sound in the mixed signal is about 7 dB. When using A3, only the frequency value of about 0.5 kHz at 2.7 m meets the above threshold. When the frequency value is 0.5 kHz, the sound pressure levels of the spurious sound at these two distances are 98.6 dB and 66.5 dB, respectively, and the sound pressure levels of the demodulated difference frequency sound are 75.4 dB and 59.1 dB, respectively. The sound pressure level differences between the spurious sound and the demodulated difference frequency sound are 23.2 dB and 7.4 dB, respectively. Their difference is −15.8 dB, which is suitable for the slope of $z^{-1.85}$. This means that the aforementioned sound pressure level difference will significantly decrease with the increase of distance after the critical distance. For larger distances, spurious sound can be ignored [112, 172].

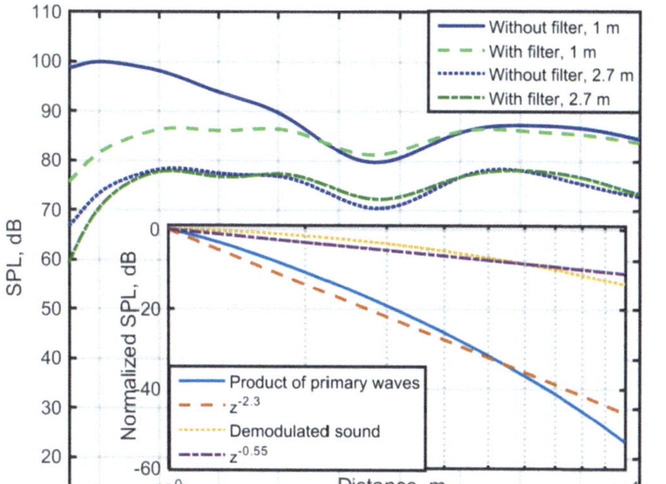

Fig. 4.36 Comparison of parametric array frequency response at 1 and 2.7 m when using A3

The simulation results normalized by the value at 1m are also shown in the inset of Fig. 4.36, using the same parameters as in the experiment, the corresponding frequency value is 0.5 kHz, and the observation distance is 1–10 m. Due to spherical expansion attenuation, the pump wave amplitude is inversely proportional to the distance after the critical distance [431], and because of the relationship between spurious sound and the pump wave, spurious sound will be inversely proportional to the square of the distance. When considering sound absorption [248], the product of the pump wave has additional attenuation. Between the two distances of 1 and 2.7 m, the total attenuation of the product of the primary waves is around about 17.6 dB. As can be seen from the inset, the spurious sound is almost proportional to $z^{-2.3}$. This means that due to the larger attenuation of the pump wave, the spurious sound will significantly decrease with distance after the critical distance. It is also noted that the demodulated difference frequency sound is inverse proportional to the distance when the distance is larger [1]. However, due to the limited observation distance, the curve $z^{-0.55}$ is more suitable for the demodulated difference frequency sound. Therefore, the relationship between the aforementioned sound pressure level difference and distance is fitted to $z^{-1.75}$, which is consistent with the experimental results.

In reference [112], Wygant et al. also provided some experimental results on the relationship between spurious sound and observation distance. In the inset of Fig. 4.37, the measurement data with and without filters are obtained from reference [112], which are the mixed signals and demodulated difference sounds at distances of 0.4–10 m. The spurious sound is calculated by $Kp_{1m}p_{2m}$, where p_{1m} and p_{2m} are also obtained from reference [112], and K is 2×10^{-4} Pa^{-1}. Then, the sound pressure level difference between the spurious sound and the demodulated difference frequency sound is calculated from the above data, and normalized by the value of 0.4 m, as shown by the solid line in Fig. 4.37. It can be clearly observed that the above normalized sound pressure level difference is fitted to the trend of $z^{-1.75}$. This means that a similar trend can be found in reference [112], which also validates the observation of this experiment.

4.5 Summary

This chapter concludes with a review of the current measurement techniques in parametric arrays. On this basis, detailed explanations are given for the cylindrical ultrasonic filter and the phononic crystal filter measurement techniques, and experimental studies are conducted on the impact of four main parameters on the measurement of parametric arrays.

This chapter first presents a simple structure acoustic filter. This filter has a low-pass characteristic, with an average attenuation of more than 15 dB for ultrasonic waves, and less than 5dB impact on audio signals. Through actual measurement of parametric array loudspeakers, the practicality of this filter is confirmed.

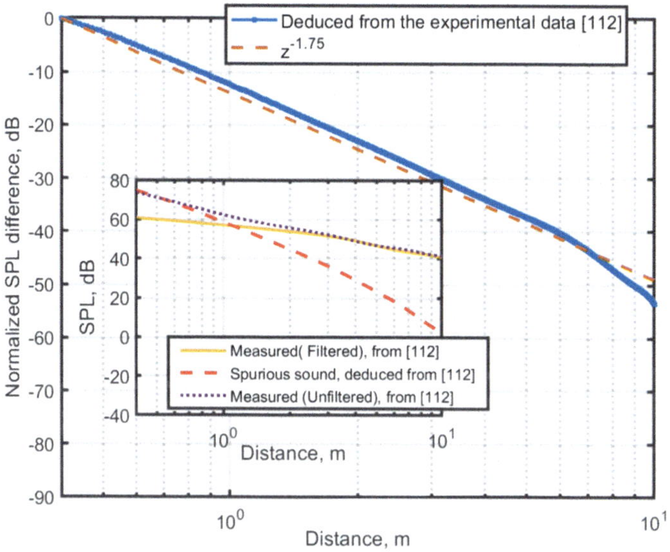

Fig. 4.37 Normalized sound pressure level curve using the value at 0.4 m (redrawn based on data from reference [112])

Secondly, this chapter presents the use of phononic crystal acoustic filters to eliminate spurious sound in parametric array loudspeaker measurements, and comparative tests were conducted in three different arrays. In Comsol, numerical studies on parameter effects were conducted through quasi-1D and 2D models. Based on the theoretical analysis, structural parameters with good performance were obtained. The experimental results show that the phononic crystal-based acoustic filter has good performance in axial propagation and directional measurement, in terms of reducing spurious sound in parametric array loudspeakers.

Finally, a more thorough explanation of the spurious sound problem in parametric array loudspeakers is given through experiments. The sound pressure level difference between the spurious sound at the observation point and the demodulated difference frequency sound (beyond the fluctuation range of the sound source array) is roughly proportional to $a^m/\left(f_d(K) \times z^l\right)$, where the coefficients m and l can be determined by simulation or experiment, and are functions related to K. In this section, it can be preliminarily obtained that, $m \approx 1.35$, $l \approx 1.85$, $f_d(K) \approx f_d^{1.7}$ (assuming that K is the same for all frequencies). It can also be observed that it is impossible to alleviate the impact of spurious sound by reducing the amplitude of the pump wave at the sound source. Similar trends were found between simulation and experimental results.

Chapter 5
FPGA Implementation of Parametric Array Loudspeakers

The current implementation of parametric array loudspeakers mainly divides into analog circuit and digital circuit implementations, and the digital circuit further divides into DSP and FPGA implementations. The following is a brief introduction.

In the design of parametric array loudspeakers, pure analog circuits were often used in the past. The overall system block diagram is shown in Fig. 5.1 (taking SRAM preprocessing as an example), which is generally composed of five parts: local oscillator circuit, preprocessing part, modulation circuit, power amplification circuit, and ultrasonic transducer array. The local oscillator uses a 40 kHz sine signal, and the preprocessing part completes the integration, gain, offset, and square root operations of the modulation signal. $x(t)$ The modulation circuit is actually a multiplier, generating the square root double sideband amplitude-modulated waves. The power amplifier circuit amplifies the power of the signal, including current and voltage, so that the output can drive the emitter unit. The ultrasonic transducer array converts voltage and current signals into ultrasonic signals and emits them. In this way, the ultrasonic amplitude-modulated wave can generate an audio frequency signal due to the self-demodulation process under the nonlinear effect of air.

In the implementation of the parametric array loudspeaker, it is necessary to adopt an appropriate preprocessing method to ensure the low distortion of the demodulated difference frequency signal. Due to the complexity of the related preprocessing algorithms, the design of analog circuits is quite difficult. Debugging analog circuits is also inconvenient, which prolongs the development cycle. These problems restrict the development and application of parametric array loudspeakers. In recent years, with the development of digital signal processing and integrated technology, the functions of digital signal processors are becoming more and more powerful, and their basic signal processing algorithms are becoming mature. Parametric array loudspeaker systems based on DSP and FPGA can conveniently implement algorithm debugging and experimental research through software programming, making their design relatively simple. Digital signal processors from TI or ADI have been used to design parametric array loudspeaker systems, and their structure diagrams are

J. Yang and P. Ji, *Parametric Array Loudspeakers*,
https://doi.org/10.1007/978-981-96-3548-1_5

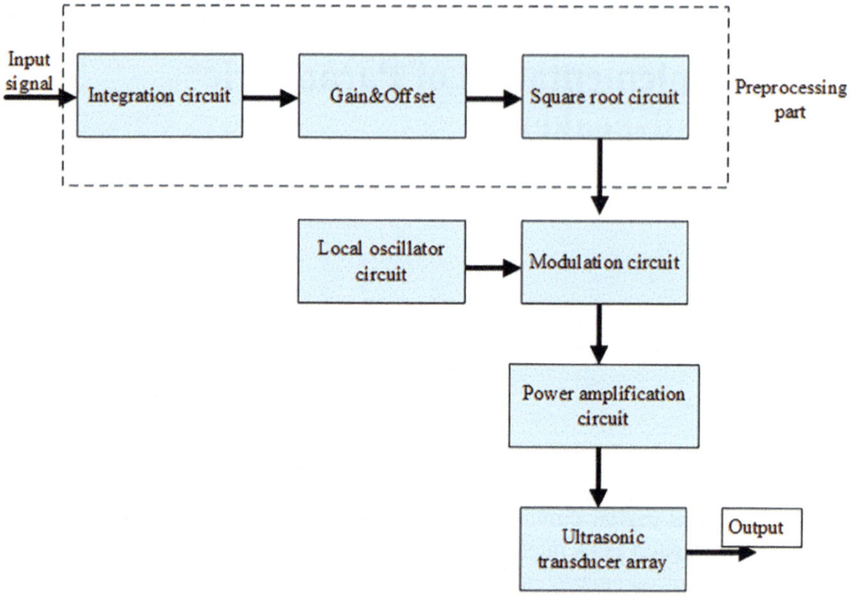

Fig. 5.1 Block diagram of parametric array loudspeaker design implemented by analog circuit

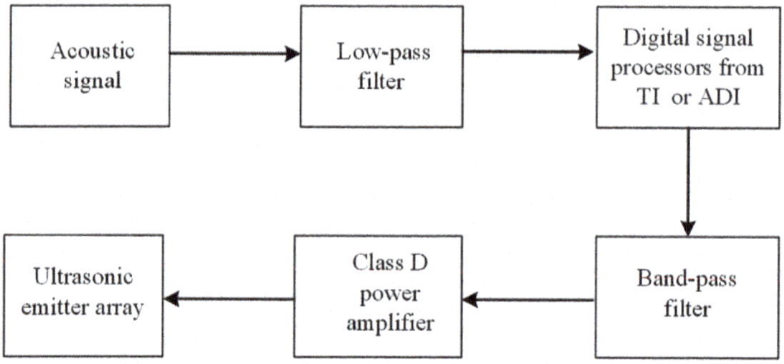

Fig. 5.2 Block diagram of parametric array loudspeaker system based on DSP

shown in Fig. 5.2. Low-pass filtering for anti-aliasing signal processing boards, with TI or ADI digital signal processors as the core is used to implement pre-processing algorithms for directional sound sources. Band-pass filtering is used for band-pass filtering of amplitude-modulated waves output by the signal processing board. Ultrasonic power amplifiers are used for power amplification of amplitude-modulated signals. The transducer array is used to convert amplitude-modulated waves into amplitude-modulated ultrasonic waves. The amplitude-modulated ultrasonic waves undergo nonlinear self-demodulation in the air, generating highly directional audio frequency beams.

With the advancement of semiconductor technology, embedded processing chips such as FPGA not only have improved integration, but also have controlled costs. FPGA adopts parallel processing, which greatly reduces the delay of signal processing, thus having better real-time performance. In engineering implementation, a top-down design method is adopted. This design method has the following advantages. Starting from the functional description to the completion of physical implementation, it conforms to the design thinking of the designer. It is easy to change the design. The design cycle is shortened, and productivity is greatly improved. Therefore, the FPGA chip is one of the best choices for small batch systems to improve system integration and reliability.

This chapter mainly implements the parametric array loudspeaker system based on the FPGA processor, which includes hardware and algorithm parts. The hardware part mainly includes the selection of chips, formulation of plans, and final implementation for four modules: power management circuits, Codec circuits, FPGA processor circuits, and D-class power amplifier circuits along with transducer matching circuits. The software part mainly includes the simulation and implementation of six modules: Codec chip control, serial-parallel conversion, carrier generation, voice activity detection, dynamic range control, and single sideband modulation. The algorithm was simulated through software, and based on this, various signal processing algorithms were implemented on the hardware platform using DSP Builder and Quartus II tools, and the resource consumption was evaluated. Finally, the performance indicators of the system were measured in the anechoic chamber.

5.1 Structure of the Parametric Array Loudspeaker System Based on FPGA

The function that the parametric array loudspeaker needs to complete is: to process the input audio signal appropriately, drive it through the amplifier, and emit it into the air by the ultrasonic transducer. According to the above basic functions, the parametric array loudspeaker system mainly includes signal acquisition, signal processing, power amplification, and transducer matching, as shown in Fig. 5.3. Therefore, the entire signal flow process is.

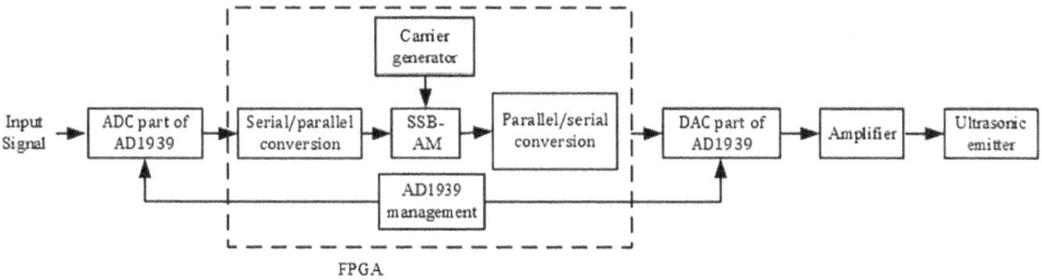

Fig. 5.3 Block diagram of parametric array loudspeaker system

1. The emitted signal is first passed through an anti-aliasing filter to remove out-of-band signals and then sampled by the ADC section of the AD1939.
2. AD1939 transmits the digital signal to the FPGA chip in a single-line serial manner. After the FPGA completes signal processing, it outputs the digital signal to AD1939 in a serial manner.
3. The DAC part of AD1939 converts the acquired digital signal into an analog signal and smooths the signal through a low-pass filter.
4. The processed analog signal is amplified by the power amplifier module to drive the ultrasonic transducer to emit ultrasonic waves, which after demodulation in the air, result in a difference frequency sound.

The entire platform is composed of hardware and software parts. The hardware includes system power hardware design, FPGA processor hardware design, AD1939 hardware design, and amplifier hardware design. The software part includes serial-parallel data conversion, carrier generation, voice activity detection, dynamic range control, single sideband modulation, and AD1939 clock and register management inside the FPGA.

5.2 Hardware Implementation

The design process of the parametric array loudspeaker hardware is device selection, schematic design, PCB design, and final debugging. Device selection is to screen from the product manuals of various semiconductor manufacturers according to needs. Then EDA tools is used to simulate the selected chips and draw the schematic. After the packaging of the devices in the schematic is completed, layout and routing of the devices. The final steps are circuit fabrication and debugging in conjunction with software.

5.2.1 Power Management

Power management mainly generates different levels of voltage required by the system. The DC voltages required by the system are 1.2, 2.5, 3.3, 5, and 12 V. The external power supply voltage of the system is DC 24 V, which generates the above different voltages through three-stage voltage reduction. The voltage reduction process and voltage regulator chip are shown in Fig. 5.4. The advantage of using a three-stage voltage reduction is that it prevents large voltage drops from generating excessive heat in the voltage regulator chip, thereby reducing the cooling requirements for the chip. Among them, the 12 V voltage is provided to the amplifier module, the 5 V voltage is provided to most of the operational amplifiers and AD1939 on the circuit board, and the remaining three voltages are provided to the FPGA chip.

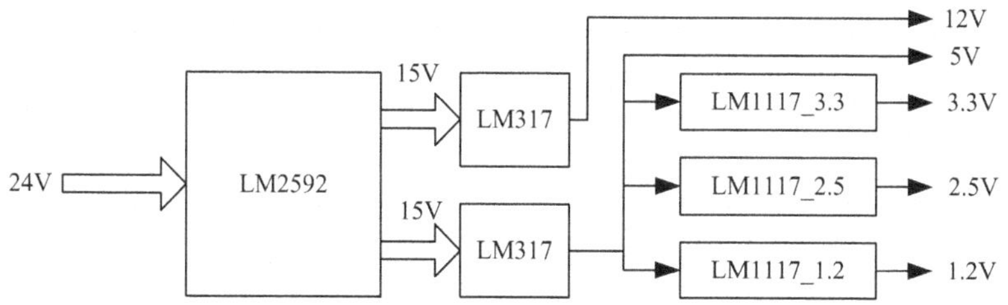

Fig. 5.4 Power management schematic diagram

5.2.2 *FPGA Processor*

The advantages of FPGA lie in its parallel processing mode, top-down modular design, rich user-defined IO ports, and integration of many modules. The main manufacturers of FPGA are Altera, Xilinx, and Lattice. In the low-end market, Altera's FPGA is relatively cheap, while the prices of the other two companies are relatively high. In addition, Altera's software is simple to use, while the other two companies' software is complex. Therefore, for not very complex systems, it is more appropriate to use Altera's Cyclone series FPGA. This system uses the Cyclone III EP3C25Q240C8 chip as the main control processor. The chip features an integration of 24,624 logic cells, 66 M9K blocks, 608,256 bits of memory, 66 18-bit multipliers, 4 PLLs, 20 clock ports, and 149 user-defined I/O ports. The chip requires three DC voltages: 1.2, 2.5, and 3.3 V.

5.2.3 *AD1939 Codec Chip*

The AD1939 is a high-performance audio Codec chip developed by Analog Devices. Its main features are: 4-channel ADC and 8-channel DAC, maximum sampling frequency of 192 kHz, data width up to 24 bits; sigma-delta structure, built-in high-performance digital filter; analog port adopts differential mode, improving signal-to-noise ratio; divided into active and passive working modes, flexible operation; uses 3.3 V single power supply, and is compatible with 5 V voltage; built-in phase-locked loop (PLL), which can set the multiplication factor according to the system clock and sampling rate, the initial value is 256 times; the chip has 16 control registers, controlled by the microprocessor through the SPI interface.

The task of the AD1939 is to complete the analog-to-digital conversion and digital-to-analog conversion. An anti-aliasing filter is needed before the ADC input, and a smoothing filter is needed after the DAC output. Also, because the analog port uses a differential mode, the anti-aliasing filter must have a single-ended to differential function, and the smoothing filter must have a differential to single-ended function.

Both filters are based on the dual op-amp AD8606ARZ design, the anti-aliasing filter is a first-order reverse input filter, with a cutoff frequency of 12 kHz. The smoothing filter uses a second-order multi-feedback structure, with a cutoff frequency of 96 kHz.

5.2.4 Power Amplifier Design and Transducer Impedance Matching

With the development of semiconductors, Class D amplifiers are becoming more mature and are gradually being adopted in the electroacoustic field. The principle of the Class D amplifier is shown in Fig. 5.5. The input sine wave and the high-frequency triangle wave pass through the comparator. When the amplitude of the sine wave is greater than the amplitude of the triangle wave, the comparator outputs a high level, otherwise, it outputs a low level, resulting in a square wave whose duty cycle varies with the amplitude of the sine wave, also known as a Pulse Width Modulation wave (PWM). The PWM wave is shaped by the GATE driver device to reduce the dead time and generate the switch tube drive signal. The switch tube is a large current switch amplifier, which only operates in two states: saturation and cutoff, so the power consumption of the switch tube is very small, improving the efficiency of the entire amplifier.

The power amplifier of the parametric array loudspeaker is designed based on the TAS5630DKD chip from TI, which has high integration and fewer peripheral devices. The main features are: the working efficiency has reached more than 90%, the internal resistance is 60 mΩ, and the maximum output power can reach 600 W@4 Ω.

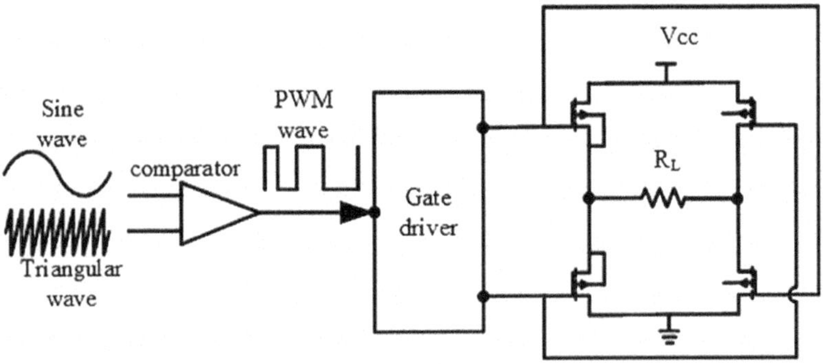

Fig. 5.5 Class D amplifier schematic diagram

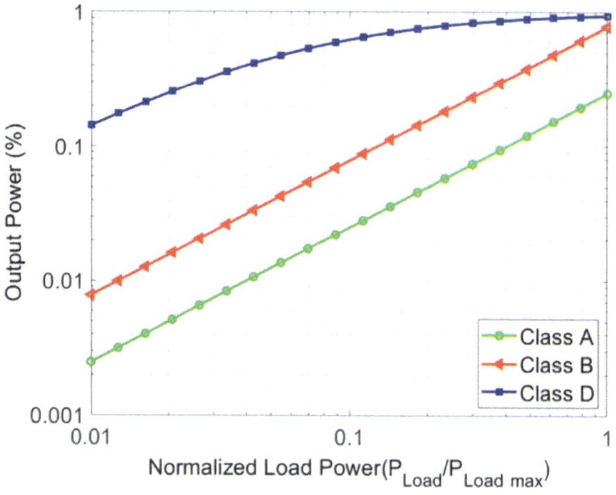

Fig. 5.6 Efficiency of three class amplifiers

It has comprehensive overcurrent, overvoltage, and high-temperature protection functions. The bandwidth can reach 80 kHz, meeting the band requirements of parametric array loudspeakers. It has three output modes: PTL, BTL, and SE, corresponding to single-channel output, dual-channel output, and four-channel output respectively. It only requires a single power supply, greatly reducing the design difficulty of the power module.

The efficiency results calculated based on the parameters of TAS5630DKD are shown in Fig. 5.6. As can be seen from the figure, the highest efficiency of Class A amplifiers is 25%; the highest efficiency of Class B amplifiers is 78.5%, and the actual working efficiency is around 50%; the efficiency of Class AB amplifiers is between Class A and Class B. The highest efficiency of Class D amplifiers based on TAS5630DKD is close to 100%, and the actual working efficiency is around 90%, which is much higher than the efficiency of Class A and Class B amplifiers.

Total harmonic distortion is also one of the important indicators of an amplifier. If the total harmonic distortion is too large, it will affect the sound quality. Some products often give the total harmonic distortion size when the power is 1 W, but this parameter is not comprehensive, because the product cannot always work at 1 W power, so it is necessary to give the harmonic distortion of the entire output power range. Here, the change curve of the total harmonic distortion of the designed amplifier tested by the audio analyzer with power is shown in Fig. 5.7. It can be seen from the figure that in the output power range of 10–100 W, the total harmonic distortion is below 0.1%, basically meeting the requirements of parametric array loudspeakers.

For resistive ordinary moving coil loudspeakers, the amplifier can directly output power, with all the power output being converted into active power. If the amplifier directly outputs to the capacitive load of the ultrasonic transducer array, it will generate a large amount of reactive power, so impedance matching must be carried out. Impedance matching is simply divided into parallel matching and series matching, both of which can achieve good matching effects. However, currently in

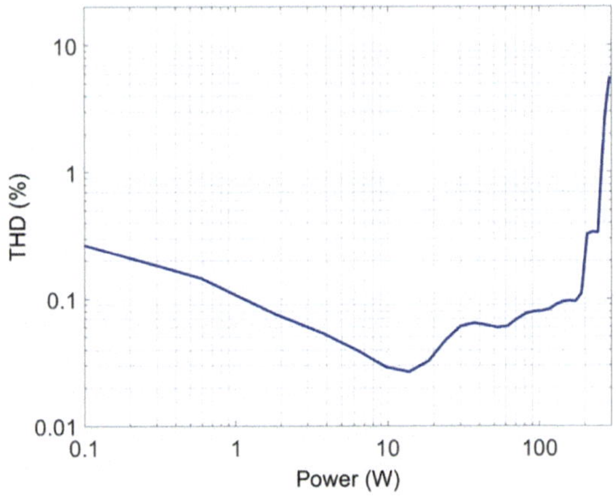

Fig. 5.7 Total harmonic distortion curve with output power variation

power ultrasound, series matching is frequently used, because when a short circuit occurs in the load, the series inductance will generate a certain impedance, which will not cause the amplifier load to be completely short-circuited, and it will protect the amplifier.

Figure 5.8 shows the impedance curve of the transducer load at 30–50 kHz. The inductance value exhibits better linearity in the 30–40 kHz range, while it shows poorer linearity in the 40–50 kHz range. In addition, the effective bandwidth of single sideband modulation is also below 40 kHz, so impedance matching is carried out in the 30–40 kHz band. According to the resonance formula, the inductance value of series matching $L = \frac{X}{2\pi f}$, where X is the reactance of the transducer array, and f is the signal frequency.

As can be seen from Fig. 5.8, the reactance is not absolutely linear with the signal frequency, so the inductance value obtained is in the range of 18–59 μH. From the transducer manual [432], it can be known that the transducer has a frequency response

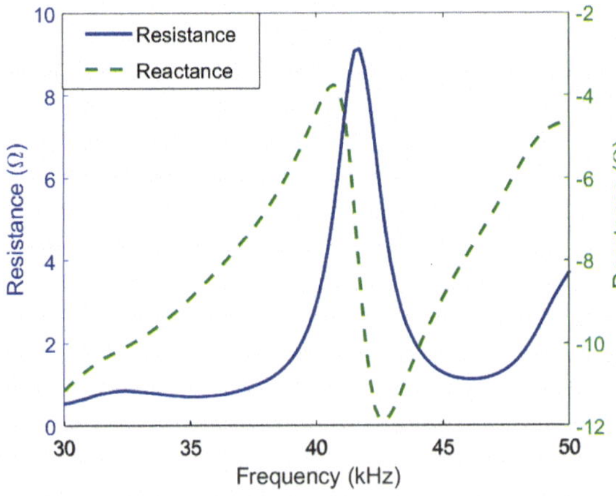

Fig. 5.8 Impedance curve of ultrasonic transducer array

Table 5.1 Power measurement results before and after transducer array matching

	Active power (W)	Reactive power (W)	Total power (W)
Before matching	10.38	9.47	14.05
After matching	10.63	0.72	10.65

peak at 40 kHz, and it decays to both sides with 40 kHz as the center frequency. Therefore, matching is chosen at 38 kHz, which allows for an improvement in the frequency response of the sidebands. For a frequency of 38 kHz, the peak-to-peak output voltage is 25 V. The power measurement results before and after matching are shown in Table 5.1. It can be seen from the table that the useless power is greatly reduced.

5.3 Algorithm Implementation

The software design process starts with the establishment of a DSP circuit model using simulation software. The establishment of the circuit model can be graphical, and the rich function blocks and IP cores provided in Simulink and DSP Builder can be conveniently used for design. After the circuit model design is completed, system-level model simulation can be performed, which is a system function simulation, unrelated to the target device and hardware system, and is a simulation based on the algorithm. Then, the DSP Builder's Signal Compiler placed in the Simulink circuit is used to convert the circuit model file, i.e., the Simulink module file (.mdl), into RTL-level VHDL code representation and Tcl (Tool Command Language) scripts. The converted VHDL file can be called in Quartus II to generate modules. At the same time, if it is necessary to verify whether the functional timing can meet the requirements of the project, simulations can be conducted using relevant software. The entire process is one that can be repeatedly debugged, so it is very convenient to adjust parameters back and forth during the design to achieve the most ideal effect.

5.3.1 AD1939 Clock and Register Management

The entire parametric array loudspeaker system has three clocks, namely the system clock, the sampling clock, and the bit clock. Among them, the system clock is the working clock of the FPGA. The sampling clock and bit clock are necessary clocks for AD1939 to complete data sampling and data output. To facilitate the control of the AD1939 working state, the AD1939 operates in passive mode, with the sampling clock and bit clock provided by the FPGA. The system clock is generated by the 50 MHz crystal oscillator on the circuit board, and 192 kHz and 12.888 MHz are obtained through the phase-locked loop (PLL) of the FPGA. The 192 kHz clock

Table 5.2 AD1939 register format

	Global address	Read/write bit	Register address bits	Register data bits
Bit	23-17	16	15-08	07-00

signal serves as the sampling clock (LRCLK) for ADC and DAC; 12.888 MHz serves as the bit clock (BCLK) for ADC and DAC. Therefore, the FPGA resources consumed by the entire clock generation are 2 PLLs.

The AD1939 cannot function properly with only two clocks provided by the FPGA, it also requires the FPGA to set the AD1939's registers appropriately. The AD1939 has a total of 16 registers, 2 PPL and Clock registers, which set the clock source, clock frequency, and ADC/DAC master–slave mode respectively; 3 DAC control registers, which set the DAC sampling rate, data bit width, serial data format, and data delay relative to DBCLK; 1 DAC mute register, which independently sets whether the 8-channel DAC is muted; 8 DAC volume control registers, which individually control the output volume of each DAC channel; 3 ADC control registers, which set ADC mute, sampling rate, data bit width, data delay relative to ABCLK, serial data format, ABCLK and ALRCLK priority, and master–slave mode. The register format is shown in Table 5.2. As can be seen from the figure, the register is composed of 24 bits, with bits 23-17 being the global address, bit 16 being the read/write bit, bits 15-8 being the register address bits, and bits 7-0 being the register data bits. Since each register has an address, the setting of each bit of the register is not sequential.

The AD1939 reads and writes to the 16 registers via a 4-wire SPI bus interface, which consists of the register write port CIN, the register read port COUT, the SPI bus clock port CCLK, and the download flag port CLATCH. This system sets the ADC and DAC sampling rate of the AD1939 to 192 kHz via the FPGA, in passive mode, with left-aligned serial data format, and data delayed by 1 cycle relative to DBCLK. After the system is powered on, the register control program runs first, and after 5000 system clock cycles, other parts of the program begin. After compilation and synthesis by Quartus II 8.0, the resources consumed by the AD1939 register management are: 63 logic units, 34 registers.

5.3.2 *Serial and Parallel Conversion*

The data transmission between AD1939 and the processor uses a serial data format, hence the FPGA must perform serial-to-parallel and parallel-to-serial conversions when exchanging data with AD1939. The AD1939 has three serial data formats: Left-Justified, Right-Justified, and Time-Division Multiplexed. In Left-Justified and Right-Justified modes, a single data line contains 2 channel data, typically used for mono or stereo input/output, while in Time-Division Multiplexed mode, a single data line contains 16 channel data, suitable for 5.1 and 7.1 and other multi-channel

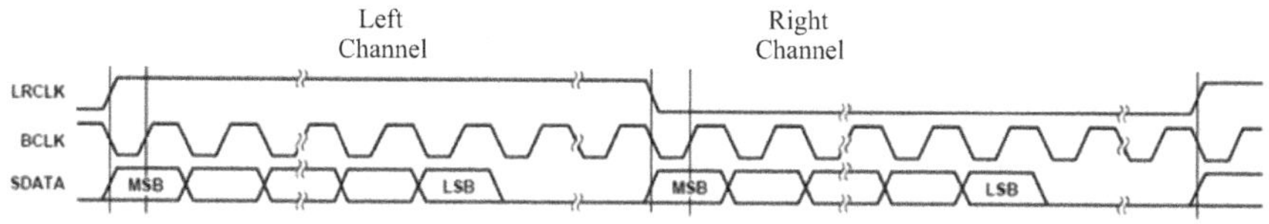

Fig. 5.9 Timing diagram in left-aligned mode

input/output. As can be seen, for multi-input and multi-output applications, Time-Division Multiplexed mode is more suitable, but for mono or stereo situations, Left-Justified and Right-Justified modes are clearly simpler and more convenient. In general recording systems, Left-Justified mode is used more frequently. Taking into account the aforementioned factors, the data transfer format between AD1939 and FPGA adopts the left-justified mode, and the timing diagram of Left-Justified mode is shown in Fig. 5.9. As can be seen from the figure, when LRCLK = 1, the left channel data is valid; when LRCLK = 0, the right channel data is valid, where $f_{LRCLK} = \frac{1}{32} f_{BCLK}$. Each channel's BCLK start clock corresponds to the highest bit of data. If the data bit width is 24 bits, then the first 24 clocks of BCLK correspond to valid data.

The serial-to-parallel and parallel-to-serial conversions in the system both utilize the shift register algorithm, as shown in Fig. 5.10. The VHDL language is used for writing, and the FPGA resource consumption is given after compilation and synthesis, as shown in Table 5.3.

5.3.3 *Carrier Generator*

The carrier generator, also known as the numerical controlled oscillator (NCO), is used to generate the sine and cosine ultrasonic carrier signals required for modulation. Due to its high frequency accuracy, short conversion time, high spectral purity, and easy frequency phase programming, the NCO is widely used in digital radio up and down conversion, as well as in various digital modulation and demodulation systems.

The main implementation methods of NCO are: lookup table method, polynomial approximation method, phase accumulation method, and coordinate axis rotation method. The lookup table method calculates the amplitude corresponding to the phase of the sine and cosine waves and stores it in the register, and the values in the table are read according to the system clock. To improve the frequency resolution of the NCO, it is often necessary to expand the capacity of the waveform memory, causing a large consumption of storage resources. For the limited memory resources of FPGA, when designing a high-speed and high-precision NCO, it is not suitable to use the lookup table method. The polynomial approximation method will consume a large number of multipliers, which is also not suitable for FPGA with limited number of multipliers. The phase accumulation method also requires large storage capacity, which is obviously not suitable for FPGA. The Coordinate Rotation Digital

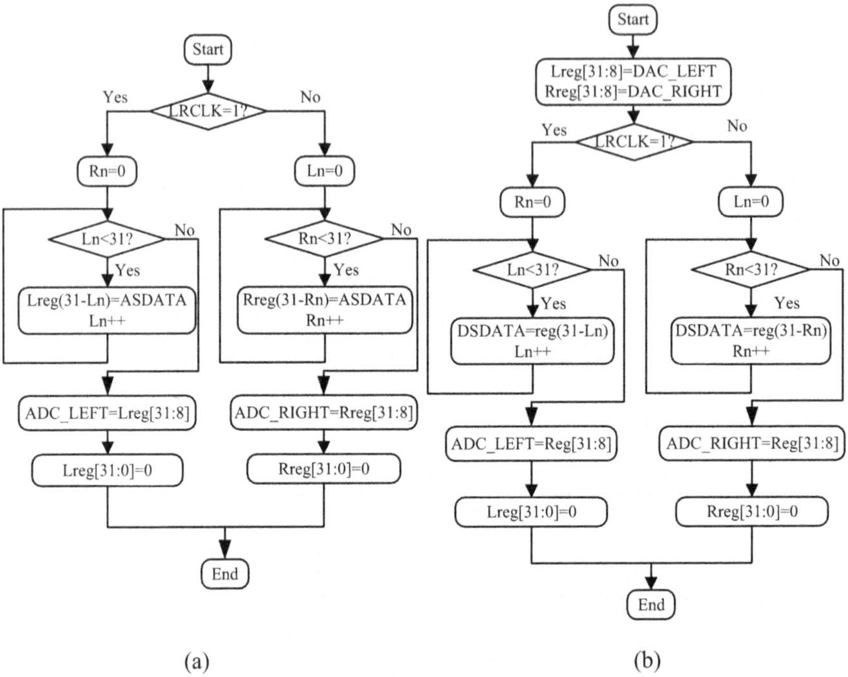

Fig. 5.10 Serial and parallel conversion program flowchart **a** serial to parallel conversion **b** parallel to serial conversion

Table 5.3 Serial-to-parallel and parallel-to-serial conversion resource consumption

	Logic unit	Register	Memory	9 bits multiplier
Serial-to-parallel conversion (data_receiver)	158	58	0	0
Parallel-to-serial conversion (data_emitter)	47	11	0	0

Computer (CORDIC) only requires a small rotation angle table. By using simple shift and addition operations, it can generate high-precision sine and cosine waveforms, which are very suitable for FPGA implementation.

The CORDIC algorithm includes three rotation systems: circular, linear, and hyperbolic. This section uses the circular system for generating sine and cosine waveforms. The circular system of the CORDIC algorithm completes a plane coordinate rotation, as shown in Fig. 5.11, where R is the radius of the circle.

As can be seen from Fig. 5.11, the vector (X_i, Y_i) is rotated by θ to get the vector $\left(X_j, Y_j\right)$, the expression is as follows

$$\begin{aligned} X_j &= R\cos(\theta+\beta) = \cos(\theta)(X_i - Y_i\tan(\theta)) \\ Y_j &= R\sin(\theta+\beta) = \cos(\theta)\left(X_i\tan^{-1}(\theta) - Y_i\right) \end{aligned} \tag{5.1}$$

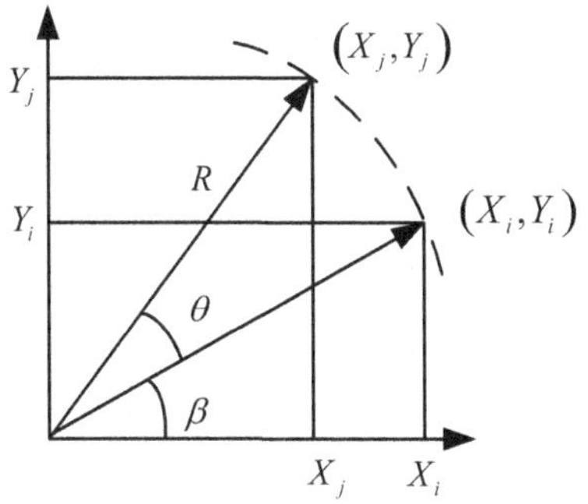

Fig. 5.11 Rotating coordinate diagram

If each step's rotation angle satisfies $\tan(\theta_i) = \pm 2^{-i}$, then the multiplication operation in the formula becomes a shift operation, and any angle rotation can be achieved through the accumulation of a series of infinitesimal rotations. Then the ith rotation can be represented as

$$\begin{aligned} X_{i+1} &= X_i - Y_i \cdot d_i \cdot 2^{-i} \\ Y_{i+1} &= X_i \cdot d_i \cdot 2^{-i} + Y_i \\ Z_{i+1} &= Z_i - d_i \tan^{-1}\left(2^{-i}\right) \end{aligned} \tag{5.2}$$

where $K_i = \cos\left(arctan\left(2^{-i}\right)\right) = \frac{1}{\sqrt{1+2^{-2i}}}$, $d_n = \{-1; +1\}$, where when $Z_i < 0$, $d_i = -1$,when $Z_i \geq 0$, $d_i = 1$. After n rotations

$$\begin{aligned} X_{n+1} &= P[X_0 \cos(Z_0) - Y_0 \sin(Z_0)] \\ Y_{n+1} &= P[Y_0 \cos(Z_0) + X \sin(Z_0)] \end{aligned} \tag{5.3}$$

where $P = \prod_{i=0}^{n} K_i \approx 0.6072$, $Z_{n+1} = 0$. If the following parameters are set: $X_0 = \frac{1}{P}$, $Y_0 = 0$, $Z_0 = \alpha$, then after n iterations, the result becomes

$$\left[X_{n+1}, Y_{n+1}, Z_{n+1}\right] = [\cos\alpha, \sin\alpha, 0] \tag{5.4}$$

From Eq. (5.4), it can be observed that the CORDIC algorithm can calculate the cosine and sine of the angle α in the circular system mode. According to Eq. (5.2), the flowchart of the CORDIC algorithm is shown in Fig. 5.12, where the content stored by $table(i)$ is $\tan^{-1}\left(2^{-i}\right)$, and the storage depth is N.

The NCO is designed according to the CORDIC flowchart, where the angle table is stored in the logic unit. After compilation and synthesis by Quartus II 8.0 software, the resource consumption is shown in Table 5.4.

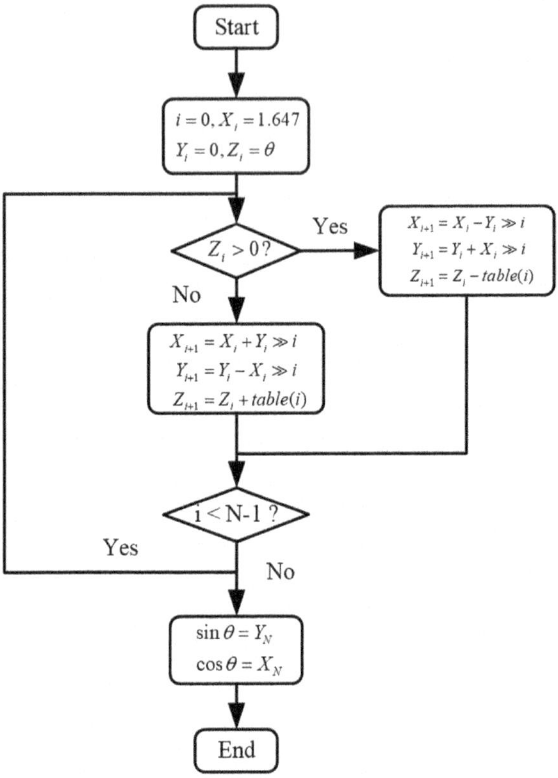

Fig. 5.12 CORDIC algorithm flowchart

Table 5.4 Carrier generator resource consumption

	Logic units	Registers	Memory	9 bits multipliers
Carrier generator	1457	0	0	0

5.3.4 *Voice Activity Detection*

In parametric array loudspeaker systems, due to the characteristics of the transducer array itself, some high-frequency noise is also generated in the absence of input audio signals. Therefore, it is necessary to add voice activity detection to exclude the possibility of noise interference before proceeding with various modulations. Voice activity detection (VAD) is a technique that detects the effective part of the voice by judging the pauses and silence intervals in the voice through specific judgment criteria. Here, by judging the size of the energy of a segment of the input signal, noise suppression can be achieved when the energy is below a certain threshold.

VAD is a crucial part in speech signal processing, and here it is analyzed in detail for its application in speech processing. Its purpose is to find the start and end points of the voice from a segment of signal containing voice, so as to only process the effective voice signal. Effective endpoint detection can not only reduce data storage and calculation time, but also avoid noise interference. The difficulty of endpoint detection lies in the noise generated by human breathing before and after the silent segment or voice segment, the weak fricative or weak plosive at the beginning of the voice, and the nasal sound at the end point, which makes the voice endpoint vague. It is necessary to comprehensively utilize various features of speech signals to ensure the accuracy of localization, avoiding the inclusion of noise signals and the loss of speech signals.

As shown in Fig. 5.13, by calculating the size of a frame energy and comparing it with a set threshold, the sound caused by circuit noise is subjected to voice activity detection. If it is below the threshold, the signal output by the FPGA is only the carrier after attenuation. The carrier, as an ultrasonic wave, attenuates rapidly and is imperceptible to the human ear. When an audio frequency signal is input, if the energy judgment is higher than the threshold, it is directly output to the next level of processing. The final subjective listening test is quite ideal, without the interference of harsh high-frequency sounds.

Short-term energy judgment is a commonly used method for voice activity detection.

$$E = \sum_{1}^{N} |s_{\omega}(n)|^2 \tag{5.5}$$

where E represents the energy value of the signal within a frame, N is the frame length, and $s_{\omega}(n)$ is the amplitude of the speech after framing and windowing. Here,

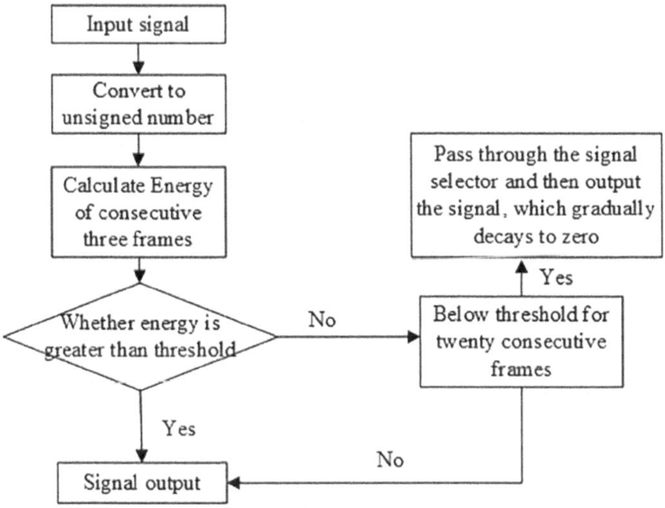

Fig. 5.13 VAD implementation process

the short-term energy method is used to implement real-time voice activity detection. When it is below the threshold, no processing is done.

The main principle of the voice activity detection module to solve the high-frequency noise problem of the system is: it calculates the energy of the audio signal in units of "frames" at the audio input port, and compares it with the noise threshold. If the energy of the audio signal is lower than the noise threshold, it is considered that there is no audio input, or the input is noise, and the ultrasonic module of the parametric array loudspeaker is shut down; if the energy of the audio signal is higher than the noise threshold, it is considered normal audio input, and the ultrasonic module is turned on. To prevent frequent switching caused by short-term silence, it is set that only when several consecutive frames (which can be set according to needs) are all below the threshold level, it is considered to input silent frames. The switch of the ultrasonic module is controlled by the level of the output of the mute detection module. To prevent sudden opening or shutting down caused by signal transition, the output of the module is smoothed to make the sound change smoothly and naturally.

There is currently no unified standard for the VAD algorithm internationally, and it is a research direction with strong subjectivity. Generally, VAD gives a probability of the occurrence of speech. The evaluation indicators of VAD are the misjudgement rate and the omission rate. The misjudgement rate is the probability of judging noise as speech, and the omission rate is the probability of misjudging speech as noise. The VAD method can overcome the influence of noise to a certain extent, and is often used in conjunction with noise suppression in practice. Therefore, considering the needs of signal processing in different acoustic environments, it is necessary to set the VAD threshold accordingly. The specific implementation is shown in Fig. 5.14.

A simulation of this module was performed using an audio clip of approximately 25 s. The simulation results are shown in Fig. 5.15. When the input signal is a signal with a higher amplitude, the output is at a high level, and the ultrasonic generation module is turned on. When there is no signal input, the output signal is "0", and the ultrasonic generation module is shut down. During short-term silence, the ultrasonic module is not shut down to avoid frequent switching between the on and off states. The output signal state changes smoothly between "0" and "1" through smoothing processing.

The introduction of this module can effectively address the auditory discomfort caused by high-frequency noise, compensate for the shortcomings of traditional dynamic range control algorithms, and to some extent, achieve the effect of low power consumption.

Figure 5.16 is the verification result. The waveform in the upper figure is the audio time-domain waveform played by the parametric array loudspeaker without the VAD module. The waveform in the lower figure is the audio time-domain waveform played after adding the VAD module. In the upper figure, when the mute detection module is not incorporated, the ultrasonic module of the parametric array loudspeaker still operates when the silent part of the input signal arrives, resulting in a small amount of noise. In the lower figure, after incorporating the mute detection module, when the silent part of the input signal is reached, the ultrasonic module is turned off after working for a period, reducing the noise level. Since the transition between output

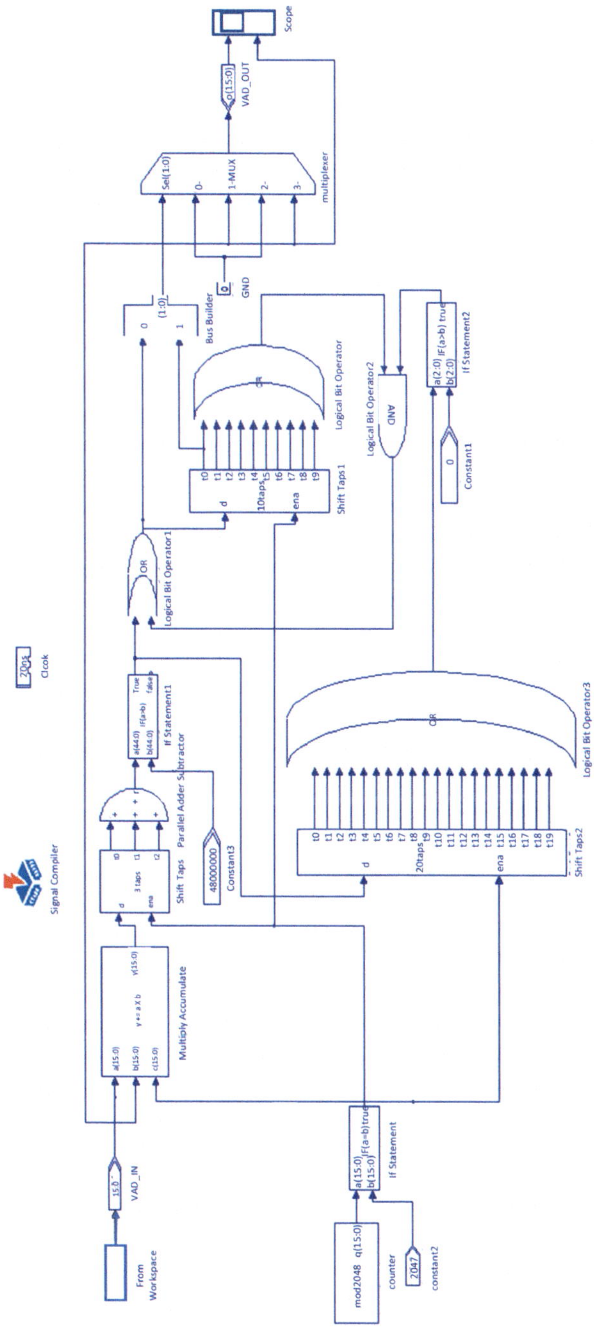

Fig. 5.14 Voice activity detection implementation

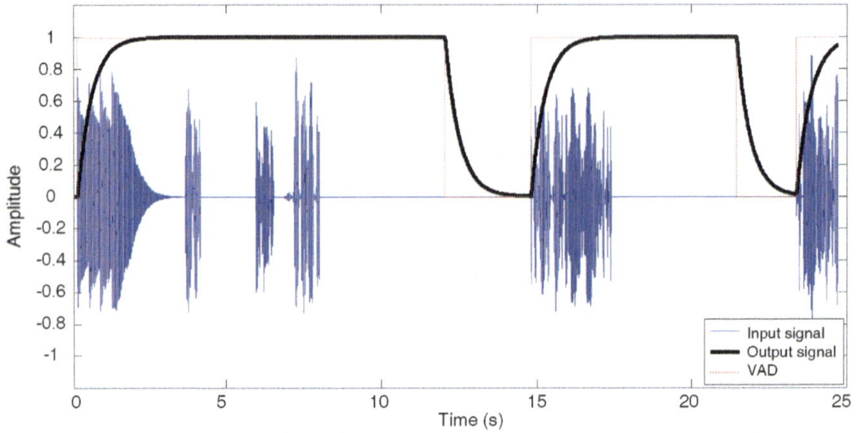

Fig. 5.15 Test results of the silence detection module

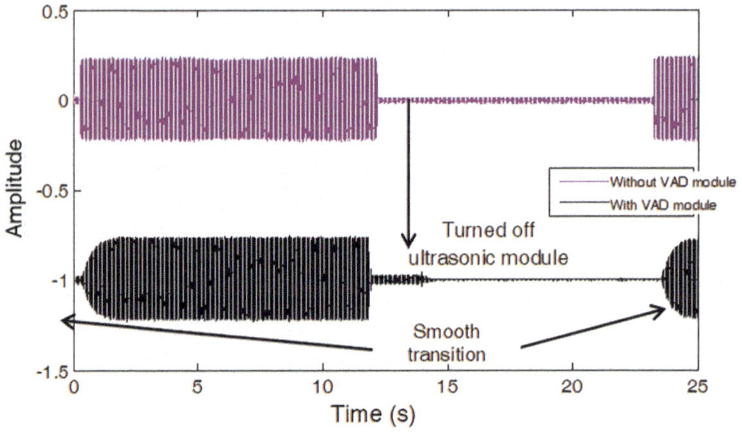

Fig. 5.16 Test sound frequency signal amplitude

levels 0 and 1 has been smoothly handled, the transition between playing and muting is more gradual, avoiding the harsh switching noise that occurs when the audio is suddenly turned on or off.

5.3.5 Dynamic Range Control

The dynamic range of the signal is defined as the logarithmic ratio of the maximum value to the minimum value of the signal amplitude, in decibels. The dynamic range of the audio signal is between 40 and 120 dB. The combination of level measurement

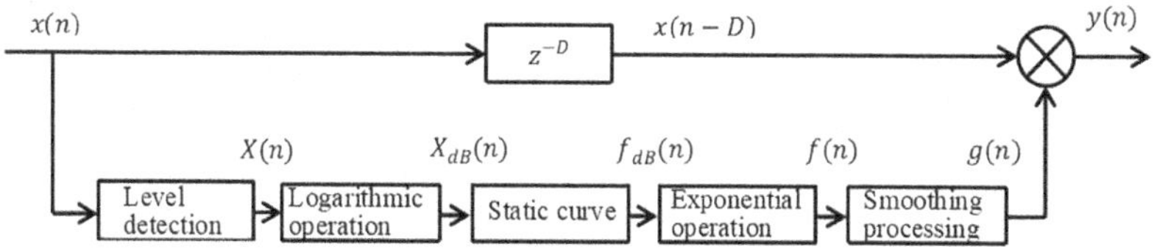

Fig. 5.17 Dynamic range control algorithm module structure diagram

and adaptive signal level adjustment is called dynamic range control. Dynamic range control of audio signals is used in many applications to match the dynamic behavior of audio signals to different requirements. During recording, dynamic range control can protect the AD converter from overload or be used in the signal path to optimally use the recording system. To suppress low-level noise, the so-called noise gate is used, and the audio signal only passes from a certain level. When playing music and speech in environments such as cars, shopping malls, restaurants, or discos, the dynamic effect must be tailored to match the specific noise characteristics of each setting. Therefore, the signal level is measured from the audio signal to derive the control signal, and then the signal level is changed to control the loudness of the audio signal, and the loudness control is applicable to the input level.

The main structure of the dynamic range control (DRC) algorithm when implemented in a field programmable gate array is shown in Fig. 5.17. As can be seen from the figure, the entire algorithm is mainly divided into the following five modules: level detection, logarithmic operation, static curve, exponential operation, and gain factor smoothing processing. The main issue to be discussed in this section is to introduce their respective designs and implementations in the FPGA platform for the above five modules.

For the FPGA platform and the characteristics of the audio dynamic range control algorithm, the basic issues that need to be paid attention to in the design process are mainly:

(1) Due to the performance characteristics of the FPGA chip, the occurrence of multiplication operations should be minimized as much as possible.
(2) Try to use fewer data bit widths while ensuring that the error is within an acceptable range.
(3) For more complex operations such as logarithms and exponentials, choose the most optimized calculation method to reduce system errors.

Level detection is the key to determining the speed of dynamic range control. There are mainly two methods for implementing level detection, peak measurement and root mean square (RMS) measurement. This section chooses the peak detection method, since it can control the smoothness of the envelope in the two states of rise and fall by adjusting the follow-up time and release time, which is more flexible.

The principle of peak detection is to compare the absolute value of the input signal with the previously detected level value. If the absolute value of the input signal is larger, the output value is as shown in Eq. (5.6)

$$x_{PEAK}(n) = (1 - \mathbf{AT}) \cdot x_{PEAK}(n-1) + \mathbf{AT} \cdot |x(n)| \tag{5.6}$$

Otherwise, the new peak level is as shown in the following formula:

$$x_{PEAK}(n) = (1 - \mathbf{RT}) \cdot x_{PEAK}(n-1) \tag{5.7}$$

where, the attack time (AT) determines the loudness of the sound and the compactness of the envelope, and the release time (RT) determines the size of the decay tail. These two parameters have a decisive impact on the speed of peak level detection.

The main role of the logarithmic operation module in the audio dynamic range control system is to convert audio signal data into decibel values. The implementation of logarithmic operations in hardware platforms has always been a widely concerned issue.

Currently, the logarithm algorithms applied in commonly used hardware platforms mainly include the following four types.

(1) Look-up table method (look-up table, LUT)
(2) CORDIC
(3) Binary Logarithm
(4) Polynomial fitting method

When using the look-up table method, it is inevitable to use a very large look-up table to obtain a smaller system error, which will consume too much hardware resources. The coordinate rotation method is not the first choice for this system due to its convergence to too small and extremely high algorithm complexity. The polynomial fitting method requires at least five multiplication operations, which is not suitable for the characteristics of the FPGA platform, so it is not adopted.

After comprehensively considering the content of resource consumption and system error, the binary logarithm algorithm proposed by Mitchell in 1962, also known as the Mitchell algorithm is selected. This part selected the error compensation for this method in reference [433], forming a binary logarithm algorithm with four-domain linear approximation error compensation. The block diagram is shown in Fig. 5.18.

The main principle of this method is to find the position of the leading 1 in the binary sequence, and process the binary digits after it into decimal form through Mitchell approximation and error compensation, and finally sum the two parts to get the logarithm result with base 2.

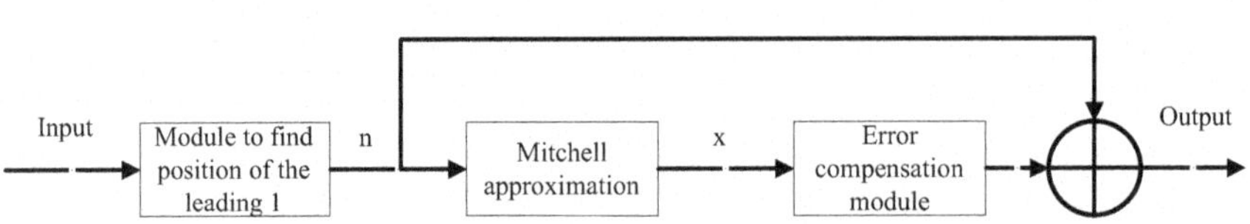

Fig. 5.18 Schematic diagram of binary logarithm algorithm with four-domain linear approximation error compensation

The static curve module, as the core module of the audio dynamic range control algorithm, its design and implementation are related to whether the entire system can meet the requirements. Figure 5.19 is a static curve representing the relationship between input and output signals. It is obvious that the properties of the static curve are closely related to the selection of four parameters: limiter threshold (LT), compressor threshold (CT), expander threshold (ET), and noise gate threshold (NT). The limiter keeps the output signal amplitude stable within a level amplitude; the compressor compresses the dynamic range of the signal level; the expander expands the dynamic range of low-level signals; the noise gate can reduce noise.

Figure 5.20 is the schematic block diagram of implementing the static curve module, judging the amplitude range of the input level, and processing the input signal differently according to different ranges. In the figure, LS, CS, ES, and G_L, G_S, G_E represent the function slopes and intercepts of the limiter, compressor, and expander parts, respectively, and M_gain represents the gain obtained by medium amplitude signals.

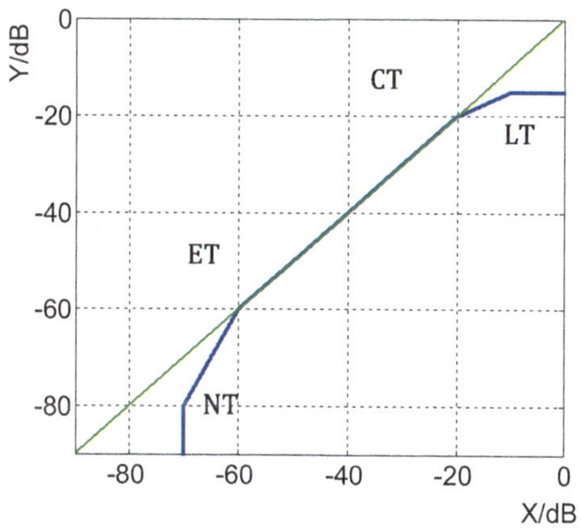

Fig. 5.19 Static curve of output input signal relationship

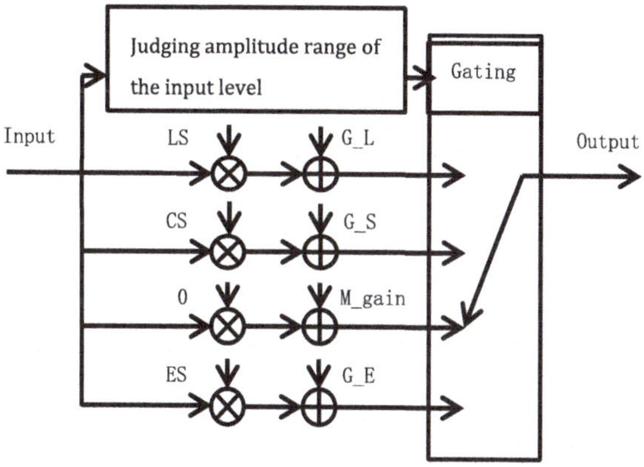

Fig. 5.20 Schematic diagram of the static curve module

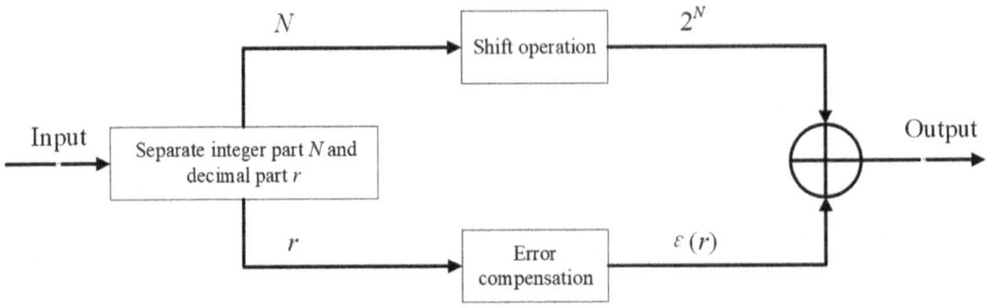

Fig. 5.21 Schematic diagram of binary exponential algorithm

The main role of the exponential operation in the audio dynamic range control algorithm is to convert the gain factor in decibel value form into a general numerical form, so it can also be understood as inverse logarithm operation. Similar to logarithm operation, this part selects the binary logarithm algorithm to implement this module.

This error can also be compensated by a linear approximation method. The complete binary exponential operation method schematic block diagram is shown in Fig. 5.21. The integer part N and the decimal part r of the input value are separated. The integer part can obtain the power of 2 through shift operation 2^N, and the decimal part can obtain the approximate value of the power of 2 through error compensation $\varepsilon(r)$. multiplying the two terms yields the power of 2 of the input value.

The implementation formula of the gain factor smoothing processing module is as follows: $g(n) = (1 - k) \cdot g(n - 1) + k \cdot f(n)$, where, the parameter k is determined by comparing the input gain factor with the size of the previous factor: when the input gain factor is greater than the previous output factor, k the value of is taken as the attack time parameter AT. Otherwise, k the value of is taken as the release time parameter RT.

The main function of this system is to smooth the control signal, avoid drastic changes in the gain factor, and thus alleviate the signal distortion caused by different gains in different parts of the audio signal.

The implementation of the dynamic range control algorithm can be completed by integrating the above five main modules according to the structure of Fig. 5.17. To save hardware resources and reduce processing delay, this section has carried out floating-point to fixed-point processing of the entire system's operations. The methods and the significance of data pointization are detailed in references [434, 435].

To compare the binary method used in this section when choosing logarithmic and exponential algorithms with other methods, this section will compare the system implemented by this method with the system implemented by the coordinate rotation algorithm, and evaluate the absolute error and hardware resource consumption of the two methods.

Figure 5.22 shows the absolute error between the output audio signal obtained by processing the same audio segment with two methods and the standard value obtained by running the algorithms. Among them, Fig. 5.22a is the absolute error of

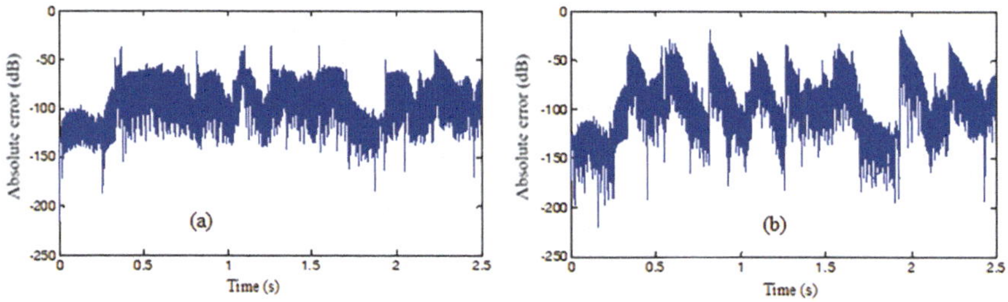

Fig. 5.22 System absolute error of coordinate rotation method and binary method implementation algorithm

the algorithm implemented by the coordinate rotation method, and Fig. 5.22b is the absolute error of the algorithm implemented by the binary method. As can be seen from the figure, the absolute error of the two methods is below -50 dB in most cases. In some cases, the system implemented by the binary method can obtain a smaller absolute error, but the coordinate rotation method can obtain a more stable absolute error range.

The Resource Usage tool in DSP builder is used to automatically evaluate the consumption of FPGA hardware resources. The main resources analyzed are logic units (Logic), Random Access Memory (RAM), and Multiplier. The comparison of resources consumed by two methods of implementing the algorithm is shown in Table 5.5.

The consumption of RAM and Multiplier by the two methods is roughly equal, but in terms of Logic consumption, the binary method is only a quarter of the CORDIC.

In Fig. 5.23, the upper waveform is the audio frequency time domain waveform played by the parametric array loudspeaker without the DRC module, and the lower waveform is the audio frequency time domain waveform played after adding the DRC module. The two waveforms show normalized amplitude values. By comparing the upper and lower waveforms, it is evident that after adding the dynamic range control module, the distortion phenomenon in the loudspeaker's audio frequency playback is significantly eliminated. On the other hand, in the small signal part, after adding the dynamic range control module, the audio frequency played is also relatively improved compared to when there is no dynamic range control module.

Table 5.5 Comparison of resource consumption for two methods of implementing dynamic range control algorithm

Implementation method	Coordinate rotation method (%)	Binary method (%)
Logic	8	2
RAM	<1	<1
Multiplier	8	9

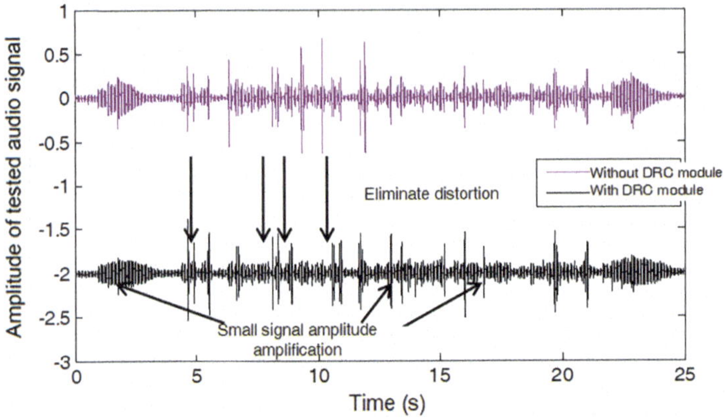

Fig. 5.23 Test sound frequency signal amplitude

5.3.6 Single-Sideband Modulation and Its Implementation on FPGA

Currently the main modulation methods for parametric array loudspeakers are double-sideband and single sideband modulation. Compared with double-sideband modulation, single sideband modulation reduces harmonic distortion and intermodulation distortion, so single sideband amplitude modulation is chosen on the platform. The implementation methods of single sideband modulation generally include filter method, phase shift method, and Weaver method. It should be noted that these methods for implementing single sideband are all methods in the communication field. The single sideband signal does not carry a carrier signal, but the parametric array loudspeaker must have a carrier in the demodulation process, so the single sideband signal in this section includes the carrier signal.

The filter method is the simplest method to implement single sideband modulation, but the problem is that the filter is not easy to implement. Because the ideal filter cutoff frequency does not exist, the actual filters all have a transition band from the passband to the stopband. The difficulty of implementing the transition band is related to the normalized value of the transition band relative to the carrier frequency. The smaller the value, the more difficult it is to implement the filter. A high-pass filter with a cutoff frequency of 500 Hz is added before the signal input of the parametric array loudspeaker, so the proportion of the low-frequency part in the signal bandwidth is not significant, which makes the design of the filter less challenging. The algorithm of the filter method for single sideband modulation is shown in Fig. 5.24, and the time domain expression is

$$\begin{aligned} S_{VSSB}(t) &= (1 + mg(t))\cos(\omega_c t) * h_v(t) \\ &= \frac{m}{2}\cos\big(\omega_g + \omega_c\big)t + \cos(\omega_c t) \end{aligned} \tag{5.8}$$

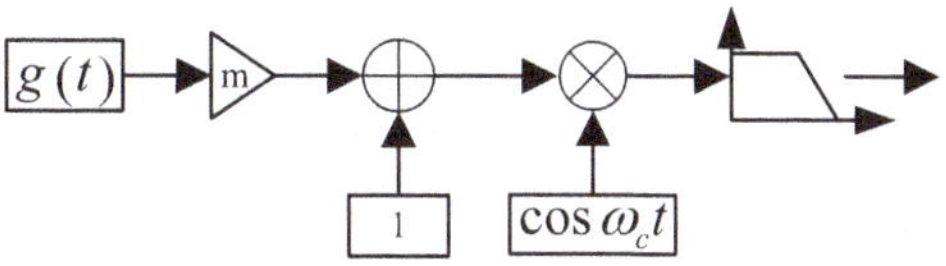

Fig. 5.24 Filter method single sideband modulation algorithm diagram

where m is the modulation index, $h_v(t)$ is the impulse responses of the filter, ω_c and ω_g are the carrier angular frequency and the input signal angular frequency, respectively.

According to the principle of the filter method for single sideband, a carrier of 40 kHz frequency needs to be generated at a sampling rate of 192 kHz. After completing double-sideband amplitude modulation, the lower sideband is retained through a low-pass filter, and the upper sideband is suppressed, thus obtaining a single sideband signal. The low-pass filter uses an elliptical structure IIR filter. The passband cutoff frequency is 40 kHz, the stopband cutoff frequency is 40.5 kHz, making the stopband amplitude attenuation not less than 60 dB, and the order of the filter is calculated to be 12. According to the bandwidth of the input signal, assuming the signal is a mixed frequency signal of 500 Hz, 2 kHz, 5 kHz, and 10 kHz, the simulation results of single sideband modulation are shown in Fig. 5.25. It can be seen from the figure that the upper sideband signal amplitudes corresponding to 500 Hz, 2 kHz, 5 kHz, and 10 kHz are −63 dB, −62 dB, −65 dB, and −68 dB, respectively, meeting the design requirements.

The phase shift method for single sideband modulation method is to shift the phase of the input signal by 90° through Hilbert transform. When the signal frequency is positive, it shifts −90°, and when the signal frequency is negative, it shifts + 90°. When the signal undergoes Hilbert transform and then undergoes orthogonal modulation with the input signal, a single sideband signal with only one sideband is obtained. Using the traditional analog method, the 90-degree phase delay of the signal requires a filter network to implement. However, the filter network is not easy

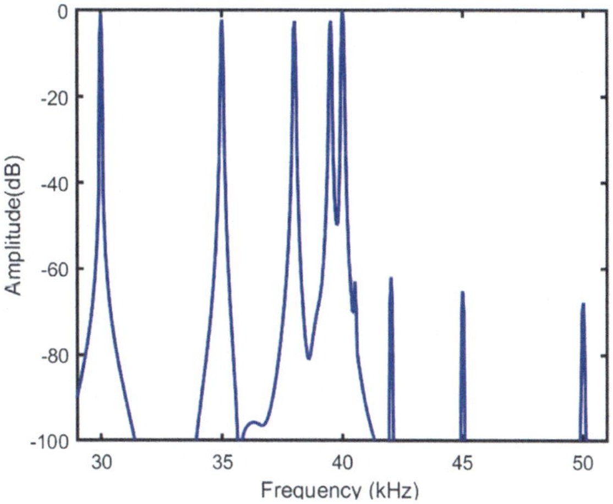

Fig. 5.25 Simulation results of filter method single sideband modulation

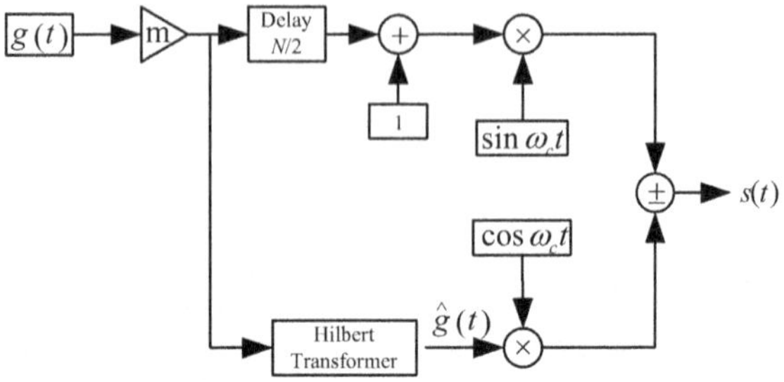

Fig. 5.26 Phase shift method single sideband modulation algorithm diagram

to implement, and its performance is difficult to meet the requirements. In contrast, it is much easier to implement in the digital domain. Hilbert also has its inherent shortcomings. The Hilbert transform requires that the signal after the transform has a strict 90-degree phase difference with the input signal, and the amplitude must remain consistent. Otherwise, the modulation result will not be a standard single sideband signal. The time domain expression of single sideband modulation by phase shift method is

$$\begin{aligned} S_{HSSB}(t) &= \left[1 + m\left(g(t) \pm \hat{g}(t)\right)\right] \cos(\omega_c t) \\ &= \cos(\omega_c t) + m \cos\left(\omega_c \pm \omega_g\right) t \end{aligned} \tag{5.9}$$

where $g(t)$ and $\hat{g}(t)$ are respectively the input signal and its orthogonal signal. The block diagram of the phase shift method single sideband modulation algorithm is shown in Fig. 5.26, where module "1" represents the added carrier signal. The module "Delay $N/2$" is to achieve synchronization of $g(t)$ and $\hat{g}(t)$, where N is the order of the Hilbert filter. In order for the delay of the module "Delay $N/2$" to be an integer, N needs to be an even number.

According to the principle of phase shift method, a pair of 40 kHz orthogonal carriers need to be generated at a sampling rate of 192 kHz. Since the order of the Hilbert FIR filter corresponding to the 192 kHz sampling rate is too high, the signal needs to be Hilbert transformed at a lower sampling rate, which is defined as the Hilbert sampling rate. According to the frequency band range of the input signal, 24 kHz is used as the Hilbert sampling rate. Finally, it is restored to 192 kHz through an interpolation factor of 8, and orthogonal modulation is performed at a sampling rate of 192 kHz. To make the upper sideband attenuation greater than 60 dB, the order of the Hilbert FIR filter is calculated to be not less than 100. The simulation uses an input signal that is a mixed-frequency signal of 500 Hz, 2 kHz, 5 kHz, and 10 kHz, and the simulation results are shown in Fig. 5.27. The upper sideband signal amplitudes corresponding to 500 Hz, 2 kHz, 5 kHz and 10 kHz are −60 dB, −78 dB, −82 dB and −80 dB, and the upper sideband amplitude is all less than −60 dB, so the single sideband modulation obtained by this parameter basically meets the requirements.

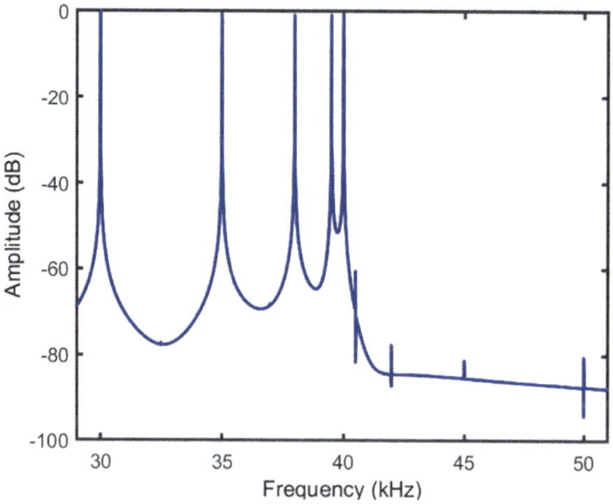

Fig. 5.27 Simulation results of phase shift method single sideband modulation

The Weaver method for the single sideband modulation is completed through two steps of orthogonal modulation. Assuming the input signal $g(t) = A_0 \cos(\omega_0 t)$, the first orthogonal modulation uses orthogonal carrier $\omega_m/2$ (where ω_m is the highest frequency of the input signal). After modulation, these two terms are obtained, i.e., X_{11} and X_{21}

$$
\begin{aligned}
X_{11} &= A_0 \cos(\omega_0 t)\cos(\omega_m t/2) = \frac{1}{2}A_0 \cos(\omega_m/2 + \omega_0)t \\
&+ \frac{1}{2}A_0 \cos(\omega_m/2 - \omega_0)t \\
X_{21} &= A_0 \cos(\omega_0 t)\sin(\omega_m t/2) = \frac{1}{2}A_0 \sin(\omega_m/2 + \omega_0)t \\
&+ \frac{1}{2}A_0 \sin(\omega_m/2 - \omega_0)t
\end{aligned}
\tag{5.10}
$$

Using a low-pass filter with bandwidth $B \geq \frac{\omega_m}{2}$, to obtain X_{12} and X_{22}

$$
\begin{aligned}
X_{12} &= \frac{1}{2}A_0 \cos(\omega_m/2 + \omega_0)t \\
X_{22} &= \frac{1}{2}A_0 \sin(\omega_m/2 + \omega_0)t
\end{aligned}
\tag{5.11}
$$

Then use another pair of orthogonal carriers with frequencies $\omega_c \pm \omega_m/2$ to shift the spectrum to the appropriate position, obtaining signals X_{13} and X_{23}, adding and subtracting the two can obtain the single sideband modulated signal. Both carrier and superposition take the plus sign to obtain the upper sideband, and both take the minus sign to obtain the lower sideband.

$$
S_{WSSB}(t) = X_{12}\cos(\omega_c \pm \omega_m/2)t \pm X_{22}\sin(\omega_c \pm \omega_m/2)t
$$

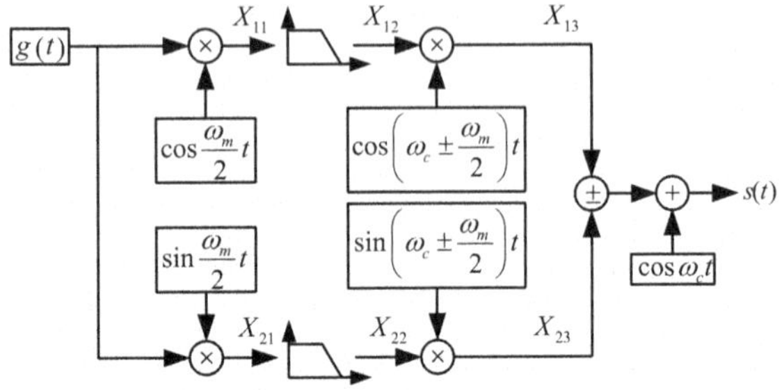

Fig. 5.28 Weaver single sideband modulation algorithm diagram

$$= \frac{1}{2}A_0(\cos(\omega_m/2 - \omega_0)t \cos(\omega_c \pm \omega_m/2)t$$
$$\pm \sin(\omega_m/2 - \omega_0)t \sin(\omega_c \pm \omega_m/2)t) \tag{5.12}$$

As can be seen from the derivation process, the Weaver method does not require a 90° phase shift of the input signal, thus eliminating the Hilbert transform. However, it needs to generate two orthogonal carrier signals of different frequencies, and after the first modulation, it needs 2 low-pass filters for filtering. In addition, the single sideband signal obtained from Eq. (5.12) does not have a carrier frequency of 40 kHz, so this carrier signal must be added before the signal is output. The entire block diagram of the Weaver algorithm is shown in Fig. 5.28.

According to the principle the Weaver method for single sideband modulation and the bandwidth of the input signal, it is necessary to generate two pairs of orthogonal carriers and a carrier of frequency 40 kHz. The frequency of the first pair of orthogonal carriers is 6 kHz, the frequency of the second pair of orthogonal carriers is 34 kHz. The low-pass filter uses an elliptical structure IIR filter. The passband cutoff frequency is 6 kHz, the stopband cutoff frequency is 6.5 kHz, making the stopband amplitude attenuation not less than 60 dB, and the filter order is calculated to be 8th-order. The input signal still uses mixed signals of 500 Hz, 2 kHz, 5 kHz, and 10 kHz, and the simulation results are shown in Fig. 5.29. From the Fig. 5.29, it is obvious that the upper sideband signal amplitudes corresponding to 500 Hz, 2 kHz, and 5 kHz are −60 dB, −64 dB, −64 dB respectively, and the upper sideband signal amplitude corresponding to 10 kHz is already less than −100 dB, meeting the design requirements.

Although all three methods can achieve single sideband modulation, the complexity and required resources vary greatly. The following discusses the computational load of the three methods. The computational load is shown in Table 5.6 under the Quartus II 8.0 environment. From the Table 5.6, it is evident that the filter method consumes the least logic units, registers, and memory, followed by the phase shift method, and the Weaver method consumes the most. The phase shift method consumes the least 9bits multipliers, followed by the filter method, and the Weaver method consumes the most.

After comparison, it is known that: in terms of computational load, the filter method requires the least resources, and the Weaver method requires the most; in

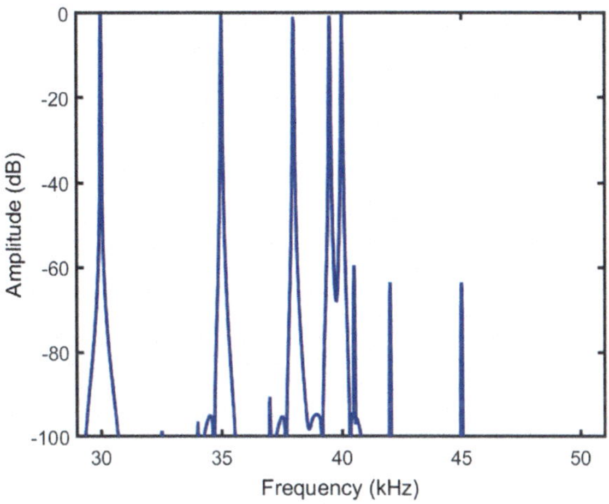

Fig. 5.29 Simulation results of Weaver method single sideband modulation

Table 5.6 Resource consumption of three single sideband modulation methods

	Logic units	Registers	Memory	9bits Multipliers
Filter method	2300	360	0	14
Phaseshift method	6560	1044	384	4
Weaver method	13,805	1724	384	20

terms of module design, the filter method is the simplest, and the Weaver method is the most complex; in terms of performance, the three algorithms do not differ much; for single-stage design, the filter method is most suitable; for multi-stage design, the phase shift method and the Weaver method are suitable.

Since the modulation technology of the parametric array loudspeaker does not exist in the intermediate frequency and radio frequency of the radio, a single-stage design can be well completed, so considering comprehensively, the filter method is chosen as single sideband modulation method of this system.

5.4 Integration and Test Results of Parametric Array Loudspeaker System

To realize the miniaturization of the parametric array loudspeaker system, the entire system is integrated, and the circuit physical diagram and overall appearance are shown in Figs. 5.30 and 5.31. The power input is a DC-05 terminal with positive inside and negative outside, powered by a 24 V power adapter. The working status of the entire circuit is displayed by 3 LEDs (yellow: top; red: middle; green: bottom): the green LED indicates that the system is working normally; the yellow LED indicates that the entire system is in a warning state due to overcurrent, overvoltage, or high

Fig. 5.30 Photo of the integrated circuit

Fig. 5.31 Appearance of the parametric array loudspeaker system

temperature; the red LED indicates that the system has a short circuit or other faults and the power must be turned off immediately. The audio input uses dual-channel input, which is added together in the FPGA to form a signal for modulation. When using, pay attention to ensure that the power supply, circuit board, and transducer array are properly connected before powering up. Then, observe the status indicator lights to determine if the system is operating normally. If a yellow or red light is on, the power should be turned off promptly.

The performance indicators of this parametric array loudspeaker system are: maximum sound pressure level: 85 dB@1 m/1 kHz; power consumption: 25 W; input sensitivity: 6 dBm (1.5 Vrms); signal bandwidth: 500 Hz–12 kHz; directivity: 5° @3 dB/1 kHz, 10° @6 dB/1 kHz, 24° @12 dB/1 kHz; harmonic distortion @ modulation index $m = 0.8$, see Fig. 5.32; frequency response curve, see Fig. 5.33.

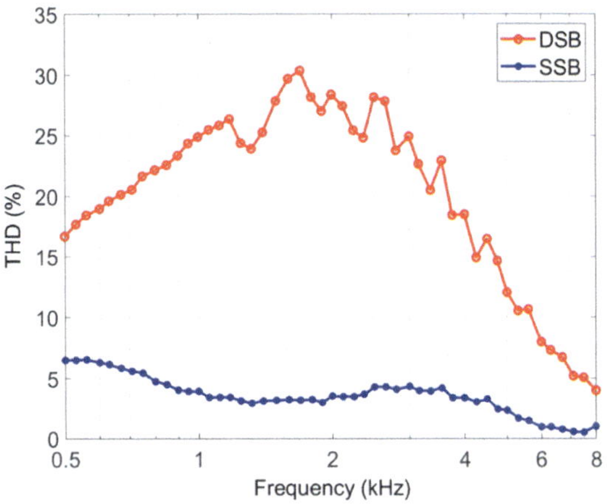

Fig. 5.32 Total harmonic distortion of double sideband modulation and single sideband modulation

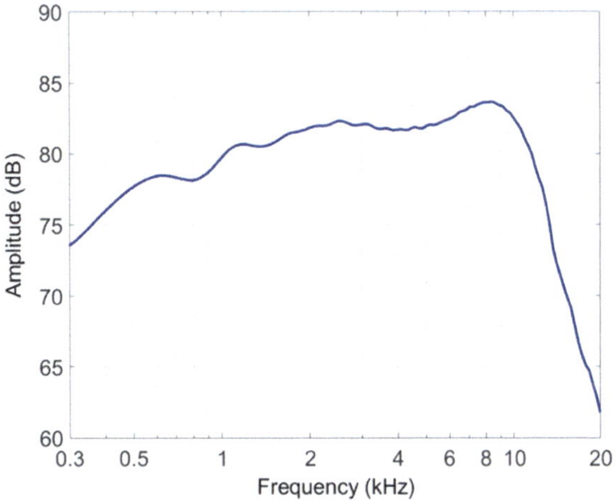

Fig. 5.33 Frequency response curve of the parametric loudspeaker array

5.5 Summary

This chapter concludes with a brief review of the methods of implementing parametric array loudspeakers in analog circuits and DSP, with a focus on FPGA implementation.

In terms of FPGA implementation, a parametric array loudspeaker system design and test based on the FPGA processor is presented, including hardware and algorithm parts. In terms of hardware, the design process of the entire system is described from four modules: power management, processor hardware, Codec chip hardware, and amplifier. After discussing the advantages and disadvantages of Class A, Class B, Class AB, and Class D amplifiers, a low-distortion Class D amplifier based on the TAS5630 chip is designed as the power drive unit of the system, and the impedance matching of the amplifier and the transducer array is completed, reducing useless power and improving amplifier efficiency. In terms of software, the internal serial and parallel data conversion, carrier generation, voice activity detection, dynamic range control, single sideband modulation, and AD1939 clock and register management are implemented on the FPGA platform. After analyzing the advantages and disadvantages of three single sideband modulation methods in terms of resource consumption and design difficulty, the filter method is adopted as the design method of the system's single sideband. On this basis, the integration of software and hardware is completed, and the performance indicators of the integrated parametric array loudspeaker are measured.

Chapter 6
Beam Control of Parametric Array Loudspeakers

According to the principle of the parametric array, the difference frequency sound of the parametric array loudspeaker has high directivity. This high directivity can make the difference frequency sound only emit to the target area, without polluting the sound environment outside the target area. Currently, there are two projection modes for using parametric array loudspeakers, namely ① directional mode and ② reflection mode. In addition to directly focusing on the audience (directional mode), the parametric array loudspeaker can also project the sound beam to the surface, thereby forming a virtual loudspeaker at the reflection point (reflection mode). In these two modes, if the direction of the sound beam produced by the parametric array loudspeaker can be further adjusted mechanically or electronically, it will be easier to apply to applications with tracking functions.

There are also many studies in this area. Since Woodward et al. [150] applied phased array technology to underwater parametric arrays for seabed mapping in 1994, which can meet the beam steering requirements within $\pm$ 20°, the current research focus mainly includes three aspects: directivity control methods, directivity models, and directivity design. This chapter will first review these three aspects, and then focus on the theory, simulation, and experimental research of beam control using fractional delay filters.

6.1 Review of Directivity Control Methods in Parametric Array Loudspeakers

This section mainly reviews the directivity control methods, directivity models, and directivity design.

J. Yang and P. Ji, *Parametric Array Loudspeakers*,
https://doi.org/10.1007/978-981-96-3548-1_6

6.1.1 Directivity Control Methods of Parametric Array Loudspeakers

Currently, in the design of parametric array loudspeakers, there are mainly two types of directivity control methods: mechanical methods and signal processing methods. The mechanical method is built with an electric rotating table and reflective objects, while the signal processing method is implemented with analog circuits or digital processors. The biggest disadvantage of the mechanical method is that its overall size is too large, and the extra reflection caused by the mobile rotating table and reflective objects is too large.

In 2005, Olszewski et al. [151] used a mechanical motor to rotate each channel's transducer, and then compensated for the phase of each channel by delaying, eliminating the sound path difference caused by rotation, thereby achieving the purpose of beam steering. However, this method has a slow beam steering speed and requires motor assistance, which adds extra equipment. The use of digital signal processing to achieve beam steering is low-cost, energy-saving, fast, and does not require additional auxiliary equipment, which is the trend of future development. The delay sum algorithm is a common method for array beamforming. This method is simple and practical, and can conveniently and quickly realize the beam steering of the parametric array loudspeaker. In 2006, Yang et al. [152] proposed to steer the parametric array difference frequency sound beam through the zero-order Bessel function weighting method. In addition, Yang et al. [153–155] Using the delay sum method, the Chebyshev weighting method combined with the SSB modulation method, the carrier and sideband frequencies are weighted separately to obtain a sound beam with a constant beam width. In actual implementation, the steering angle of the parametric array beam is limited by the sampling rate of the digital-to-analog converter. In the case of a low sampling rate, if an integer sampling period delay is used, although the parametric array beam can be steered, there is a minimum steering angle of the beam, and the steering angle of the beam is only distributed at a few discrete angles. In 2006, Gan et al. [156] pointed out that at a sampling rate of 192 kHz, the minimum steering angle is 26°. For this reason, Gan was able to achieve a very small steering angle with the smallest amount of calculation without increasing the sampling frequency by delaying the carrier and sideband signals separately, but this method requires the generation of an omnidirectional carrier beam, which greatly reduces the demodulation of the energy of the difference frequency sound. In 2010, Takeokal et al. [157] first used a high-pass filter to upsample and then delay the integer, and then used an inverter to replace the power amplifier to drive the ultrasonic transducer to emit sound waves. Although this method solves the problem of delay accuracy, the signal-to-noise ratio of this method depends heavily on the stability of the clock, and secondly, the output dynamic range is small, resulting in a small difference frequency sound pressure level.

For sound steering using array sensors, delay and beamforming are used [436], because it is a simple and powerful array signal processing algorithm. The idea of this processing algorithm is to apply appropriate delays to each transducer, so that

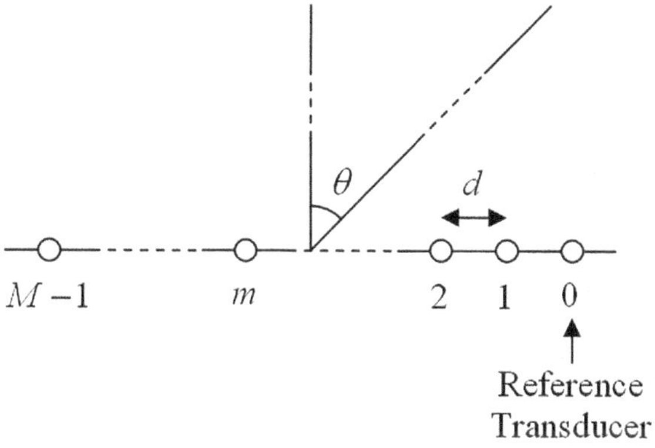

Fig. 6.1 Uniform linear array

the entire system has a maximum response in a certain arrival direction (Direction of arrival, DOA). Different weighting functions can also be added in this algorithm to control the beam width and the sidelobe of the array.

Consider a group of ultrasonic transducers arranged in a uniform linear array (ULA), with the spacing between units as shown in Fig. 6.1. Set an observation point in the far field of the array, its angle is θ, relative to the normal of the transducer array aperture. If each transducer is weighted w_m $(m = 0, 1, 2, \ldots, M-1)$, the array response function can be derived as

$$H(\omega\tau) = \frac{1}{M}\sum_{m=0}^{M-1} w_m e^{jm\omega\tau} \tag{6.1}$$

where $\tau = d/c \cdot \sin\theta$ is the time delay caused by the array geometry. By choosing appropriate weights, the level of the sidelobes can be adjusted at the expense of the beamwidth of the main lobe. As can be seen from Eq. (6.1), the maximum value of the main lobe occurs on the wide side of the ULA ($\theta = 0$). However, the maximum value of the main lobe can be changed by adding a phase offset or delay to each sensor. A time delays ($m\tau_0$) must be added to the transducers before the signal is transmitted to the medium in order to steer the ULA. The delay can be calculated as $\tau_0 = d/c \cdot \sin\theta_0$ and the array response of delay and beamforming becomes

$$H(\omega\tau) = \frac{1}{M}\sum_{m=0}^{M-1} w_m e^{-jm\omega(\sin\theta_0 - \sin\theta)d/c} \tag{6.2}$$

Then the far-field directivity of weighted array $D_{1a}(\theta)$ becomes

$$D_{1a}(\theta) = D_1(k_a, \theta)H(k_a, \theta) \tag{6.3}$$

where $D_1(k_a, \theta)$ is the aperture directivity of the frequency ω_a given in Eq. (1.75), and in Eq. (6.3) for far-field array response $H(k_a, \theta)$, w_{am} and ω_a are used to replace

w_m and ω. Similarly, the far-field directivity $D_{1b}(\theta)$ of the primary wave ω_b can be written as

$$D_{1b}(\theta) = D_1(k_b, \theta)H(k_b, \theta) \tag{6.4}$$

Therefore, by substituting Eqs. (6.3) and (6.4) into the far-field directivity function of the demodulated difference frequency, the beam pattern of the difference frequency sound can be obtained.

$$D_{-}(\theta) = D_1(k_a, \theta)H(k_a, \theta)D_1(k_b, \theta)H(k_b, \theta) \tag{6.5}$$

It shows that the directivity of the far-field difference frequency can be roughly estimated by the product of four terms: the aperture directivity of a single transducer for frequency ω_a, the aperture directivity of a single transducer for frequency ω_b of the aperture directivity, the far-field array response of the weighting function w_{am} and the far-field array response of the weighting function w_{bm}.

6.1.2 *Directivity Model of the Parametric Array Loudspeakers*

This section mainly introduces the directivity model of the parametric array loudspeaker. Currently, there are three main types of directivity models: product model, improved product model, and convolution model. They will be briefly introduced below.

6.1.2.1 Product Model

For the case where the non-linear effect of the pump wave occurs in the far field, the directivity of the pump wave in the far field must be considered, and the pump wave cannot be approximated as a collimated beam sound wave. Assume that the transducer is placed on the YOZ plane, and the sound source axis coincides with the X-axis, as shown in Fig. 6.2, (R, θ, η) and (r, γ, ϕ) are the spherical coordinates of the far-field point and the sound source point, respectively. The far-field sound pressure expressions of the pump waves with frequencies ω_1 and ω_2 are as follows:

$$p_{1,2} = P_{1,2}D_{1,2}(\gamma, \varphi)\exp\left[-\left(\alpha_{1,2} + ik_{1,2}\right)r\right]/r \tag{6.6}$$

where $P_{1,2}$ is the sound pressure amplitude of the two pump waves, $D_{1,2}(\gamma, \phi)$ is the normalized directivity function, $\alpha_{1,2}$ and $k_{1,2}$ are the attenuation coefficients and wavenumbers of the two pump waves. Ignoring the near-field nonlinearity and higher-order nonlinearity, then according to the Berktay far-field solution, the difference frequency sound at a farfield point (R, θ, η) can be expressed as

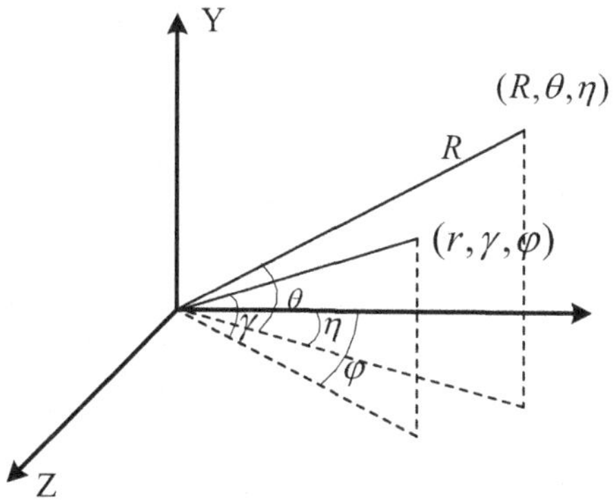

Fig. 6.2 Schematic diagram of the sound source coordinate system

$$p_d(R,\theta,\eta)=\frac{\omega_d^2P_1P_2\beta S_0}{4\pi\rho_0c_0^4}\iiint\frac{D_1(\gamma,\varphi)D_2(\gamma,\varphi)}{r^2s}\times\exp[-(\alpha_1+\alpha_2+ik_d)r-(\alpha_d+ik_d)s]dV \tag{6.7}$$

where the integral element $dV = r^2\cos\gamma d\gamma d\phi dr$, $s = |\mathbf{R}-\mathbf{r}|$. If the far-field point is sufficiently distant from the region of integration, the result can be derived

$$s = R - ru \tag{6.8}$$

where $u = \cos\gamma\cos\theta\cos(\phi-\eta)+\sin\gamma\sin\theta$. Assuming that the pump wave and the difference frequency sound have high directivity, the volume integral involves a small angle, and by approximating u to the second order, it can be derived:

$$u\approx 1-\frac{1}{2}(\gamma-\theta)^2-\frac{1}{2}(\varphi-\eta)^2 \tag{6.9}$$

Substitute Eqs. (6.8) and (6.9) into Eq. (6.7) to yield

$$p_d(R,\theta,\eta)=\frac{\omega_d^2P_1P_2\beta S_0}{4\pi\rho_0c_0^4R}\exp[-(\alpha_d+ik_-)R]\int_0^\infty\int_{-\pi/2}^{\pi/2}\int_{-\pi/2}^{\pi/2}D_1(\gamma,\varphi)D_2(\gamma,\varphi)\times\exp\{-[(\alpha_1+\alpha_2-u\alpha_d)+ik_d(1-u)]r\}\cos\gamma d\gamma d\varphi dr \tag{6.10}$$

Because $\alpha_d << \alpha_1+\alpha_2$, then an approximation can be made without losing accuracy

$$\alpha_1+\alpha_2-u\alpha_d\approx\alpha_1+\alpha_2-\alpha_d=\alpha_T \tag{6.11}$$

After integrating Eq. (6.10) with respect to r, the difference frequency sound becomes

$$\begin{aligned} p_d(R,\theta,\eta) &= \frac{\omega_d^2 P_1 P_2 \beta S_0}{4\pi \rho_0 c_0^4 R \alpha_T} exp[-(\alpha_d + ik_d)R] \\ &\times \int\limits_{-\pi/2}^{\pi/2} \int\limits_{-\pi/2}^{\pi/2} \frac{D_1(\gamma,\varphi) D_2(\gamma,\varphi) cos\gamma}{1 + i(k_d/2\alpha_T)((\gamma-\theta)^2 + (\varphi-\eta)^2)} d\gamma d\varphi \\ &= \frac{\omega_d^2 P_1 P_2 \beta S_0}{4\pi \rho_0 c_0^4 R \alpha_T} exp[-(\alpha_d + ik_d)R] \\ &\times D_1(\theta,\eta) D_2(\theta,\eta) cos\theta \otimes D_W(\theta,\eta) \end{aligned} \tag{6.12}$$

where $D_W(\theta, \eta)$ is the Westervelt directivity function. The directivity of the difference frequency sound is extracted from the above formula as the convolution of the product of the pump wave directivity and $D_W(\theta, \eta)$

$$D_d(\theta,\eta) = D_1(\theta,\eta) D_2(\theta,\eta) \cos\theta \otimes D_W(\theta,\eta) \tag{6.13}$$

Assuming that the Westervelt directivity is very sharp, close to the shock function, then the directivity of the difference frequency sound can be changed to the product form of the pump wave directivity

$$D_d(\theta,\eta) \approx D_1(\theta,\eta) D_2(\theta,\eta) \tag{6.14}$$

6.1.2.2 Improved Product Model

In order to improve the accuracy of the product directivity model, Shi et al. introduced the equivalent Gaussian source array [159]. In the equivalent Gaussian source array, the directivity of each Gaussian source is approximately a lobe in the directivity of the main frequency. The coefficients of the equivalent Gaussian source array are determined by nonlinear regression analysis based on the following model.

$$D(\theta) = \sum_{n=1}^{N} p_n \exp\left[-\left(\frac{\omega a_n}{2c_0}\right)^2 \tan^2(\theta - \theta_n)\right] \tag{6.15}$$

where θ is the independent variable, ω is the angular frequency of the pump wave, N is the number of Gaussian sources, p_n, a_n and θ_n are the amplitude, radius, and rotation angle of the nth Gaussian source. The directivity of the main frequency provides observational data in nonlinear regression analysis. Using an equivalent Gaussian source array can make the Gaussian source assumption valid in parametric array loudspeakers.

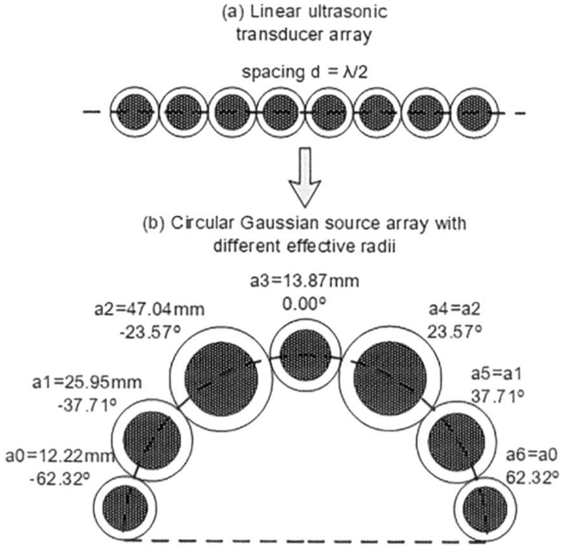

Fig. 6.3 **a** A uniform linear ultrasonic sensor array ($M = 8$, spacing half wavelength) deformed into **b** an equivalent circular Gaussian sound source array ($N_{\text{lobe}} = 7$)

Taking a uniform linear ultrasonic sensor array with 8 half-wavelength spacings as an example, as shown in Fig. 6.3a. A total of 7 lobes were observed in its beam pattern, as shown in Fig. 6.4. The configuration of the equivalent Gaussian source array is obtained by using Gaussian beams to approximate the response of the measured ultrasonic transducer. When the pump wave is set to 40 kHz, a Chebyshev weight with 20 dB attenuation is used in the ultrasonic sensor array. As shown in Figs. 6.3b and 6.4, the size of the Gaussian source varies according to the beam width of each lobe, and the axis of the Gaussian source is located at the peak of the corresponding lobe. A good match accuracy is observed between the beam pattern of the ultrasonic sensor array and the equivalent Gaussian source array.

Based on this equivalent Gaussian source array method, three improved forms have been proposed: advanced product directivity, exponential product directivity, and combined product directivity, among which the advanced product directivity model currently performs well. The model is represented as:

$$D_d(\theta) = \sum_{m=1}^{N_1}\sum_{n=1}^{N_2} \frac{p_{1,m}p_{2,n}a_{1,m}^2 a_{2,n}^2}{a_{1,m}^2 + a_{1,n}^2} e^{\frac{-a_{1,m}^2\omega_1^2\tan^2\left(\theta-\theta_{1,m}\right)-a_{2,n}^2\omega_2^2\tan^2\left(\theta-\theta_{2,n}\right)}{4c_0^2}} \tag{6.16}$$

In Eq. (6.16), $N1$ and $N2$ represent the number of Gaussian sources of the two pump waves, respectively. The principle of product directivity is applied to each pair of Gaussian sources in the equivalent Gaussian source array. The directivity of the difference frequency is obtained by superimposing all product directivities. In the experimental verification in reference [159], the advanced product directivity model showed higher accuracy compared to the traditional product directivity principle,

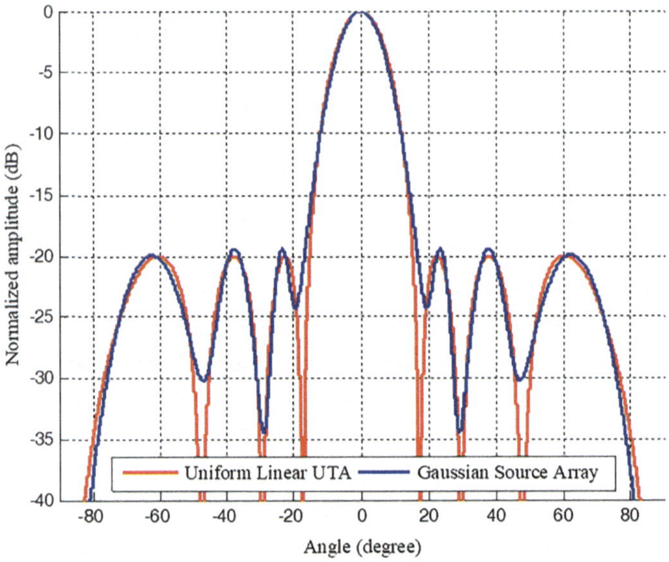

Fig. 6.4 A beam diagram of an ultrasonic sensor array with 8 channels using Chebyshev weighting (20 dB attenuation) and its equivalent Gaussian sound source array

while the other two models showed similar accuracy, but with higher computational complexity.

6.1.2.3 Convolution Model

Recently, a convolution model based on Westervelt directivity has been proposed by scholars, which provides an accurate directivity model for parametric array loudspeakers. For a detailed derivation of this model, please refer to reference [160]. The directivity of the difference frequency sound is expressed as follows:

$$D_d(\theta) = [D_1(\theta)D_2(\theta)] \otimes D_W(\theta) \tag{6.17}$$

where $\otimes$ represents the discrete convolution operation, $D_1(\theta)$ and $D_2(\theta)$ are the directional functions of the two pump waves respectively, $D_W(\theta)$ is the Westervelt directional function.

In Eq. (6.17), the product directivity and Westervelt's directivity are convolved to obtain the directivity of the difference frequency sound. Therefore, it is named as the convolution model. Reference [160] has already provided some experimental verification of the convolution model. Here is another experimental verification [161]. Figure 6.5 shows a picture of a parametric array loudspeaker composed of 8 channels. When the steering angle is set to 0°, all channels of this parametric array loudspeaker are driven by the same modulated carrier. At the difference frequencies of 1, 2, 4,

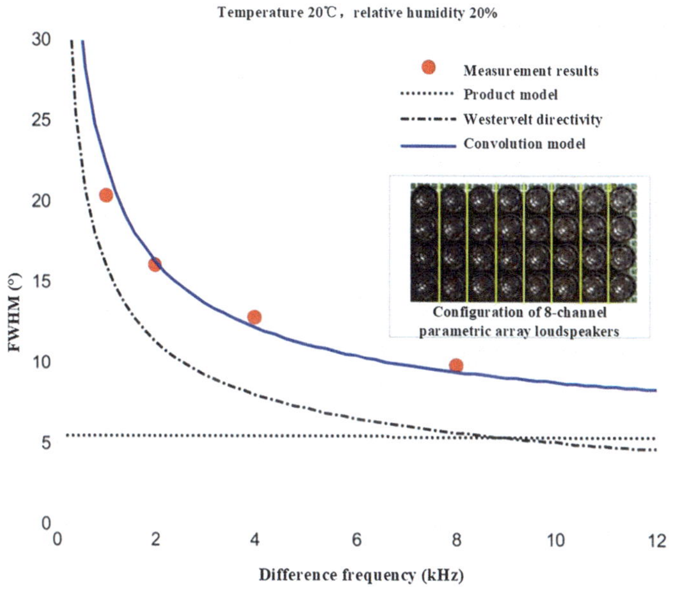

Fig. 6.5 Experimental verification of the convolution model

8 kHz, the directivity of the full width of the beam at half its maximum amplitude (FWHM) is measured. The measurement results are compared with the product directivity model, Westervelt's directivity model, and the convolution model. As shown in Fig. 6.5, the convolution model basically overlaps with the measurement results, and the effect is the best.

6.1.3 Directional Design of Parametric Array Loudspeakers

6.1.3.1 Constant Beamwidth Beamformer

The work on beam width control of parametric array loudspeakers is relatively less. In the related research [155], an algorithm with Chebyshev weighting, based on delay and beamforming technology, is proposed to control the beam width and sidelobe of the difference sound generated by the pump wave. In practical applications, the audio signal modulated by the ultrasonic carrier cannot be considered narrowband, so a wideband beamformer is very desirable, which is also the current technical interest. In the beamformer with constant beam width in the parametric array loudspeaker, the sidelobe attenuation of the carrier frequency is determined by the beamwidth from zero to zero, and the sidelobe attenuation of the sideband frequency is adjusted to make the first sidelobe of the sideband frequency coincide with the first zero point of the carrier frequency.

According to the commonly used convolution model, the beam width of the demodulated difference sound will basically decrease with the increase of frequency, which is mainly caused by the Westervelt beam pattern. Therefore, to establish a wideband beamformer with constant beam width, it is necessary to compensate for the Westervelt beam pattern. Here, to meet a larger dynamic range, the Dolph-Chebyshev weighting function is used. The product beam pattern of the ultrasonic frequency can be mainly controlled by the sideband attenuation. Shi et al. [437] gave a simulation result. They fixed the sidelobe attenuation of the carrier frequency at 6dB, and considered the upper sideband and lower sideband modulation methods. The sidelobe attenuation of the sideband varies between 6 and 50 dB. The working frequency is from 500 to 8000 Hz, and the constant half-power bandwidth is 25°. Here, the difference between the two modulation methods is not significant. In this way, a beamformer with constant beam width can be implemented in the parametric array loudspeaker, as shown in Fig. 6.6. It is evident that in the range of 500 to 8000 Hz, the parametric array loudspeaker basically maintains a constant half-power bandwidth of about 25°.

Recently, Gou et al. [438] proposed to approximate the directivity of Westervelt with a mixed Gaussian directivity method, so that the convolution model can be decomposed into solvable equations. Taking the steering angle of 20° as an example, the comparison result is shown in Fig. 6.7, which proves the effectiveness of this method in the beam control design method of the parametric array loudspeaker by comparing with the simulation results of the *k*-wave method [439, 440].

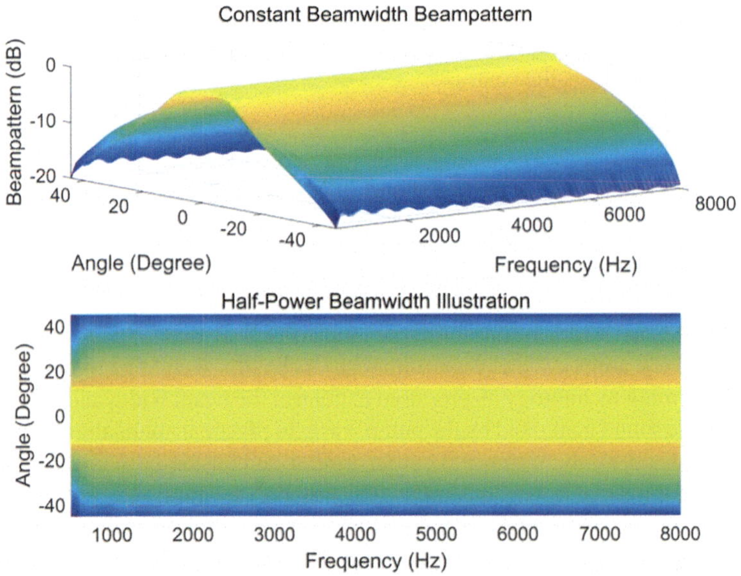

Fig. 6.6 Constant beamwidth

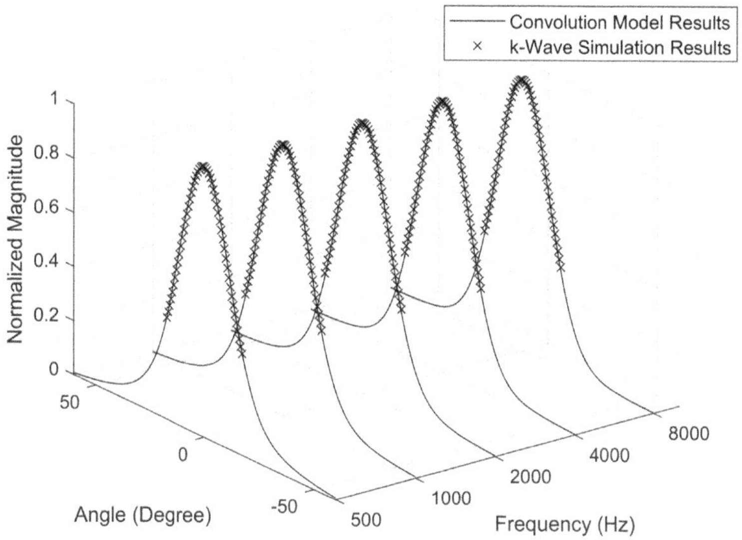

Fig. 6.7 Comparison of results using the mixed Gaussian directivity method with k -wave results (20°)

6.1.3.2 Digital Beamsteering of Parametric Array Loudspeakers

The implementation of the parametric array loudspeaker system in digital domain has many advantages. First, the performance of the digital signal processing (DSP) system is not affected by hardware aging. Secondly, DSP has high flexibility and can implement different algorithms without changing the hardware configuration. But there are also some limitations. A major limitation of this system is that the delay depends on the sampling rate of the DSP. For example, in a DSP board with a sampling rate of 160 kHz, the sampling period can be calculated by Eq. (6.18).

$$T = 1/F_S = 1/160 \text{ kHz} = 6.25 \ \mu s \tag{6.18}$$

The minimum value of delay τ_0 is equal to the sampling period T, and the delay τ_0 is rewritten as

$$\theta_0 = \sin^{-1}(\frac{\tau_0 c}{d}) \tag{6.19}$$

In practical situations, to avoid grating lobes, some small transducers need to be closely arranged. The transducer elements are arranged in a hexagonal pattern to minimize the distance between two adjacent groups of transducers. For a circular transducer with a diameter of 9.7 mm, the spacing between its elements is d = 4.9 mm. From Eq. (6.19), it is obvious that the minimum steering angle is $\theta_0 = \sin^{-1}\frac{6.25\ \mu s \times 344\ \text{ms}^{-1}}{4.9\ \text{mm}} = 26.03^\circ$. This means that it is impossible to steer the difference frequency to a smaller angle than using the sampling delay method. Similarly, if the delay time is twice the sampling period ($\tau_0 = 2T$), the steering angle is 61.35°.

One method to improve the steering angle is to increase the sampling frequency. However, this method is costly and inefficient. For example, if the sampling frequency is doubled (320 kHz), the minimum sampling delay, τ_0 is 3.125 μs. The minimum steering angle is 12.67°. For smaller steering angles, the computation load doubles. Moreover, the higher the sampling rate, the higher the cost of ADC and DAC.

An alternative method to improve the steering angle is to implement fractional delay [441, 442] or perform beamsteering in the frequency domain. There are several ways to implement fractional delay. The first and simplest method is interpolation. This is achieved by upsampling the modulated signal and then applying a low-pass interpolation filter, which requires high computational power. A delay is imposed on the upsampled signal. The delayed signal is then downsampled before being sent to the driver.

Another way to achieve fractional delay is to obtain a continuous-time version by interpolating the modulated signal. This process is usually accomplished by infinite convolution, which is impossible to implement. In practical applications, it is necessary to truncate the convolution length, but this introduces some errors. After obtaining the continuous modulated signal, it is precisely delayed. Finally, the delayed signal is resampled and sent to the drive circuit. The methods of fractional delay will be detailed in Sect. 6.2 below.

One drawback of the above two time-domain interpolation methods is that they introduce errors, and the reconstructed signal is inaccurate. These methods also require high computational power, and need a low-pass interpolation filter, where the latter requires long convolution.

When performing beamsteering in the frequency domain, an accurate delay can be obtained. This is achieved by performing a discrete Fourier transform on the modulated signal. An equivalent phase shift (delay) is added to the transformed signal. Then the signal is transformed back to the time-domain signal by inverse discrete Fourier transform.

The purpose of the carrier frequency weighting function w_{am} is to control the beam width of the difference frequency and attenuate the sidebands of the carrier frequency. As for the sideband frequency weighting function w_{bm}, it generates a flat directional response (similar to a low-pass filter) within a certain angle range, making the amplitude gain of the difference frequency the same at different steering angles. The weighting function w_{bm} can also attenuate the sidelobes of the sideband frequencies, resulting in lower side lobe directivity for the generated difference frequencies.

The FIR filter coefficients can be calculated using the Parks-McClellan optimal equivalent FIR filter design method [443]. It is used to design an M length FIR filter, which has the best approximation to the required frequency response. During simulation, the carrier frequency of the lower sideband modulation is set to 40 kHz. The sampling frequency of the DSP board is 160 kHz. The input frequency of the beam steering system ranges from 500 Hz to 19.5 kHz, with an increment of 500 Hz. It forms $M = 32$ units in total, with a spacing between units of $d = 4.9$ mm.

Table 6.1 Weight function of carrier and sideband frequencies

m	Weight function		
	w_{am}	w_{bm} (LSB)	w_{bm} (USB)
0	0.2843	−0.0246	0.1047
1	0.1906	−0.1607	0.0014
2	0.2484	0.0078	−0.0445
3	0.3124	0.0495	0.0242
4	0.3817	0.0563	0.0363
5	0.4549	−0.0168	−0.0329
6	0.5303	−0.0850	−0.0436
7	0.6062	−0.0530	0.0661
8	0.6806	0.0646	0.0262
9	0.7515	0.1297	−0.0916
10	0.8168	0.0290	−0.0096
11	0.8746	−0.1642	0.1447
12	0.9231	−0.2103	−0.0472
13	0.9609	0.0718	−0.2397
14	0.9868	0.5885	0.2300
15	1.0000	1.0000	1.0000

The effective sound source radius is set to $a = 3.85$ mm. The speed of sound $c = 344\ \text{ms}^{-1}$, the beam width of the difference frequency θ_- is 10°. According to the design procedure of the weight function, the calculated weight function of the carrier frequency w_{am}, and the calculated weight function of the sideband frequency w_{bm}, are shown in Table 6.1. Since the weight function is symmetrical, only the first half (truncated) of the weight function is listed in Table 6.1.

By adopting the given beam steering algorithm, Fig. 6.8 is the beam pattern of the carrier frequency 40 kHz (solid line), sideband frequency 30 kHz (dotted line) and difference frequency 10 kHz (dashed line). The steering angle of the difference frequency is set to 10°, with a delay of

$$\tau_{a0} = \frac{4.9\ \text{mm}}{344\ \text{ms}^{-1}} \sin^{-1} 10^\circ = 2.47\ \mu s,\ \tau_{b0} = 0 \times T = 0\ \text{s} \tag{6.20}$$

One characteristic of the given beam steering algorithm is that the directivity of the sideband frequency is almost flat (small ripples). This means that the directivity of the sideband frequency is a weak function of the main lobe angle. Therefore, the directivity of the difference frequency inherits the directivity of the carrier at the main lobe angle of the sideband frequency. The steering angle and main lobe width of the difference frequency also inherit the directivity of the carrier frequency.

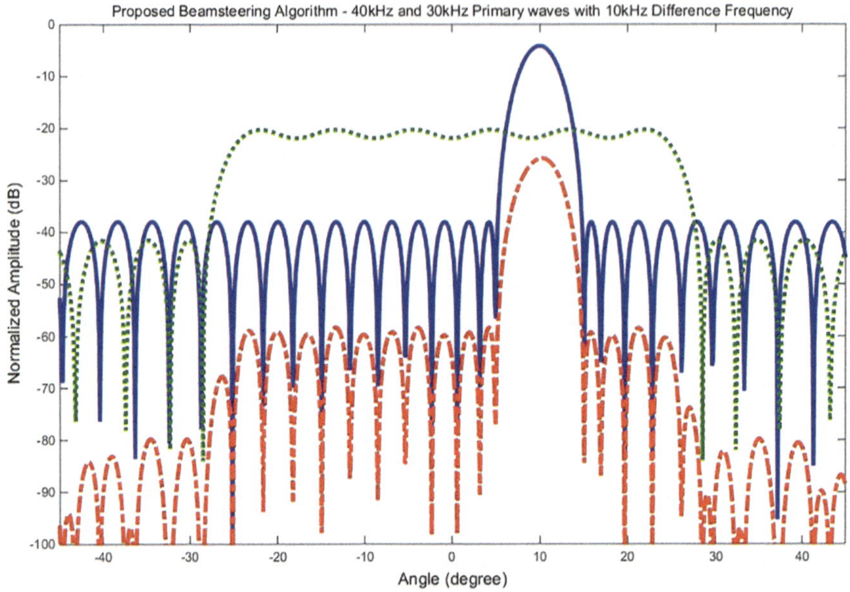

Fig. 6.8 Steering of difference frequency (10 kHz) at $\theta_{-} = 10°$

6.2 Application of Fractional Delay Filters in Beamsteering of Parametric Array

In the above applications of parametric array beamsteering, integer delay algorithms are used in all cases, so the accuracy of the beam steering angle is not high. The fractional delay algorithm can produce non-integer delays, thereby improving the accuracy of the beamsteering angle. First, various fractional delay algorithms are reviewed, and then the fractional delay filter is applied to the beamsteering of the parametric array, and theoretical and experimental comparisons are made.

6.2.1 Review of Fractional Delay Algorithm

Suppose the input signal is $x(n)$, the output after a linear delay system is $y(n)$

$$y(n) = L[x(n)] = x(n - D) \tag{6.21}$$

where the delay D is split into an integer part $Int(D)$ and a decimal part d, such as $D = Int(D) + d$. If D is an integer, then the output is the data of a previous sampling point, which is relatively simple to handle. But if D is not an integer, the output falls between two previous sampling points. For the case where D is not an integer, there are usually two solutions: one is to convert the digital signal into a band-limited continuous signal, and then resample after precise delay; the other is to achieve the required decimal delay through the approximation of linear filter coefficients, which is also known as the fractional delay filter. Fractional delay filters are divided into IIR-based all-pass filters and FIR filters.

Generally speaking, for the same frequency domain characteristics, the resources (multipliers and memory) consumed by IIR filters are less than those of FIR filters. However, the disadvantage of this type of filter is that the phase response is non-linear, and the design of IIR to meet the requirements of amplitude and phase response (group delay) is much more complex than FIR. In addition, there are stability issues with IIR filters, because the system function of IIR filters has zeros and poles, and when the poles are not within the unit circle, the system will become unstable. Therefore, for real-time processing systems, it is very difficult to ensure that the poles are within the unit circle. In general, the IIR structure is convenient for designing conventional filters (such as low-pass, high-pass, band-pass filters), but it is more complex for filters with irregular amplitude responses. For these reasons, the FIR fractional delay filter is the main focus of this section. Depending on the coefficient design method, FIR fractional delay filters can be roughly divided into three types: least squares, maximum flatness, and equiripple methods.

6.2.1.1 Least Squares Fractional Delay Filter

The ideal fractional delay filter has an impulse response of

$$h_{id}(n) = \frac{\sin[\pi(n-D)]}{\pi(n-D)} = \mathrm{sinc}(n-D) \tag{6.22}$$

This is obviously impossible to achieve, and the design idea of the least squares method is to minimize the mean square error between the designed fractional delay filter and the ideal fractional filter. That is

$$E_1 = \frac{1}{\pi}\int_0^{\pi}\left|E\left(e^{j\omega}\right)\right|^2 d\omega = \frac{1}{\pi}\int_0^{\pi}\left|H\left(e^{j\omega}\right) - H_{id}\left(e^{j\omega}\right)\right|^2 d\omega \tag{6.23}$$

where $H\left(e^{j\omega}\right)$ is the transfer function of the designed filter, $H_{id}\left(e^{j\omega}\right)$ is the transfer function of the ideal filter. Converting from the frequency domain to the time domain yields

$$e_1 = \sum_{n=-\infty}^{\infty} |h(n) - h_{id}(n)|^2 \tag{6.24}$$

$h(n)$ can be specifically implemented through the following methods.

(a) Truncation of ideal impulse response

Because the impulse response of the ideal fractional delay filter is infinitely long and is a non-causal system, it is impossible to implement. By directly truncating $h_{id}(n)$, an approximate truncated impulse response can be obtained.

$$h(n) = \begin{cases} \mathrm{sinc}(n - D) & M \leq n \leq M + N \\ 0 & \text{others} \end{cases} \tag{6.25}$$

where M is the starting point of truncation, N is the length of truncation. Although this method is simple and feasible, the amplitude response and phase response of the filter both have serious Gibbs effects. In addition, the available bandwidth is very narrow, which is not a good solution for broadband signals.

(b) Window function method

The expression for the impulse response of the window function method is:

$$h(n) = \begin{cases} W(n - D)\mathrm{sinc}(n - D) & 0 \leq n \leq N \\ 0 & \text{others} \end{cases} \tag{6.26}$$

where N is the length of the impulse response, $W(n - D)$ is the applied window function. It is obvious that the window function method is a general form of the truncation method, because the truncation method can be regarded as truncation using a rectangular window. In 1978, Harris [444] summarized various window function methods. For example, the advantage of the Kaiser window is that it can control the amplitude of the ripple through a parameter, while the Dolph-Chebyshev window can achieve the smallest sidelobe amplitude. The window function method can improve the Gibbs phenomenon to a certain extent. Another advantage of the window function method is that it requires less computational effort and storage resources for generating filter coefficients online, making it suitable for real-time processing applications.

(c) Smooth transition transfer function method

The discontinuity of the window function will inevitably lead to the Gibbs effect. If the required frequency response is a smooth function rather than a discontinuous rectangle, the decay of the impulse response accelerates, which may cause the window function response that is not needed to be truncated. Burrus and Selesnick et al. [445, 446] used spline function models to generate fractional delay filters with arbitrary transition bands. They designed the passband within $[0, \omega_p]$ to have an ideal frequency response $e^{-jD\omega}$, the stopband within $[\omega_s, \pi]$ is 0, and the transition region is a P order spline function. Then the transfer function becomes

$$h(n) = \left\{ \frac{\sin[\frac{n-D}{2P}(\omega_s-\omega_p)]}{\frac{n-D}{2P}(\omega_s-\omega_p)} \right\}^P \frac{\sin[\frac{n-D}{2}(\omega_s+\omega_p)]}{\frac{\pi(n-D)}{2P}} \quad n = 0, 1, 2, \ldots, N \tag{6.27}$$

The smooth transition transfer function method maintains a constant amplitude response when the fractional delay changes, but this method makes the phase response very poor.

(d) General least squares FIR

When designing a fractional delay filter, the main focus is on the error within the passband of the designed filter and the ideal filter, so it is achieved by multiplying the frequency response by a weighting function. Then the error can be transformed into the form:

$$E_1 = \frac{1}{\pi}\int_0^{\alpha\pi} W(\omega)\left|E\left(e^{j\omega}\right)\right|^2 d\omega = \frac{1}{\pi}\int_0^{\alpha\pi} W(\omega)\left|H\left(e^{j\omega}\right) - H_{id}\left(e^{j\omega}\right)\right|^2 d\omega \tag{6.28}$$

where $W(\omega)$ is a non-negative frequency weighting function. This method mainly depends on the selection and weighting method of band approximation, but the cost is the increase of out-of-band error.

6.2.1.2 Maximum Flat Fractional Delay Filter

Maximum flatness refers to the Lagrange interpolation method. The Lagrange interpolation method was applied to the recovery of finite bandwidth analog signals during sampling in the 1950s [447] and the transformation of sampling rates [448]. According to the requirement of maximum flatness,

$$\left.\frac{d^n\left(e^{j\omega}\right)}{d\omega^n}\right|_{\omega=\omega_0} = 0 \quad n = 0, 1, 2, \ldots, N \tag{6.29}$$

Given that $E\left(e^{j\omega}\right)$ is a complex error function, transforming the frequency domain differential equation into a time-domain difference equation yields

$$\sum_{k=0}^{N} k^n h(k) = D^n \quad n = 0, 1, 2, \ldots, N \tag{6.30}$$

After determinant transformation, the filter coefficients can be calculated as

$$h(n) = \prod_{\substack{k=0 \\ k \neq n}}^{N} \frac{D-k}{n-k} \quad n = 0, 1, 2, \ldots, N \tag{6.31}$$

The advantage of the maximum flat fractional delay filter is that the calculation of coefficients is simple, the low frequency has a good response and a very flat amplitude response, but the disadvantage is that it is troublesome to update the coefficients in real time online. The first to use Lagrange interpolation to achieve fractional delay was Strube in 1975 [449] who approximated it using the Taylor series, but he did not know that the result he derived was the Lagrange interpolation algorithm. In 1988, Laine [450] applied Lagrange interpolation to fractional delay and obtained the maximum flat characteristics, but he did not give a strict mathematical derivation. In addition, to perform rate conversion, Liu and Wei [451] derived the maximum flat fractional delay filter using Lagrange interpolation, and obtained the general formula for the N order general FIR filter.

6.2.1.3 Equiripple Fractional Delay Filter

The disadvantage of the least squares and maximum flat fractional delay filters is that they cannot accurately control the peak error, for example, to design a fractional filter based on a given tolerance, these two methods are not suitable, while the equiripple fractional delay filter designed using the maximum and minimum criterion is very suitable. The error function constraint of the maximum and minimum criterion is

$$E_{\max} = \min\left\{ \max_{\omega \in [\alpha\pi]} \left\{ \left| E\left(e^{j\omega}\right) \right| \right\} \right\} \tag{6.32}$$

where $E\left(e^{j\omega}\right)$ is the error function. Under certain conditions, Eq. (6.32) will take on an equal ripple form. This method is complex because the values of the approximation function of the fractional delay filter are usually complex numbers, making it more complex than the linear phase FIR filter. If iterative calculations are involved in the design process, it is not suitable for real-time coefficient updates in applications. In 1987, Pyfer and Ansari [452] proposed a simple method for designing complex numbers based on the maximum and minimum criteria, which simplifies the design of complex numbers by decomposing complex calculations into two real calculations. However, because the coefficient calculation requires the use of the Remez alternating algorithm, it is still difficult to apply in real-time coefficient calculation. However, in 1979, Oetken [453] proposed a method that meets the requirements of real-time calculation, using a polynomial structure to design an even-length interpolation filter. This interpolation/extraction polynomial structure used a fixed step size to complete the fractional delay.

The above methods can independently generate the coefficients of the required fractional delay filter, but for delay-sensitive applications, these methods may become inappropriate because each method will produce filter network delay. In addition, for adaptive or variable FIR filters, the coefficients may be unknown, so it is necessary to control the delay independently of the filter coefficients. There are two common types of arbitrary delay control FIR filters: based on Fourier transform and Farrow structure. The Fourier transform method is discussed in detail in reference [454]. In 1987, Adams [455] proposed a similar method, using two discrete Fourier transforms for experimental control. The Fourier transform-based method requires a certain amount of computation and has high requirements for real-time processing processors. The Farrow structure is a method proposed by Farrow [456] in 1988 to continuously change the delay. By polynomial approximation, the filter coefficients are separated from the delay, so the delay can be arbitrarily changed without changing the coefficients. The arbitrary fractional delay FIR filter of the Farrow structure is just a change in the filter structure, not a filter design method, so this structure can be arbitrarily combined with three design methods: least squares, maximum flatness, and equiripple.

Due to the virtual sources formed by nonlinear effects, i.e., the end-fire array structure, the parametric array loudspeaker can emit a high directional audio frequency beam. This section combines the parametric array with the linear array to form a phased parametric array loudspeaker, achieving the purpose of steering the difference frequency sound beam. The directivity formula of the difference frequency sound beam steering can be obtained through the KZK equation and the Berktay far-field solution. However, due to the principle of the parametric array loudspeaker, the channel delay cannot be directly calculated based on the steering angle of the difference frequency sound, so it is necessary to use the linear array method to calculate the relationship between the steering angle of the pump wave and the delay of the pump wave, and then guide the steering of the difference frequency sound beam according to the results of the Berktay far-field solution.

The method of delay addition is a common algorithm for beamsteering, and the size of the delay depends on the steering angle. Because the frequency of the pump wave is high, the beamsteering angle accuracy obtained by the integer sampling period delay under the condition of limited sampling rate is poor. The fractional delay filter can realize the delay of non-integer sampling period, thereby improving the delay accuracy under the condition of not high sampling rate. This section introduces the fractional delay filter of the Farrow structure, which can realize the arbitrary angle steering of the wideband difference sound. Combined with the design advantages of FPGA, the phased array fractional delay algorithm is implemented on the FPGA platform, and experiments are conducted on the entire system in the anechoic chamber, and compared with the simulation results.

6.2.2 Implementation of Phased Parametric Array Loudspeaker Model

Although the directivity of the difference sound of the phased parametric array loudspeaker model has been given by Eq. (6.14), this formula cannot determine the relationship between the delay of each channel of the phased parametric array loudspeaker and the steering angle of the difference sound beam. The phased parametric array loudspeaker is based on the line array, and the difference sound can be steered to the target direction only after the pump wave beam is steered to the target direction, so the delay of each channel needs to be calculated according to the parameters of the pump wave. The following derives the directivity formula of the pump wave and discusses the influence of array parameters on beam steering.

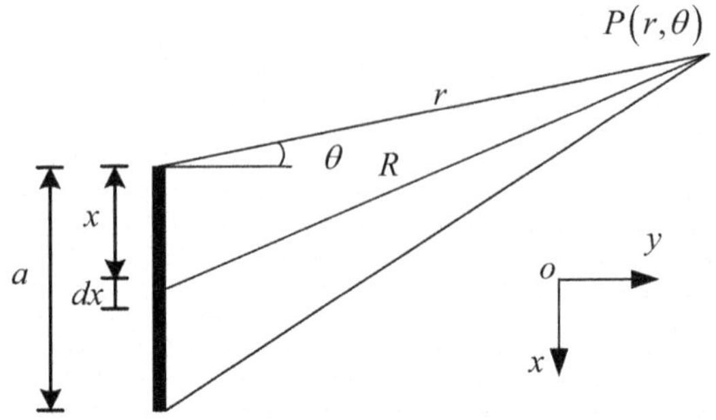

Fig. 6.9 Schematic diagram of far-field radiation of a column of acoustic units

Before calculating the linear superposition of the line array sound wave, first calculate the sound pressure of a column of units in the far field, and its two-dimensional plane schematic diagram is shown in Fig. 6.9, where a is the unit width, r and R are the far field $P(r, \theta)$ point and the differential variable dx distance between the upper edge of the unit. Then the sound pressure of the differential unit at $P(r, \theta)$ can be expressed as

$$dp = \frac{p_0}{R} e^{j(\omega t - kR)dx} \tag{6.33}$$

where p_0 is the sound pressure amplitude per unit length, and $r \gg a$ in the far field situation

$$R = \sqrt{r^2 + x^2 - rx\sin\theta} = r\left(1 - \frac{x}{r}\sin\theta + \cdots\right) \approx r - x\sin\theta \tag{6.34}$$

In Eq. (6.34), θ is the angle between the line connecting the upper edge and $P(r, \theta)$ and the direction of y. Therefore, the sound pressure level at $P(r, \theta)$ can be obtained through integration.

$$\begin{aligned} p(r, \theta, t) &= \int_0^a dp = \int_0^a \frac{p_0}{r} e^{j[\omega t - k(r - x\sin\theta)]} \\ &= \frac{p_0 a}{r} \frac{\sin\left(\frac{ka\sin\theta}{2}\right)}{\frac{ka\sin\theta}{2}} e^{-\frac{jka\sin\theta}{2}} e^{j(\omega t - kr)} \end{aligned} \tag{6.35}$$

Equation (6.35) already represents the sound pressure of a unit in the far field at $P(r, \theta)$. Therefore, the far-field sound pressure level of the pump wave of the phase-controlled parameter array can be obtained through the linear superposition of each unit, see Fig. 6.10. Similarly, first approximate the distance from each channel unit to $P(r, \theta)$.

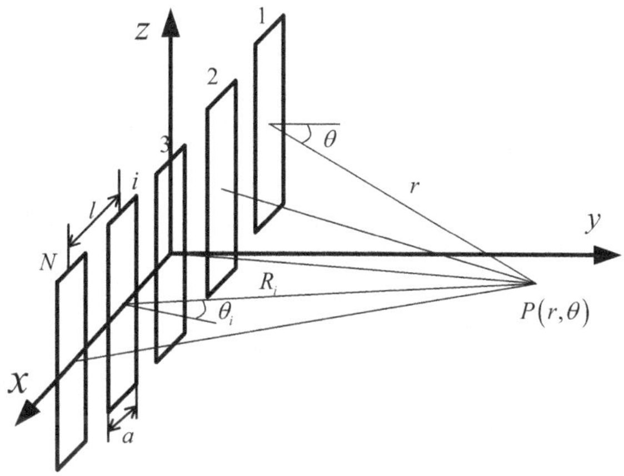

Fig. 6.10 Schematic diagram of the phased array

$$
\begin{aligned}
R_i &= \sqrt{r^2 + [(i-1)l]^2 - 2r(i-1)l\sin\theta} \\
&= r\left(1 - (i-1)\frac{l}{r}\sin\theta + \cdots\right) \\
&\approx r - (i-1)l\sin\theta
\end{aligned}
\tag{6.36}
$$

In Eq. (6.36), θ has the same meaning as θ defined in Eq. (6.34). r is the distance from the first channel unit to the point $P(r, \theta)$, l is the distance between two adjacent channels, and the ellipsis represents higher order small quantities. Then the sound pressure of the far field $P(r, \theta)$ is

$$
\begin{aligned}
p(r,\theta,t) &= \sum_{i=1}^{N} p_i(r,\theta,t) \\
&= \frac{p_0}{r}\frac{\sin\left(\frac{ka\sin\theta}{2}\right)}{\frac{k\sin\theta}{2}}\frac{\sin\left[\left(\frac{\omega\Delta\tau - kl\sin\theta}{2}\right)N\right]}{\sin\left(\frac{\omega\Delta\tau - kl\sin\theta}{2}\right)} \\
&\quad \times e^{-j\left(\frac{ka\sin\theta}{2}\right)} e^{-j\left(\frac{\omega\Delta\tau - kl\sin\theta}{2}\right)(n-1)} e^{j(\omega t - kr)}
\end{aligned}
\tag{6.37}
$$

where N is the number of channels, a is the aperture of the channel unit, ω is the angular frequency of the sound wave, $k = \omega/c_0$ is the wavenumber of the sound wave, c_0 is the speed of sound at small amplitude, $\Delta\tau$ is the delay between adjacent channels, which is related to the steering angle θ_s as follows

$$
\sin\theta_s = \frac{c_0\Delta\tau}{l}
\tag{6.38}
$$

According to Eq. (6.37), the directivity function of the beam steering of the phased parametric array loudspeaker for the pump wave can be obtained as

$$D(\theta) = \left|\frac{p(r,\theta,t)}{p(r,\theta_s,t)}\right| = \left|\frac{\sin\left(\frac{ka\sin\theta}{2}\right)}{\frac{ka\sin\theta}{2}}\frac{\sin\left[\left(\frac{\omega\Delta\tau - kl\sin\theta}{2}\right)N\right]}{N\sin\left(\frac{\omega\Delta\tau - kl\sin\theta}{2}\right)}\right| \tag{6.39}$$

Equation (6.38) is substituted into Eq. (6.39), and the directivity function of the beam steering of the phased parametric array loudspeaker for the pump wave becomes

$$D(\theta) = \left|\frac{\sin\left(\frac{ka\sin\theta}{2}\right)}{\frac{ka\sin\theta}{2}}\frac{\sin\left[\left(\frac{\pi l(\sin\theta_s - \sin\theta)}{\lambda}\right)N\right]}{N\sin\left(\frac{\pi l(\sin\theta_s - \sin\theta)}{\lambda}\right)}\right| = H_a(\theta)H_l(\theta) \tag{6.40}$$

where λ is the wavelength of the pump wave. Equation (6.40) includes the directivity of each channel unit $H_a(\theta)$ and the directivity of the array $H_l(\theta)$.

According to reference [457], let $D(\theta) = 0$, that is,

$$\frac{\pi l(\sin\theta_s - \sin\theta)N}{\lambda} = m\pi \tag{6.41}$$

where m is an integer. The main lobe width can be derived as when $m = 1$

$$\Delta\theta_m = \sin^{-1}\left(\sin\theta_s + \frac{\lambda}{Nl}\right) - \sin^{-1}\left(\sin\theta_s - \frac{\lambda}{Nl}\right) \tag{6.42}$$

The main lobe position of the phased parametric array loudspeaker can be determined by setting $D(\theta) = 1$. The condition for $D(\theta) = 1$ is

$$\frac{\pi l(\sin\theta_s - \sin\theta)}{\lambda} = m'\pi \tag{6.43}$$

where m' is an integer. let $m' = 1$ and solve for θ represents the position of the first grating lobe.

$$\theta_{g1} = \sin^{-1}\left(\sin\theta_s - \frac{\lambda}{l}\right) \tag{6.44}$$

Combining Eqs. (6.41) and (6.44), the relationship between the channel spacing without grating lobes and the maximum steering angle is

$$l_{max} = \frac{\lambda}{1 + \sin(\theta_s)_{\max}}\frac{N-1}{N} \tag{6.45}$$

According to Eqs. (6.42) and (6.45), the curves of the main lobe width of the pump wave with the steering angle (Fig. 6.11) and the steering angle with the adjacent channel spacing (Fig. 6.12) are given, respectively. From Fig. 6.11, it becomes clear that the main lobe of the pump wave widens with the increase of the steering angle, and narrows with the increase of the number of channels and the increase of the

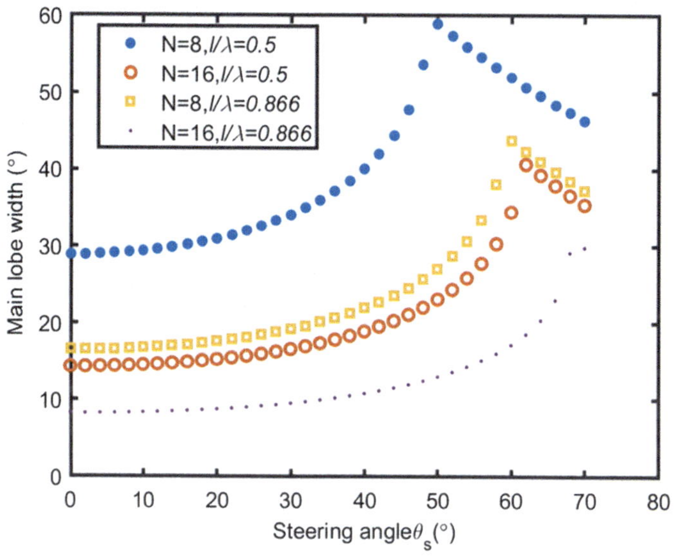

Fig. 6.11 Curve of the main lobe width of the primary wave of the phased array with steering angle

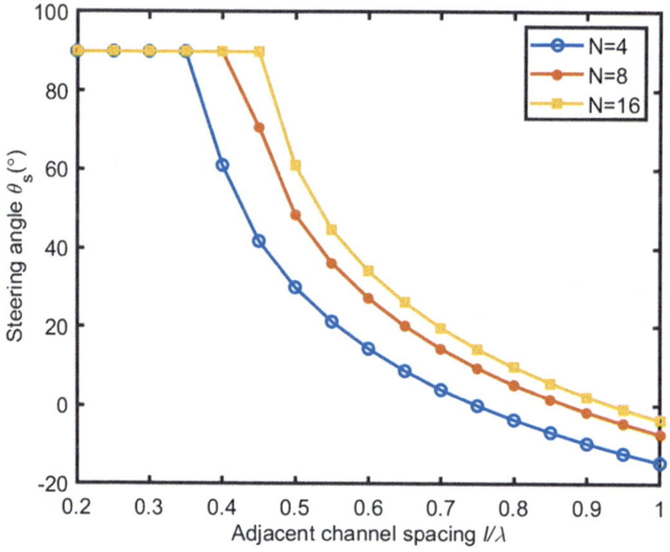

Fig. 6.12 Curve of the steering angle of the grating-free lobe with the spacing between adjacent channels

channel spacing. From Fig. 6.12, it can be observed that the steering angle of the grating-free lobe is inversely proportional to the channel spacing. The larger the channel spacing, the smaller the steering angle of the grating-free lobe.

6.2.3 Phased Parametric Array Loudspeaker System

The block diagram of the phased parametric array loudspeaker system can be seen in the Fig. 6.13, and the entire block diagram can be divided into five parts according to the signal flow direction: signal input, signal processing, signal output, signal driving, and sound wave emission. The signal analog-to-digital conversion and signal digital-to-analog conversion are completed by AD1939. FPGA is responsible for signal processing, as well as generating the sampling clock required by AD1939 and managing the registers of the AD1939 chip. The signal driving is completed by the 8-channel D-class amplifier based on TAS5630, and finally radiated out through the 8-channel ultrasonic transducer array. This section mainly introduces the implementation of fractional delay, and the implementation of most other modules has been introduced in Sect. 5.2.

6.2.3.1 Hardware Platform

The hardware of the phased parametric array loudspeaker system is designed as three circuit boards and a multi-channel transducer array board, as shown in Fig. 6.14. The three circuit boards are the FPGA development board, the AD1939 development board, and the amplifier board. The FPGA development board, with EP3C80 as the core, is responsible for the operation of the entire system algorithm and the control of AD1939 sampling. The AD1939 development board has 4-channel input and 8-channel output, responsible for acquisition of voice signal and signal of steering angle, as well as the output of 8-channel modulated signals after delay. The FPGA development board communicates with the AD1939 development board through user-defined IO ports. The amplifier board is designed based on the TAS5630 chip, working in SE mode. Each TAS5630 can provide 4-channel drive, and 8-channel drive requires 2 TAS5630 circuit boards.

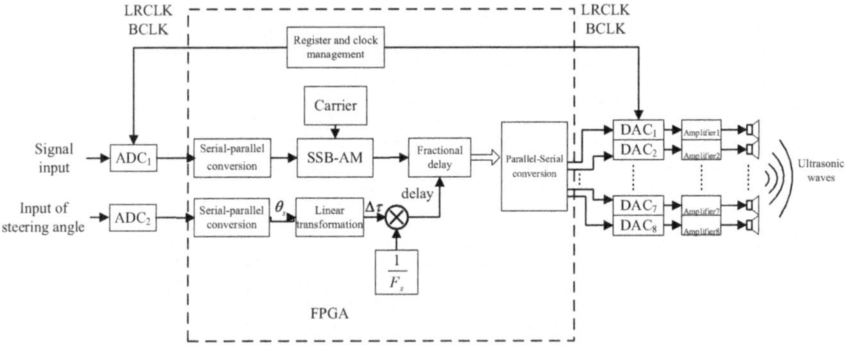

Fig. 6.13 Block diagram of the phased parametric array loudspeaker system

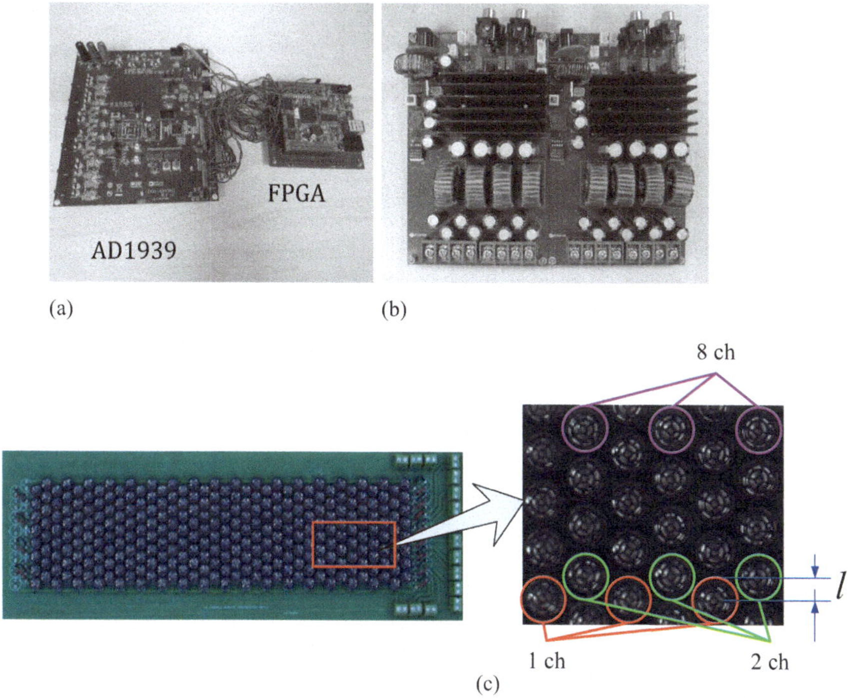

Fig. 6.14 Hardware part of the phased parametric array loudspeaker system **a** FPGA+AD1939 circuit board **b** amplifier board **c** transducer array board

6.2.3.2 Implementation of Fractional Delay

Equation (6.38) has given the relationship between the steering angle and the channel delay. For the parametric array loudspeaker system with a carrier of 40 kHz, under the condition of channel spacing $l = \lambda/2$, if $\Delta\tau$ is the unit sampling interval, the steering angle can only be a finite number of discrete angles, and the accuracy is limited. As can be seen from Table 6.2, even if the sampling rate reaches 1 MHz, the smallest steering angle is only 4.22 degrees. Although higher steering accuracy can be achieved by further increasing the sampling rate, the cost of multi-channel, high-bit-width, high-sampling-rate analog-to-digital and digital-to-analog conversion chips will also be higher. Therefore, it is necessary to use fractional delay to delay each channel on the analog-to-digital and digital-to-analog conversion chips based on a low sampling rate, to achieve low-cost and high-steering accuracy.

Various fractional delay algorithms have been detailed in Sect. 6.2.1. The fractional FIR filter of the ordinary structure (Fig. 6.15) needs to update the filter coefficients according to the change of delay. The characteristic of the Farrow structure (Fig. 6.16) is that the filter coefficients are separated from the delay, so changing the delay does not need to recalculate the filter coefficients or consume a large amount of memory to store the filter coefficients, which makes the fractional delay filter

Table 6.2 Relationship between sampling rate and steering angle

Sampling rate	Steering angle (degrees)				
96 kHz	50	–	–	...	n × 50
192 kHz	22.56	45.02	67.58	...	n × 22.56
500 kHz	8.47	16.94	25.41	...	n × 8.47
1 MHz	4.22	8.44	12.66	...	n × 4.22

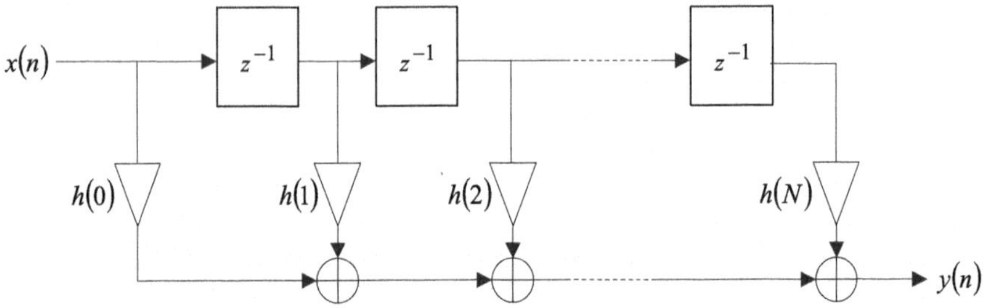

Fig. 6.15 Structure of a common FIR filter

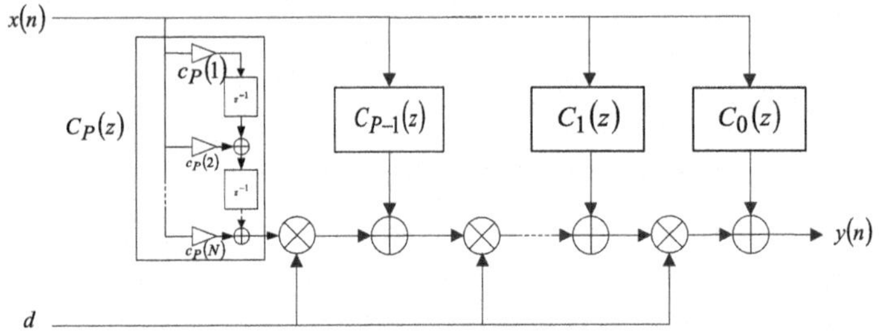

Fig. 6.16 Structure of a Farrow FIR filter

more convenient to use in real-time systems. Because of the above advantages, the fractional delay used by the phased array is implemented by the FIR filter of the Farrow structure.

Assume the fractional delay filter coefficient is $h_d(n)$, it is the function of the delay $d = \frac{\Delta\tau}{T} - \text{int}\left(\frac{\Delta\tau}{T}\right)$ (int() represents the rounding operation)

$$h_d(n) \approx \sum_{m=0}^{P} c_m(n) d^m,\ n = 0, 1, 2, \ldots, N \tag{6.46}$$

where $c_m(n)$ is the real approximation coefficient, and P is the highest power of the polynomial. Perform $h_d(n)$ transformation on z

$$
\begin{aligned}
H_d(z) &= \sum_{n=0}^{N} h_d(z)z^{-n} = \sum_{n=0}^{N}\left[\sum_{m=0}^{P} c_m(n)d^m\right]z^{-n} \\
&= \sum_{m=0}^{P}\left[\sum_{n=0}^{N} c_m(n)z^{-n}\right]d^m = \sum_{m=0}^{P} C_m(z)d^m
\end{aligned}
\tag{6.47}
$$

where $C_m(z) = \sum_{n=0}^{N} c_m(n)z^{-n}$. By the method of polynomial approximation, the coefficients of the filter can be written as

$$
h_{d,q} = \sum_{m=0}^{P} c_m(n)d_q^m,\ q = 0, 1, 2, \ldots, Q;\ n = 0, 1, 2, \ldots, N \tag{6.48}
$$

The transfer function of the filter is represented as a polynomial of delay d, and $C_m(z)$ is the coefficient of the polynomial. The filter is transformed into $Q + 1$ sub-filters, where the coefficients of the sub-filters $c_m(n)$ are independent of the delay d. The filter coefficients are designed offline using the least squares phase method. The higher the order of the filter, the closer the phase response and amplitude response of the filter are to the ideal impulse response. However, when the actual system is working, it is not necessary to achieve the ideal amplitude and phase response in all frequency band ranges, so the appropriate parameters can meet the design requirements and reduce the consumption of system resources.

Define the total in-band amplitude error E_m and the total phase error E_p. The total error represents the sum of errors within the channel, characterizing the overall fluctuation within the band. In addition, the maximum amplitude error in the band $e_{m\,\max}$ and the maximum phase error in the channel $e_{m\,\max}$ are defined to characterize the instantaneous error values, that is, the maximum fluctuation values. The error expression is defined as follows

$$
E_m = \sqrt{\int_0^{\omega_s}\int_0^1 |H_{i\Delta}(\omega, \Delta)_m - H_\Delta(\omega, \Delta)_m|^2 d\Delta d\omega} \tag{6.49}
$$

$$
E_p = \sqrt{\int_0^{\omega_s}\int_0^1 \left|H_{i\Delta}(\omega, \Delta)_p - H_\Delta(\omega, \Delta)_p\right|^2 d\Delta d\omega} \tag{6.50}
$$

$$
e_{m\max} = \max\left[\left|H_{i\Delta}(\omega, p)_m - H_\Delta(\omega, p)_m\right|, 0 \le \omega \le \omega_s, 0 \le p \le 1\right] \tag{6.51}
$$

$$
e_{p\max} = \max\left[\left|H_{i\Delta}(\omega, p)_p - H_\Delta(\omega, p)_p\right|, 0 \le \omega \le \omega_s, 0 \le p \le 1\right] \tag{6.52}
$$

To avoid confusion between the fractional delay unit d and the differential symbol d, the Δ is temporarily used to replace the delay unit d. ω_s represents the passband cutoff frequency, $H_{i\Delta}(\omega, \Delta)_m$ and $H_{i\Delta}(\omega, \Delta)_p$ respectively represent the ideal amplitude response and phase response of fractional delay, $H_\Delta(\omega, \Delta)_m$ and $H_\Delta(\omega, \Delta)_p$

represent the fractional delay amplitude response and phase response of the Farrow structure. The selection of fractional filter coefficients refers to the total error, maximum error, and the total delay size.

First, determine the channel frequency range of the fractional delay. The carrier of the parametric array loudspeaker system is 40 kHz, and the modulation method uses the lower sideband of SSB-AM, so the highest frequency of the signal is 40 kHz. The normalized highest frequency of the signal is $\omega = \frac{40.5\text{k}}{192\text{k}} * 2\pi \approx 0.42\pi$ at a sampling rate of 192 kHz, so the normalized cutoff frequency of the filter is set to $\omega_s = 0.5\pi$. The performance of the filter are compared with different filter order N and polynomial power Q according to Eq. (6.48), see Table 6.3. From the table, it is evident when Q is constant, the phase response fluctuation of even N is less than that of odd N, but the amplitude response fluctuation of even N is greater than that of odd N. When N is constant, the amplitude response and phase response become smoother with the increase of Q, and the amplitude response and phase response fluctuations improve significantly when Q changes from 2 to 3 and from 3 to 4, but the improvement is already very small when it changes from 4 to 5. The integer delay of the filter is half of the filter order, the corresponding amplitude and phase fluctuations when $N = 6$, $Q = 3$ are already less than −40 dB, which fully meets the needs, and the performance improvement brought by further increasing the order and power is not significant, so choose $N = 6$, $Q = 3$ as the order of the filter and the power of the polynomial. The corresponding impulse response, amplitude response, and phase response are shown in Figs. 6.17, 6.18 and 6.19.

To verify the delay effect of the Farrow structure fractional delay filter, it was compared with the Lagrange interpolation algorithm, as shown in Eq. (6.31). First,

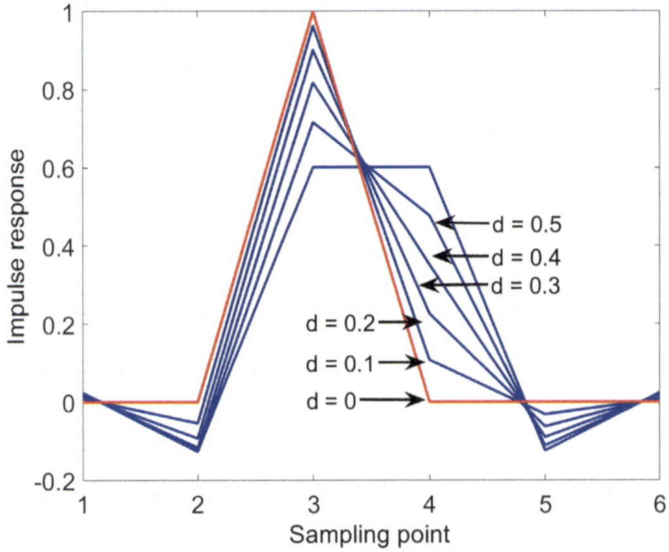

Fig. 6.17 Corresponding impulse response for $N = 6$, $Q = 3$, $\omega_s = 0.5\pi$

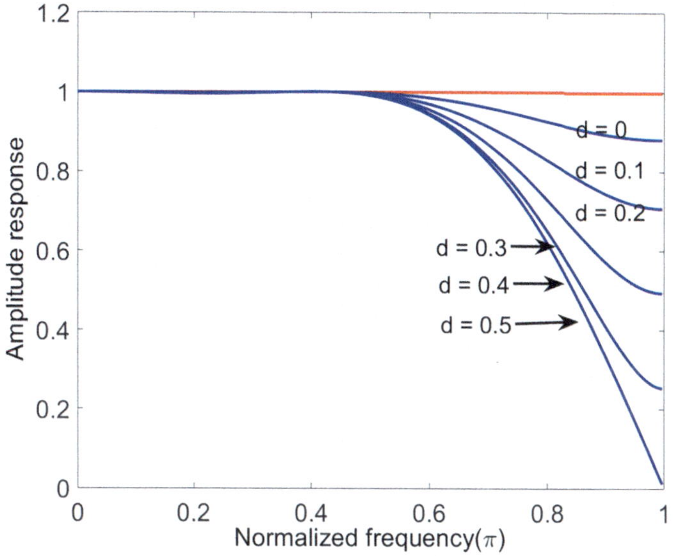

Fig. 6.18 Corresponding amplitude frequency response for $N = 6$, $Q = 3$, $\omega_s = 0.5\pi$

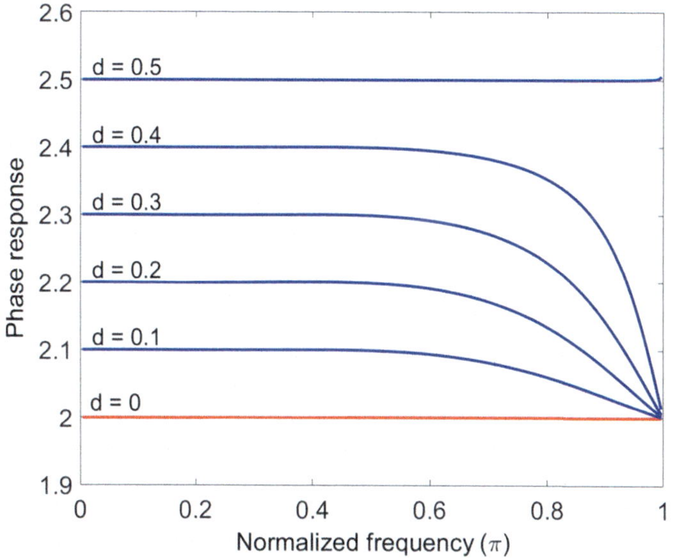

Fig. 6.19 Corresponding phase frequency response for $N = 6$, $Q = 3$, $\omega_s = 0.5\pi$

Table 6.3 Performance of different parameters of the filter

	E_m	E_p	$e_{m\max}$	$e_{p\max}$	Integer delay
N = 4, Q = 3	0.4670	0.0528	0.0285	0.0062	2
N = 4, Q = 4	0.4670	0.0530	0.0293	0.0066	2
N = 5, Q = 2	0.0848	0.2863	0.0159	0.0199	3
N = 5, Q = 3	0.0713	0.2609	0.0126	0.0199	3
N = 5, Q = 4	0.0692	0.2610	0.0116	0.0202	3
N = 5, Q = 5	0.0692	0.2610	0.0116	0.0202	3
N = 6, Q = 2	0.0915	0.0089	0.011	0.0096	3
N = 6, Q = 3	0.0673	0.0095	0.0095	0.001	3
N = 6, Q = 4	0.0653	0.0087	0.0079	0.0012	3
N = 6, Q = 5	0.0652	0.0087	0.0079	0.0012	3
N = 7, Q = 2	0.0450	0.1273	0.0071	0.0113	4
N = 7, Q = 3	0.0181	0.0457	0.0034	0.0040	4
N = 7, Q = 4	0.0098	0.0455	0.0019	0.0041	4
N = 7, Q = 5	0.0098	0.0455	0.0019	0.0041	4
N = 8, Q = 2	0.0406	0.1190	0.0059	0.0105	4
N = 8, Q = 3	0.0179	0.0055	0.0024	5.6×10^{-4}	4
N = 8, Q = 4	0.0099	0.0039	7.5×10^{-4}	3.4×10^{-4}	4
N = 8, Q = 5	0.0099	0.0036	7.5×10^{-4}	3.0×10^{-4}	4

the delays corresponding to steering angles of 1° and 10° were calculated according to Eq. (6.38), and then the Lagrange interpolation algorithm and the Farrow structure FIR fractional delay filter were used to process the 40 kHz carrier, as shown in Figs. 6.20 and 6.21. The delay results of the Farrow structure are in good agreement with the delay results of the Lagrange interpolation algorithm.

Through simulation comparison, the order and polynomial power of the Farrow structure fractional delay filter $N = 6$ and $Q = 3$ fully meet the delay accuracy requirements of the phased parametric array loudspeaker. The fractional delay algorithm was implemented on the FPGA platform based on the above parameters.

The steering angle input in Fig. 6.13 is a voltage signal, which is controlled by a potentiometer to control the amplitude of the voltage signal. The range of voltage amplitude changes is [0 V, 2.5 V], corresponding to the steering angle range [0°, 90°]. The voltage signal representing the angle is converted into a digital signal through ADC, and then a delay compatible with the fractional delay filter is generated through linear transformation. The relationship between the voltage signal and the delay is

$$\text{delay} = \frac{\Delta\tau}{T_s} = \frac{\sin(\theta_s)F_s l}{c_0} = \frac{\sin\left(\frac{V_{in}}{V_{ref}} \times \frac{\pi}{2}\right)F_s l}{c_0} \tag{6.53}$$

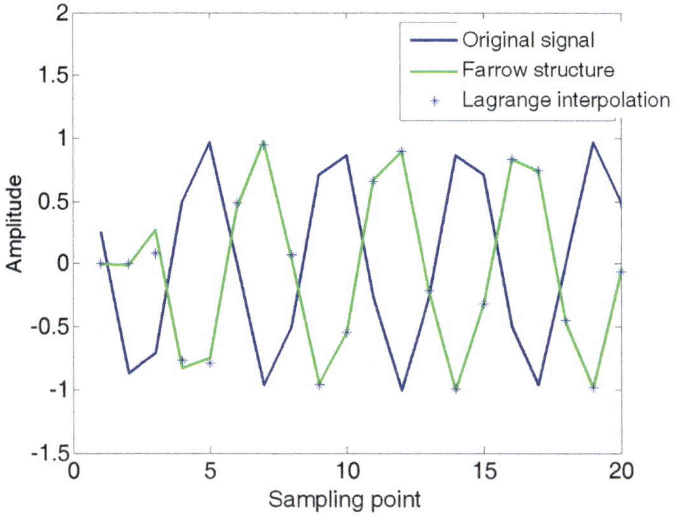

Fig. 6.20 Comparison of signals before and after 1° steering at 40 kHz

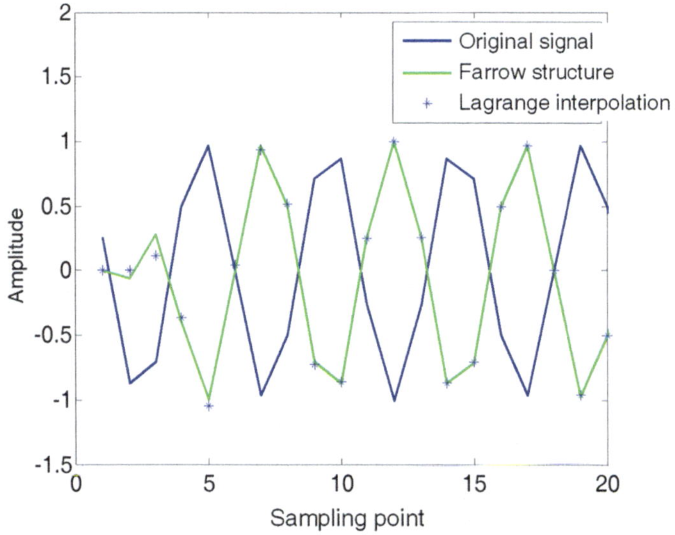

Fig. 6.21 Comparison of signals before and after 10° steering at 40 kHz

where, V_{in} is the amplitude of the voltage signal representing the steering angle, the reference voltage $V_{ref} = 2.5$ V, θ_s is the steering angle, $\Delta\tau$ is the delay interval between adjacent channels, T_s is the sampling interval, which is inversely proportional to the sampling frequencyF_s, c_0 is the speed of sound, l is the distance between adjacent channels of the array. According to the parameters$l =$

5mm,$c_{\circ} = 344\frac{m}{s}$, $\theta_s = \left[-90^{\circ}, 90^{\circ}\right]$ in Sect. 6.2.3, the calculations yield $\Delta\tau = [-14.53\,\mu s,\ 14.53\,\mu s]$, $\text{delay} = \left[-2.7907,\ 2.7907\right]$.

The entire algorithm is implemented through DSP Builder. The 8-channel delay system is shown in Fig. 6.22. In the figure, subsystemI mainly completes the extraction of the sign bit and the delay of each channel relative to the first channel caused by the delay of adjacent channels. SubsystemII is the delay module of 8 channels, as shown in Fig. 6.23. The inputs to the subsystem module *delay* are divided into integer and decimal parts. The integer part is implemented through the delay line, and the decimal part is completed by the Farrow structure decimal delay filter. Subsystem III determines the steering direction based on the sign bit extracted by subsystemI, and then outputs the signal after the delay of each channel. If *delay* is positive, it represents steering to the left of the transducer array axis, and if negative, it represents steering to the right of the transducer array axis.

The module is converted into VHDL language through the Signal Compiler tool of DSP Builder, and after being compiled and synthesized with other modules in Quartus II 8.0 software, a download file is generated. The FPGA input signal, carrier signal, and modulated delay output signal waveform can be observed through JTAG download to the FPGA chip and SignalTap, as shown in Fig. 6.24. For convenience, only the output waveforms of 4 channels are shown in the figure. The resources consumed by the 8-channel Farrow decimal delay algorithm are: 5285 logic units, 3292 registers, 2 PLLs, 193,848 bits of memory, and 22 9-bit multipliers.

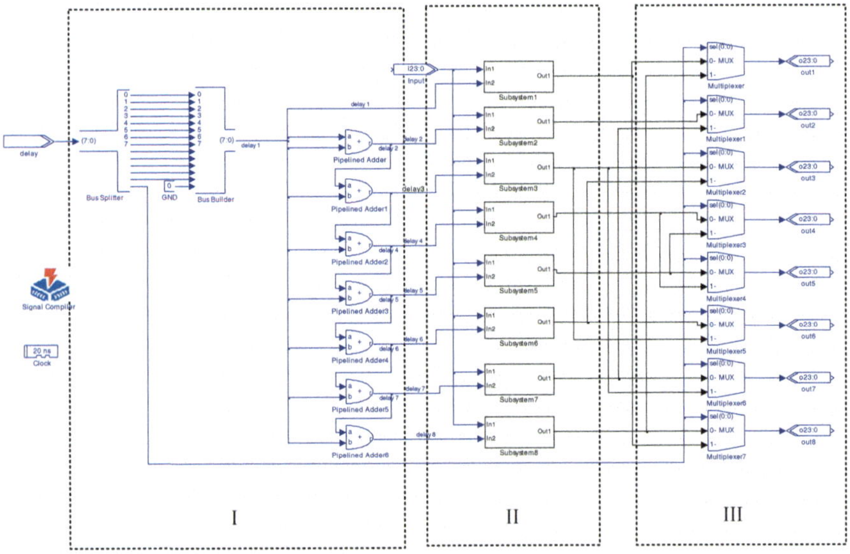

Fig. 6.22 Block diagram of FPGA implementation of fractional delay filter

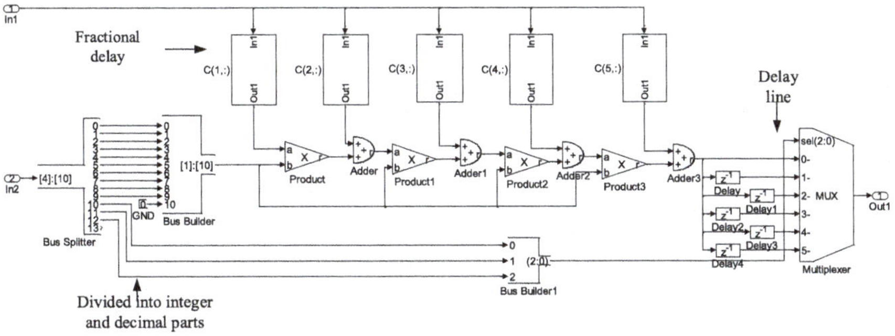

Fig. 6.23 Internal structure of the Subsystem module

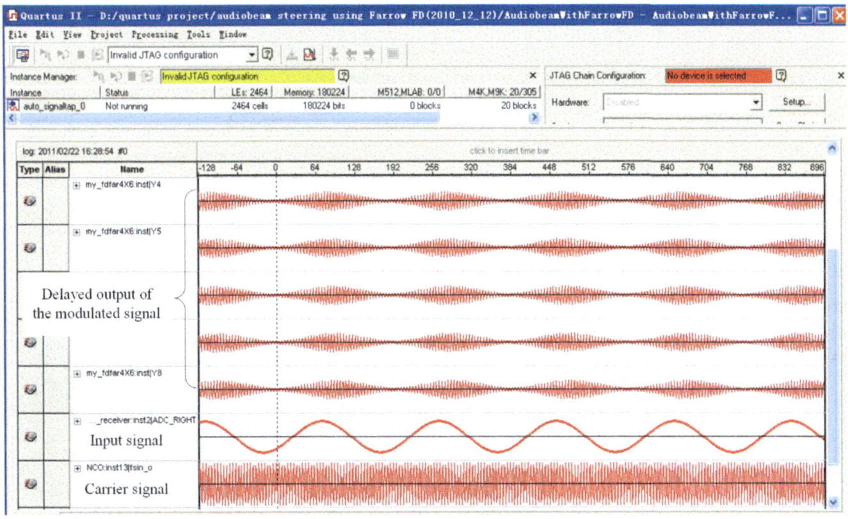

Fig. 6.24 SignalTap displays real-time processing results

6.2.4 *Beam Steering Comparison Between Simulation and Experiment*

6.2.4.1 Simulation and Experimental Settings

To verify the beam steering effect, based on the phased parametric array loudspeaker system, experimental measurements were carried out in the anechoic chamber, and compared with simulation results. The schematic diagram of the test and the measurement site of the anechoic chamber are shown in Figs. 6.25 and 6.26. In Fig. 6.25, microphones1 and 2 respectively measure the pump wave sound pressure level and the difference frequency sound pressure level. To eliminate the influence of

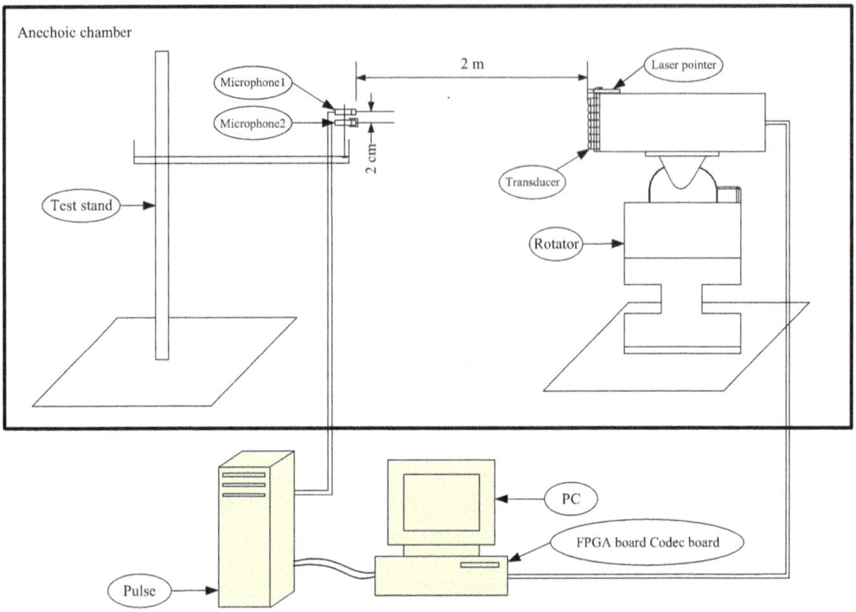

Fig. 6.25 Measurement schematic diagram

Fig. 6.26 Scene of the experiment in the anechoic chamber

spurious sound on the difference frequency sound, a sound filter [191] is installed in front of microphone 2. The signal at each angle is measured by rotating the sound source through the digital cloud platform. In $[\theta_s - 5°, \theta_s + 5°]$ range, measure once every 1°, in $[-50°, \theta_s - 5°]$ and $[\theta_s + 5°, +50°]$ range, measure once every 2°, in $[-90°, -50°]$ and $[+50°, +90°]$ range, measure once every 5°.

The input signal is generated by a computer and is a mixed frequency signal of 900 and 2300 Hz. The purpose of using this input signal is to test the algorithm's applicability to broadband, and the two frequencies are not multiples of each other, so the measurements will not be affected by their respective harmonics. According to the analysis of the pump wave beam steering in Sect. 6.2.2, the channel spacing needs to be as small as possible. The array uses MA40S4S units with a diameter of 10 mm. To reduce the possibility of grating lobes, the spacing is minimized to 5 mm through staggered arrangement. For a 40 kHz carrier, the transducer array with 8 channels, this spacing corresponds to a maximum grating-free steering angle of about 30°. Although any steering angles can be generated in practice, due to the limited measurement accuracy, the smallest angle measured in the experiment is 5°.

6.2.4.2 Simulation and Experimental Results

Figures 6.27, 6.28, 6.29, 6.30 and 6.31 are the beam patterns of the pump wave steering of 5°, 10°, 20°, 30° and 40°, where (a), (b) and (c) correspond to the carrier of 40 kHz and the sidebands of 39.1 kHz after modulation at 900 Hz, and the sidebands of 37.7 kHz after modulation at 2300 Hz.

For the simulation results of the pump wave based on the KZK model, it can be seen from the figure that when the steering angle is 5° and 10°, the KZK simulation results and experimental results match well within 30° centered on the main lobe, and there is a certain deviation in the part outside the main lobe 30°. When the steering angle is 20°, the main lobe of the KZK simulation result has deviated from 20°, and the main lobe width is narrower than the experimental result, and the fluctuation at the angle far from the main lobe increases. When the steering angle is 30°, the main lobe of the KZK simulation result has a large deviation from the target direction, and the sidelobe position also deviates from the sidelobe position of the experimental result. When the steering angle is 40°, the main lobe angle of the KZK simulation result not only does not match the target angle, but also a grating lobe appears at approximately −40°. Obviously, this position will not exhibit a grating lobe, so the simulation result has completely lost its accuracy at this steering angle. When the steering angle is greater than 40°, the error of the KZK simulation result will be greater. This also verifies that the simulation results of the KZK model provide good reference when the steering angle is less than 20°. When the steering angle is greater than 20°, the time error becomes too large to be used to guide the results. This result occurs because the KZK equation uses a parabolic approximation (or near-axis approximation), and this limits its ability to describe the sound field to areas not far from the axial direction. Therefore, the simulation error is acceptable within a certain angle, but it becomes larger when the angle exceeds a certain limit.

For the beam steering simulation results under the line array model, it can be seen from the figure that within the range of 40° on both sides of the main lobe, the simulation results match the experimental results very well; outside the 40° range, the simulation results decay quickly and deviate from the experimental results. This result occurs because Eqs. (6.34) and (6.36) use a first-order approximation for distance, ignoring higher-order small quantities.

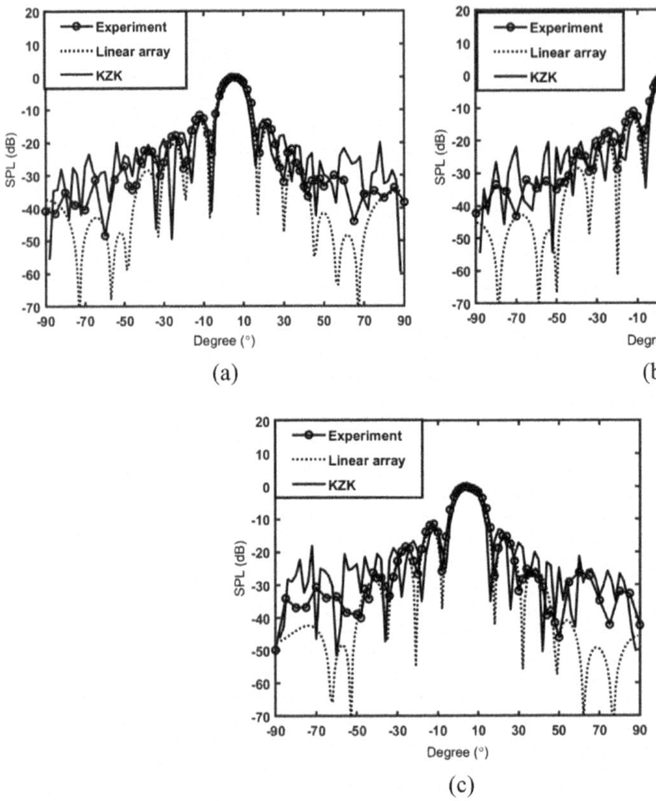

Fig. 6.27 Beam diagram with primary wave steering angle of 5° **a** carrier 40 kHz, **b** sideband 39.1 kHz, **c** sideband 37.7 kHz

Furthermore, because the sound source is not an ideal point source, the side lobe amplitude on the side near the transducer axis will be higher than the amplitude on the far axis side. This can be seen in both the experiment and simulation results of the pump wave. According to the simulation and experimental results, the main lobe width of the pump wave changes with the steering angle, as shown in Table 6.4. The table shows that the higher the frequency, the narrower the beam; the larger the steering angle, the wider the beam.

Figures 6.32, 6.33, 6.34, 6.35 and 6.36 are the beam patterns of the difference frequency sound steering at 5°, 10°, 20°, 30° and 40° respectively, where (a) and (b) correspond to the simulation and experimental results at 900 Hz and 2300 Hz respectively.

As can be seen from the figure, for the difference frequency sound simulation results based on the KZK model, when the beam steering angle is 5° and 10°, the simulation results within 30° on the left and right of the target steering angle match well with the experimental results, but there is a difference between the simulation

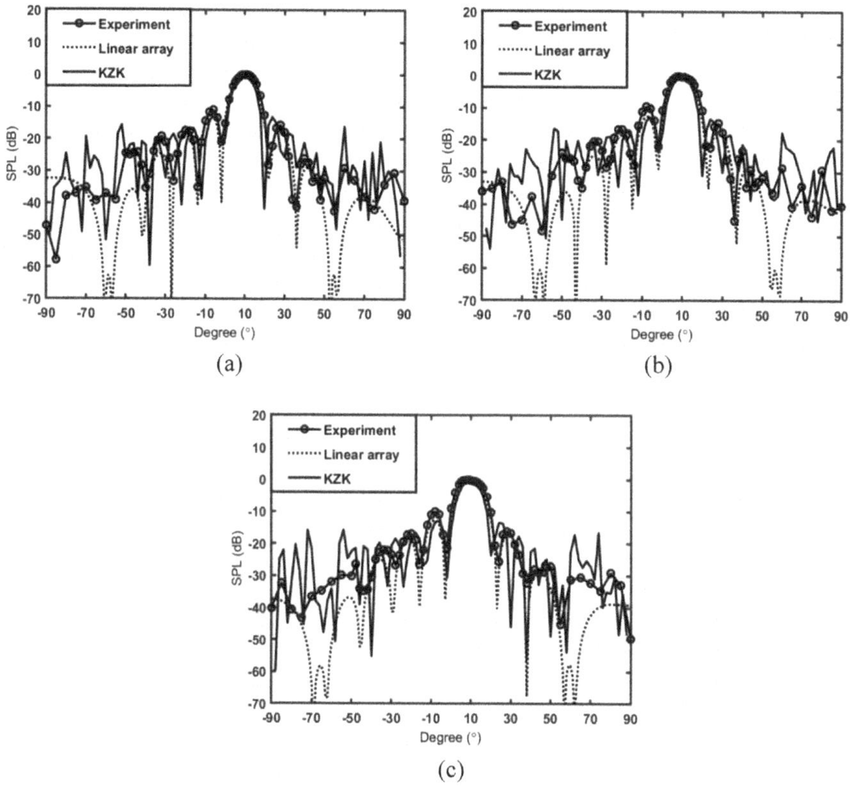

Fig. 6.28 Beam diagram with primary wave steering angle of 10° **a** carrier 40 kHz, **b** sideband 39.1 kHz, **c** sideband 37.7 kHz

results and the experimental results at angles greater than 30°. When the beam steering angle is 20°, the beam width of the simulation result is narrower than the experimental result, that is to say, it is slightly worse than the case where the beam steering angle is 5° and 10°. In addition, a grating lobe appears at the −50° position, which does not match the analysis of the grating lobe. When the beam steering angle is 30° and 40°, the beam steering angle of the simulation does not match the specified steering angle, and there is a large deviation from the experimental results. In addition, the amplitude of the grating lobe at the −50° position is larger, which does not match the analysis of the grating lobe. These results prove the near-axis characteristics of the KZK model, and the KZK model can only guide the situation where the beam steering angle is within 20°.

As can be seen from the figure, for the difference frequency sound simulation results based on the Berktay model, the simulated steering angle matches the experimental results, but the beam shape is quite different from the experimental results. The reason is precisely due to the non-fulfillment of the approximation conditions of Eq. (6.14).

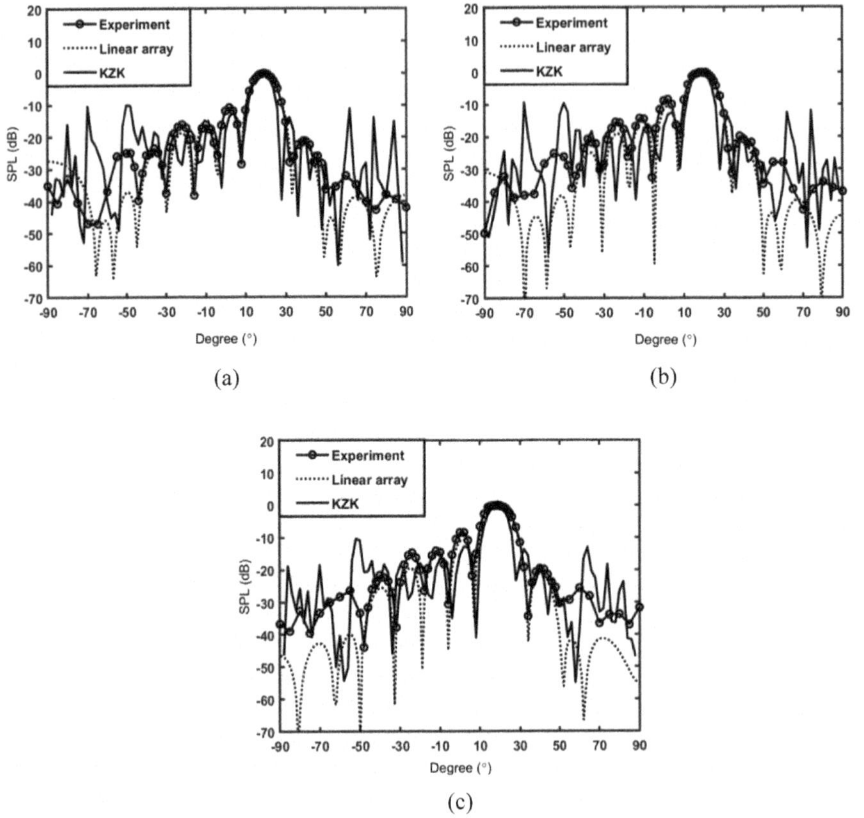

Fig. 6.29 Beam diagram with primary wave steering angle of 20° **a** carrier 40 kHz, **b** sideband 39.1 kHz, **c** sideband 37.7 kHz

According to the experimental results of beam steering, the 3 dB width of the difference frequency sound beam is given, see Table 6.5. From the table, it is obvious that the 3 dB beam width of the difference frequency sound becomes narrower as the frequency increases and wider as the angle increases.

6.3 Summary

This chapter concludes with a brief review of directional control methods, directional models, and directional design, followed by a focus on the implementation of difference frequency sound beam steering in parametric array loudspeaker using fractional delay filters. Based on the pump wave product model of the directivity of the difference frequency sound, a simulation study was conducted on the beam steering of the secondary sound field of the parametric array loudspeaker, and then the delay of

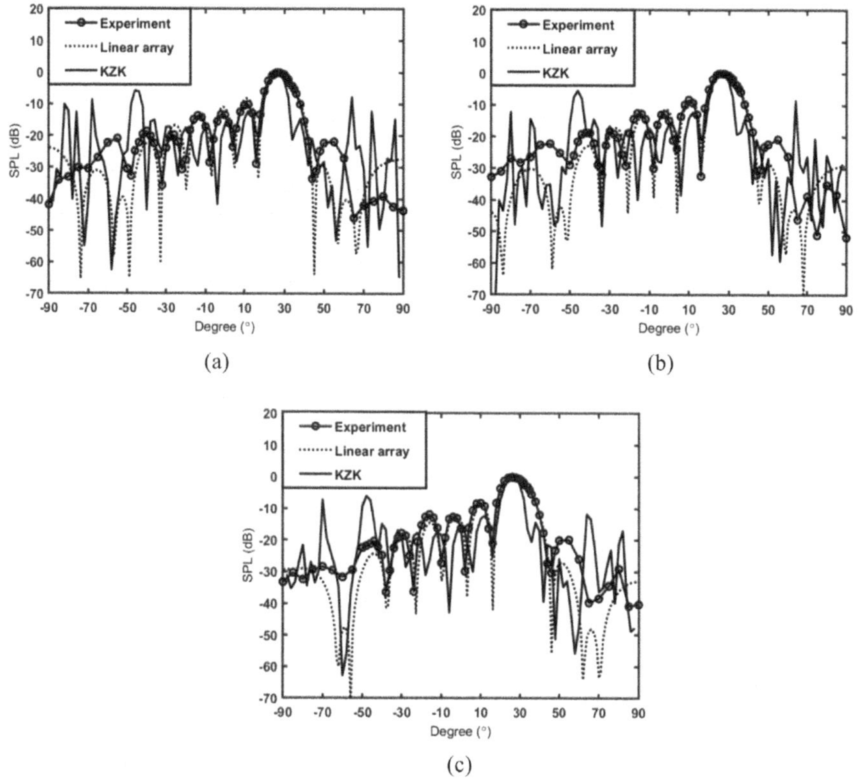

Fig. 6.30 Beam diagram with primary wave steering angle of 30° **a** carrier 40 kHz, **b** sideband 39.1 kHz, **c** sideband 37.7 kHz

the phased parametric array loudspeaker at different steering angles was obtained. By comparing the total passband error and the instantaneous maximum error, the parameters most suitable for the performance and cost of the Farrow structure small filter were determined. The selected parameters were compared with the Lagrange interpolation algorithm, and the results demonstrated a good level of agreement within the specified bandwidth. The Farrow fractional delay filter was built using the DSP Builder tool, and the phased parametric array loudspeaker system was implemented on the FPGA platform, and the beam steering was experimentally verified in the anechoic chamber. The parabolic approximation simulation results of the KZK model are quite accurate within 20° on both sides of the axis. Due to the large air sound absorption coefficient, the approximation conditions of the Berktay model are not satisfied, so its simulation results can only predict the steering angle, and the beam shape deviates significantly from the experimental results.

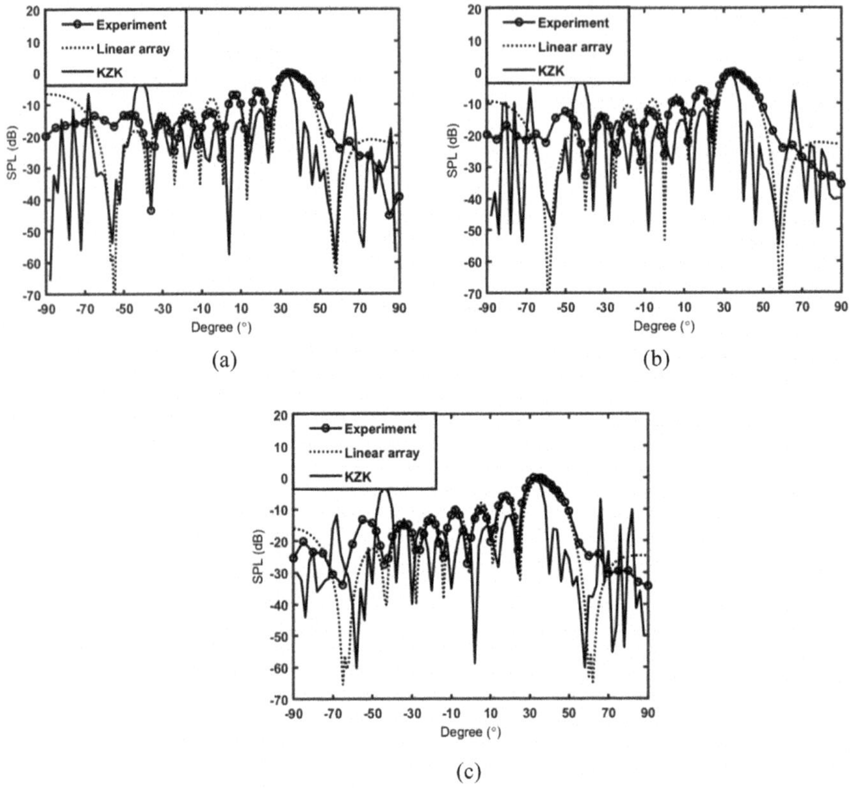

Fig. 6.31 Beam pattern with primary wave steering angle of 40° **a** carrier 40 kHz, **b** sideband 39.1 kHz, **c** sideband 37.7 kHz

Table 6.4 Relationship between pump wave main lobe width and steering angle

Steering angle (°)	Main lobe width at corresponding frequencies (°)		
	40 kHz	39.1 kHz	37.7 kHz
5	24	26	26
10	24	26	26
20	24	26	28
30	28	28	30
40	36	36	36

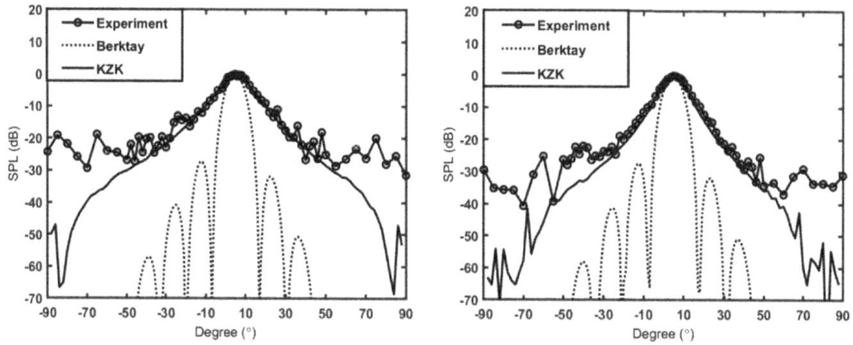

(a) (b)

Fig. 6.32 Beam pattern with difference frequency wave steering of 5° **a** 900 Hz, **b** 2300 Hz

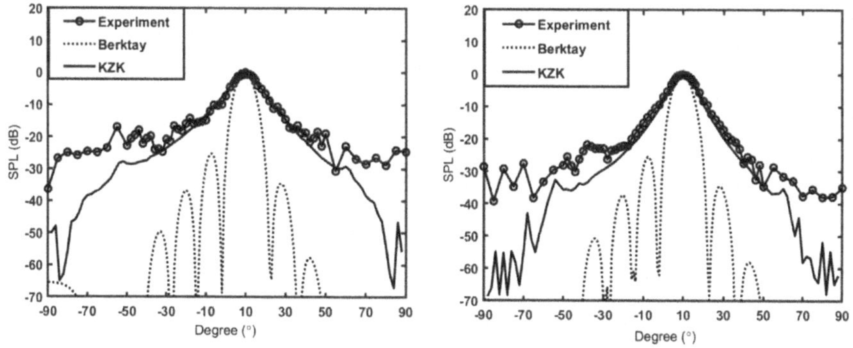

(a) (b)

Fig. 6.33 Beam pattern with difference frequency wave steering of 10° **a** 900 Hz, **b** 2300 Hz

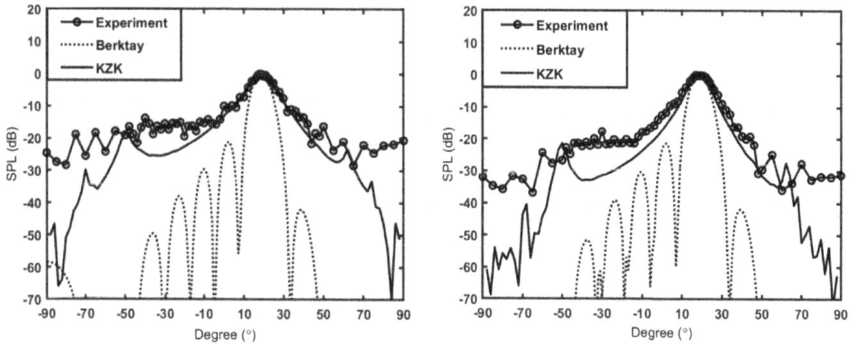

(a) (b)

Fig. 6.34 Beam pattern with difference frequency wave steering of 20° **a** 900 Hz, **b** 2300 Hz

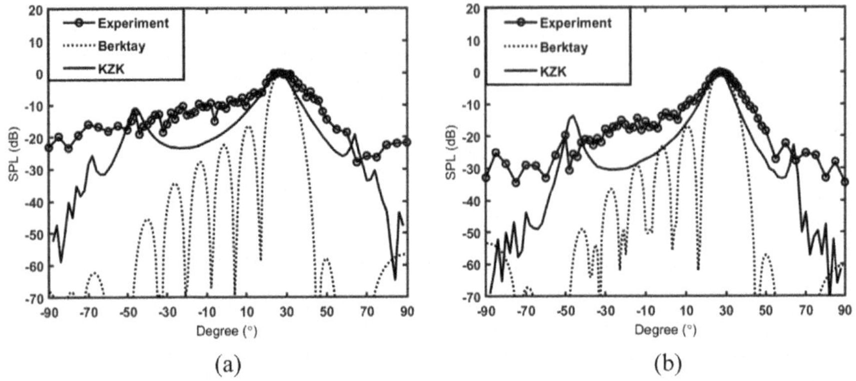

Fig. 6.35 Beam pattern with difference frequency wave steering of 30° **a** 900 Hz, **b** 2300 Hz

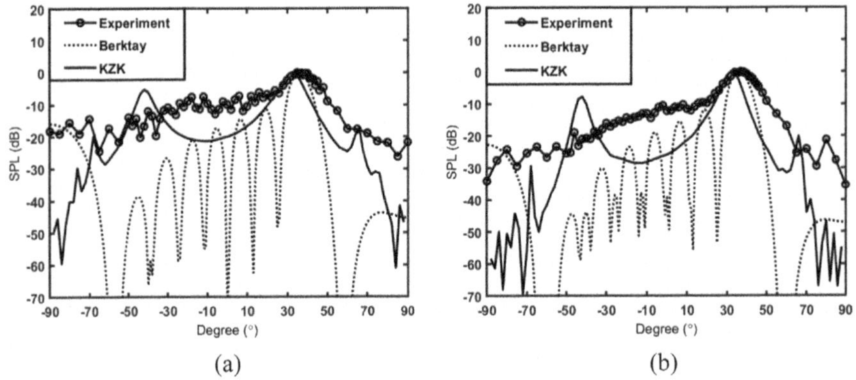

Fig. 6.36 Beam pattern with difference frequency wave steering of 40° **a** 900 Hz, **b** 2300 Hz

Table 6.5 Relationship between difference frequency sound beam 3 dB width and steering angle

Steering angle (°)	3 dB beam width of corresponding frequency (°)	
	900 Hz	2300 Hz
5	12	10
10	12	11
20	13	12
30	14	13
40	16	14

Chapter 7
Typical Applications of Parametric Array Loudspeakers

Parametric array loudspeakers, as a new type of directional sound source, not only have broad application prospects in noise control, personal communication, acoustic detection, and material measurement, but can also be used for intelligent sound field control. Especially in active noise control, using parametric array loudspeakers as secondary sound sources can not only reduce secondary path acoustic feedback, but also reduce the impact of secondary sound sources on non-control areas in the sound field. In 2005, Brooks et al. from University of Adelaide, Australia [458] initially explored the possibility of using parametric array loudspeakers as secondary sound sources for active noise control systems, believing that although parametric array loudspeakers can produce highly directional sound frequencies, due to the involvement of nonlinear effects, they cannot precisely control the phase and amplitude of the sound frequency. In addition, due to the low conversion efficiency of parametric array loudspeakers, to obtain the appropriate sound frequency, the input signal amplitude needs to be increased, which may produce high sound pressure level ultrasonic waves, possibly raising safety issues related to ultrasonic waves [301, 302]. In 2006, the research group studied an active noise control system combining parametric array loudspeakers and virtual loudspeaker technology, but it was limited to theoretical exploration and did not provide a practically feasible solution [459]. In 2010, Tanaka et al. from Tokyo Metropolitan University in Japan [460, 461] used phased array technology to steer the beam of parametric array loudspeakers and used it as a secondary sound source for active noise control systems. In 2011, the research group used the reflection of an arc-shaped parametric array loudspeaker to radiate spherical waves to the primary sound source sound field, overlapping as much as possible with the radiated sound waves of the primary sound source, achieving global noise reduction in the entire sound field. However, the implementation and control of this method are quite cumbersome and not conducive to practical application [462]. Komatsuzaki and Iwata [463] simulated and experimentally studied the sound field of parametric array loudspeakers and primary sound sources in three different relative positions (same direction emission, opposite direction emission,

J. Yang and P. Ji, *Parametric Array Loudspeakers*,
https://doi.org/10.1007/978-981-96-3548-1_7

and perpendicular intersection), achieving control of the local sound field. In terms of personal communication, Nakashima et al. [195, 196] installed two parametric array loudspeakers composed of 16 piezoelectric ceramic transducers on a mobile phone prototype, and the experimental measurement found that the sound pressure level exceeded 70 dB at 50 cm from the mobile phone. Using the high directivity of parametric array loudspeakers, the sound level difference between the two ears is about 15 dB, so binaural sound reproduction can be achieved. Since parametric array loudspeakers also have high directivity in the low-frequency range, they can be used for material acoustic property measurement. Castagnede et al. [201, 464] used parametric array loudspeakers to excite pulse signals to measure the absorption coefficient of materials under normal incidence. Ahn et al. [186] considered the key factors of applying parametric array loudspeakers to mobile devices: power consumption and device size, and provided corresponding solutions. This section mainly demonstrates the breadth and effectiveness of its applications in active noise control in free space, active noise control in pipelines, and measurement absorption coefficients in situ.

Active noise control is a method of artificially generating a secondary sound signal in a designated area to control the primary sound signal, achieving the purpose of noise reduction. According to the principle of sound wave cancellation interference, if the secondary sound source produces a sound wave with equal amplitude and opposite phase to the primary sound source, it will cancel each other out after superimposing with the original sound field in that area, thereby reducing noise. In 1933, German physicist Lueg [465, 466] applied for a patent called "Process for Eliminating Sound Oscillations". He used the spatial and temporal coherence of the sound field and Young's interference principle to propose the concept of artificially causing sound field cancellation interference to eliminate noise. In 1953, Olson et al. [467] published an article on "Electronic Sound Absorbers". In the frequency range of 20~200Hz, the optimal noise reduction in local space reached 25 dB. Subsequently, Olson used his system for research in areas such as pipeline noise control, noise-resistant helmets, and vibration and sound reverberation control [468]. In the following twenty years, electronic circuits and signal processing technology have made significant advancements. Not only have transistor circuits become quite mature, but large-scale integrated circuits and digital circuits have also been rapidly integrated into various actual control systems. Active noise control technology has developed rapidly in theory and experiment [469–473], and has industrial applications in areas such as active noise reduction earmuffs (headphones), pipeline active noise cancellation, and local space active noise cancellation.

The active noise control system reduces the sound energy in the target area. In the energy conservation system, the energy mechanism problem is more worth studying. Since the change in sound energy in the sound field is closely related to the change in radiation from the sound source, the energy mechanism of active noise control is often closely linked to the explanation method of sound source radiation impedance. Early studies on the energy mechanism of active noise control were usually combined with pipeline active noise cancellation. Theoretical analysis and experimental results show [471, 474–477] that the physical mechanism of pipeline noise active control is generally divided into two types: energy suppression mechanism and energy absorption mechanism. The former indicates that the introduction of the secondary sound source changes the radiation impedance of the primary sound source, reducing the radiation sound power of the primary sound source as a whole. The latter indicates that the secondary sound source acts as a sound energy absorber. In the pipeline sound field, if the secondary sound source is a monopole, the energy suppression mechanism plays a dominant role, and if the secondary sound source is a dipole, the energy absorption mechanism plays a dominant role. The energy suppression mechanism can be further divided into the spatial transfer mechanism of energy and the resistance storage mechanism of the multipole field [474]. In this mechanism, the secondary sound source reflects the sound wave back to the upstream of the pipeline, reducing the radiation impedance of the primary sound source, thereby reducing its radiation sound power. At this time, the secondary sound source plays a "unloading" role for the primary sound source. In the energy absorption mechanism, the radiation sound power of the secondary sound source becomes negative, and the absorbed energy is used to overcome its own mechanical resistance, driving the secondary sound source to make a sound, and the energy required from the outside will be greatly reduced. In addition, although the secondary sound source acts as an energy absorber in the energy absorption mechanism, not all types of secondary sound sources absorb energy in all directions, and the energy absorbed in different directions is different.

In the active noise control system, traditional omnidirectional loudspeakers are generally used as secondary sound sources, resulting in noise reduction in the control area, but noise will increase in other areas. When the control area is a certain direction or specific local space, the omnidirectional traditional loudspeakers cannot meet the requirements well. However, a parametric array loudspeaker with high directivity can be used as a new type of audio source for active noise control systems.

The on-site measurement of the sound absorption coefficient of materials is an important topic in the field of architectural acoustics. The existing reflection method using ordinary loudspeakers has high requirements for material size and testing environment, and due to the limitations of spherical waves, it needs to use sound field models and iterative approximation algorithms to approximate, which can easily generate errors. Parametric array loudspeakers, due to the directional sound frequency generated by nonlinear effects, have plane wave characteristics in the near field, which can better meet the requirements of on-site measurement. However, there are still many practical problems that need to be analyzed and solved when applying parametric array loudspeakers to the measurement field. Based on this, this chapter has carried out theoretical and experimental research on the on-site measurement of the sound absorption coefficient of materials using parametric array loudspeakers. In the measurement of air acoustics, the measurement of the sound absorption coefficient has mainly relied on the impedance tube method and the reverberation chamber method for many years. The on-site measurement method for materials under actual conditions has not been well resolved due to the limitations of the sound source conditions [478], which plays a very important role in actual acoustic design. Currently, the most commonly used is the reflection method of ordinary loudspeakers. The omnidirectional characteristics of ordinary sound sources are greatly affected by material size and measurement environment in the actual measurement process, and have high requirements for measurement conditions. Because the sound source emits sound waves in the form of spherical waves, there is a certain difference between the impedance obtained in the actual measurement process and the plane wave impedance under ideal conditions, which needs to use sound field models and iterative approximation algorithms to approximate [479], which can easily introduce errors. Parametric array loudspeakers have sharp directivity, which can minimize the impact of the testing environment, and by using their plane wave characteristics within the near field range, they can solve a series of problems brought about by spherical wave approximation. Bernard's group [201, 464] first used parametric array loudspeakers to measure the sound absorption coefficient of materials using the pulse method, but they only conducted preliminary experiments and did not effectively analyze this issue. The issues of nonlinear secondary sound field, reflection coefficient impact, and spurious sound were not mentioned, which are the core issues in the acoustic measurement using parametric array loudspeakers, directly affecting the accuracy and validity of the results.

7.1 Application of Parametric Array Loudspeakers in Free Space Active Noise Control

This section first analyzes the energy transfer mechanism of parametric array loudspeakers as secondary sound sources for active noise control systems, analyzes from the aspects of global energy control and local space control, and summarizes the advantages and disadvantages of parametric array loudspeakers for active noise control systems [480].

7.1.1 Global Energy Control

7.1.1.1 Ordinary Loudspeaker as Secondary Sound Source

Assume the sound source intensity vector of the primary sound source is $\mathbf{q}_p$, and the sound pressure vector is $\mathbf{p}_p$ The intensity vector and pressure vector of the secondary sound source are respectively $\mathbf{q}_s$ and $\mathbf{p}_s$, and then $\mathbf{p}_p = \mathbf{Z}_{pp}\mathbf{q}_p + \mathbf{Z}_{ps}\mathbf{q}_s$, $\mathbf{p}_s = \mathbf{Z}_{sp}\mathbf{q}_p + \mathbf{Z}_{ss}\mathbf{q}_s$, where $\mathbf{Z}_{pp}$ and $\mathbf{Z}_{ss}$ are the self-impedance matrices of the primary and secondary sound sources, respectively, $\mathbf{Z}_{sp}$ and $\mathbf{Z}_{ps}$ are mutual impedances, representing the sound transmission impedance matrices between the primary and secondary sound sources, respectively. The sound power of the primary sound source is $W_p = \frac{1}{2}Re\left[\mathbf{q}_p^H\mathbf{p}_p\right]$, and the sound power of the secondary sound source is $W_s = \frac{1}{2}Re\left[\mathbf{q}_s^H\mathbf{p}_s\right]$. The total output power of the sound field is

$$
\begin{aligned}
W_T = W_p + W_s &= \tfrac{1}{2}\mathrm{Re}\left[\mathbf{q}_s^H\mathbf{Z}_{sp}\mathbf{q}_p + \mathbf{q}_s^H\mathbf{Z}_{ss}\mathbf{q}_s + \mathbf{q}_p^H\mathbf{Z}_{pp}\mathbf{q}_p + \mathbf{q}_p^H\mathbf{Z}_{ps}\mathbf{q}_s\right] \\
&= \tfrac{1}{2}\left[\mathbf{q}_s^H\mathbf{R}_{ss}\mathbf{q}_s + \mathbf{q}_s^H\mathbf{R}_{sp}\mathbf{q}_p + \mathbf{q}_p^H\mathbf{R}_{ps}\mathbf{q}_s + \mathbf{q}_p^H\mathbf{R}_{pp}\mathbf{q}_p\right] \\
&= \mathbf{q}_s^H\left[\tfrac{1}{2}\mathbf{R}_{ss}\right]\mathbf{q}_s + \mathbf{q}_s^H\left[\tfrac{1}{2}\mathbf{R}_{sp}\mathbf{q}_p\right] + \left[\tfrac{1}{2}\mathbf{q}_p^H\mathbf{R}_{ps}\right]\mathbf{q}_s + \tfrac{1}{2}\mathbf{q}_p^H\mathbf{R}_{pp}\mathbf{q}_p \\
&= \mathbf{q}_s^H\mathbf{A}\mathbf{q}_s + \mathbf{q}_s^H\mathbf{b} + \mathbf{b}^H\mathbf{q}_s + c
\end{aligned} \tag{7.1}
$$

where $\mathbf{R}_{ss} = Re[\mathbf{Z}_{ss}]$, $\mathbf{R}_{sp} = Re\left[\mathbf{Z}_{sp}\right] = \mathbf{R}_{ps}^T$, $\mathbf{R}_{pp} = Re\left[\mathbf{Z}_{pp}\right]$, $\mathbf{A} = \frac{1}{2}R_{ss} = Z_{ss}^H Z_{ss}$, $b = \frac{1}{2}R_{sp}q_p$, (here $R_{sp} = Z_s^H Z_p$), $c = \frac{1}{2}q_p^H R_{pp}q_p$. Due to the scalar c is a real number, so $\mathbf{q}_s^H\mathbf{A}\mathbf{q}_s$ should also be a real number, therefore the matrix $\mathbf{A}$ is a positive definite symmetric matrix.

The two most typical control strategies for active noise control are: minimizing the radiated sound power of the primary sound source and maximizing the sound absorption of the secondary sound source, which are respectively referred to as minimizing output sound power and the maximum sound absorption criterion [477].

The minimum value of W_T can be obtained by using the unconstrained optimization method. Therefore, under the condition of global energy minimization, the optimal secondary sound source intensity vector is obtained

$$
\mathbf{q}_{so} = -\mathbf{A}^{-1}\mathbf{b} = -\mathbf{R}_{ss}^{-1}\mathbf{R}_{sp}\mathbf{q}_p. \tag{7.2}
$$

Substituting the above formula into Eq. (7.1), the expression of W_T for the output sound power of global energy minimization is obtained

$$
W_T = \frac{1}{2}\mathbf{q}_p^H\left[\mathbf{R}_{pp} - \mathbf{R}_{sp}^T\mathbf{R}_{ss}^{-1}\mathbf{R}_{sp}\right]\mathbf{q}_p. \tag{7.3}
$$

The sound power of the secondary sound source is

$$
W_{so} = \frac{1}{4}j\mathbf{q}_p^H\left[\mathbf{X}_{sp}^T R_{ss}^{-1} R_{sp} - R_{sp}^T R_{ss}^{-1} X_{sp}\right] q_p \tag{7.4}
$$

where $j\left[X_{sp}^T R_{ss}^{-1} R_{sp} - R_{sp}^T R_{ss}^{-1} X_{sp}\right]$ is a Hermitian matrix, so W_{so} is a real number, its sign can be positive or negative, depending on $\mathbf{R}_{sp}$ and $\mathbf{X}_{sp}$. When W_{so} is positive, it indicates that the secondary sound source radiates energy outward; when it is negative, it indicates that the secondary sound source absorbs energy.

In the case of global energy minimization, the radiation impedance of the primary and secondary sound sources before and after active control is

$$
\mathbf{Z}_p = \mathbf{Z}_{pp} - j\mathbf{Z}_{ps}\frac{\mathbf{R}_{sp}}{\mathbf{R}_{ss}} = \left(\mathbf{R}_{pp} - \mathbf{R}_{ss}^{-1}\mathbf{R}_{sp}\mathbf{R}_{ps}\right) + j\mathbf{R}_{ss}^{-1}\mathbf{R}_{sp}\mathbf{X}_{ps}. \tag{7.5}
$$

$$
\mathbf{Z}_s = j\left(\mathbf{X}_{ss} - \mathbf{X}_{ps}\frac{\mathbf{R}_{ss}}{\mathbf{R}_{sp}}\right). \tag{7.6}
$$

At this time, the sound pressure vector of the secondary sound source is $\mathbf{p}_{so} = \mathbf{Z}_{sp}\mathbf{q}_p + \mathbf{Z}_{ss}\mathbf{q}_{so} = j[\mathbf{X}_{sp}\mathbf{q}_p + \mathbf{X}_{ss}\mathbf{q}_{so}]$,, and it is obvious n that if the N units of the primary sound source are in phase, then $\mathbf{q}_p$ and $\mathbf{q}_{so}$ are all real numbers, then $\mathbf{Z}_s$ is a pure imaginary number, the radiation resistance of the secondary sound source is zero, only the radiation reactance exists, that is to say, the secondary sound source is reactive. The radiation impedance of the primary sound source has also changed, the radiation resistance has decreased, which leads to a decrease in the total output sound power of the system, this is the mechanism of active noise control in the whole space, and it is also a typical reactive sound energy storage mechanism [474, 477]. The primary and secondary sound sources are coupled to each other and work on each other. Under the active control state, the work done by the secondary sound source on the medium cancels out the work done by the medium on it.

The maximum sound absorption criterion is to make the sound power absorbed by the secondary sound source the maximum. W_s part can be obtained from Eq. (7.1).

$$W_s = \frac{1}{4}\mathbf{q}_s^H \mathbf{R}_{ss}\mathbf{q}_s + \frac{1}{4}\mathbf{q}_s^H \mathbf{Z}_{sp}\mathbf{q}_p + \frac{1}{4}\mathbf{q}_p^H \mathbf{Z}_{sp}^H \mathbf{q}_s. \tag{7.7}$$

Similarly, the optimal secondary source intensity $\mathbf{q}_{sa}$ can be calculated to minimize W_s to the minimum value W_{sa}

$$\mathbf{q}_{sa} = -\frac{1}{2}\mathbf{R}_{ss}^{-1}\mathbf{Z}_{sp}\mathbf{q}_p. \tag{7.8}$$

$$W_{sa} = -\frac{1}{8}\mathbf{q}_p^H \mathbf{Z}_{sp}^H \mathbf{R}_{ss}^{-1}\mathbf{Z}_{sp}\mathbf{q}_p. \tag{7.9}$$

Equation (7.9) indicates that the secondary sound source is absorbing sound energy and the output power of the primary sound source is

$$W_{pa} = \frac{1}{4}\mathbf{q}_p^H \left[2\mathbf{R}_{pp} + \mathbf{X}_{sp}^T \mathbf{R}_{ss}^{-1}\mathbf{X}_{sp} - \mathbf{R}_{sp}^T \mathbf{R}_{ss}^{-1}\mathbf{R}_{sp}\right]\mathbf{q}_p \tag{7.10}$$

Analysis of the single-channel system shows that the third term in the above equation can be greater than or less than the second term, indicating that the output sound power of the primary sound source may increase or decrease after active control.

7.1.1.2 Parametric Array Loudspeaker as Secondary Sound Source

When the parametric array loudspeaker is used as the secondary sound source, due to its directivity, when the primary sound source is not within the beam angle of the secondary sound source, it can be considered that the primary sound source has no effect on the sound field of the secondary sound source, that is $\mathbf{Z}_{ps} = 0$, at this time the total output sound power of the sound field is

$$W_T = W_p + W_s = \mathrm{Re}\Big[\mathbf{q}_s^H\mathbf{Z}_{sp}\mathbf{q}_p + \mathbf{q}_s^H\mathbf{Z}_{ss}\mathbf{q}_s + \mathbf{q}_p^H\mathbf{Z}_{pp}\mathbf{q}_p\Big]/2 \tag{7.11}$$

In fact, the last term of the above equation is W_p a constant, which is irrelevant to $\mathbf{q}_s$. Therefore, the minimum value of W_T is the minimum value of W_s, at which point the total output power of the system is minimized, which is also the maximum sound absorption of the secondary sound source. The W_s can also be transformed into the form of a Hermitian quadratic equation [477].

$$\begin{aligned} W_s &= \tfrac{1}{2}\mathrm{Re}\big[\mathbf{q}_s^H\mathbf{Z}_{sp}\mathbf{q}_p + \mathbf{q}_s^H\mathbf{Z}_{ss}\mathbf{q}_s\big] \\ &= \mathbf{q}_s^H\left(\tfrac{1}{2}\mathbf{R}_{ss}\right)\mathbf{q}_s + \mathbf{q}_s^H\left(\tfrac{1}{4}\mathbf{Z}_{sp}\mathbf{q}_p\right) + \left(\tfrac{1}{4}\mathbf{Z}_{sp}\mathbf{q}_p\right)^H\mathbf{q}_s + c \\ &= \mathbf{q}_s^H\mathbf{A}\mathbf{q}_s + \mathbf{q}_s^H\mathbf{b} + \mathbf{b}^H\mathbf{q}_s + c \end{aligned} \tag{7.12}$$

where, $\mathbf{A} = \frac{1}{2}\mathbf{R}_{ss}$, $\mathbf{b} = \frac{1}{4}\mathbf{Z}_{sp}\mathbf{q}_p$, $c = 0$.

In this way, the optimal solution is still in the form of $-\mathbf{A}^{-1}\mathbf{b}$, therefore it is obtained

$$\mathbf{q}_{sa} = -\frac{1}{2}\mathbf{R}_{ss}^{-1}\mathbf{Z}_{sp}\mathbf{q}_p \tag{7.13}$$

The radiation impedances of the primary and secondary sound sources are $\mathbf{Z}_p = \mathbf{Z}_{pp}$ and $\mathbf{Z}_s = -\mathbf{R}_{ss} + j\mathbf{X}_{ss}$. It is evident that the radiation impedance of the secondary sound source becomes negative, indicating that it absorbs the energy of the sound field. Therefore, when the parametric array loudspeaker acts as a secondary sound source, the minimization of the total output power globally is equivalent to the maximum sound absorption of the secondary sound source, which is equivalent to the parametric array loudspeaker absorbing the energy of the primary sound source.

7.1.1.3 Radiation Impedance of Parametric Array Loudspeaker

It can be seen from Eq. (7.13), the optimal complex resonance speed of the secondary sound source is related to the radiation impedance, and the parametric array loudspeaker cannot directly solve the radiation impedance due to its nonlinear mechanism. Therefore, it can only be obtained indirectly through the sound power method [481, 482], the sound power W_s is obtained by the far-field integration of the sound pressure expression. The equivalent radiation resistance of the sound source is given by $W_s = \frac{1}{2}R|U|^2$.

When the nonlinear effect is completed in the near field of the sound source, that is, the parametric array length is less than the Rayleigh distance ($L_a < z_0$), the quasi-linear solution of the difference frequency sound is [2]

$$q_-(\theta, z) \cong -\frac{jp_{0a}p_{0b}\beta k^2 a^2}{4\rho_0 c_0^2\alpha_T}\frac{e^{-\alpha_- z}}{z}D(\theta)\exp\left(-\frac{1}{2}jkz\tan^2\theta\right) \tag{7.14}$$

where $D(\theta) = D_W(\theta)D_A(\theta)$ is the directional function

$$D_W(\theta) = \frac{1}{1 + j(k/2\alpha_T)\tan^2\theta}, D_A(\theta) = \frac{2J_1(ka\tan\theta)}{ka\tan\theta} \tag{7.15}$$

The axial sound pressure of the difference frequency is

$$q_-(0, z) \cong \frac{p_{0a}p_{0b}\beta k^2a^2}{4\rho_0 c_0^2\alpha_T}\frac{e^{-\alpha_- z}}{z} \tag{7.16}$$

Due to the sharp directivity of the parametric array loudspeaker, the sound power is not uniformly distributed in space, so it can be considered that the difference frequency sound is a non-directional scattering wave with a sound intensity of I_s, and its total radiated sound power is

$$W_s = 4\pi z^2 I_s = 4\pi z^2 \frac{q_-^2(0, z)}{2\rho_0 c_0} \tag{7.17}$$

If the parametric array loudspeaker is assumed to be equivalent to a pulsating sphere source with the vibration speed of $u_{0a}u_{0b}$, and the radiation resistance of R_E, (where u_{0a} and u_{0b} are the oscillation speeds of the pump wave), then its average radiated sound power is

$$\overline{W_E} = \frac{1}{2}R_E(u_{0a}u_{0b})^2 \tag{7.18}$$

The equivalent radiation impedance of the parametric array loudspeaker by setting $W_s = \overline{W_E}$

$$R_E = \frac{\pi\rho_0\beta^2k^4a^4}{4\alpha_T^2 c_0} \tag{7.19}$$

7.1.1.4 Variation of Sound Field Energy

Considering the simplest case, the primary sound source is a point sound source, and the secondary sound source is a parametric array loudspeaker. Then, the optimal complex resonance speed formula of the secondary sound source, i.e., Eq. (7.13) becomes as follows

$$q_{sa} = -\frac{1}{2}\frac{Z_{sp}}{R_{ss}}q_p \tag{7.20}$$

From this, the relationship between the energy of the primary and secondary sound sources and the total energy of the sound field before and after control can be obtained [477], which is.

(1) The ratio of the energy of the primary sound source after control to the energy of the primary sound source before control

$$W_p/W_{pp} = 1 + \left(X_{sp}^2 - R_{sp}^2\right)/2R_{ss}R_{pp} \tag{7.21}$$

(2) The ratio of the energy of the secondary sound source after control to the energy of the primary sound source before control

$$W_s/W_{pp} = -\left(X_{sp}^2 + R_{sp}^2\right)/4R_{ss}R_{pp} \tag{7.22}$$

(3) The ratio of the total energy of the sound field after control to the energy of the primary sound source before control

$$W_T/W_{pp} = 1 + (X_{sp}^2 - 3R_{sp}^2)/4R_{ss}R_{pp} \tag{7.23}$$

The primary sound source is assumed to be a pulsating sphere source with a radius of 0.03 m, and the secondary sound source is a parametric array loudspeaker with a radius of 0.05 m. Moreover, the primary sound source is outside the beam angle of the secondary sound source, i.e., the primary sound source has no effect on the sound field of the secondary sound source. When the distance between the primary and secondary sound sources is 1.5 m, the relationship between the ratios of the three energies and the noise frequency is shown in Fig. 7.1. When the noise frequency is 2 kHz, the relationship between the ratios of the three energies and the change in distance between the primary and secondary sound sources is shown in Fig. 7.2.

From the above figures, the following can be observed.

(1) The energy of the secondary sound source is less than zero, indicating that it has absorbed the energy of the primary sound source. Moreover, both the optimal resonance speed formula of the secondary sound source. Equation (7.13) and the energy formula of the secondary sound source, Eq. (7.22) are less than zero, indicating that the secondary sound source is in a state of sound absorption.
(2) When the noise frequency changes, the energy of the primary sound source increases and decreases below 1.5 kHz. As can be seen from Eq. (7.21), this depends on the impact of the secondary sound source on the primary sound source, i.e., the real part R_{sp} and imaginary part X_{sp} of their mutual impedance Z_{sp}.
(3) Although the secondary sound source absorbs the energy of the primary sound source, due to the oscillation of the energy of the primary sound source, the total energy of the sound field also oscillates.
(4) Because the secondary sound source has directivity, the primary sound source has no effect on the sound field of the secondary sound source, and the primary

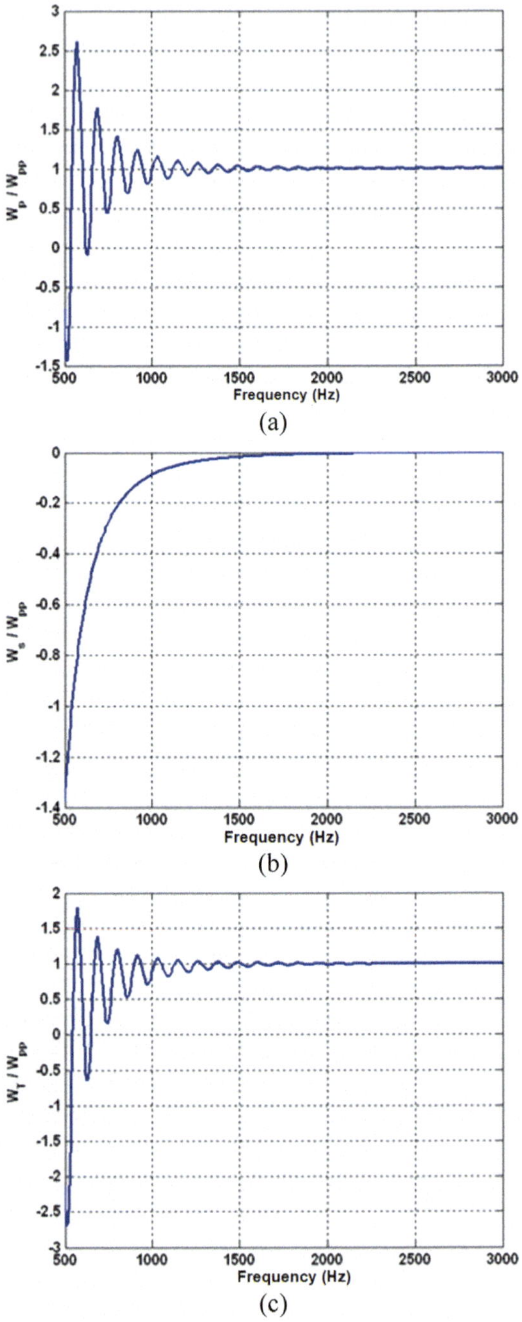

Fig. 7.1 Relationship between the ratio of sound energy and noise frequency **a** ratio of primary sound source energy after control to before control **b** ratio of secondary sound source energy to primary sound source energy before control **c** ratio of total sound field energy to primary sound source energy before control

Fig. 7.2 Relationship between the ratio of sound energy and the change in distance between primary and secondary sound sources **a** ratio of primary sound source energy after control to before control **b** ratio of secondary sound source energy to primary sound source energy before control **c** ratio of total sound field energy to primary sound source energy before control

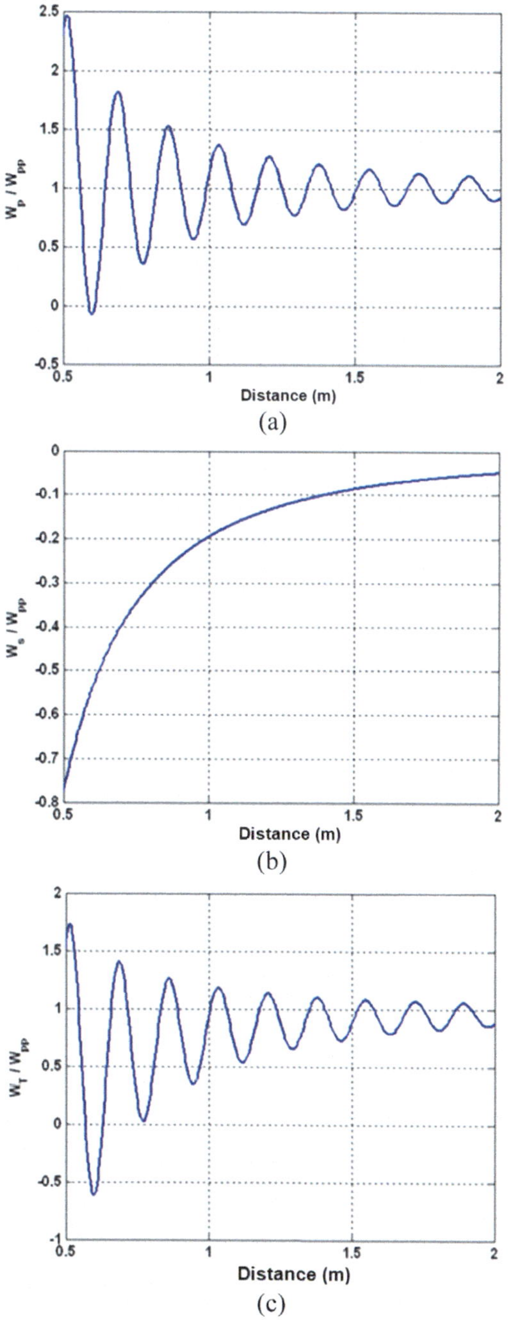

sound source is assumed to be a pulsating sphere source radiating spherical waves, so the change in distance between the primary and secondary sound sources does not significantly affect the control effect.

7.1.2 *Local Space Control*

When a common omnidirectional loudspeaker is used as the secondary sound source for local spatial control, and the distance between the primary and secondary sound sources is greater than half the wavelength, the sound energy in some areas of the sound field will increase, and excellent noise reduction effect cannot be achieved [473]. On the other hand, even though active noise control across an entire space can be theoretically feasible, in practical scenarios, the cost of implementing such a system is often prohibitive. Moreover, it is not always necessary to reduce noise across the entire area, and only part of the space requires noise reduction, which is called local active noise control.

From the perspective of noise control, if the average sound pressure level of a spatial area is reduced by more than 3 dB after control, then noise control in that area has been effective. If the average sound pressure level is reduced by more than 10 dB, then that area can be called a "quiet zone". The purpose of local spatial noise control is to place the secondary sound source and select the observation point in the local spatial area where noise reduction is needed, set the control target, and implement active control to create a quiet zone [483–487]. There are many specific expressions of the objective function, but they need to take into account two requirements: (1) to obtain as large a quiet zone as possible, (2) to minimize the increase in sound power outside the quiet zone.

7.1.2.1 Local Active Quiet Zone

The primary sound source is a sound source array is assumed to be composed of L monopoles. N secondary sound sources are arranged in a row in front of the primary sound source, and M error sensors are also arranged in a row in front of the secondary sound source, as shown in Fig. 7.3 [477]. In the figure, only one monopole sound source is selected as the primary sound source, and its vertical distances from the secondary sound source and the error sensor are r_{ps} and r_{pe}. The vertical distance between the secondary sound source and the error sensor is r_{se}, and the distances between adjacent secondary sound sources and error sensors are respectively r_{ss} and r_{ee}.

The intensities of the primary sound sources are represented by L and N order column vectors $\mathbf{q}_p$ and $\mathbf{q}_s$, respectively, so any observation point in space r_{em}. $(m = 1, 2, \ldots M)$ at the sound pressure can be expressed as

$$p(\mathbf{r}_{em}) = p_p(\mathbf{r}_{em}) + p_s(\mathbf{r}_{em}) = \mathbf{Z}_p\mathbf{q}_p + \mathbf{Z}_s\mathbf{q}_s \tag{7.24}$$

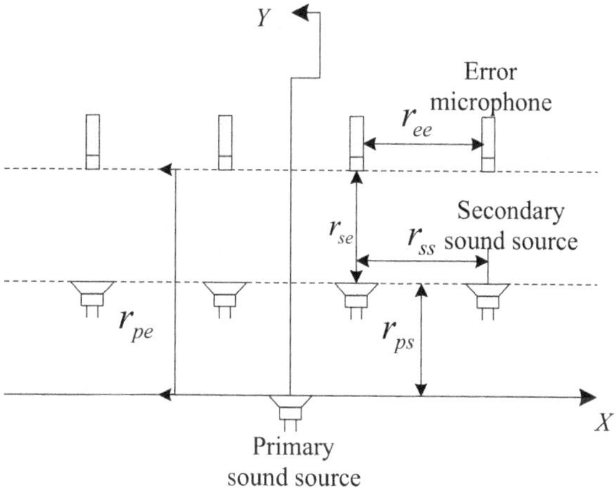

Fig. 7.3 Schematic diagram of the local space active noise control system

In the formula, $\mathbf{Z}_p$, $\mathbf{Z}_s$ are L and N order column vectors representing the sound transmission impedance from the primary and secondary sound sources to the observation point, respectively. Therefore, the sound pressure vector at M error sensors is

$$\mathbf{P}_e = \mathbf{Z}_{pe}\mathbf{q}_p + \mathbf{Z}_{se}\mathbf{q}_s \tag{7.25}$$

where $\mathbf{Z}_{pe}$ and $\mathbf{Z}_{se}$ are $L \times M$ and $N \times M$ order sound transmission impedance matrices, respectively. If the goal of active control is to minimize the sum of the squares of the sound pressure in the local space, then the objective function of the control system is

$$J_A(\mathbf{q}_s) = \mathbf{P}_e^H \mathbf{P}_e \tag{7.26}$$

Substituting Eq. (7.25) into the above equation, the relationship between the objective function and the intensity vector of the secondary sound source is obtained, which is quadratic, so the secondary sound source intensity that minimizes the objective function can be found.

$$\mathbf{q}_{sA} = -[\mathbf{Z}_{se}^H \mathbf{Z}_{se}]^{-1}\mathbf{Z}_{se}^H \mathbf{Z}_{pe}\mathbf{q}_p \tag{7.27}$$

When the number of error sensors equals the number of secondary sound sources, that is, $N = M$, a large quiet area can be obtained, and it can be ensured that the sound pressure at the error sensor after active control is zero, and the corresponding secondary sound source intensity is

$$\mathbf{q}_{sA} = -\mathbf{Z}_{se}^{-1}\mathbf{Z}_{pe}\mathbf{q}_p. \tag{7.28}$$

The sound pressure at any point in space is

$$p(\mathbf{r}) = \mathbf{Z}_p\mathbf{q}_p - \mathbf{Z}_s\mathbf{Z}_{se}^{-1}\mathbf{Z}_{pe}\mathbf{q}_p. \tag{7.29}$$

The total sound power in space is

$$\begin{aligned} W_A =& \frac{1}{2}\Big[\mathbf{q}_p^H \mathrm{Re}(\mathbf{Z}_{pp})\mathbf{q}_p + \mathbf{q}_s^H \mathrm{Re}(\mathbf{Z}_{ss})\mathbf{q}_s \\ &+ \mathbf{q}_p^H \mathrm{Re}(\mathbf{Z}_{ps}^T)\mathbf{q}_s + \mathbf{q}_s^H \mathrm{Re}(\mathbf{Z}_{ps})\mathbf{q}_p\Big]. \end{aligned} \tag{7.30}$$

In the formula, $\mathbf{Z}_{pp}$ and $\mathbf{Z}_{ss}$ are the $L \times L$ order and $N \times N$ order sound transmission impedance matrices of the primary sound source array and the secondary sound source array, respectively. Taking an active control system comprising two secondary sound sources and two error sensors as an example, this study investigates the changes in sound field power and its distribution pattern before and after implementation of the control measures.

Suppose the primary sound source is located at the coordinate origin, and the two secondary sound sources and error sensors are symmetrically arranged on both sides of the y axis, and $r_{ss} = r_{ee}$, where the distance from the secondary sound source to the adjacent error sensor is r_{se}, and the distance to the other error sensor is $r_{se'}$, as shown in Fig. 7.3. Since both the sound source and the observation point are in free space, it is easy to obtain the sound transmission impedance,

$$\mathbf{Z}_{pe} = \frac{j\omega\rho_0}{4\pi}\begin{bmatrix} e^{-jkr_{pe'}}/r_{pe'} \\ e^{-jkr_{pe'}}/r_{pe'} \end{bmatrix} \tag{7.31}$$

$$\mathbf{Z}_{se} = \frac{j\omega\rho_0}{4\pi}\begin{bmatrix} e^{-jkr_{se}}/r_{se} & e^{-jkr_{se'}}/r_{se'} \\ e^{-jkr_{se'}}/r_{se'} & e^{-jkr_{se}}/r_{se} \end{bmatrix} \tag{7.32}$$

where $r_{pe'}$ is the distance from the primary sound source to the two error sensors, which are equal, and $r_{pe'}^2 = \sqrt{r_{pe}^2 + r_{ee}^2/4}$, $r_{se'}^2 = \sqrt{r_{se}^2 + r_{ss}^2}$. By substituting into Eq. (7.28), the optimal intensity of the two secondary sound sources can be obtained.

$$q_{sA1} = q_{sA2} = Aq_p \tag{7.33}$$

where

$$A = \frac{e^{-jkr_{pe'}}(e^{jkr_{se}}/r_{se} + e^{jkr_{se'}}/r_{se'})}{r_{pe'}[1/r_{se}^2 + 1/r_{se'}^2 + 2\cos k(r_{se'} - r_{se})/r_{se}r_{se'}]} \tag{7.34}$$

The total sound power after active control is

$$W_A = W_{pp}[1 + 2(1 + \sin\mathrm{c}(kr_{ss}))|A|^2 + 4\sin\mathrm{c}(kr_{ps'})\mathrm{Re}(A)] \tag{7.35}$$

In the formula, $r_{ps'}$ is the distance from the primary sound source to the two secondary sound sources, which is $r_{ps'} = \sqrt{r_{ps}^2 + r_{ss}^2}$.

From Eqs. (7.33) to (7.35), it can be seen [483] as follows. (1) The error sensor should not be too close to the secondary sound source. If the error sensor is located in the near field of the secondary sound source, although the sound power will not increase significantly after control, the area of the quiet zone obtained is too small, losing its practical value. (2) The optimal layout depends directly on the distance between the secondary sound sources r_{ss}, and is also related to r_{se} and r_{ps}, and r_{ss} has an optimal value range [483].

When the number of secondary sound sources and error sensors is more than 2, it is impossible to obtain the optimal intensity of secondary sound sources and the minimum sound power analytical expressions like Eqs. (7.33) and (7.35), but the dependence of the quiet zone, sound power distribution on the layout of secondary sound sources, and error sensors can be obtained through numerical calculations [483–485]. The essential conclusion is as follows. The quiet zone obtained by the multi-channel system is not a simple superposition of the quiet zones obtained by a single secondary sound source. After the frequency range of the primary noise is determined, the optimal layout of the secondary sound sources and error sensors can be found to maximize the quiet zone range and minimize the increase in sound power. The quiet zone is formed around the secondary sound source, and its size depends on the distance from the error sensor array to the secondary sound source array, and the distance from the primary sound source to the secondary sound source array.

7.1.2.2 Experimental Study of Local Quiet Zone

Using parametric array loudspeakers as secondary sound sources, due to their directivity, it is possible to achieve sound attenuation at the control point while minimizing the impact of the parametric array loudspeakers on other areas of the sound field. This section studies the use of parametric array loudspeakers and ordinary loudspeakers as secondary sound sources in a single-channel active noise control system, and while obtaining the maximum sound attenuation at the microphone, the size of the quiet zone in the sound field is compared through simulation experiments and actual measurements.

Assuming that the primary sound source, secondary sound source, and control point are aligned on a straight line, the distance between the primary sound source. The control point is 240 cm, and the distances between the primary and secondary sound sources are 140 cm, 100 cm, and 60 cm respectively. The primary sound source is a common loudspeaker with a radius of 0.03 m, and a parametric array loudspeaker with a radius of 0.05 m and a common loudspeaker with a radius of 0.03 m are used as the secondary sound sources respectively. The input signal of the secondary sound source is adjusted to achieve the maximum sound attenuation at the control point, and the sound field near the control point is measured, as shown in the shaded area in Fig. 7.4. The parametric array loudspeaker is a regular hexagon

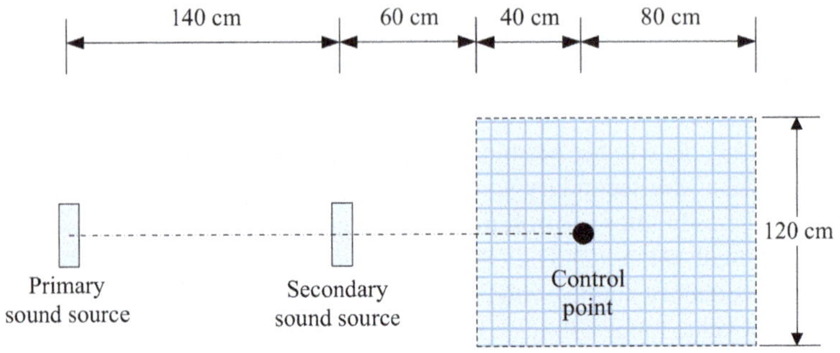

Fig. 7.4 System schematic diagram

composed of 217 ultrasonic transducers with a diameter of 10 mm in parallel. The sound pressure measurement uses a 1/4-in. microphone GRAS 46BE, and the experiment was conducted in the anechoic chamber of Institute of Acoustics, Chinese Academy of Sciences. The schematic diagrams of the system corresponding to the distances between the primary and secondary sound sources of 140 cm, 100 cm, and 60 cm are shown in Figs. 7.4, 7.11, and 7.18 respectively, and the results corresponding to different distances are shown in Figs. 7.5, 7.6, 7.7, 7.8, 7.9 and 7.10, Figs. 7.12, 7.13, 7.14, 7.15, 7.16 and 7.17, and Figs. 7.19, 7.20, 7.21, 7.22, 7.23 and 7.24 respectively. The ratio of the sound field energy after and before control is shown in Table 7.1.

The following can be seen from the above charts.

1. When the parametric array loudspeaker is used as the second sound source, the following conclusions can be drawn

While achieving the maximum sound attenuation at the control point, the axial area near the control point can also achieve noise reduction, with a maximum attenuation of 25 dB. The simulation and experimental results of areas of −10 dB and −6 dB, in the three cases are basically consistent, but the experimental measurement area for 0 dB attenuation is slightly larger than the simulation results.

(1) The simulation and experimental values of the axial sound pressure attenuation of the secondary sound source in the first case basically coincide, but the simulation values in the other two cases are 10 dB lower than the experimental values. This is because the simulation results are more precise in controlling the phase of the parametric array loudspeaker, while various factors such as the consistency and stability of the transducer in the experimental process lead to the inability to achieve theoretical attenuation. However, a maximum sound attenuation of 30 dB can still be obtained from actual measurements.
(2) The secondary sound source has almost no effect on increasing the noise in the sound field, with the maximum increase being only 2 dB.
(3) The closer the primary and secondary sound sources are, the larger the control area (areas of −10 dB, −6 dB, and 0 dB attenuation). In the third case, it is obvious that the entire sound field is almost all attenuated.

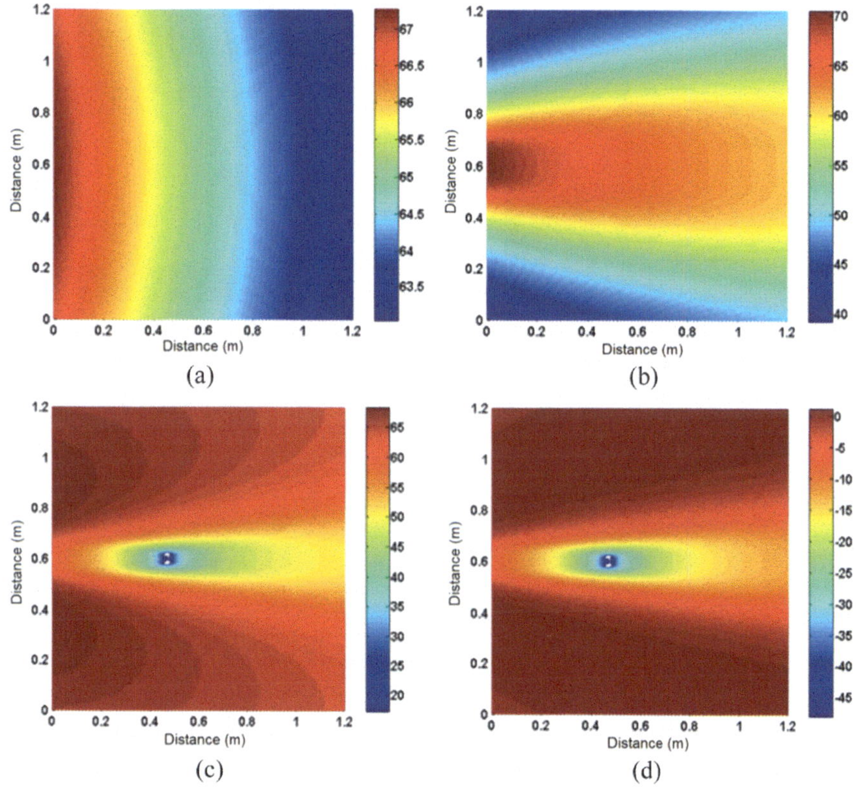

Fig. 7.5 Simulation results of the sound field before and after control using a parametric array loudspeaker as the secondary sound source **a** sound field before control, **b** secondary sound source field, **c** sound field after control, **d** sound attenuation before and after control

(4) The total energy of the sound field after control is significantly reduced, and as the distance between the primary and secondary sound sources decreases, the energy attenuation is greater, and the control effect is better.
(5) Overall, the experimental and simulation results can basically match.

2. When using ordinary loudspeakers as secondary sound sources, it can be seen as follows.
3. While obtaining the maximum sound attenuation at the control point, the axial area near the control point can also achieve noise reduction effects, with the maximum attenuation reaching 20 dB. However, there is a significant difference between the experimental results and the simulation results. The sound field of the primary and secondary sound sources measured in reality does not present the radiation sound field of the point sound source, but has a clear beam (especially it is more obvious with the secondary sound source), and the entire sound field is steered upwards. So, the attenuation area of the sound field after control is also

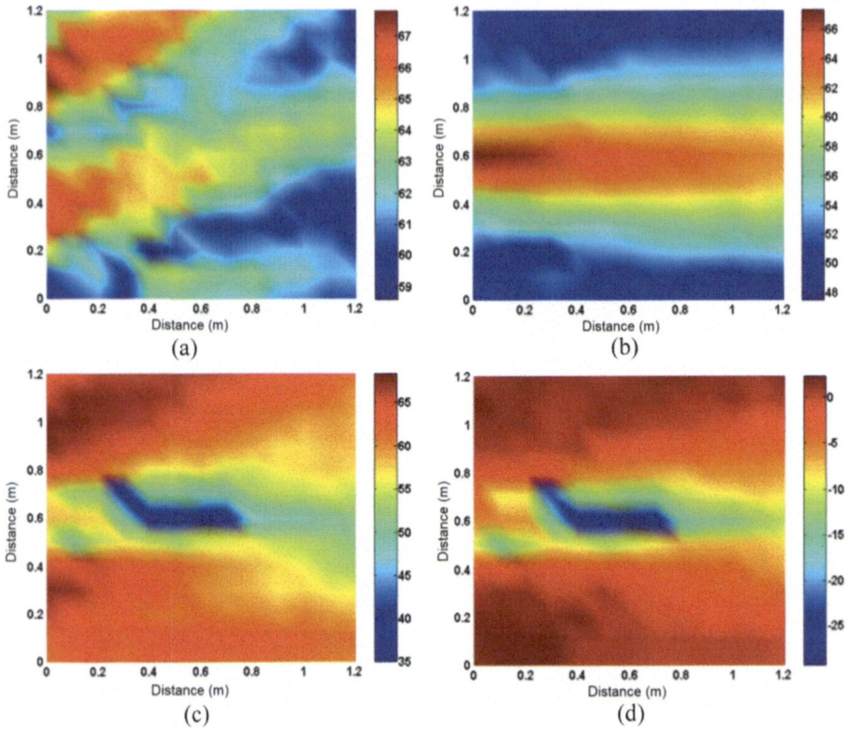

Fig. 7.6 Measurement results of the sound field before and after control using a parametric array loudspeaker as the secondary sound source **a** sound field attenuation before and after control in the 10 dB area, **b** sound field attenuation before and after control in the 6 dB area, **c** sound field attenuation before and after control in the 0 dB area, **d** axial sound pressure of the secondary sound source before and after control

significantly different from the simulation results, and the attenuation area of the simulation results is significantly larger than the experimental results.

4. The secondary sound source has a significant effect on increasing the sound field, with the maximum increase reaching 8 dB. But due to the limitation of the measurement area, which is only 120 cm × 120 cm, there are not many measurement points where the sound pressure increases, resulting in the total energy of the sound field in the second and third cases being reduced.
5. Overall, the comparison of experimental and simulation results using ordinary loudspeakers as secondary sound sources is not very good, and the difference is large, for the following reasons:

(i) The loudspeaker used in the experiment, despite its small size (radius 0.03 *m*), cannot be considered an ideal point sound source, so its radiated sound field does not present the spherical wave form of the simulation results.

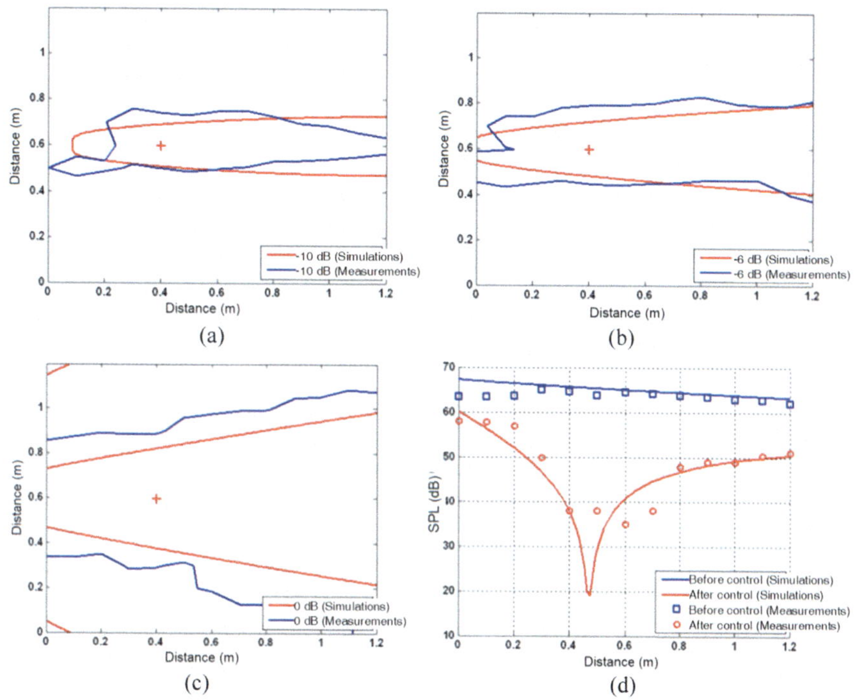

Fig. 7.7 Comparison of simulation and experimental results of sound field attenuation area using a parametric array loudspeaker as the secondary sound source **a** 10 dB attenuation region of the sound field before and after control, **b** 6 dB attenuation region of the sound field before and after control, **c** 0 dB attenuation region of the sound field before and after control, **d** axial sound pressure of the secondary sound source before and after control (**a**, **b**)

(ii) The measurement area is limited by the loudspeaker stand in the anechoic chamber. Although in some places the sound pressure increases after control, the overall energy does not increase significantly.

(iii) Comparing the results of the parametric array loudspeaker and the ordinary loudspeaker as secondary sound sources, it can be seen as follows.

(iv) While noise control is achieved at the control point, sound attenuation is also achieved in the axial area of the control point. The simulation and experimental results of the parametric array loudspeaker as a secondary sound source are in good agreement, and the sound attenuation area is on the axis of the secondary sound source. The simulation results of the ordinary loudspeaker show that the sound attenuation area is larger than that of the parametric array loudspeaker, but the actual measured attenuation area is not that large.

(v) When the parametric array loudspeaker is used as a secondary sound source, the closer the distance between the primary and secondary sound sources, the better the control effect becomes. This results in greater sound field energy attenuation and minimal impact on other areas of the sound field. Consequently, the overall

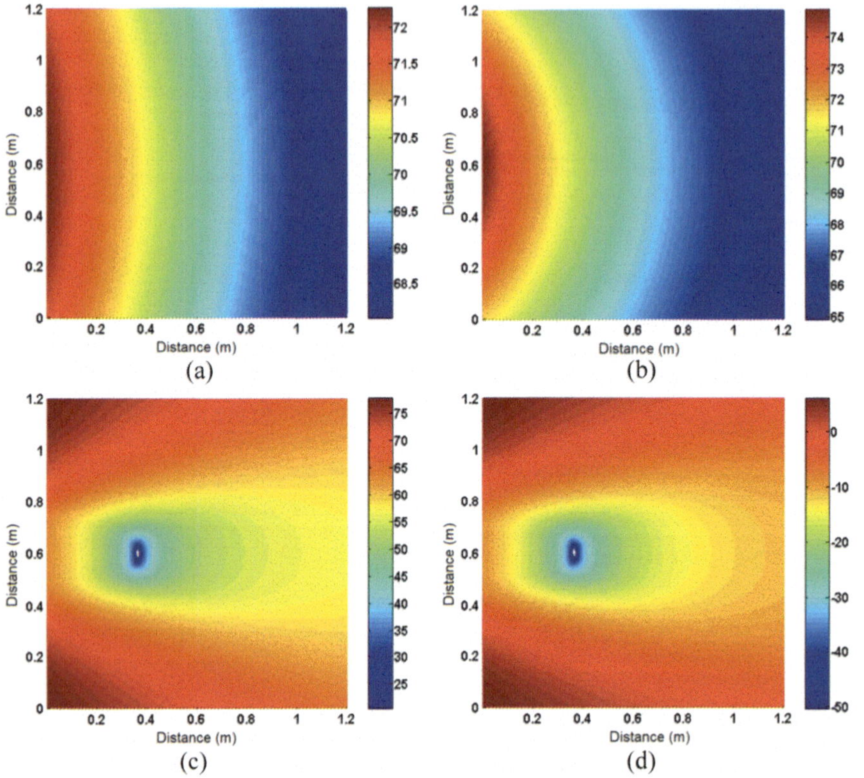

Fig. 7.8 Simulation results of the sound field before and after control using a conventional loudspeaker as the secondary sound source **a** sound field before control, **b** secondary sound source field, **c** sound field after control, **d** sound attenuation before and after control

sound field energy is always decreasing. When an ordinary loudspeaker is used as a secondary sound source, due to the limitations of the measurement area, the actual measurement results show that the sound field energy in this area also decreases.

7.2 Active Noise Control of Parametric Array Loudspeakers in Pipelines

The pipeline is widely used in industrial production and people's lives, and the noise of the pipeline seriously affects the efficiency of industrial production and the quality of people's lives, so the problem of pipeline noise urgently needs to be solved. Active noise control is gradually being used in actual noise reduction projects because of its low cost and small size. However, ordinary loudspeakers as secondary

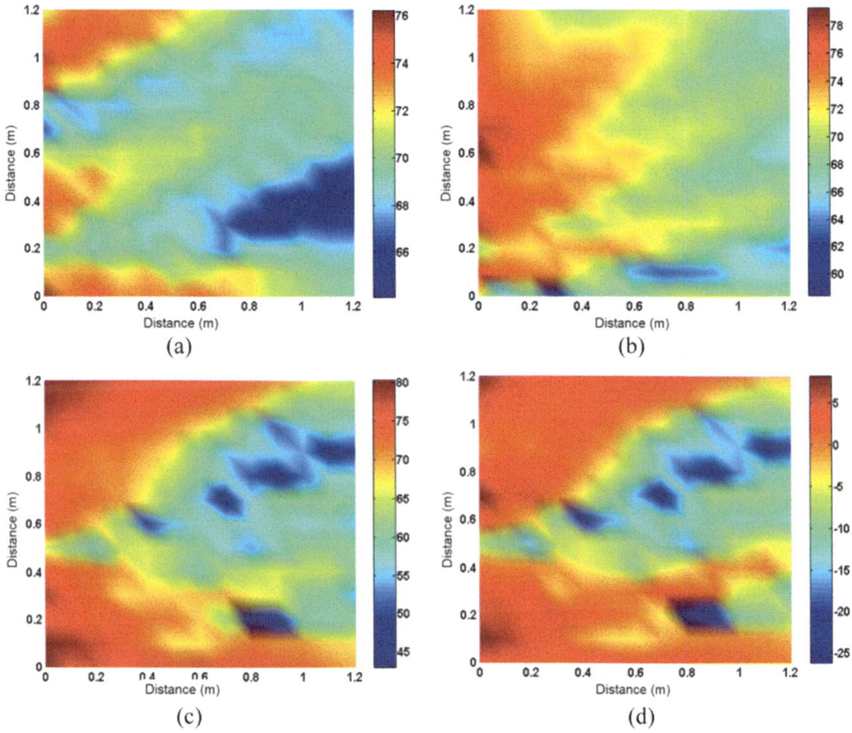

Fig. 7.9 Measurement results of the sound field before and after control using a conventional loudspeaker as the secondary sound source **a** sound field before control, **b** secondary sound source field, **c** sound field after control, **d** sound attenuation before and after control

sound sources for pipeline active noise control will produce strong acoustic feedback, which will affect the control effect and the stability of the control system. To solve this problem, scholars generally add a feedback compensation network to cancel out the acoustic feedback, but this will increase the complexity of the algorithm and the consumption of resources. The parametric array loudspeaker has the advantages of high directivity and small size, and can be used as a secondary sound source to reduce acoustic feedback, thereby reducing the complexity of the active noise control algorithm and also the amount of computation. This section first tests the size of the acoustic feedback of the two sound sources in a square waveguide. The focus is on the control of noise in straight and L pipes by the parametric array loudspeaker, and by comparing with ordinary loudspeakers, the noise control effect of the parametric array loudspeaker without an acoustic feedback supplement network is verified. Finally, by measuring the sound field of the pipeline and the amount of noise reduction, the noise control area and area of the parametric array loudspeaker are determined [488].

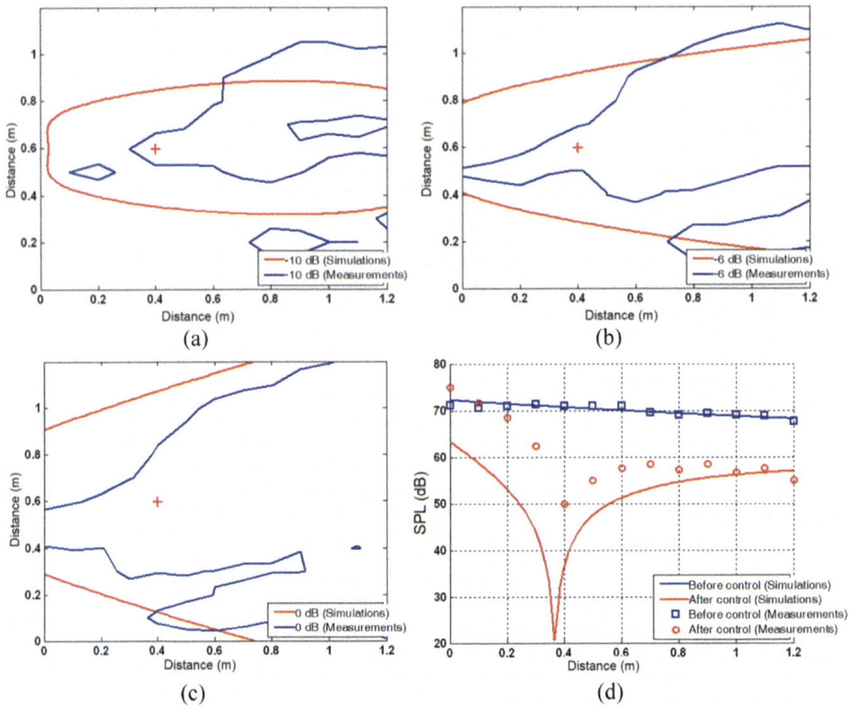

Fig. 7.10 Comparison of simulation and experimental results of sound field attenuation area using a conventional loudspeaker as the secondary sound source **a** 10 dB attenuation region of the sound field before and after control, **b** 6 dB attenuation region of the sound field before and after control, **c** 0 dB attenuation region of the sound field before and after control, **d** axial sound pressure of the secondary sound source before and after control

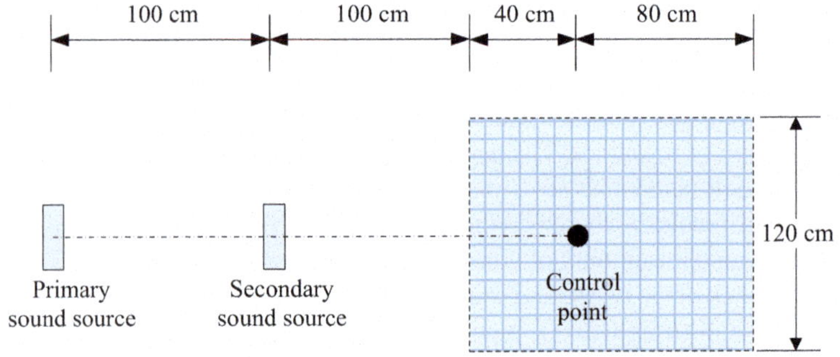

Fig. 7.11 System schematic diagram

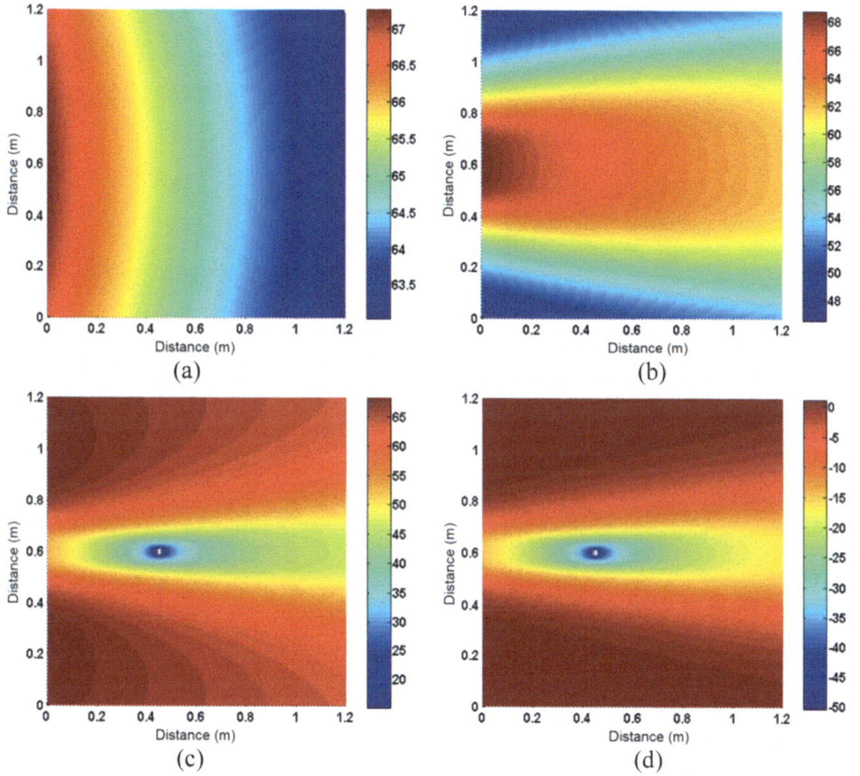

Fig. 7.12 Simulation results of the sound field before and after control using a parametric array loudspeaker as the secondary sound source **a** sound field before control, **b** secondary sound source field, **c** sound field after control, **d** sound attenuation before and after control

7.2.1 Secondary Acoustic Feedback

7.2.1.1 Secondary Acoustic Feedback Analysis

For general feedforward adaptive active noise control, the problem of acoustic feedback has always been a major problem in active noise control. Without taking appropriate measures, it will affect the control effect and even cause the system to diverge. The following analyzes the impact of the feedback path on the system. The adaptive algorithm diagram with acoustic feedback is shown in Fig. 7.5, where $P(z)$ is the primary path, $S(z)$ and $F(z)$ are the secondary path and feedback path, $x(n)$, $y(n)$, $d(n)$ and $e(n)$ are respectively the reference signal, control signal, primary noise, and error signal. $y'(n)$ is the error microphone control signal, $W(z)$ is the adaptive filter.

In Fig. 7.25, the error signal in the z domain is

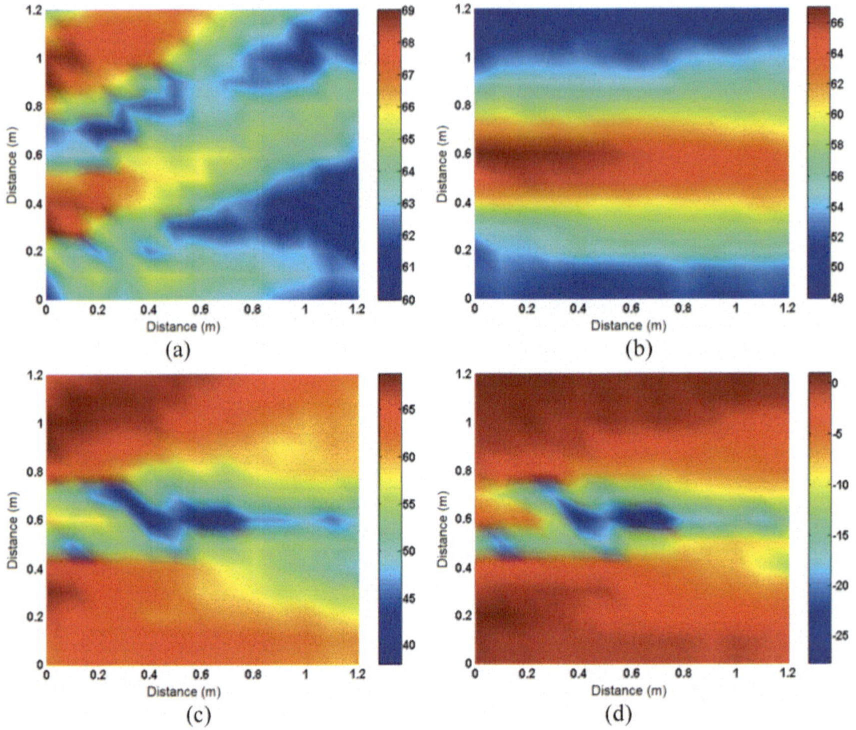

Fig. 7.13 Measurement results of the sound field before and after control using a parametric array loudspeaker as the secondary sound source **a** sound field before control, **b** secondary sound source field, **c** sound field after control, **d** sound attenuation before and after control

$$E(z) = P(z)X(z) + S(z)\frac{W(z)X(z)}{1 - W(z)F(z)} \tag{7.36}$$

Assuming adaptive convergence, the system is stable. The transfer function of the controller's weight coefficients can be calculated as

$$W^o(z) = -\frac{P(z)}{S(z) - P(z)F(z)} \tag{7.37}$$

The open-loop transfer function of the feedback loop can be derived from Fig. 7.25 as

$$H_{OL}(z) = W(z)F(z) \tag{7.38}$$

After the adaptive convergence, its open-loop transfer function becomes

$$H_{OL}(z) = -\frac{P(z)F(z)}{S(z) - P(z)F(z)} \tag{7.39}$$

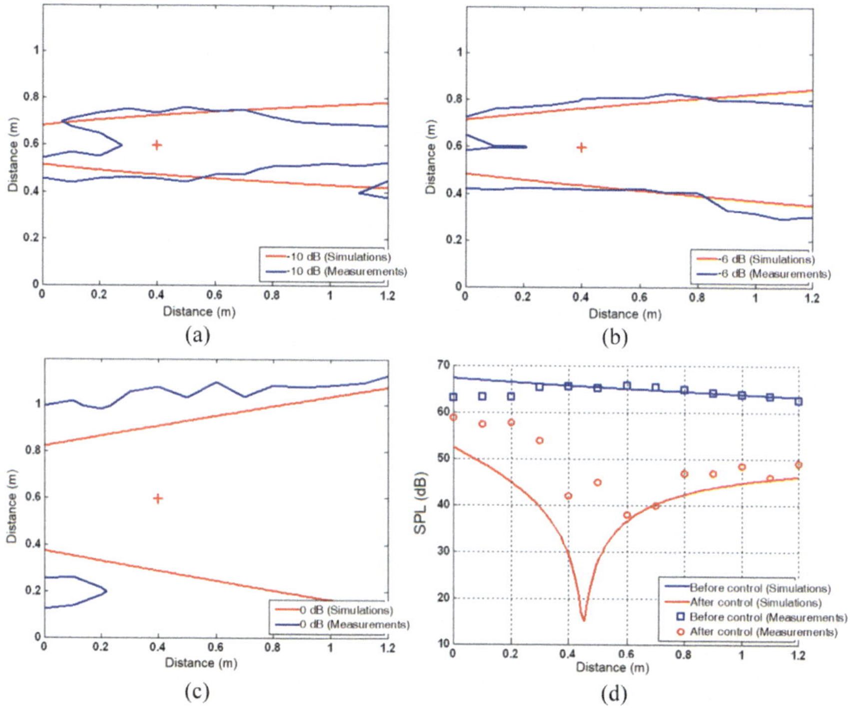

Fig. 7.14 Comparison of simulation results and experimental results of sound field attenuation area using a parametric array loudspeaker as the secondary sound source **a** 10 dB attenuation region of the sound field before and after control, **b** 6 dB attenuation region of the sound field before and after control, **c** 0 dB attenuation region of the sound field before and after control, **d** axial sound pressure of the secondary sound source before and after control

$H_{OL}(z)$ can characterize the stability of the system. When the open-loop gain is greater than 1 and the phase lag is 180°, the system will become unstable. To ensure the stability of the system, a feedback compensation filter is generally added to the controller to offset the acoustic feedback $F(z)$, or reduce the response of the feedback path $F(z)$.

7.2.1.2 Secondary Acoustic Feedback Comparison

From the existing research results it can be observed that the directivity of the parametric array loudspeaker in the free field is much better than that of the ordinary loudspeaker. It is hoped to use the high directivity of the parametric array loudspeaker to reduce the acoustic feedback in the pipeline. Therefore, this section compares the acoustic feedback of the two sound sources in the pipeline through experiments.

The schematic diagram of the experimental device of this experiment is shown in Fig. 7.26. Because the pipeline is too long, for convenience, it is presented in the

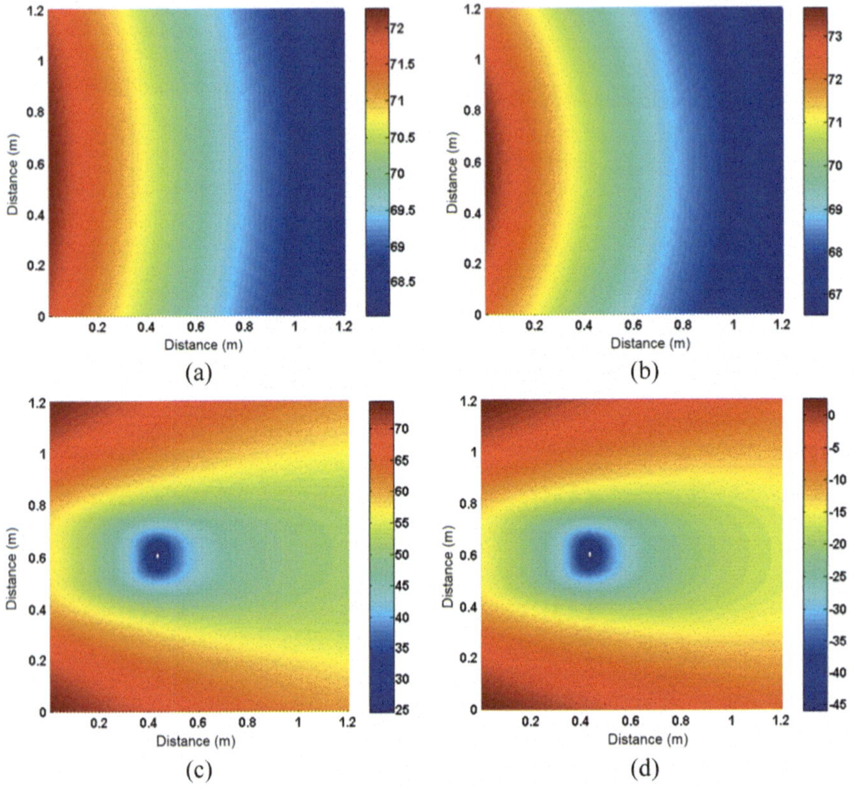

Fig. 7.15 Simulation results of the sound field before and after control using a conventional loudspeaker as the secondary sound source **a** sound field before control, **b** secondary sound source field, **c** sound field after control, **d** sound attenuation before and after control

form of a schematic diagram and a partial physical diagram. As can be seen from the figure, the ordinary loudspeaker is located on the side wall of the pipeline, and the parametric array loudspeaker is located in the center of the pipeline. The pipeline is a square waveguide tube with a diameter of 10 cm and a length of 4 m. The noise source is a loudspeaker with a diameter of 10 cm, located on one side of the pipeline, and is not turned on in this acoustic feedback comparison experiment, with this pipeline port as the coordinate 0 point. The positions of other devices are marked in the figure.

The secondary sound source emits white noise signals, and the reference loudspeaker and error loudspeaker respectively collect the feedback path and secondary path, and then use the Normalized Least Mean Square (NLMS) algorithm to identify the feedback path and secondary path. The update step size of the NLMS algorithm is set to 0.01, and the identification length is 512. To facilitate comparison, ensure that the sound pressure level of the sound waves emitted by the two sound sources is equivalent during measurement, and normalize the results with the amplitude of the secondary path as a reference. The path response measurement results of the

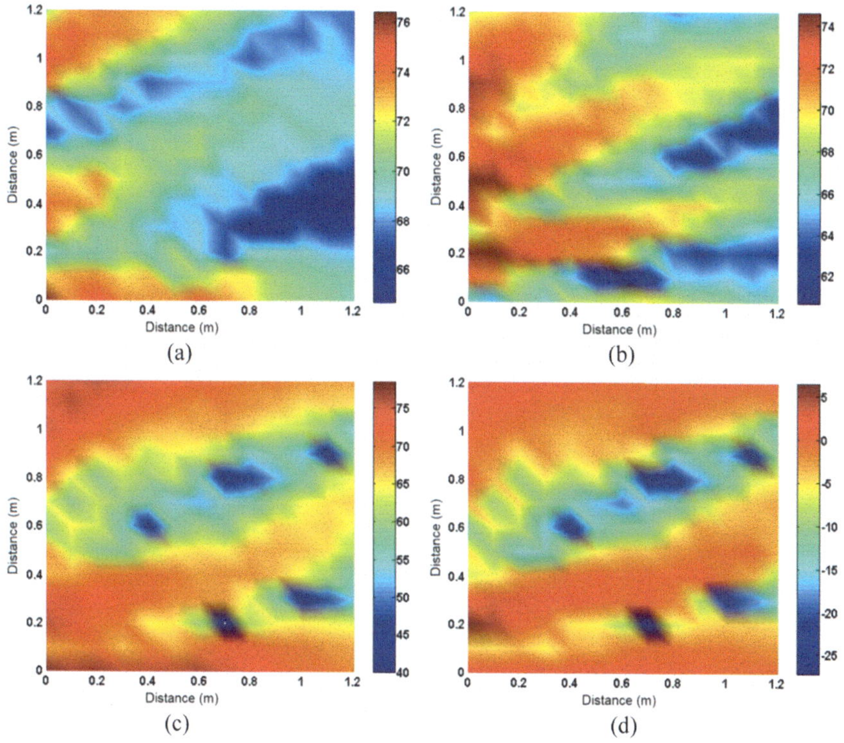

(a) (b) (c) (d)

Fig. 7.16 Measurement results of the sound field before and after control using a conventional loudspeaker as the secondary sound source **a** sound field before control, **b** secondary sound source field, **c** sound field after control, **d** sound attenuation before and after control

ordinary loudspeaker and the parametric array loudspeaker are shown in Figs. 7.27 and 7.28. As can be seen from the figure, when the ordinary loudspeaker is used as the secondary sound source, the amplitude of the secondary path is almost the same as that of the feedback path. When the parametric array loudspeaker is used as the secondary sound source, the amplitude of the feedback path is significantly smaller than the amplitude of the secondary path, with a difference of about 16 dB. This shows that the parametric array loudspeaker as a secondary sound source can reduce acoustic feedback.

Since placing the parametric array loudspeaker in the pipeline will affect the use of the pipeline, the straight pipe is changed to an L-pipe, as shown in Fig. 7.29. The length from the noise source to the corner is 1 m, and the length from the corner to the pipeline noise outlet is 2 m. Placing the parametric array loudspeaker at the corner can both take advantage of its high directivity without affecting the use of the pipeline. Since the position of the ordinary loudspeaker in the straight pipe does not affect the normal use of the pipeline, only the secondary path and feedback path of the parametric array loudspeaker were measured in the L-pipe, as shown in Fig. 7.30.

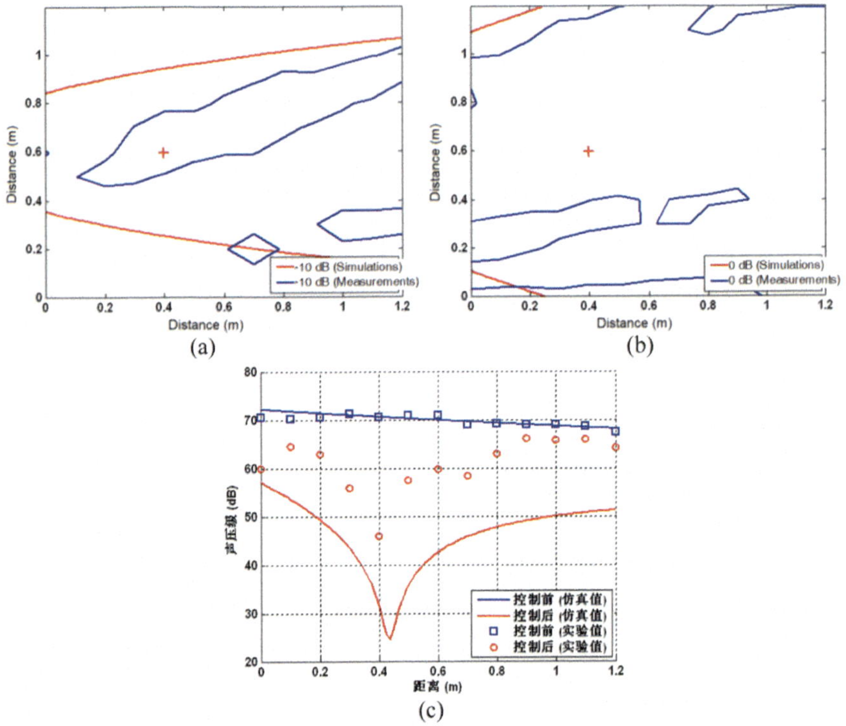

Fig. 7.17 Comparison of simulation results and experimental results of the sound field attenuation area using a conventional loudspeaker as the secondary sound source **a** 10 dB attenuation region of the sound field before and after control, **b** 0 dB attenuation region of the sound field before and after control, **c** axial sound pressure of the secondary sound source before and after control

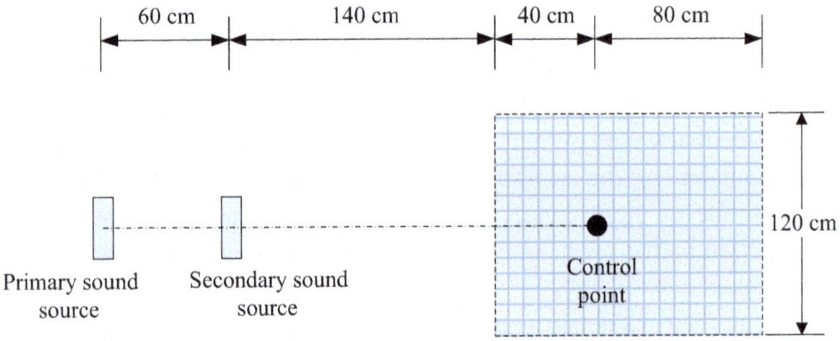

Fig. 7.18 System schematic diagram

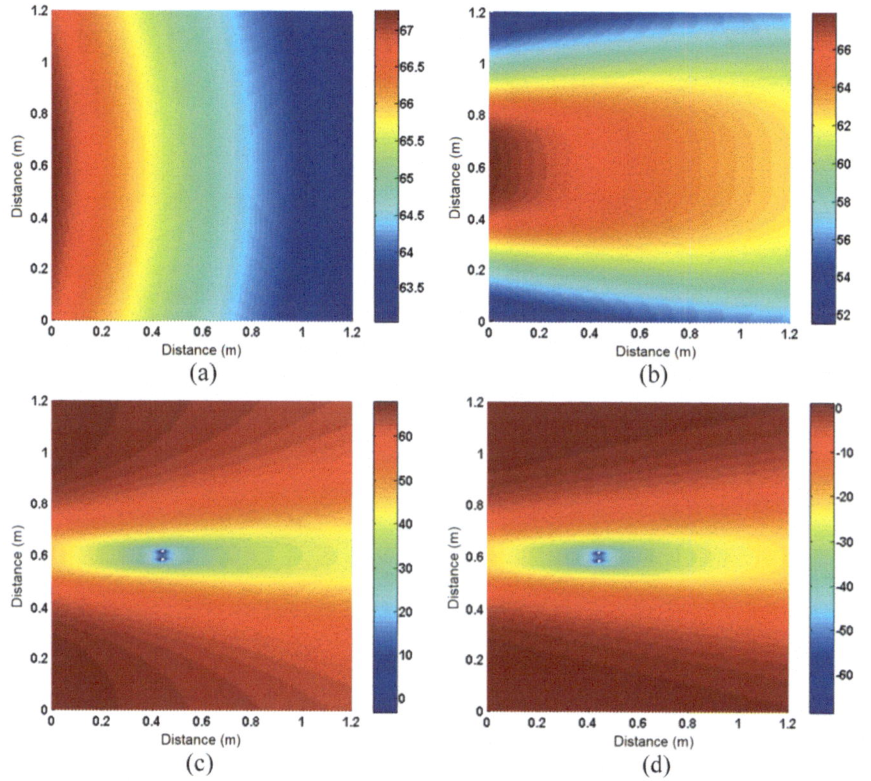

Fig. 7.19 Simulation results of controlling the front and rear sound fields using a parametric array loudspeaker as the secondary sound source **a** sound field before control, **b** secondary sound source field, **c** sound field after control, **d** sound attenuation before and after control

The method of measuring the secondary path and feedback path response in the L-pipe is the same as in the straight pipe, and will not be repeated here. As can be seen from the figure, the amplitude of the feedback path impulse response is about one third of the amplitude of the secondary path impulse response, which means it is about 10 dB smaller. Therefore, under the L-shaped structure, the parametric array loudspeaker can also reduce acoustic feedback.

7.2.2 *Active Noise Control*

7.2.2.1 Adaptive Active Noise Control

The parallel structure narrowband adaptive control algorithm was proposed by Glover [489] in 1977. According to the algorithm proposed by Glover, a block diagram of

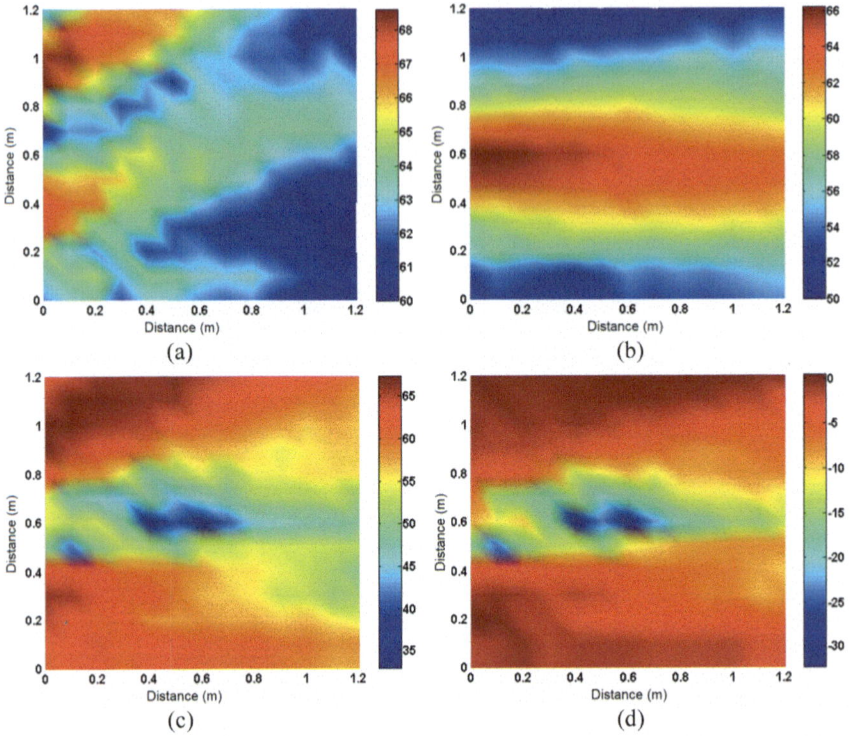

Fig. 7.20 Measurement results of the sound field before and after control using a parametric array loudspeaker as the secondary sound source **a** sound field before control, **b** secondary sound source field, **c** sound field after control, **d** sound attenuation before and after control

a narrowband adaptive algorithm with acoustic feedback in parallel is obtained, as shown in Fig. 7.31. Among them, Fig. 7.31a has a feedback compensation network, and Fig. 7.31b has no feedback compensation network.

The reference signal passed through the peak filter in Fig. 7.31 is

$$x_m(n) = A_m\cos(\omega_m n),\ m = 1, 2, \ldots, M \tag{7.40}$$

where M represents the number of single frequencies in the narrowband noise, and $M = 1$ represents that there is only one single frequency noise. The algorithm becomes a single-frequency active noise control algorithm. Each single frequency noise input corresponds to an adaptive filter $W_m(z)$. These adaptive filters form a parallel structure, and then the output signal of the filter $y_m(n)$ generates the control signal by summing

$$y(n) = \sum_{m=1}^{M} y_m(n) \tag{7.41}$$

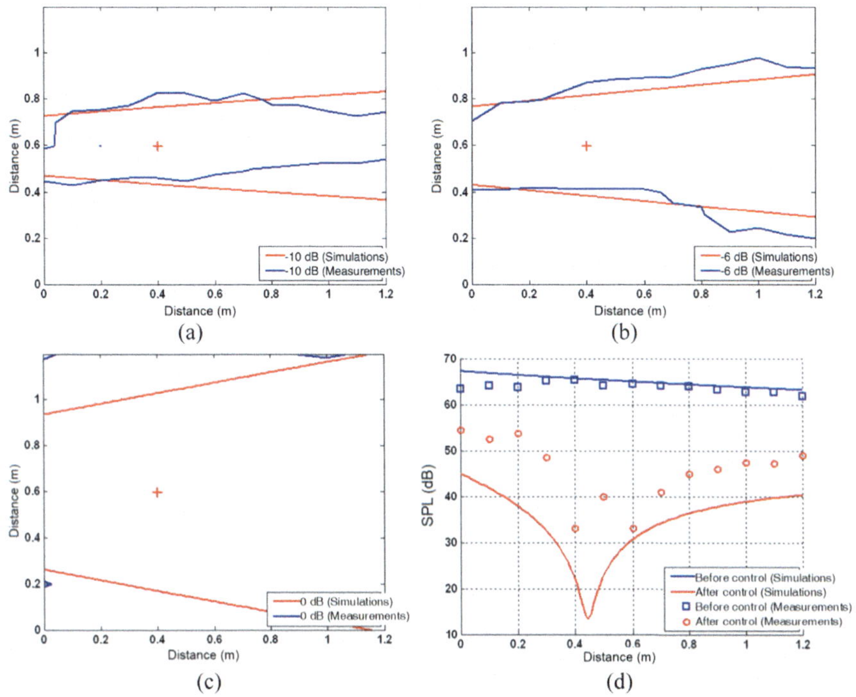

Fig. 7.21 Comparison of simulation results and experimental results of sound field attenuation area using a parametric array loudspeaker as the secondary sound source **a** 10 dB attenuation region of the sound field before and after control, **b** 6 dB attenuation region of the sound field before and after control, **c** 0 dB attenuation region of the sound field before and after control, **d** axial sound pressure of the secondary sound source before and after control

After filtering by the secondary path estimation $\hat{S}(z)$, the noise signal is

$$x'_m(n) = \sum_{l=1}^{L_1-1} \hat{s}_l x_m(n-l), \quad m = 1, 2, \dots M \tag{7.42}$$

where L_1 is the length of $\hat{S}(z)$. The update formula for the filter $W_m(z)$ is

$$w_m(n+1) = w_m(n) + \mu x'_m(n) e(n), \qquad n = 0, 1, \dots, N-1 \tag{7.43}$$

$$e(n) = d(n) - y(n) \otimes s(l) \tag{7.44}$$

where μ is the update step size.

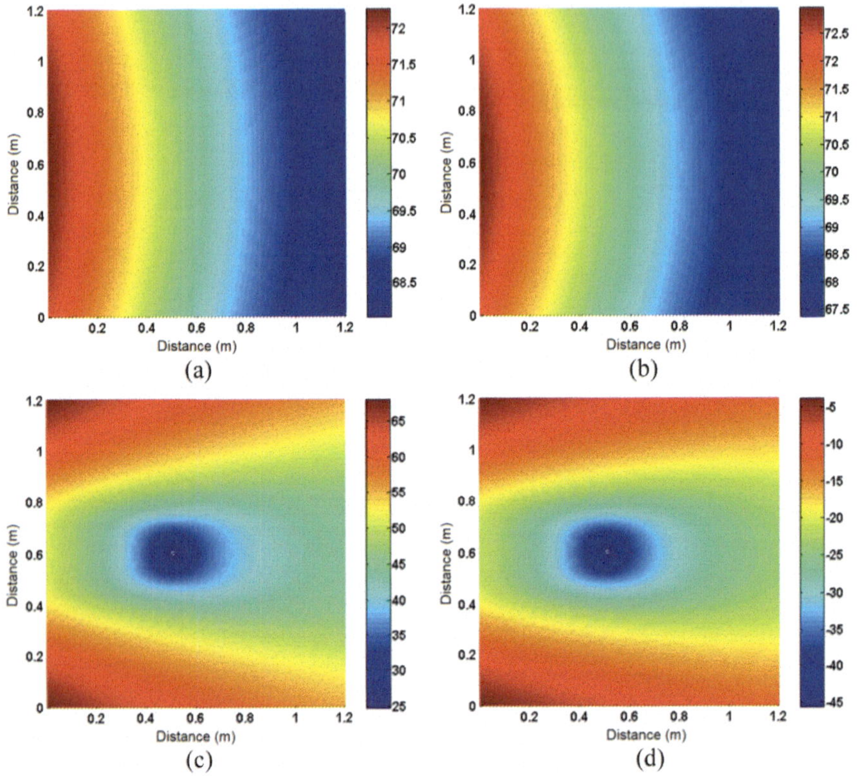

Fig. 7.22 Simulation results of the sound field before and after control using a conventional loudspeaker as the secondary sound source **a** sound field before control, **b** secondary sound source field, **c** sound field after control, **d** sound attenuation before and after control

7.2.2.2 Experimental Setup

To meet the requirements of real-time control, noise control experiments use Texas Instruments' DSP chip TMS320C6713 as the main control unit, with ADS8365 and DAC8544 serving as the analog-to-digital converter and digital-to-analog converter respectively. The maximum processing capability of the TMS320C6713 chip can reach 2400 MIPS, and ADS8365 and DAC8544 can achieve a data conversion speed of 10 MHz and a conversion precision of 16 bits, which can meet the speed and precision requirements of noise control. Considering the frequency of noise control and the computational load of the processor, the system sampling rate is designed to be 16 kHz. The FxLMS algorithm can automatically adjust the filter coefficients according to the changes in the noise source and acoustic environment parameters. For single-frequency noise control, the length of the adaptive filter is chosen to be 32. The purpose of this experiment is to compare the final noise reduction effects of

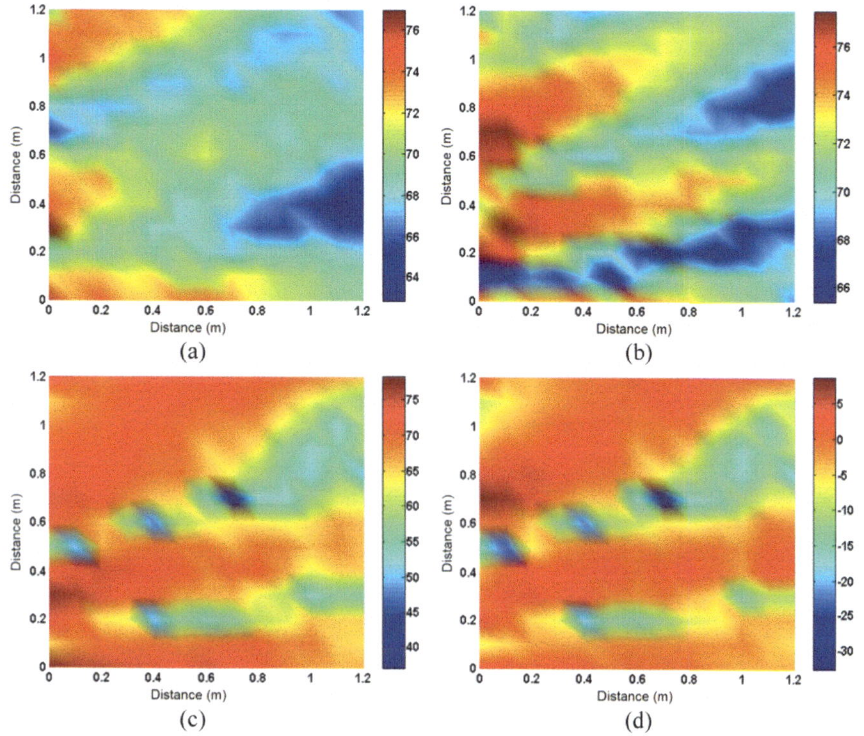

Fig. 7.23 Measurement results of the sound field before and after control using a conventional loudspeaker as the secondary sound source **a** sound field before control, **b** secondary sound source field, **c** sound field after control, **d** sound attenuation before and after control

the two sound sources, so the step size for updating the coefficients of the adaptive filter $W(n)$ is set sufficiently small.

The straight pipe experimental setup is shown in Fig. 7.32. The pipe used in the figure has an inner cavity with side length of 10 cm, a wall thickness of 1 cm, and a length of 4 m. The waveguide tube is set with the axial direction as x direction, the noise source is located at one end of the waveguide tube, and this end is set as $x = 0$. The reference microphone is located at $x = 0.45$ m. The secondary sound source is placed at $x = 1.55$ m. To ensure that the distance between the error microphone and the secondary sound source is greater than the accumulation length of the parametric array loudspeaker, the error microphone is placed at $x = 3.55$ m. To reduce the impact of external factors on the experiment, the port at $x = 4$ m is filled with sound-absorbing material. The emission unit of the parametric array loudspeaker is a square ultrasonic transducer array with a size of 6 cm by 6 cm. The parametric array loudspeaker is located inside the pipe, as shown in Fig. 7.32b. To reduce harmonic distortion, the modulation mode of the parametric array loudspeaker is single-sideband modulation. The ordinary loudspeaker uses a HiVi dynamic driver loudspeaker unit with the same

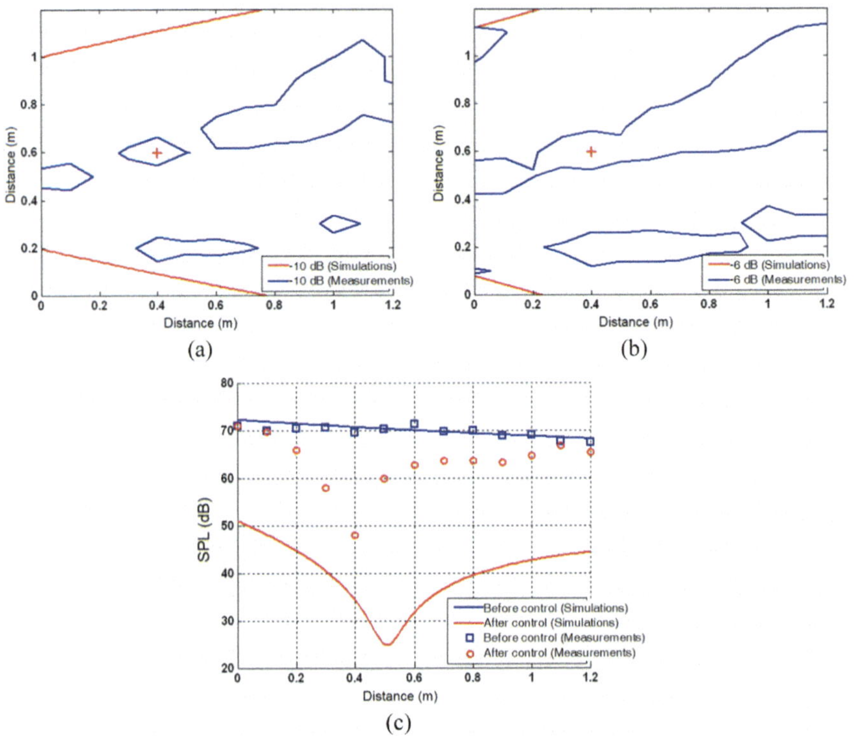

Fig. 7.24 Comparison of simulation results and experimental results of the sound field attenuation area using a conventional loudspeaker as the secondary sound source **a** 10 dB attenuation region of the sound field before and after control, **b** 6 dB attenuation region of the sound field before and after control, **c** axial sound pressure of the secondary sound source before and after control

Table 7.1 Ratio of sound field energy after and before control

	Parametric array loudspeaker (%)	Common loudspeaker (%)
Case 1	76.5	132.1
Case 2	62.1	78.3
Case 3	47.4	81.3

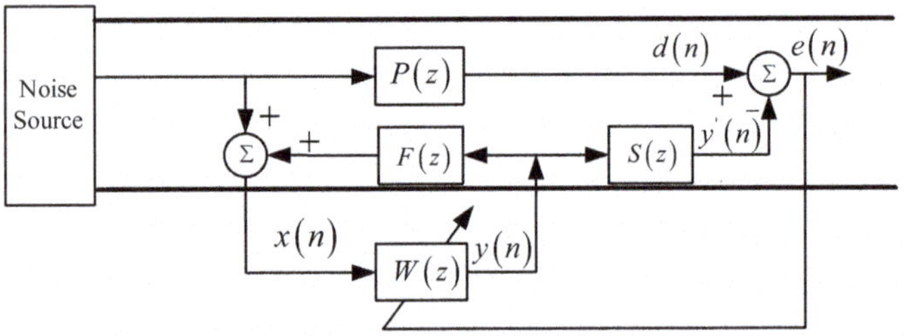

Fig. 7.25 Block diagram of feedback compensation FxLMS algorithm

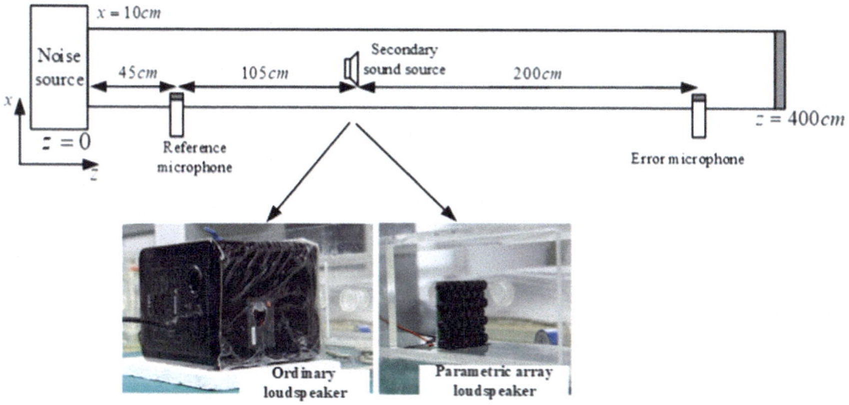

Fig. 7.26 Schematic diagram and local physical diagram of straight pipe noise control

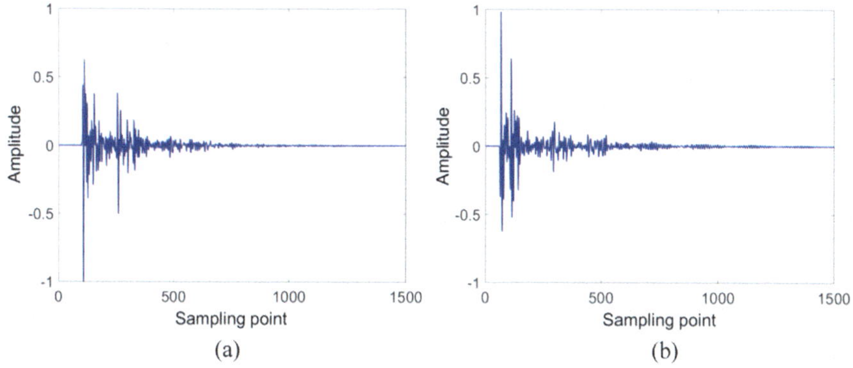

Fig. 7.27 Straight pipe ordinary loudspeaker channel response **a** secondary path impulse response, **b** feedback path impulse response

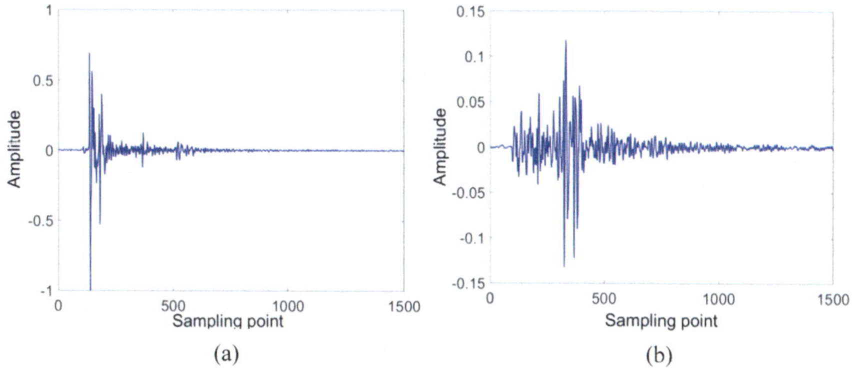

Fig. 7.28 Straight pipe parametric array loudspeaker sound channel response **a** secondary path impulse response, **b** feedback path impulse response

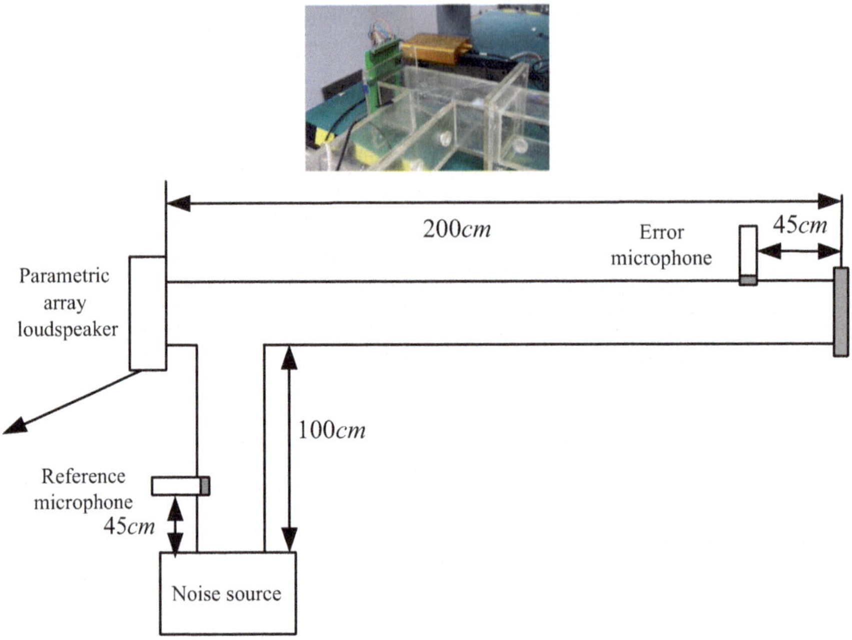

Fig. 7.29 Schematic diagram of L-pipe noise control

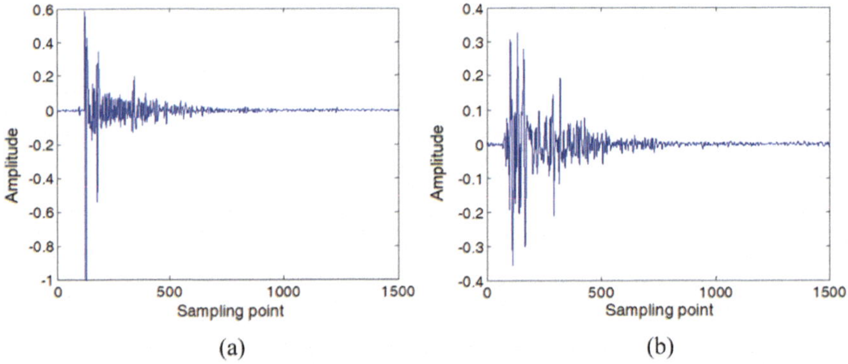

Fig. 7.30 L-pipe parametric array loudspeaker path response **a** secondary path impulse response, **b** feedback path impulse response

size as the parametric array loudspeaker, and the ordinary loudspeaker is installed on the side wall of the notch in Fig. 7.32b.

Figure 7.33 shows the physical picture of the noise control of the L-tube parametric array loudspeaker. The parametric array is placed at the corner of the tube. The distance between the reference microphone and the noise source and the distance between the error microphone and the pipe noise outlet are both 0.45 m. The upstream

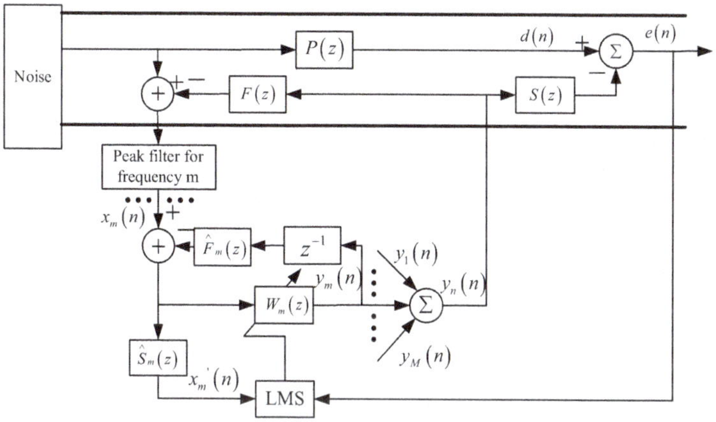

(a)

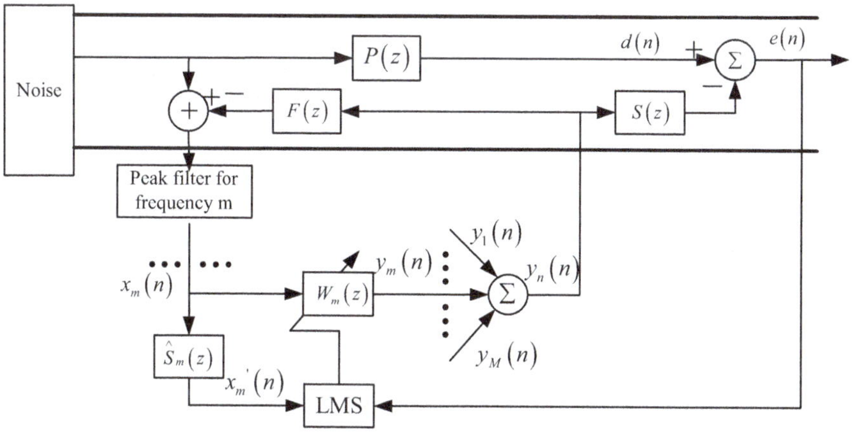

(b)

Fig. 7.31 Block diagram of parallel narrowband active noise control algorithm with sound feedback **a** with feedback compensation network, **b** without feedback compensation network

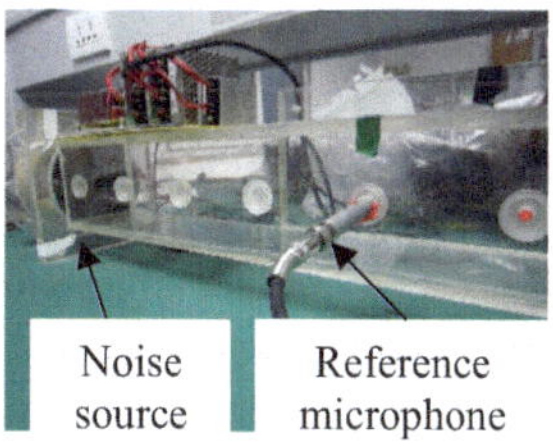

(a)

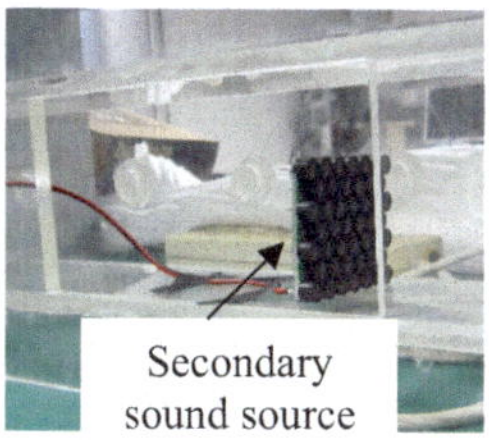

(b)

(c)

Fig. 7.32 Straight pipe experimental setup diagram **a** noise source and reference microphone, **b** parametric array loudspeaker, **c** error microphone

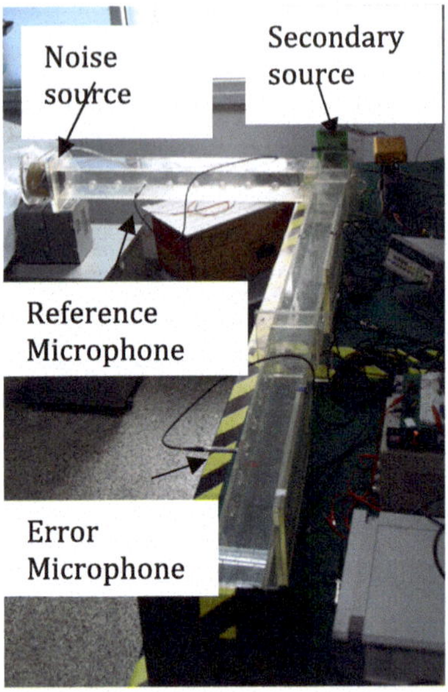

Fig. 7.33 L-pipe experimental setup diagram

pipe length of the secondary sound source is 1 m, and the downstream pipe length is 2 m. Other configurations are the same as Fig. 7.32.

According to the size of the pipe, the cutoff frequency of the pipe is 1.715 kHz. Due to the physical characteristics of the parametric array loudspeaker, its low-frequency performance is poor. To observe better, the noise frequency range is selected between 600 and 1200 Hz. The noise signal is generated by the Agilent signal source (model 33220A), amplified by the D-class amplifier, and then emitted through a dynamic loudspeaker with a diameter of 10 cm. The controller begins to process noise reduction. When the controller is stable, the error microphone signal is collected to the computer through the sound card (model FIREFACE 400). For easy comparison, the noise before and after the experimental results control was normalized.

7.2.2.3 Single-Frequency Noise Reduction Results

The ordinary loudspeakers and parametric array loudspeakers as secondary sound sources are used for adaptive active noise control of single-frequency noise in this section. When ordinary loudspeakers are used as secondary sound sources, they are only implemented in straight pipes, employing two parallel adaptive algorithms, one without a feedback compensation network and the other with a feedback compensation network. When parametric array loudspeakers are used as secondary sound sources, they are implemented in both straight pipes and L pipes, and the parallel adaptive algorithms have no feedback compensation networks. Six sets of experiments were conducted, with the frequencies of single-frequency noise being 600 Hz,

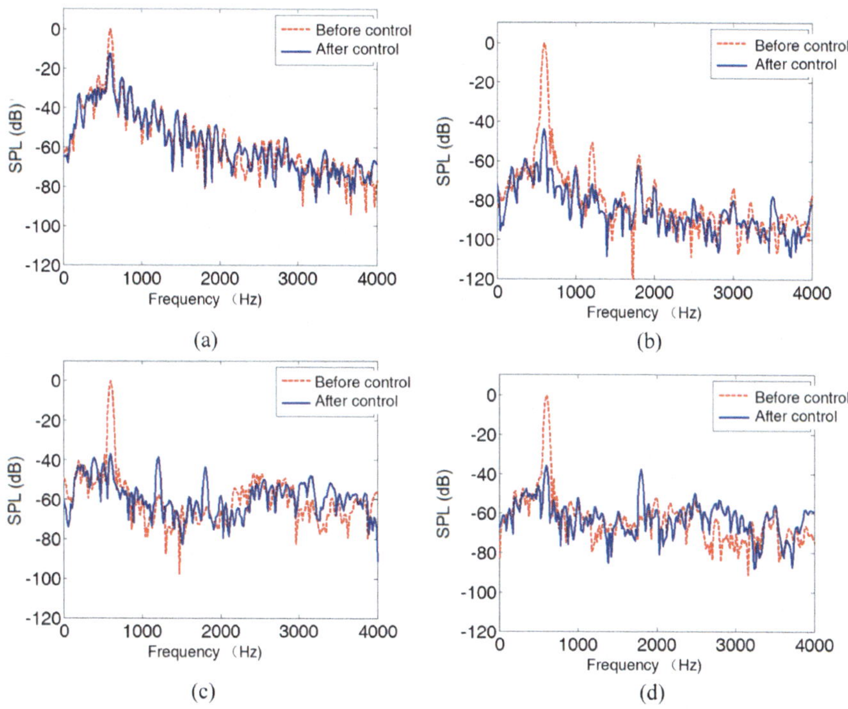

Fig. 7.34 Control results for a single-frequency noise at 600 Hz **a** control results of straight pipe without feedback compensation loudspeaker, **b** control results of straight pipe with feedback compensation loudspeaker, **c** control results of straight pipe without feedback compensation parametric array loudspeaker, **d** control results of L pipe without feedback compensation parametric array loudspeaker

800 Hz, 900 Hz, 1000 Hz, 1100 Hz, and 1200 Hz. Figures 7.34, 7.35, 7.36 and 7.37 show the experimental results of 600 Hz, 800 Hz, 1000 Hz, and 1200 Hz respectively.

Figure 7.34 shows the control results of two sound sources for a single-frequency noise at 600 Hz. As can be seen from the figure, when there is no feedback compensation network in the straight pipe, the noise reduction amplitude of the ordinary loudspeaker as a secondary sound source for noise at 600 Hz is about 12 dB. When there is feedback compensation in the straight pipe, the noise reduction amplitude of the ordinary loudspeaker is 44 dB. When the parametric array loudspeaker is used as a secondary sound source, the noise reduction amplitudes of the straight pipe and L pipe without feedback compensation network are 39 dB and 37 dB respectively.

Figure 7.35 shows the control results of two sound sources for a single-frequency noise at 800 Hz. As can be seen from the figure, when there is no feedback compensation network in the straight pipe, the noise reduction amplitude of the ordinary loudspeaker as a secondary sound source for noise at 800 Hz is about 5 dB. When there is feedback compensation in the straight pipe, the noise reduction amplitude of the ordinary loudspeaker is 38 dB. When the parametric array loudspeaker is used

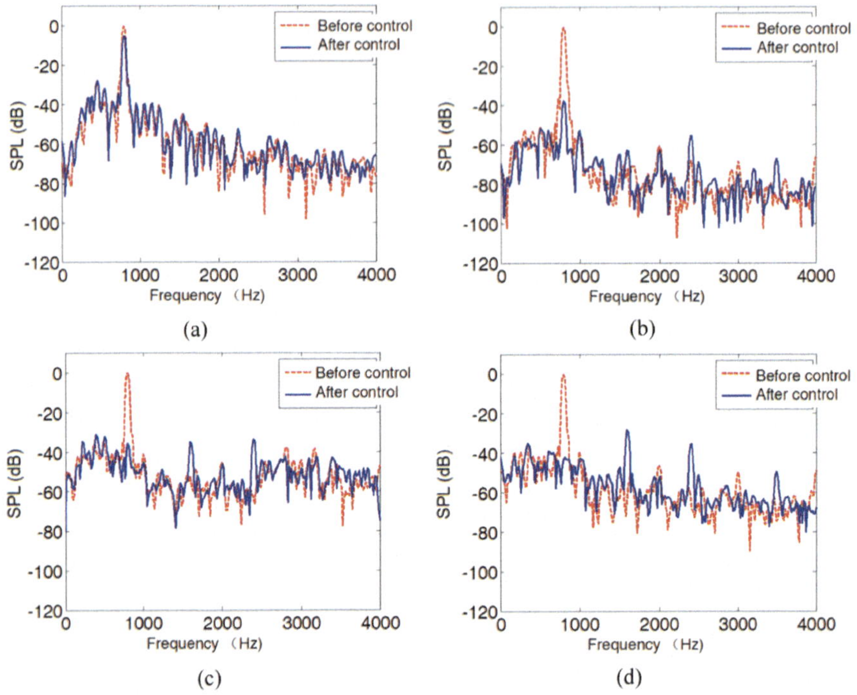

Fig. 7.35 Control results for a single-frequency noise at 800 Hz, **a** control results of straight pipe without feedback compensation loudspeaker, **b** control results straight pipe with feedback compensation loudspeaker, **c** control results of straight pipe without feedback compensation parametric array loudspeaker, **d** control results of L-pipe without feedback compensation parametric array loudspeaker

as a secondary sound source, the noise reduction amplitudes of the straight pipe and L pipe without feedback compensation network are 37 dB and 45 dB respectively.

Figure 7.36 shows the control results of two sound sources for a single-frequency noise at 1000 Hz. As can be seen from the figure, when there is no feedback compensation network in the straight pipe, the control of the ordinary loudspeaker as a secondary sound source for noise at 1000 Hz diverges. When there is feedback compensation in the straight pipe, the noise reduction amplitude of the ordinary loudspeaker is 46 dB. When the parametric array loudspeaker is used as a secondary sound source, the noise reduction amplitudes of the straight pipe and L pipe without feedback compensation network are 43 dB and 37 dB respectively.

Figure 7.37 shows the control results of two sound sources for a single-frequency noise at 1200 Hz. As can be seen from the figure, when there is no feedback compensation network in the straight pipe, the control of the ordinary loudspeaker as a secondary sound source for noise at 1200 Hz diverges When there is feedback compensation in the straight pipe, the noise reduction amplitude of the ordinary loudspeaker is 63 dB When the parametric array loudspeaker is used as a secondary

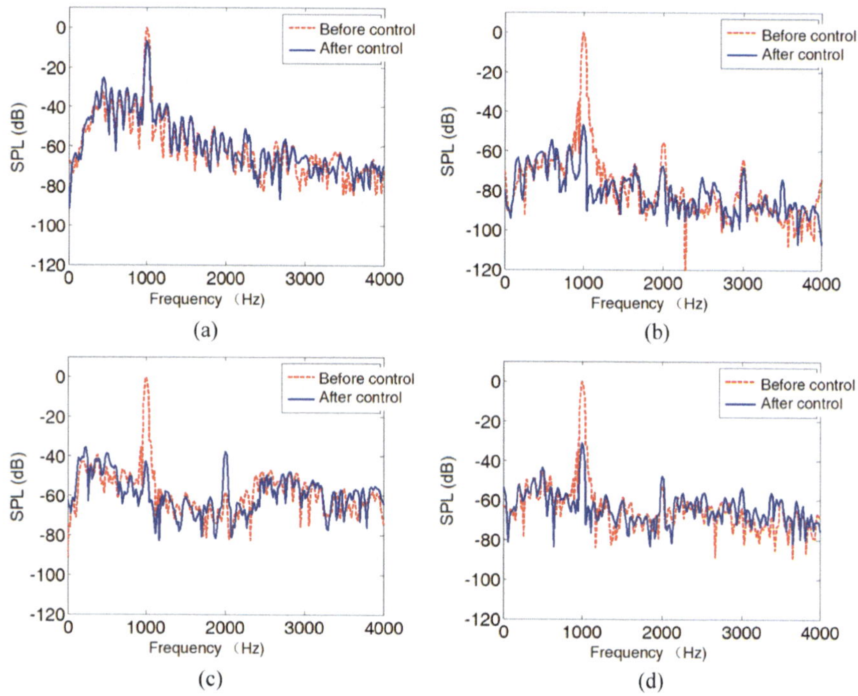

Fig. 7.36 Control results for a single-frequency noise at 1000 Hz **a** control results of straight pipe without feedback compensation loudspeaker, **b** control results straight pipe with feedback compensation loudspeaker, **c** control results of straight pipe without feedback compensation parametric array loudspeaker, **d** control results of L-pipe without feedback compensation parametric array loudspeaker

sound source, the noise reduction amplitudes of the straight pipe and L pipe without feedback compensation network are 44 dB and 56 dB respectively.

The basic frequency noise reduction and the amplitude of the second harmonic after noise reduction in the 6 sets of experimental results were extracted for comparison, see Figs. 7.38 and 7.39. Since normalization was applied, the noise reduction at fundamental frequency can represent the amplitude after noise reduction. The control effect of the ordinary loudspeaker in the straight pipe without feedback compensation network is poor. Although the noise at the control point has decreased, self-excitation may occur at other frequency points. It is meaningless to only compare the noise reduction at fundamental frequency and the amplitude of the second harmonic after noise reduction, so the control data under this condition is not given in Figs. 7.38 and 7.39.

From Fig. 7.38, it is evident that in the straight pipe, the noise reduction effect of the parametric array loudspeaker without feedback compensation network is comparable to that of the ordinary loudspeaker with feedback compensation network. At

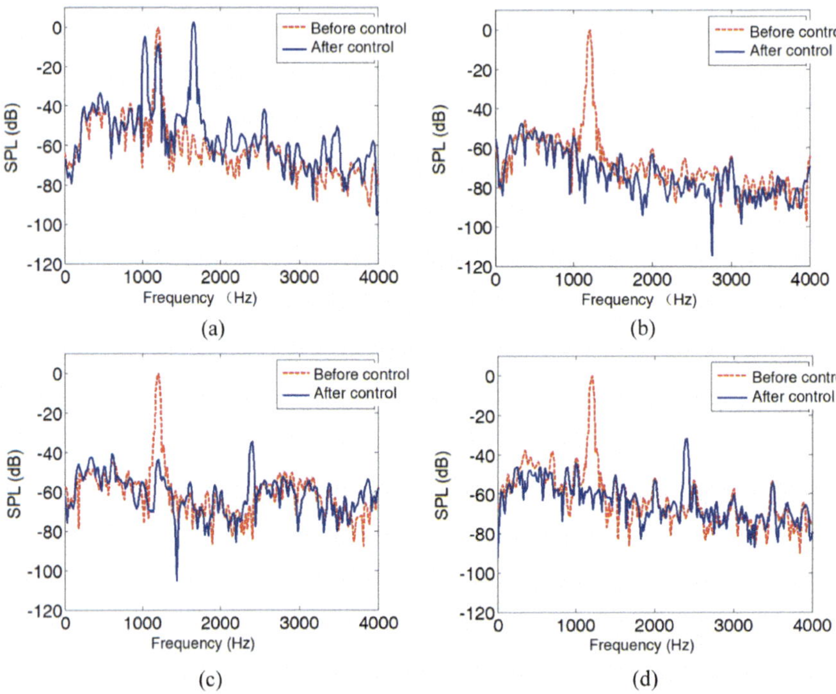

Fig. 7.37 Control results for a single-frequency noise at 1200 Hz, **a** control results of straight pipe without feedback compensation loudspeaker, **b** control results straight pipe with feedback compensation loudspeaker, **c** control results of straight pipe without feedback compensation parametric array loudspeaker, **d** control results of L-pipe without feedback compensation parametric array loudspeaker

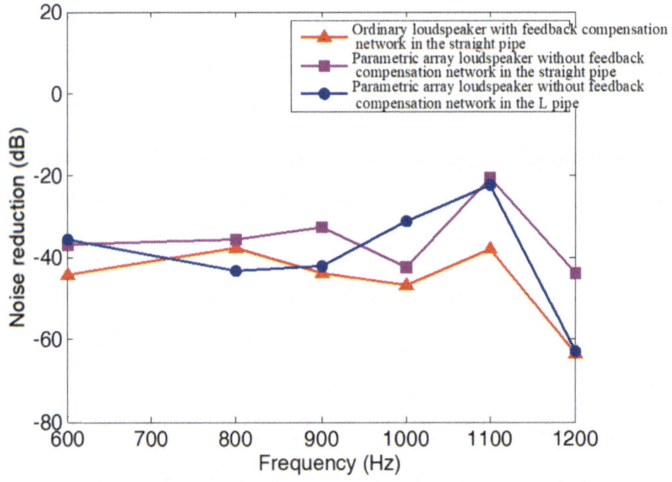

Fig. 7.38 Noise reduction at fundamental frequency for a single-frequency noise

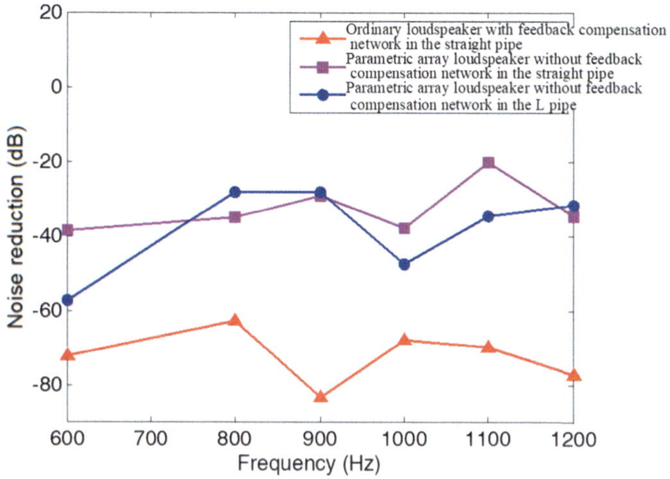

Fig. 7.39 Amplitude of the second harmonic after noise reduction for a single-frequency noise

several frequency points from 600 to 1000 Hz, the effect of the parametric array loudspeaker without feedback compensation network in the L pipe is not much different from the other two situations, but the noise reduction effect at 1100 and 1200 Hz is not as good as the noise reduction effect of the other two situations. This result is due to the nonlinearity of the parametric array loudspeaker, which can also be seen from Fig. 7.39. It is more obvious that the second harmonic of the parametric array loudspeaker is greater than the second harmonic of the ordinary loudspeaker, indicating that the harmonic distortion of the parametric array loudspeaker is larger when controlling the fundamental frequency. The noise control uses a linear adaptive algorithm, so the noise reduction effect of the parametric array loudspeaker will decrease at some frequency points.

In summary, without feedback compensation, the noise reduction effect of the parametric array loudspeaker is far stronger than that of the ordinary loudspeaker. While under the same noise reduction effect, the parametric array loudspeaker does not need feedback compensation, simplifying the control algorithm and reducing resource consumption.

7.2.2.4 Narrowband Line Spectrum Noise Reduction Results

The ordinary loudspeakers and parametric array loudspeakers as secondary sound sources for adaptive active noise control of narrowband line spectrum noise is adopted in this section. When the ordinary loudspeaker is used as the secondary sound source, it is only implemented in the straight pipe, and the algorithm has a feedback compensation network. When the parametric array loudspeaker is used as the secondary sound source, it is carried out in both the straight pipe and the L pipe, and the

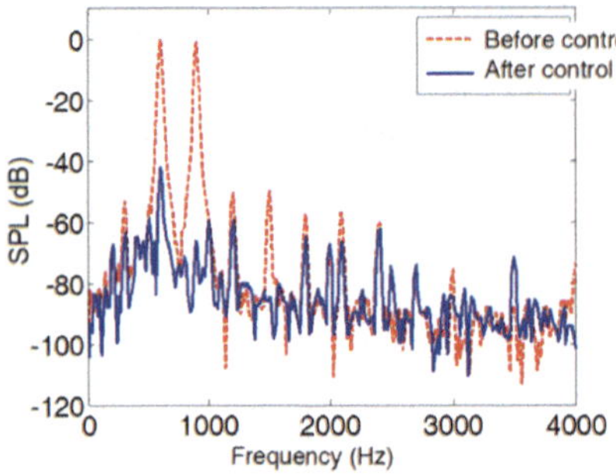

(a)

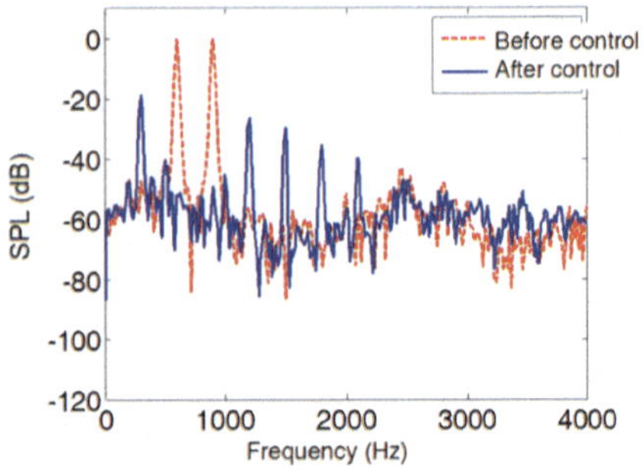

(b)

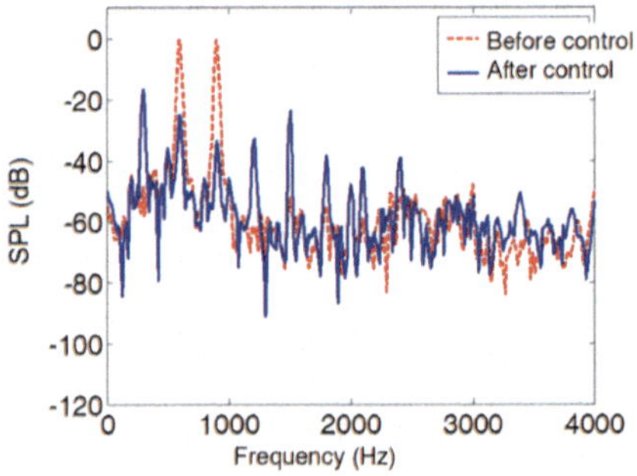

(c)

Fig. 7.40 Control results of 600 Hz + 900 Hz narrowband noise **a** straight pipe ordinary loudspeaker, **b** straight pipe parametric array loudspeaker, **c** L pipe parametric array loudspeaker

algorithm has no feedback compensation network. Five sets of experiments were conducted, using line spectrum noise of 600 Hz + 900 Hz, 600 Hz + 1000 Hz, 600 Hz + 1100 Hz and 600 Hz + 1200 Hz, and the control results are shown in Figs. 7.40, 7.41, 7.42 and 7.43.

Figure 7.40 shows the control results of 600 Hz + 900 Hz narrowband noise. The noise reduction amounts of the straight pipe loudspeaker, straight pipe parametric array, and L pipe parametric array for the 600 Hz line spectrum noise are 42 dB, 51 dB, and 24 dB respectively. For the 900 Hz line spectrum noise, the noise reduction amounts are 66 dB, 49 dB, and 33 dB respectively.

Figure 7.41 shows the control results of 600 Hz + 1000 Hz narrowband noise. The noise reduction amounts of the straight pipe loudspeaker, straight pipe parametric array, and L pipe parametric array for the 600 Hz line spectrum noise are 45 dB, 50 dB, and 40 dB respectively. For the 1000 Hz line spectrum noise, the noise reduction amounts are 66 dB, 40 dB, and 29 dB respectively.

Figure 7.42 shows the control results of 600 Hz + 1100 Hz narrowband noise. The noise reduction amounts of the straight pipe loudspeaker, straight pipe parametric array, and L pipe parametric array for the 600 Hz line spectrum noise are 34 dB, 42 dB, and 31 dB respectively For the 1100 Hz line spectrum noise, the noise reduction amounts are 34 dB, 39 dB, and 17 dB respectively.

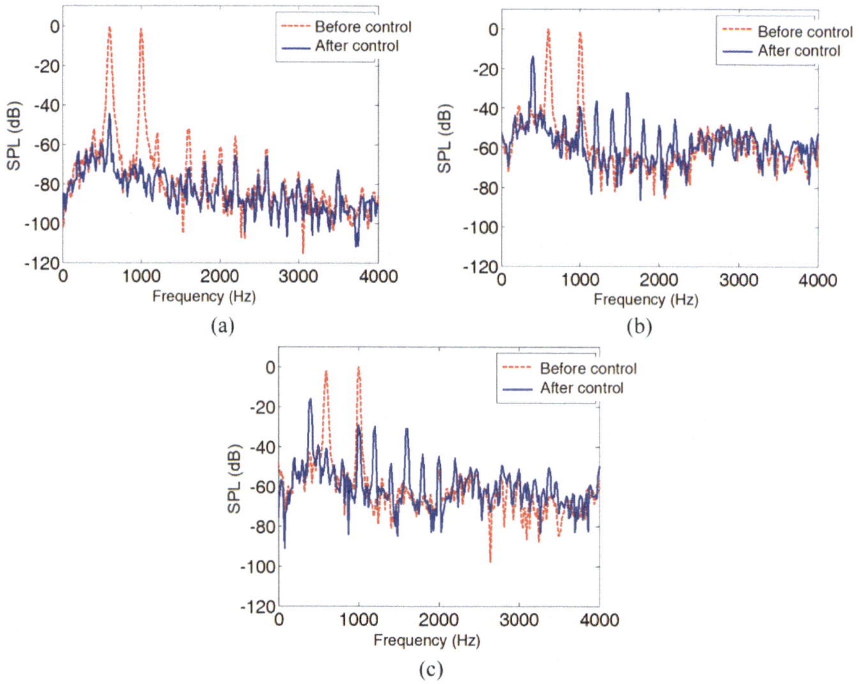

Fig. 7.41 Control results of 600 Hz + 1000 Hz narrowband noise **a** straight pipe ordinary loudspeaker, **b** straight pipe parametric array loudspeaker, **c** L pipe parametric array loudspeaker

Figure 7.43 shows the control results of 600 Hz + 1200 Hz narrowband noise. The noise reduction amounts of the straight pipe loudspeaker, straight pipe parametric array, and L pipe parametric array for the 600 Hz line spectrum noise are 41 dB, 48 dB, and 27 dB respectively For the 1200 Hz line spectrum noise, the noise reduction amounts are 54 dB, 41 dB, and 45 dB respectively.

The noise reduction experiments of narrowband dual-frequency line spectrum noise at 600 Hz + 900 Hz, 600 Hz + 1000 Hz, 600 Hz + 1100 Hz and 600 Hz + 1200 Hz are respectively coded as experiment numbers 1, 2, 3, and 4. According to the experimental results, the fundamental frequency noise reduction of spectral line 1 and the second harmonic noise reduction are given respectively, as shown in Fig. 7.44. The fundamental frequency noise reduction of spectral line 2 and the second harmonic noise reduction, as shown in Fig. 7.45. The noise reduction of the intermodulation distortion amplitude of the two spectrum lines, is shown in Fig. 7.46. In the figure, the intermodulation distortion frequencies corresponding to experiment numbers 1, 2, 3, and 4 are 300 Hz, 400 Hz, 500 Hz, and 600 Hz respectively. Since the experimental results have been normalized, the fundamental frequency noise reduction can represent the amplitude after the fundamental frequency noise reduction. From the figure, it becomes clear that for spectrum line 1, the control effect of the

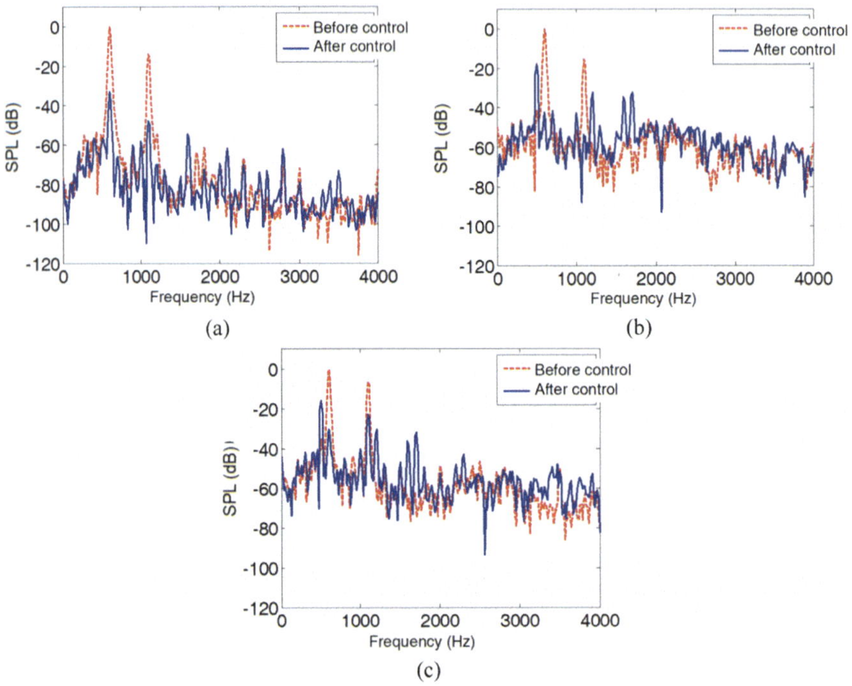

Fig. 7.42 Control results of 600 Hz + 1100 Hz narrowband noise **a** straight tube ordinary loudspeaker, **b** straight tube parametric array loudspeaker, **c** L Tube parametric array loudspeaker

parametric array loudspeaker in the straight tube is the best, followed by the ordinary loudspeaker in the straight tube, and the control effect of the parametric array loudspeaker in the L tube is the worst. For spectrum line 2, the control effect of the ordinary loudspeaker in the straight tube is relatively the best, followed by the parametric array loudspeaker in the straight tube, and the control effect of the parametric array loudspeaker in the L tube is the worst. For the second harmonic, the amplitude difference of the parametric array loudspeaker in the straight tube and the L tube is not large, and the amplitude is relatively high, while the second harmonic of the ordinary loudspeaker is smaller. It can also be seen that the amplitude of the second harmonic after noise reduction by the parametric array loudspeaker is comparable to the amplitude after the fundamental frequency noise reduction. Figure 7.46 shows that the intermodulation distortion of the parametric array loudspeaker in the first three experiments is much higher than that of the ordinary loudspeaker, while the intermodulation distortion of the parametric array loudspeaker in the fourth experiment is not much different from that of the ordinary loudspeaker. The reason is that the intermodulation distortion frequency of the fourth experiment is 600 Hz, which is the same as the fundamental frequency of spectrum line 1. If the fundamental frequency of spectrum line 1 is controlled, then the intermodulation distortion is also reduced.

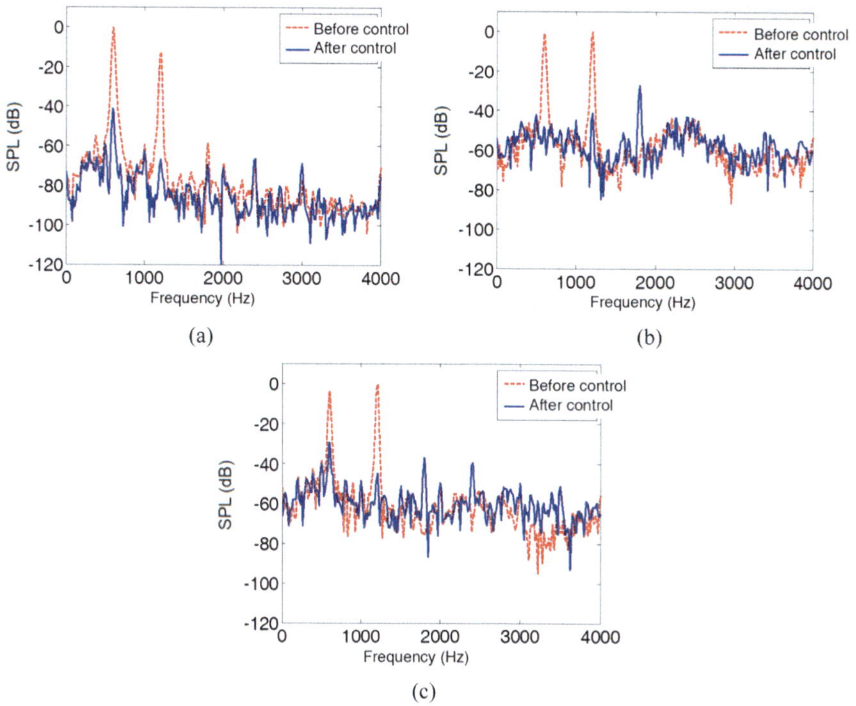

Fig. 7.43 Control results of 600 Hz + 1200 Hz narrowband noise **a** straight tube ordinary loudspeaker, **b** straight tube parametric array loudspeaker, **c** L Tube parametric array loudspeaker

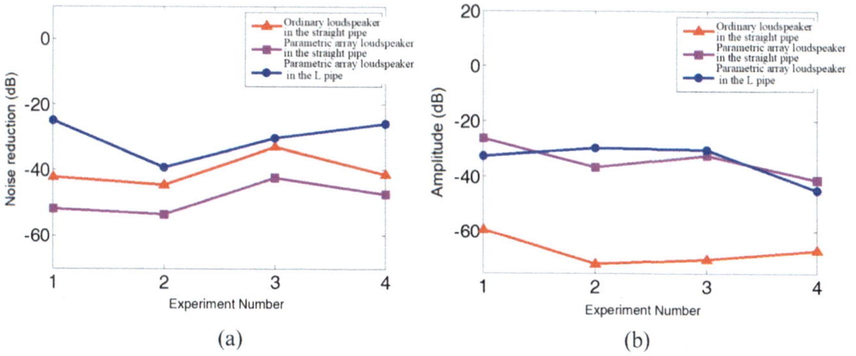

Fig. 7.44 The fundamental frequency noise reduction of spectral line 1 (**a**) and the second harmonic noise reduction (**b**)

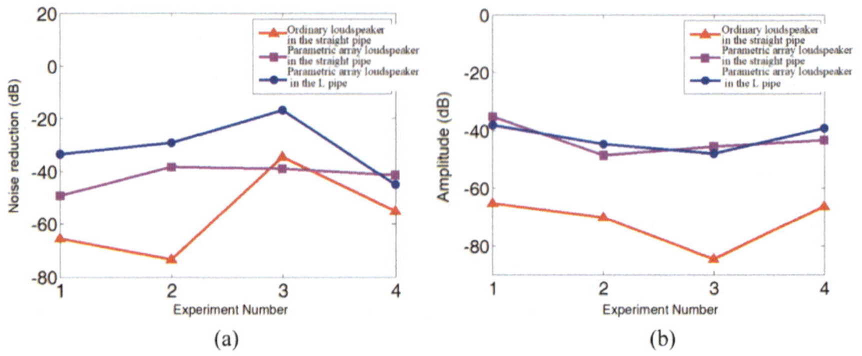

Fig. 7.45 The fundamental frequency noise reduction of spectral line 2 (**a**) and the second harmonic noise reduction (**b**)

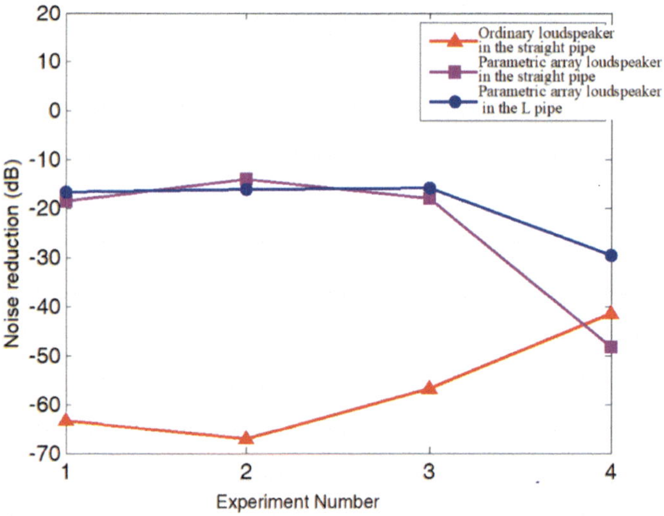

Fig. 7.46 Intermodulation distortion amplitude with noise reduction

In summary, in the straight tube, the control of noise of spectrum line 1 by the parametric array loudspeaker is better than that by the ordinary loudspeaker, while the control of spectrum line 2 is the opposite. In the L tube, the noise reduction effect of the parametric array loudspeaker on the two spectrum lines is relatively worse. This is mainly because the feedback path impulse response amplitude of the parametric array loudspeaker in the L tube is higher than that in the straight tube, that is, the acoustic feedback of the parametric array loudspeaker in the L tube is greater than that in the straight tube, thus resulting in insufficient noise reduction. In addition, the nonlinear effects of the two spectrum lines are more complex, which also leads to the weakening of the noise reduction effect of the parametric array narrowband spectrum noise.

7.2.2.5 Noise Reduction Area of the Parametric Array Loudspeaker

Based on the theory of sound wave guide theory, the sound wave below the cutoff frequency of the waveguide tube propagates in the form of plane waves. The frequency of active noise control in general pipes is lower than the cutoff frequency, so the sound wave emitted by the ordinary loudspeaker as a secondary sound source cancels out the noise in the form of plane waves. Therefore, the noise control area of the ordinary loudspeaker is the entire downstream pipe below the error microphone, and the control area is the entire cross-section of the pipe. The sound emission mechanism of the parametric array loudspeaker is different from that of the ordinary loudspeaker, so it must be proven by measurement that the noise reduction area and noise reduction area of the parametric array loudspeaker.

First, the noise reduction area is measured. According to the setup in Fig. 7.26, the error sensor is located at $x = 3.55$ m, and the secondary sound sources are all located at $x = 1.55$ m, where the parametric array loudspeaker is inside the tube, and the ordinary loudspeaker is on the side wall. The results of the sound field test are shown in Fig. 7.47. In the downstream area of the sound source, the sound waves emitted by ordinary loudspeakers propagate in the form of plane waves. Ignoring sound absorption attenuation and sound pressure fluctuations, the sound pressure amplitude remains basically unchanged. The sound waves emitted by the parametric array loudspeaker accumulate around 1.5 m–2.5 m, and remain basically stable around 2.5 m–4 m, which are consistent with the sound pressure of ordinary loudspeakers. This indicates that after the accumulation of sound waves from the parametric array loudspeaker is completed, the sound field is consistent with the sound pressure of the ordinary loudspeaker.

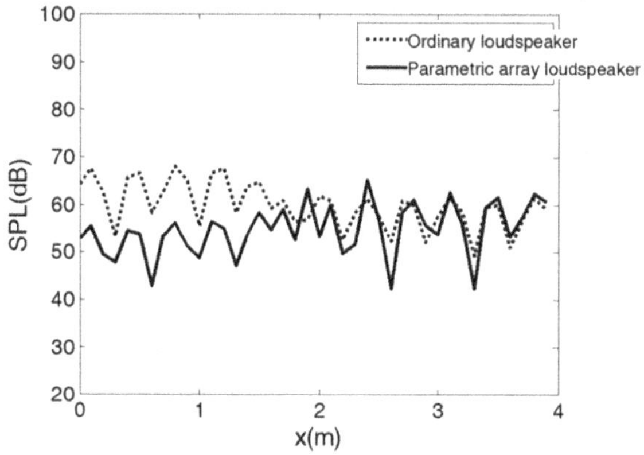

Fig. 7.47 In-tube sound field measurement results at 600 Hz

In addition, by measuring the sound pressure before and after noise reduction at 10 cm, 20 cm, 30 cm, and 40 cm downstream of the error sensor, the noise reduction situation in the downstream area of the error microphone is confirmed, as shown in Table 7.2. Although the noise reduction at 600 Hz downstream is slightly less than the noise reduction at the control position, the noise reductions at 800, 1000, and 1200 Hz downstream are not much different from the noise reductions at the control point, so all frequencies have basically achieved the noise reduction effect. It can be considered that the control area of the parametric array loudspeaker is the same as the control area of the ordinary loudspeaker, both are the entire downstream part below the error sensor.

Next, the noise reduction area is measured, with a measurement frequency of 600 Hz. Refer to Fig. 7.48 for the cross-sectional measurement diagram, which shows the section where the error microphone is located for noise reduction area measurement. The error sensor is connected to the controller in the figure, reflecting the error situation at the control point. The sensor is connected to the Pulse instrument for measuring the sound pressure level at different positions on the cross-section after noise reduction. The duct is a square duct with a side length of 10 cm, so only the results on one side of the duct center need to be measured due to symmetry. L represents the distance from the front end of the sensor to the center position. $L = 0$ cm means the front end of the sensor is at the center, and $L = 5$cm indicates that the front end of the speaker is flush with the inner wall of the pipe. T The feedback sensor is fixed in a position that is flush with the inner wall of the pipe, and the measuring microphone determines the sound pressure level at various distances by altering its own position. The results of the cross-sectional noise reduction and sound field are shown in Table 7.3.

The results show that the noise reduction at different distances on the cross-section of the pipe is basically the same, with an average of 40.3 dB and a fluctuation of less than ± 1 dB. The sound pressure level at different distances on the cross-section is also basically the same, with an average of 74.4 dB and a fluctuation within ± 0.1 dB. This indicates that the noise reduction area of the difference frequency sound of the parametric array loudspeaker is the same as that of the ordinary loudspeaker, and the noise reduction of the entire cross-section can be achieved.

Table 7.2 Noise reduction in the downstream area of the error microphone

	10 cm	20 cm	30 cm	40 cm
600 Hz	28 dB	26 dB	29 dB	32 dB
800 Hz	35 dB	43 dB	37 dB	38 dB
1000 Hz	34 dB	31 dB	31 dB	38 dB
1200 Hz	40 dB	36 dB	40 dB	39 dB

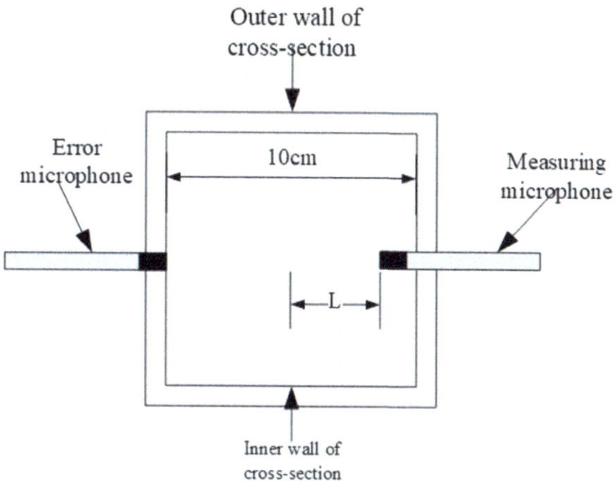

Fig. 7.48 Noise reduction area measurement schematic diagram

Table 7.3 Cross-sectional noise reduction of error microphone and sound field

L (cm)	Noise reduction (dB)	Sound field (dB)
0	42	74.3
1	40	74.4
2	39.5	74.4
3	40.5	74.5
4	39.5	74.5
5	40	74.4

7.3 Application of Parametric Array Loudspeakers in Sound Absorption Coefficient Measurement

There are two types of methods for measuring the sound absorption coefficient of acoustic materials: laboratory measurement and field measurement. Laboratory measurements mainly include impedance tube method and reverberation chamber method, but the results obtained under laboratory conditions cannot fully reflect the sound absorption performance of sound-absorbing materials in actual field use.

How to accurately measure the sound absorption coefficient of materials on site has become an important issue in acoustic design and engineering. This section reviews the existing methods that can be used for field measurement of sound absorption coefficients. In combination with the actual application of this section, the two-microphone transfer function method is selected, combined with the reflection model for key analysis. The impact of different measurement conditions on the measurement results is investigated [490].

The experimental setup is shown in Fig. 7.49, and the experiment was conducted in the anechoic chamber of Institute of Acoustics, Chinese Academy of Sciences. Where P_1 is the sound pressure at microphone 1, P_2 is the sound pressure at microphone 2, d

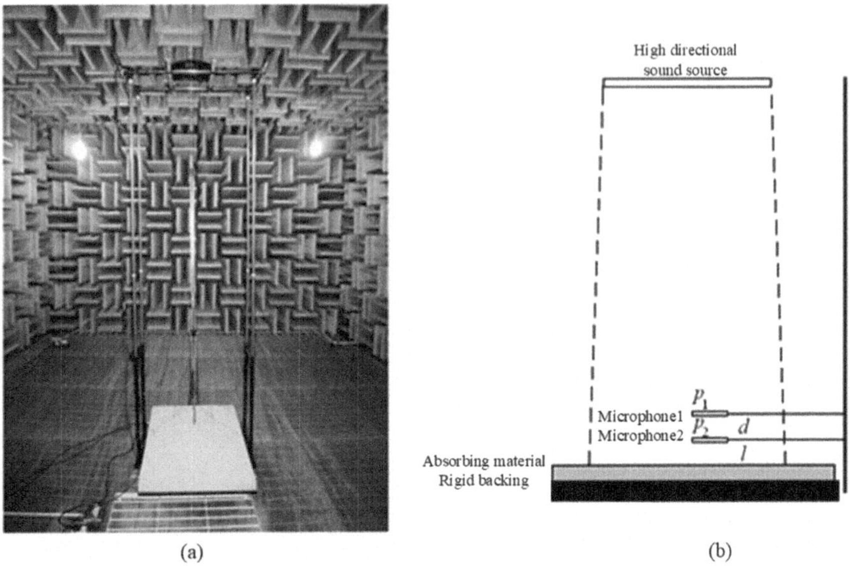

Fig. 7.49 Duplicate **a** experimental setup of transfer function method, **b** schematic diagram

is the distance between the two microphones, and l is the distance from microphone 2 to the material being tested.

7.3.1 Methods of On-Site Measurement of Sound Absorption Coefficient

7.3.1.1 Principle of Pulse Echo Method

The principle of the pulse echo method is a technique that has been applied to on-site measurement of sound-absorbing materials in recent years. Based on the propagation model of the free field, it separates the incident sound wave and the reflected sound wave for spectral analysis to obtain parameters such as the sound absorption coefficient.

The principle of this method is shown in Figs. 7.50 and 7.51 [201]. The sound source emits a pulse signal, which is received by a single microphone placed between the sound source and the sound-absorbing material to obtain the incident signal. The pulse sound signal continues to propagate, reflects after contacting the material, and is collected by the microphone to obtain the reflected signal. The incident and reflected signals are Fourier transformed to obtain the spectrum, as shown in Figs. 7.50 and 7.51. The reflection coefficient corresponding to different frequencies is obtained

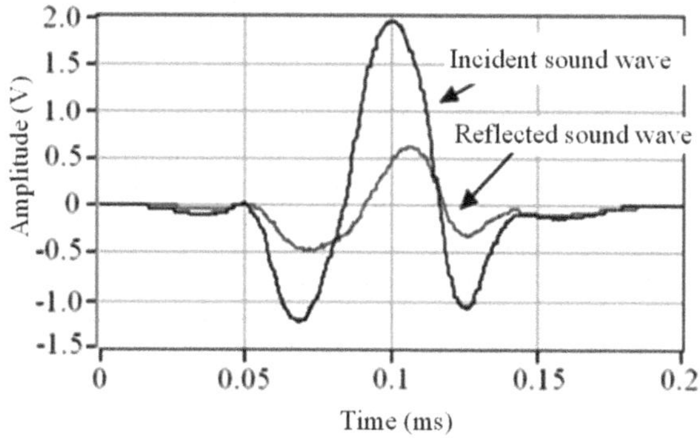

Fig. 7.50 Time domain diagram of incident and reflected waves by impulse method

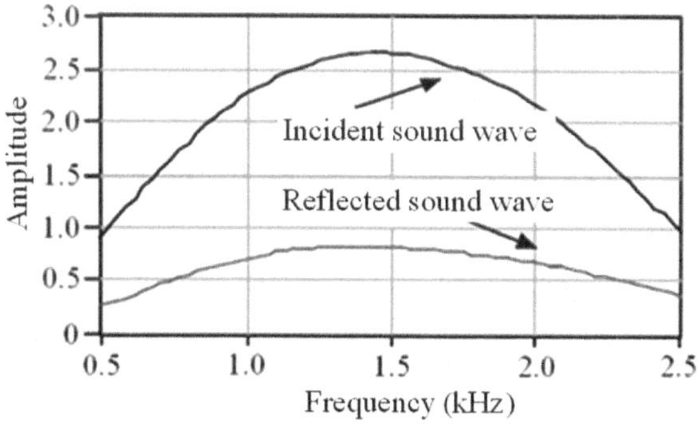

Fig. 7.51 Frequency domain diagram of incident and reflected waves of impulse method

by division, from which the sound absorption coefficient of the material can be calculated.

The advantage of this method is that the experimental setup is simple and convenient, and there is no need for microphone matching and synchronous collection processes. The requirements for the test system are relatively low. However, the actual test signal is far more complex than the signal shown in the figure above. It is very difficult to judge the starting point of the incident sound wave and the reflected sound wave and to effectively separate the two signals. Under the influence of high-intensity ultrasound, the situation is even more complex. This method has high requirements for the test environment, and the amplitude of the background noise has a direct impact on the test results. In addition, the reflection method requires increasing the distance between the microphone and the material to distinguish the incident wave and the reflected wave in time. But when there is sound-absorbing material, and the reflection surface is not rigid, the difference frequency sound frequency will produce a certain attenuation, resulting in a large error in the results [490].

7.3.1.2 Principle of the Impedance Method

The principle of the impedance method is to measure the sound pressure signal generated by a common loudspeaker on the material surface through two microphones, from which the sound impedance of the material surface is inferred to calculate the sound absorption coefficient of the material [479]. Assuming that both the incident wave and the reflected wave are plane waves, microphones 1 and 2 are placed at two positions close to the test sample to measure the sound pressure. Let the midpoint of the two microphones be M, p_1 and p_2 represent the sound field sound pressure signals collected by microphones 1 and 2, respectively, and the sound pressure at point M can be approximated by the following formula:

$$p_M \approx \frac{p_1 + p_2}{2} \tag{7.45}$$

The particle velocity at point M can be approximated as:

$$v_M = -\frac{\partial p}{i\omega\rho\partial z} \approx \frac{p_2 - p_1}{i\omega\rho d} \tag{7.46}$$

where ρ represents air density, d is the distance between microphones 1 and 2.The acoustic impedance at point M can be expressed as:

$$Z_M \approx \frac{i\omega\rho d(p_1 + p_2)}{2(p_2 - p_1)} \tag{7.47}$$

Then the sound impedance of the material surface can be obtained from Eq. (7.48)

$$Z_S = \frac{Z_M + i\rho c \tan\left(\frac{\omega d}{c}\right)}{\rho c + iZ_M \tan\left(\frac{\omega d}{c}\right)} \tag{7.48}$$

The reflection coefficient of the material r is

$$r = \frac{Z_S - \rho c}{Z_S + \rho c} \tag{7.49}$$

The sound absorption coefficient can be obtained as follows

$$\alpha = 1 - |r|^2 \tag{7.50}$$

There are mainly three problems with this method.

(1) The surface acoustic impedance of the material obtained from actual measurements is the result under spherical wave conditions, which differs from the surface acoustic impedance under plane wave conditions.

(2) The sound pressure and particle velocity at point M are approximated, not accurate values. Therefore, the accuracy of the measurement results is closely related to the distance between the microphones. The measurement error increases with the increase in the distance between the microphones and with the increase in frequency.
(3) The problem of microphone mismatch cannot be solved. Since the calculation process requires the absolute measurement values of two microphones rather than the relative ratio, it is impossible to avoid the error caused by microphone mismatch by swapping the microphones. It can only be matched through an impedance tube, which increases the complexity of the measurement.

7.3.1.3 Principle of Transfer Function Method

The two-microphone transfer function method was developed by Chung and Blaser [491,492] in the 1980s for the measurement of the normal sound absorption coefficient and impedance rate of materials. This method replaces the moving measurement points in the traditional standing wave ratio method with two fixed points on the impedance tube, making full use of the phase information of the sound pressure, greatly simplifying the testing equipment and process. This method can use sweep frequency signals or broadband noise signals to quickly obtain the sound absorption coefficient and complex reflection coefficient of the tested material over a wide frequency band, and then calculate the normal acoustic impedance rate. By swapping the microphones for repeated measurements, the problem of microphone mismatch is further solved, reducing the requirements for the microphones.

The experimental setup is shown in Fig. 7.52 [493]. Assume p_{in} is the incident sound wave, p_{re} is the reflected sound wave, and then, the sound pressure at microphone 1 can be expressed as

$$p_1 = P_{in}e^{ik(d+l)} + P_{re}e^{-ik(d+l)} \tag{7.51}$$

where, d is the distance between the two microphones, l is the distance from microphone 2 to the tested material, and P_{in}, P_{re} are the amplitudes of the incident and reflected sound waves on the surface of the tested material, respectively.

The sound pressure at microphone 2 can be expressed as

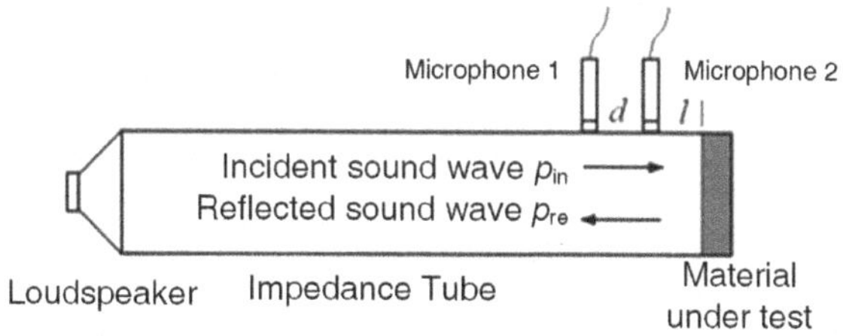

Fig. 7.52 Experimental setup diagram of impedance tube method

$$p_2 = P_{in}e^{ikl} + P_{re}e^{-ikl} \tag{7.52}$$

The transfer function of the incident wave from microphone 1 to microphone 2 is

$$H_{in} = e^{-ikd} \tag{7.53}$$

The transfer function of the reflected wave from microphone 1 to microphone 2 is

$$H_{re} = e^{ikd} \tag{7.54}$$

Substituting Eqs. (7.51) and (7.52) into Eq. (7.55), the sound transfer function between the two microphones is

$$H_{12} = \frac{p_2}{p_1} = \frac{e^{ikl} + re^{-ikl}}{e^{ik(d+l)} + re^{-ik(d+l)}} \tag{7.55}$$

The reflection coefficient can be obtained from Eq. (7.55)

$$r = \frac{H_{12} - H_{in}}{H_{re} - H_{12}} e^{2ik(d+l)} \tag{7.56}$$

Normalized surface acoustic impedance

$$z_s = \frac{1+r}{1-r} \tag{7.57}$$

According to Eq. (7.50), the sound absorption coefficient of the material can be obtained, and the following relationships can also be obtained:

Incident sound pressure amplitude on the material surface

$$\hat{p}_{in\_s} = \frac{p_1 e^{-ikl} - p_2 e^{-ik(l+d)}}{2i\sin(kd)} = \frac{p_1\left(e^{-ikl} - H_{12}e^{-ik(l+d)}\right)}{2i\sin(kd)} \tag{7.58}$$

Reflected sound pressure amplitude on the material surface

$$\hat{p}_{re\_s} = \frac{p_1 e^{ikl} - p_2 e^{jk(l+d)}}{-2i\sin(kd)} = \frac{p_1\left(e^{ikl} - H_{12}e^{ik(l+d)}\right)}{-2i\sin(kd)} \tag{7.59}$$

Sound pressure at the material surface

$$p_s = \hat{p}_1 + \hat{p}_2 = \frac{-p_1(\sin(kl) + H_{12}\sin(k(l+d)))}{\sin(kd)} \tag{7.60}$$

Sound quality point velocity at the material surface

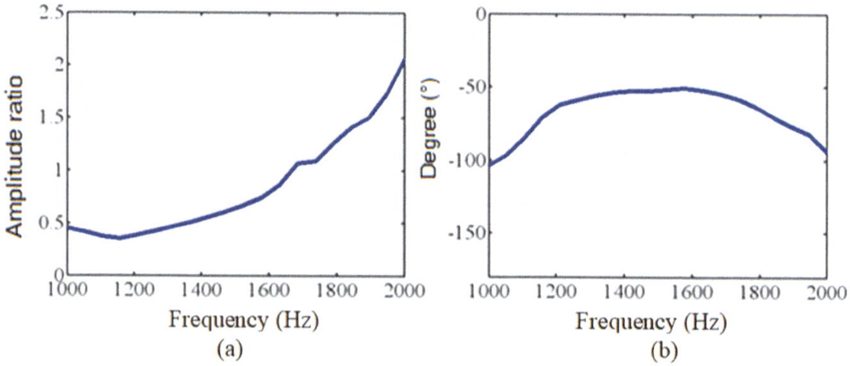

Fig. 7.53 Amplitude frequency response of the sound transfer function between two microphones **a** amplitude response, **b** phase response

$$U_s = \frac{\hat{p}_1 - \hat{p}_2}{\rho c} = \frac{p_1(\cos(kl) - H_{12}\cos(k(l+d)))}{i\rho c\sin(kd)} \tag{7.61}$$

Considering that the two-microphone method is simple to test and has strong anti-interference ability, it is selected as the research object in the following tests. Figure 7.53 shows the sound transfer function measured between the two microphones.

7.3.1.4 Effectiveness of Using the Transfer Function Method

To verify the effectiveness of using the transfer function method to measure the sound absorption coefficient in the reflected sound field of the parametric array loudspeaker, two extreme cases are first selected for testing, namely the full absorption case without any reflectors and the full reflection case of rigid reflection from the hard backboard.

In the case of no reflectors, the ideal result should be no echo, that is, $P_{re} = 0$.

Therefore, the sound absorption coefficient should be fully absorbed, that is, the sound absorption coefficient is 1. Figure 7.54 shows the actual measurement results using a sweep signal.

In the case of rigid reflection from the hard backboard, the ideal result should be that the incident sound wave is completely reflected, with no absorption, that is, $P_{re} = P_{in}$. Therefore, the sound absorption coefficient should be no absorption, that is, the sound absorption coefficient is 0. Figure 7.55 shows the actual measurement results.

The above curves are all single measurement results, using a 20-s, 400-point sweep signal, without multiple measurement averages. It is evident that the results obtained by the parametric array loudspeaker in two extreme cases are generally in good agreement with the theoretical values. However, the results measured in the low frequency part are inaccurate and have a certain deviation from the actual value. This

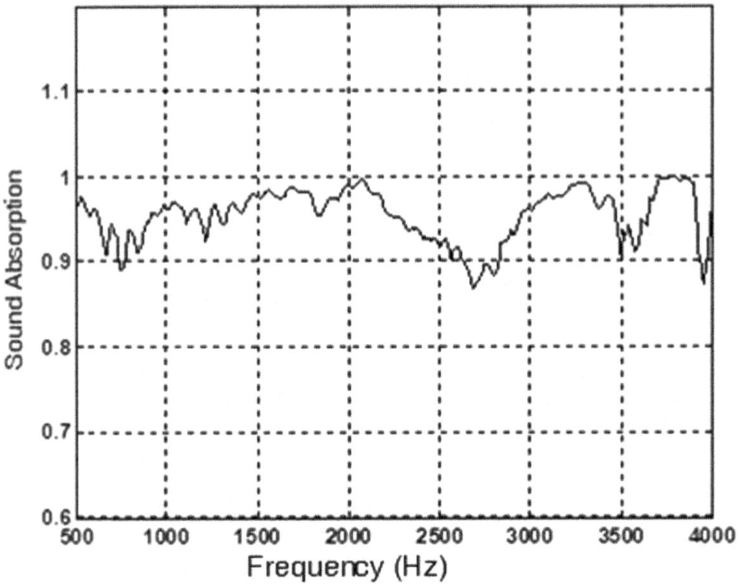

Fig. 7.54 Results obtained when there is no reflection surface

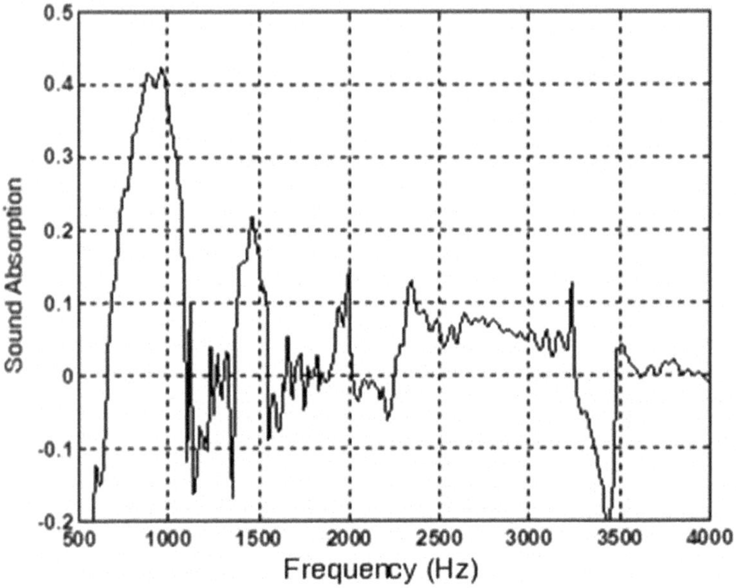

Fig. 7.55 Results obtained when there is rigid reflection

is caused by the characteristics of the parametric array loudspeaker itself. Due to the low efficiency of the nonlinear effect of generating the difference frequency sound frequency in the low frequency situation, the sound pressure level of the difference frequency sound frequency is low, the signal-to-noise ratio of the measurement is low, and the result is difficult to meet the requirements.

7.3.2 *Influencing Factors of the Model in Sound Absorption Coefficient Measurements*

This section will analyze various parameters that affect the measurement of the sound absorption coefficient, combine the normal incidence reflection model of the parametric array loudspeaker, analyze the impact of each variable on the test results in the model, and compare it with the actual measurement results, find out the influencing factors in the actual test, and provide guidance for the improvement of the test process.

7.3.2.1 Influence of the Distance Between the Microphones

According to the requirements of ISO10534 [485], when using the transfer function method to measure the sound absorption coefficient, there are certain requirements for the test frequency. The working frequency is determined by the following formula

$$f_l < f < f_u \tag{7.62}$$

where f_l is the lowest working frequency, determined by the test system, f_u is the highest effective frequency, which is related to two aspects: (1) The diameter of the impedance tube D determines, $f_u \cdot D < 0.5c$; (2) The distance between the microphones d, both need to satisfy the relationship $f_u \cdot d < 0.45c$. In the testing process of the parametric array loudspeaker, there is no limitation of the impedance tube, so the distance between the microphones directly determines the highest frequency of the effective results. The results of this part mainly use 3.5 cm, taking the speed of sound $c = 340m/s$, and the effective frequency is

$$f_u < \frac{0.45c}{d} = 4371 \text{ Hz} \tag{7.63}$$

The standing wave tube results used as a reference value are limited by size, and the upper cut-off frequency of the test results is 2 kHz. Therefore, in the comparison results in the following text, the upper limit is mostly 2 kHz.

7.3.2.2 Influence of the Distance Between the Microphone and the Material

In the experiment the high-directivity difference frequency sound beam is used. If the ultrasonic primary wave after reflection is ignored as a new parameter array loudspeaker generating sound frequency, the sound frequency generated before reflection is as shown in Fig. 7.56 after reflection. As the distance from the reflection plane increases, the sound pressure level gradually decreases.

Three sets of signals were measured in the anechoic chamber. The tested material was a polyurethane foam sponge with a thickness of 4.2 cm and a size of 1.2m × 1.2m. The distances between microphone 2 and the surface of the test material, denoted as l, were 15 mm, 30 mm, and 50 mm, respectively. The results are shown in Fig. 7.57. In actual measurements, as the distance between the microphone and the material increases, the absorption coefficient tends to increase, reaching a maximum of about 15%. Around 1 kHz, due to the influence of spurious sound, there was an overestimation of about 10%, which has been analyzed in reference [490]. Therefore, if conditions permit, reducing the distance between the microphone and the test material as much as possible can effectively reduce errors. In this part of the experiment, this distance was controlled to be less than 30 mm, and it becomes clear that the error is basically less than 5%.

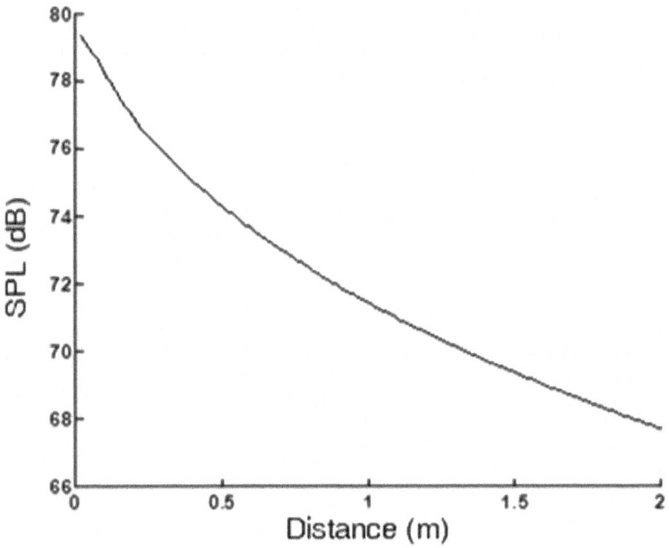

Fig. 7.56 Schematic diagram of sound attenuation of the sound frequency generated before reflection after reflection

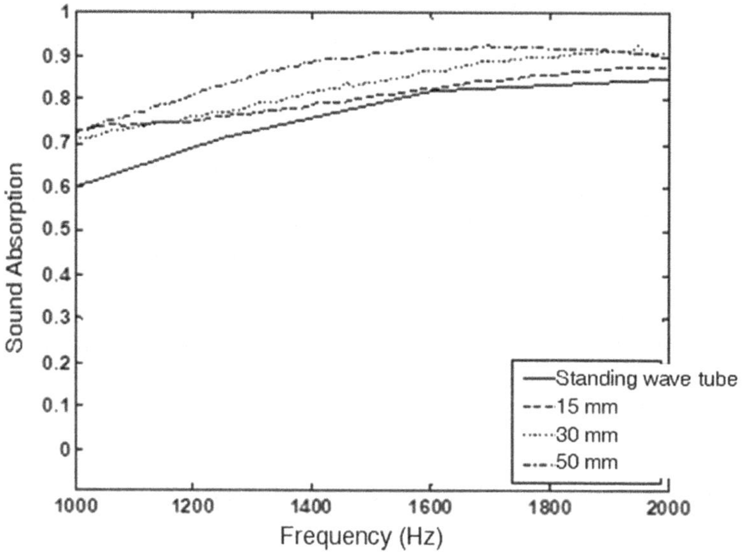

Fig. 7.57 Effect of the distance between the microphone and the material on the results

7.3.2.3 Influence of the Measurement Position

The center point positioning is an important aspect in the actual measurement process. The ideal situation is that the parametric array loudspeaker emits sound waves directly at the center of the material, and the measurement position of the microphone is at the sound axis position. However, this is difficult to achieve in actual measurements, which requires the test system to have a certain robustness to the placement error.

According to reference [490], the distribution curve of the absorption coefficient obtained in different positions of the microphone sample position can be simulated. As shown in Fig. 7.58, the horizontal axis is the distance between the test position and the acoustic center of the parametric array loudspeaker, and the vertical axis is the error size of the results measured at the offset position compared to the results at the center axis.

It can be observed that the basic trend of error distribution gradually expands as the distance from the center increases. There is a slight fluctuation at 0.2 m, and although the distance increases afterwards, the error change is basically stable.

Figure 7.59 is a schematic diagram of the results measured in the experiment when the microphone measurement position is at different distances from the sound axis of the parametric array loudspeaker. It can be observed that when the deviation from the center position is within 6 cm, the measured result is in good agreement with the central result, with the maximum error around 3%. However, as the distance increases, the actual measured result is significantly smaller than the central result, with the maximum deviation reaching about 10%. After that, the decrease remains basically unchanged as the distance increases.

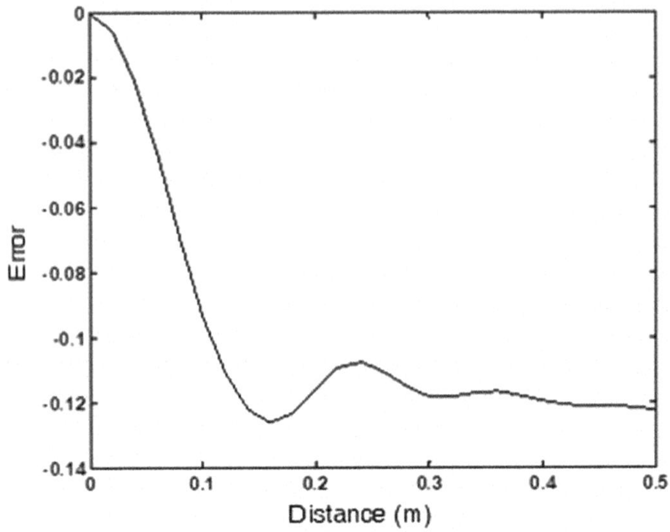

Fig. 7.58 Relative error size when the measurement position is different

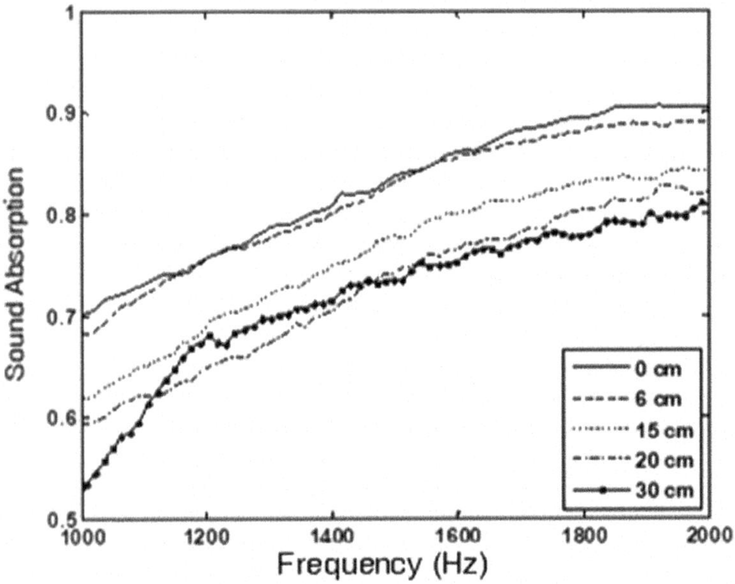

Fig. 7.59 Schematic diagram of results obtained at different distances from the center

When the distance from the center is large, the deviation of low-frequency data begins to expand. This is mainly because the size of the material used in the experiment is limited. When the deviation is large, the distance to the edge of the material is closer, which has a more significant impact on low frequencies.

7.3.2.4 Impact of Microphone Matching on the Results

When using the two-microphone transfer function method to measure the absorption coefficient, it is an important step to swap the positions of the microphones during the test to avoid errors between the two channels introduced by the microphone frequency response and transmission line. This part of the error will enter the actual test transfer function, which has a decisive impact on the stability and reliability of the results obtained.

The transfer function actually used in the test process is calculated by the following formula

$$H_{12} = \sqrt{\frac{H_{12}^{I}}{H_{21}^{II}}} = |H_{12}|e^{i\phi} \tag{7.64}$$

where, H_{12}^{I} is the transfer function measured before the exchange, H_{12}^{II} is the transfer function obtained by only swapping microphone 1 and microphone 2 while keeping the measurement position and calculation method unchanged. ϕ is the phase of the transfer function.

As shown in Fig. 7.60, the results obtained after balancing H_{12} are calculated to obtain the absorption coefficient results, as well as the comparison of the absorption coefficient calculated from the unbalanced results obtained using the transfer function before and after the exchange. It is obvious that the difference in results is quite large. If no matching is implemented, the results obtained are essentially meaningless.

7.3.3 Impact of Test Conditions

7.3.3.1 Impact of the Size of the Material

In on-site measurement of sound absorption coefficient, a key element is the requirement for the size of the material being tested. In actual situations, the layout of the material may have already been completed, with a fixed area that cannot be significantly modified. This requires the proposed method to have strong adaptability, enabling accurate measurement of the sound absorption coefficient even when the size of the material being tested is small.

When studying the reflection and scattering of a certain size plane, a common situation is to consider a usual structure from the source to the plane to the receiving point,

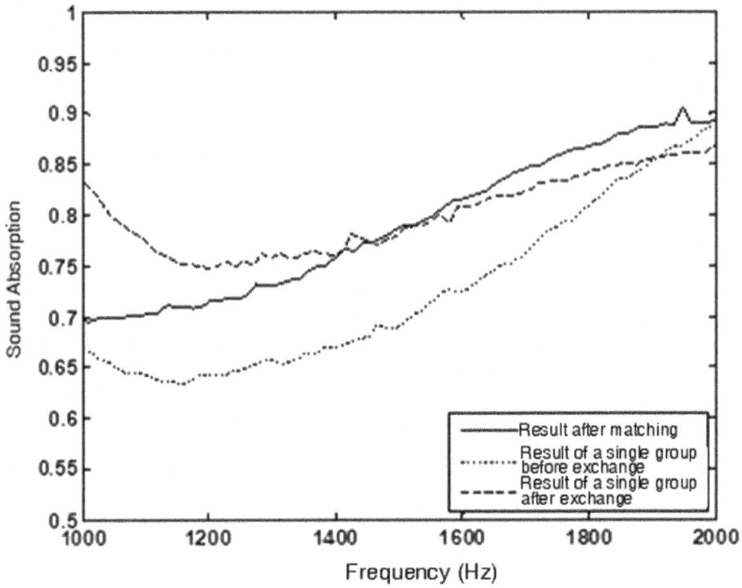

Fig. 7.60 Effect of microphone matching on the results

as shown in Fig. 7.61. Since the size of the material being tested in the experiment is limited, it can be considered as a special high-pass filter in actual situations. Due to scattering and diffraction, the intensity of the reflected sound gradually decreases as the frequency decreases. For such a problem, the first Fresnel zone theory can be used for analysis [494, 495].

Considering the reflection on the $x-y$ plane, outside the reflection center P point, the P'. The role of the point in the reflection structure can be obtained as follows

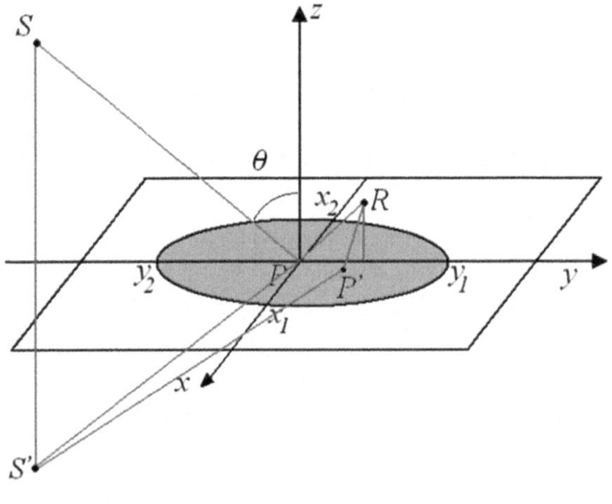

Fig. 7.61 Schematic diagram of reflection sound field

$$S'P' + RP' - R_2 = F'\lambda \tag{7.65}$$

$$R_2 = SP + RP \tag{7.66}$$

Here, F' is a constant less than F, S is the location of the sound source, R is the receiving point, S' is the image sound source, λ is the wavelength, P' is any point in the set of points where the difference between the path length and the direct reflection path length is less than $F\lambda$. Here, F is between 1/4 and 1/2, so in this section, F is set to be 1/2.

The set of P' can be represented by an ellipse, where the length of the semi-major axis is

$$a = \frac{R_2 + F\lambda}{2} \tag{7.67}$$

The length of the semi-minor axis is

$$b = \sqrt{\frac{R_2 F\lambda}{2} + \left(\frac{F\lambda}{2}\right)^2} \tag{7.68}$$

The area within the ellipse corresponds to the Fresnel zone in this study, and the reflection sound field of each point in the area significantly affects the receiving point. Therefore, it is necessary to determine the ellipse formula corresponding to the boundary of the Fresnel zone. This formula depends on the constant F, the wavelength of the sound wave λ, and the geometric distribution of the sound source and receiving point during reflection.

$$\frac{x^2}{b^2} + \frac{(y\sin\theta - c)^2}{a^2} + \frac{y^2\cos^2\theta}{b^2} = 1 \tag{7.69}$$

$$c = \frac{R_2}{2} - SP \tag{7.70}$$

The coordinates of the four vertices of the ellipse corresponding to the equation can be determined by the following formula

$$x_{1,2} = \pm b\sqrt{1 - \frac{(y_m \sin\theta - c)^2}{a^2} + \frac{y_m^2 \cos^2\theta}{b^2}} \tag{7.71}$$

$$y_{1,2} = -y_m \pm \sqrt{\frac{1}{A} - \left(\frac{c\cos\theta}{Aab}\right)^2} \tag{7.72}$$

where

$$y_m = \frac{B}{A}$$

$$A = \left(\frac{\sin\theta}{a}\right)^2 + \left(\frac{\cos\theta}{b}\right)^2$$

$$B = c\frac{\sin\theta}{a^2}$$

In this section, since the research subject is the measurement of the absorption coefficient under normal incidence, hence $\theta = 0°$.

Therefore, the expression can be rewritten as

$$x^2 + y^2 = b^2\left(1 - \frac{c^2}{a^2}\right) \tag{7.73}$$

Figure 7.62 shows a schematic diagram of the size of the material required for reflection testing under different frequency conditions.

In this experiment, with the distance between the microphone and the material under test being 1.9 m, the most extreme case is considered, where it is assumed that the sound source has no directivity and is treated as a point sound source. The microphone 1 is about 0.07 m away from the material to be tested, so the distance from the microphone to the material to be tested is taken as 0.07 m. As can be seen

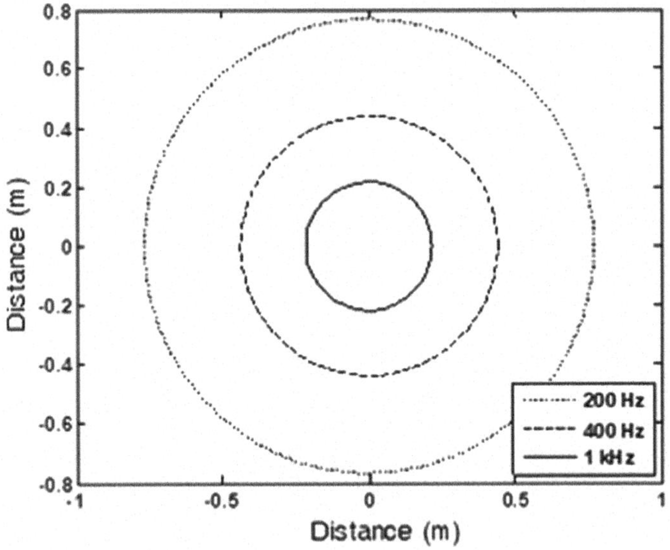

Fig. 7.62 Schematic diagram of the size of the corresponding area at different frequencies

from the figure, if the lowest frequency to be measured is 200 Hz, the minimum size of the square material required is 1.54 m × 1.54 m, 0.88 m × 0.88 m in the case of 400 Hz, and 0.44 m × 0.44 m in the case of 1 kHz.

In this experiment, three different sizes of polyurethane sponges with a thickness of 5 cm were used, namely 0.20 m × 0.20 m, 0.45 m × 0.45 m, 0.70 m × 1.0 m, and the corresponding lower limits of the frequency that can be fully reflected are shown in Table 7.4.

The experimental results are shown in Figs. 7.63 and 7.64. As can be seen from the figures, when the material is small, the accuracy of low-frequency measurement is seriously affected, and there is a large error with the results of the standing wave tube. For materials of 0.45 m and 0.70 m, the test results show good consistency, which is in line with the predicted lower limit of the reflection frequency. Among them, the material of 0.45 *m* has the smallest error with the measurement results of the standing wave tube, and the maximum relative error is only 5.4%, and the relative error of 85% of the sampling points is below 3.0%. The maximum relative error of the 0.70 m material is 20.8%, 65% of the sampling points have a relative error within the range of 5.0%, 20% of the sampling points have a relative error within the range of 5.0~10.0%, and 10% of the sampling points have a relative error within the range of 10.0~20.0%.

The main reason for the error is that the polyurethane sponge is an elastic material and cannot be accurately cut. When measuring in the standing wave tube, the material needs to be cut into a standard cylindrical shape, and there must be no gap between the material and the tube wall, which is difficult to meet in practice. A slightly larger material is placed in the standing wave tube, and the density changes slightly. There

Table 7.4 Lower limit of reflection frequency corresponding to different sizes of materials

Width	0.20 m	0.45 m	0.70 m
$f_{\min}$ (Hz)	3260	1025	591

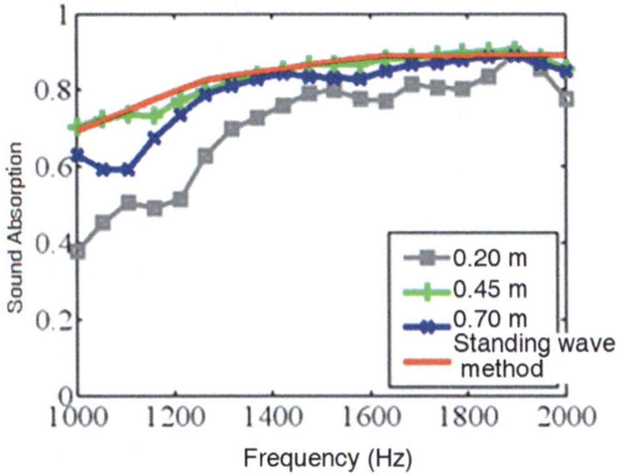

Fig. 7.63 Comparison of sound absorption coefficient measurement results for materials of different sizes

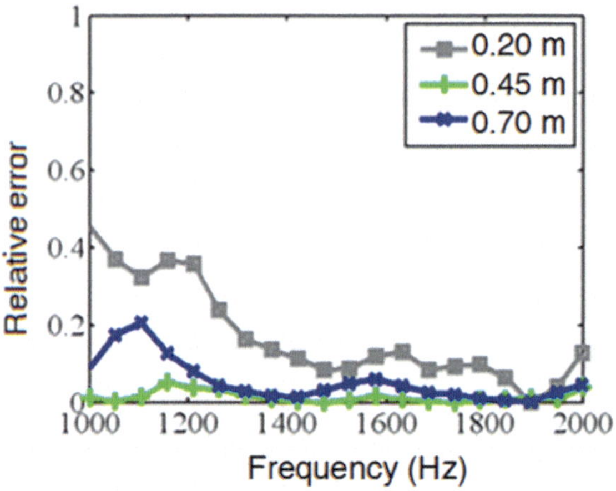

Fig. 7.64 Relative error between the sound absorption coefficient measurement results of different size materials and the standing wave tube results (with the standing wave tube results as the reference value)

may be some differences when measuring with the parameter array. On the other hand, because the method used in this part requires placing the microphone on the surface of the material to be tested, it will inevitably affect the sound field on the surface of the material. The microphone used in this experiment is relatively large, with a diameter of 1.3 cm and a length of 10 cm, plus the attached fixing device, and it will definitely have some impact on the sound field on the surface of the material to be tested. The interference intensity for different frequencies is not the same. As can be seen from Fig. 7.64, the error is significantly higher than other frequencies in the range of 1100~1200 Hz. Further work needs to consider the impact of the microphone on the measurement of the sound field, and include the impact between the two closely placed microphones into the analysis range to improve the accuracy of the measurement.

7.3.3.2 Impact of the Test Environment

To test the performance of the method under different environments, this section conducted sound absorption measurements in a regular room and compared the results with those in a fully anechoic room to test the impact of reflections and other issues in the regular room on the results.

Measurements were taken in a room measuring 4.5 m × 4 m × 3 m. The room layout is shown in Fig. 7.65. The room environment is a regular suspended ceiling, rigid floor, and the room's background noise is between 30 and 40 dB.

The measurement results are shown in Fig. 7.66. The measurement results of the regular room and the fully anechoic room are very consistent. In the case of 0.40 m material, only one sampling point has a relative error of 11.1%, all other sampling points have a relative error of less than 5.0%, and 85% of the sampling points have a relative error of less than 3%, showing good stability. In the measurement of 0.70 m material, the maximum relative error is 6.9%, and 85% of the sampling points have

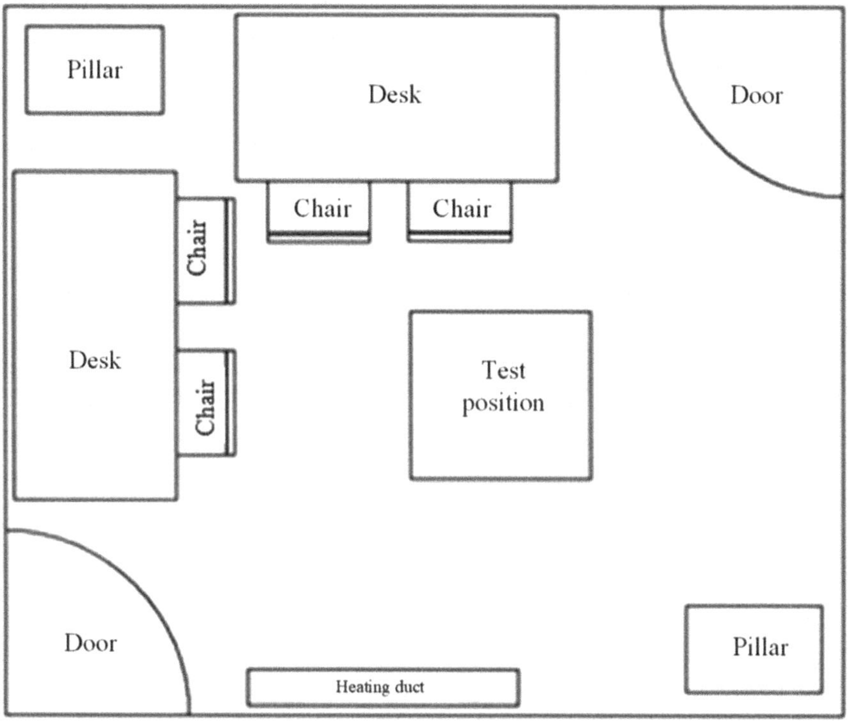

Fig. 7.65 Ordinary room layout

a relative error of less than 5.0%. As the test frequency increases, the error shows a downward trend. The intensity of the nonlinear self-demodulation sound frequency nonlinear effect of the parametric array is directly proportional to the square of the frequency. As the frequency increases, the sound pressure level will also increase, the signal-to-noise ratio of the obtained signal is higher, and the anti-interference ability is stronger.

Due to the high directivity of the parametric array loudspeaker, most of the energy of the sound waves is absorbed by the material after contact, and the signal entering the microphone through the reflection and scattering of the wall and the objects placed in the room is very weak. Therefore, the size of the room, the characteristics of the walls, and the placement of items in the room have limited impact on the test results. This method has very good effects on the measurement of the sound absorption coefficient under complex sound environments, and has high application value for quick measurement on construction sites and online automatic detection in material production.

The current test method has certain errors at some frequency points. The reasons for the errors may be various, including the impact of the microphone on the sound field, the interference between the microphones, and the inconsistency of the material under different measurement methods, etc.

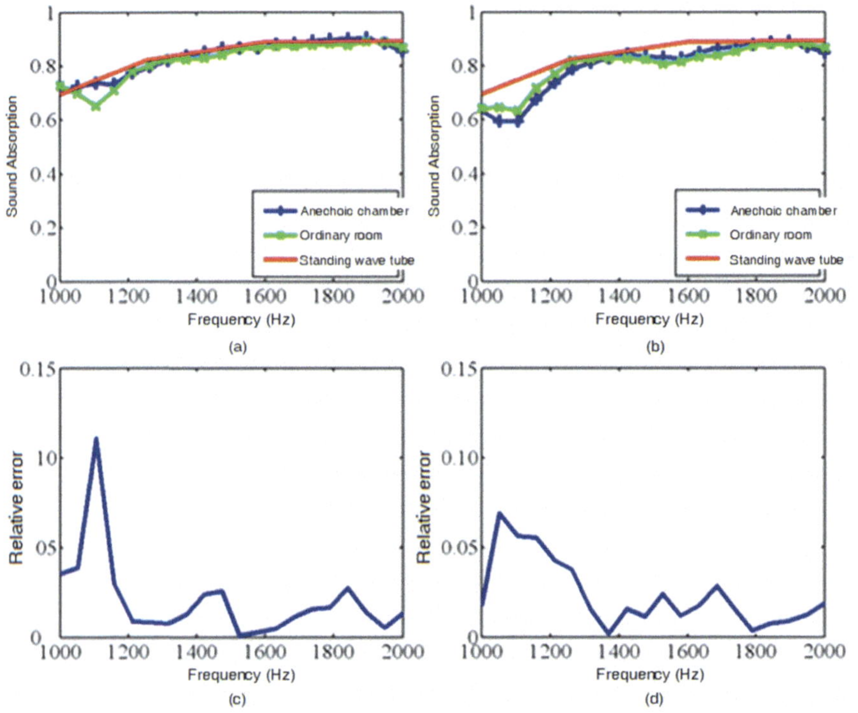

Fig. 7.66 Comparison of results between anechoic chamber and ordinary room **a** sound absorption coefficient of the material with a size of 0.45 m, **b** sound absorption coefficient of the material with a size of 0.70 m, **c** relative error between the measurement results of the material with a size of 0.45 m in the anechoic chamber and the ordinary room, **d** relative error between the measurement results of the material with a size of 0.70 m in the anechoic chamber and the ordinary room (with the standing wave tube results as the reference value)

7.3.3.3 Stability Analysis of the Test

The stability analysis of the test is repeated eight times under normal incidence, and eight sets of results are obtained. Figure 7.67 shows the results of the three trials.

The standard deviation between the eight sets of repeated measurement results is given in Fig. 7.68. The values corresponding to different frequency points are the standard deviation values between all eight results at that frequency point.

The value of the standard deviation is less than 0.007 across the entire frequency band, and less than 0.005 for the vast majority of test frequency points. The differences between different test results are small, and they match well with each other. It becomes clear that this method has high robustness for ordinary tests.

As the frequency increases, the value of the standard deviation shows a gradual decreasing trend. As the frequency increases, the amplitude response of the parametric array loudspeaker becomes greater, therefore, the signal-to-noise ratio has been significantly improved.

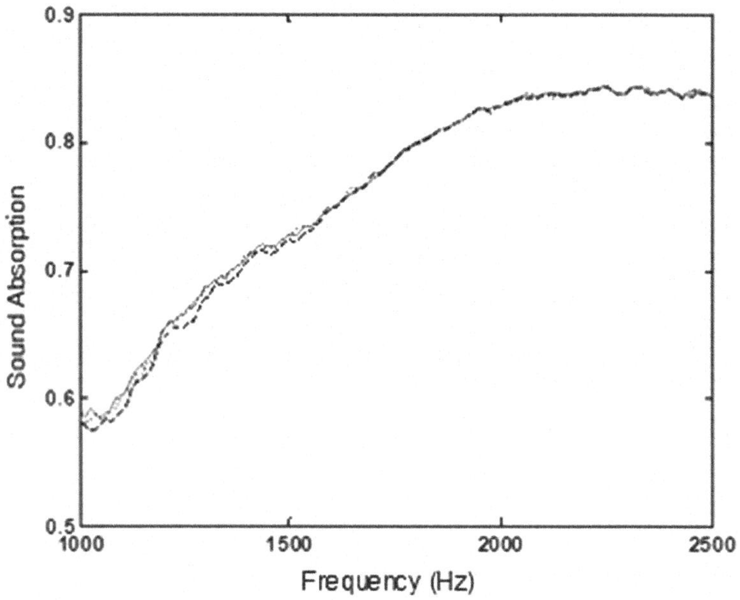

Fig. 7.67 Comparison of three different measurement results

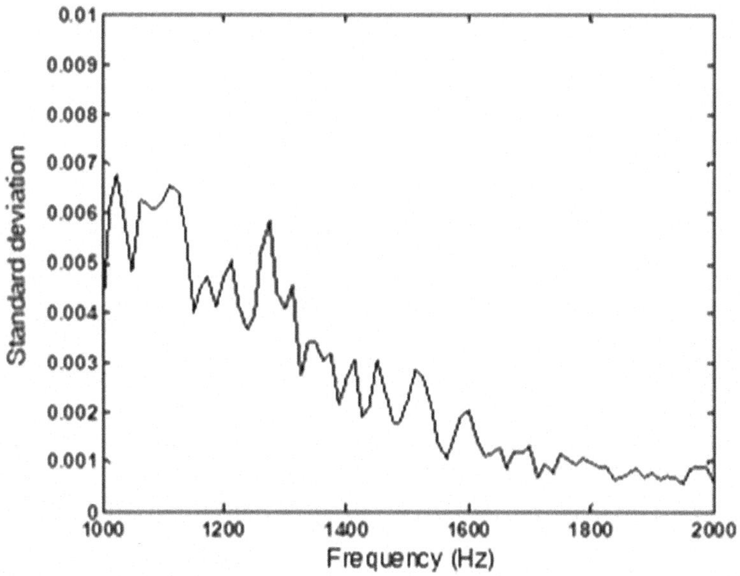

Fig. 7.68 Standard deviation of test results

7.4 Summary

This chapter reviews the current main applications of parametric array loudspeakers, and focuses on detailed explanations of its applications in active noise control in free space, pipeline active noise control, and field measurement coefficients.

First, a study was conducted on the active noise control system with the parametric array loudspeaker as the secondary sound source, and the energy transfer mechanism of the sound field was analyzed. Under the condition of minimizing global energy, the parametric array loudspeaker absorbs the sound field energy, reducing the total energy of the entire sound field after control. When local space control is performed, a significant quiet zone can be generated in the axial direction of the parametric array loudspeaker. Due to the directivity of the parametric array loudspeaker, while creating a quiet zone in the local space, the impact on other areas of the sound field is minimal.

Next, a method for reducing acoustic feedback in parametric array loudspeakers is presented, achieving active noise control in the duct of the parametric array loudspeakers. The parametric array loudspeaker with smaller acoustic feedback was determined by measuring the response of the feedback path. Active noise control in the duct was carried out using the parametric array loudspeaker under the condition without a feedback compensation network, and compared with the active noise control in the duct under two conditions: with and without a feedback compensation network using a regular loudspeaker. The results show that the noise reduction effect of the parametric array under the condition without a feedback compensation network is not much different from that of the regular loudspeaker with a compensation network, thereby reducing the complexity of the adaptive algorithm and reducing the amount of algorithm computation.

Finally, a method for measuring the absorption coefficient of materials in the field using a parametric array loudspeaker based on the two-microphone transfer function method is presented. The effects of factors such as measurement location, distance between the microphone and the material, and microphone matching on the measurement results were studied. The impact of conditions such as material size and measurement environment in actual measurements was analyzed, and the stability of the measurement results was analyzed. Based on this foundation, by measuring the sound absorption coefficient of the material under different angles of sound wave incidence, utilizing the relationship of the Paris formula, and employing a discretization approximation method to simulate the diffuse field absorption in a reverberation room within a free field, the diffuse field sound absorption coefficient of the material in the free field was successfully measured. This method was then compared with actual measurements taken in a reverberation room, proving its effectiveness of this method.

Bibliography

1. Hamilton MF, Blackstock DT (1998) Nonlinear acoustics. Academic Press, San Diego
2. Blackstock DT (1998) History of nonlinear acoustics: 1750s–1930s. In: Hamilton MF, Blackstock DT (eds) Nonlinear acoustics. Academic Press, San Diego, pp 233–261
3. Stokes GG (1848) On a difficulty in the theory of sound. Phil Mag 33:349–356
4. Earnshaw S (1860) On the mathematical theory of sound. Philos Trans R Soc Lond 150:133–148
5. Cleveland RO, Hamilton MF, Blackstock DT (1995) Time-domain modeling of finite-amplitude sound in relaxing fluids. J Acoust Soc Am 98(5):3312–3318
6. Beyer RT (1984) Nonlinear acoustics in fluids (Benchmark Papers in Acoustics). Van Nostrand Reinhold, New York
7. Rankine WJM (1870) On the thermodynamic theory of waves of finite longitudinal disturbance. Philos Trans R Soc Lond 160:277–288
8. Rayleigh L (1910) Aerial plane waves of finite amplitude. Proc Roy Soc Lond Ser A 84(570):247–284
9. Taylor GI (1910) The conditions necessary for discontinuous motion in gases. Proc Roy Soc Lond Ser A 84(571):371–377
10. Fubini E (1935) Anomalies in the propagation of an acoustic wave of large amplitude. Alta Freq 4:532–581
11. Fay RD (1931) Plane sound waves of finite amplitude. J Acoust Soc Am 3(2A):222–241
12. Thuras AL, Jenkins RT, O′ Neil HTO (1935) Extraneous frequencies generated in air carrying intense sound waves. J Acoust Soc Am 6(3):173–180
13. Eckart C (1948) Vortices and streams caused by sound waves. Phys Rev 73(1):68–76
14. Lighthill MJ (1952) On sound generated aerodynamically. I. General theory. Proc Roy Soc Lond Ser A 211(1107):564–587
15. Mendousse JS (1953) Nonlinear dissipative distortion of progressive sound waves at moderate amplitudes. J Acoust Soc Am 25(1):51–54
16. Khokhlov RV, Naugolnych KA, Soluyan SI (1964) Waves of moderate amplitude in absorbing media. Acustica 14:248–253
17. Blackstock DT (1964) Thermovicous attenuation of plane, periodic, finite-amplitude sound waves. J Acoust Soc Am 36:534–542
18. Fenlon FH (1972) An cxtension of the bessel-fubini series for a multiple-frequency CW acoustic source of finite amplitude. J Acoust Soc Am 51(1B):284–289
19. Westervelt PJ (1957) Scattering sound by sound. J Acoust Soc Am 29(2):199–203
20. Muir TG, Korman MS (2006) Professor Peter Westervelt and the parametric array. J Acoust Soc Am 119(5):3231–3231
21. Westervelt PJ (1960) Parametric end-fire array. J Acoust Soc Am 32(7):934–935
22. Westervelt PJ (1963) Parametric acoustic array. J Acoust Soc Am 35(4):535–537

J. Yang and P. Ji, *Parametric Array Loudspeakers*,
https://doi.org/10.1007/978-981-96-3548-1

23. Bellin JLS, Beyer RT (1962) Experimental investigation of an end-fire array. J Acoust Soc Am 34(8):1051–1054
24. Berktay HO (1965) Possible exploitation of non-linear acoustics in underwater transmitting applications. J Sound Vib 2(4):462–470
25. Berktay HO (1965) Parametric amplification by the use of acoustic non-linearities and some possible applications. J Sound Vib 2(4):462–470
26. Berktay HO (1967) Some proposals for underwater transmitting applications of non-linear acoustics. J Sound Vib 6(2):244–254
27. Tjøtta S (1967) Some non-linear effects in sound fields. J Sound Vib 6(2):255–267
28. Zverev VA, Kalachev AI, Stepanov NS (1968) Utilization of nonlinear effects in underwater acoustics. Soviet Phys Acoust 13:324–326
29. Zabolotskaya EA, Khokhlov RV (1969) Quasi-plane waves in the nonlinear acoustics of confined beams. Soviet Phys Acoust 15:25–40
30. Kuznetsov VP (1971) Equations of nonlinear acoustics. Soviet Phys Acoust 16:467–470
31. Tjøtta JN, Tjøtta S (1981) Nonlinear equations of acoustics, with application to parametric acoustic arrays. J Acoust Soc Am 69(6):1644–1652
32. Garrett GS, Tjøtta JN, Tjøtta S (1982) Nearfield of a large acoustic transducer, Part I: linear radiation. J Acoust Soc Am 72(3):1056–1061
33. Garrett GS, Tjøtta JN, Tjøtta S (1983) Nearfield of a large acoustic transducer, Part II: parametric radiation. J Acoust Soc Am 74(3):1013–1020
34. Garrett GS, Tjøtta JN, Tjøtta S (1984) Nearfield of a large acoustic transducer. Part III: general results. J Acoust Soc Am 75(3):769–779
35. Aanonsen AI, Barkve T, Tjøtta JN et al (1984) Distortion and harmonic generation in the nearfield of a finite amplitude sound beam. J Acoust Soc Am 75(3):749–768
36. Berntsen J, Tjøtta J N, Tjøtta S (1984) Nearfield of a large acoustic transducer. Part IV: second harmonic and sum frequency radiation. J Acoust Soc Am 75(5):1383–1391
37. Beyer RT (1974) Nonlinear acoustics. Naval Sea Systems Command, Washington DC
38. Rudenko OV, Soluyan SI (1977) Theoretical foundations of nonlinear acoustics. Consultants Bureau, New York
39. Qian Z (1992) Nonlinear acoustics. Science Press, Beijing
40. Qian Z (1976) About sound of scattering by sound. Acta Physica Sinica 25(6):472–480
41. Qian Z (1981) A new method for calculating the radiation field of a linear acoustic array. Acta Physica Sinica 30(11):1479–1487
42. Qian Z (1981) On the near field of a parametric array. Acta Physica Sinica 30(11):1559–1561
43. Shao D, Qian Z, Zhang B et al (1982) Parametric array underwater communication. J Appl Acoust 1:33–34
44. Qian Z, Shi X, Zhao C et al (1983) Finite amplitude reflected wave of a circular piston sound source. Acta Physica Sinica 9:1109–1117
45. Qian Z, Shao D (1986) On the application of shallow sea sound parametric array. Acta Physica Sinica 10:1374–1377
46. Qian Z (1987) Nonlinear (acoustic) parameters and their application in ultrasound imaging. J Appl Acoust 3:1–7
47. Qian Z (1988) Acoustic interaction of plane wave and plane traveling wave pulse. Acta Physica Sinica 37(2):221–228
48. Qian Z (1988) Research on broadband parametric arrays (1)—non-axial field of pulse self-demodulation problem. Acta Acustica 2:119–123
49. Qian Z (1988) Research on broadband parametric arrays (2)—impact of original wave signal envelope. Acta Acustica 4:271–278
50. Shao D, Qian Z (1988) Experimental research on broadband parametric arrays. Acta Acustica 4:279–283
51. Qian Z (1991) Overview of nonlinear acoustics. Physics 5:261–266
52. Qian Z (1999) Progress in nonlinear acoustics research in China. Physics 10:593–600
53. Gong X, Zhu Z Du G (1979) Research on nonlinear interaction of finite amplitude and small amplitude plane sound waves. J Nanjing Univ (Natural Sciences Edition) 3:19–28

54. Yang D, Li Z, Fang E (2016) Research on parametric array sound field formed by three columns of different phase original waves. J Harbin Eng Univ 37(1):7–12
55. Yang D, Lan C, Li S et al (2010) Analysis of energy changes during nonlinear interaction of sound waves in water. J Harbin Eng Univ 31(8):995–999
56. Yang D, Lan C, Shi S et al (2012) Research on sound absorption of nonlinear interaction of sound waves in water. Vibr Impact 31(8):52–56
57. Yang D, Liu Q, Shi J et al (2017) Nonlinear interaction rules of noise and single frequency sound waves. J Harbin Eng Univ 38(9):1345–1350
58. Chotiros NP, Mautner AM, Lovik A et al (1997) Acoustic penetration of a silty sand sediment in the 1–10-kHz band. IEEE J Oceanic Eng 22(4):604–615
59. Williams KL, Satkowiak LJ, Bugler DR (1989) Linear and parametric array transmission across a water–sand interface—theory, experiment, and observation of beam displacement. J Acoust Soc Am 86(1):311–325
60. Kopp L, Cano D, Dubois E et al (2000) Potential performance of parametric communications. IEEE J Oceanic Eng 25(3):282–295
61. Humphrey VF, Robinson SP, Smith JD et al (2008) Acoustic characterization of panel materials under simulated ocean conditions using a parametric array source. J Acoust Soc Am 124(2):803–814
62. Zhou H, Huang SH, Li W (2020) Parametric acoustic array and its application in underwater acoustic engineering. Sensors (Basel, Switzerland) 20(7):2148
63. Zhu J (2014) Research on shallow stratum detection technology of parametric array. Ph.D. Thesis of Harbin Engineering University
64. Darvennes CM, Hamilton MF (1990) Scattering of sound by sound from two Gaussian beams. J Acoust Soc Am 87(5):1955–1964
65. Westervelt PJ (1983) The nonscattering of sound by sound. J Acousti Soc Am 73(S1):S82–S82
66. Westervelt PJ (1993) Nonscattering of sound by sound: exact second-order transient solutions. J Acoust Soc Am 94(3):1774–1774
67. Westervelt PJ (1994) Scattering of sound by sound within the interaction zone. J Acoust Soc Am 96(5):3320–3320
68. Westervelt PJ (1995) Scattering of sound by sound within the interaction zone: approximate solutions. J Acoust Soc Am 97(5):3375–3375
69. Westervelt PJ (1995) General proof of the nonscattering of sound by sound in the lossless dispersionless case. J Acoust Soc Am 97(5):3375–3375
70. Trivett DH, Rogers PH (1982) Scattering of a cw plane wave by a pulse. J Acoust Soc Am 71(5):1114–1117
71. Lighthill MJ (1954) On sound generated aerodynamically. II. Turbulence as a source of sound. Proc Roy Soc Lond Ser A 222(1148):1–32
72. Ingard U, Pridmore-Brown DC (1956) Scattering of sound by sound. J Acoust Soc Am 28(3):367–369
73. Bellin JLS, Beyer RT (1960) Scattering of sound by sound. J Acoust Soc Am 32(3):339–341
74. Dean LM (1962) Interactions between sound waves. J Acoust Soc Am 34(8):1039–1044
75. Tjøtta JN, Tjøtta ST (1987) Interaction of sound waves. Part I: basic equations and plane waves. J Acoust Soc Am 82(4):1425–1428
76. Tjøtta JN, Tjøtta S (1987) Interaction of sound waves. Part II: plane wave and real beam. J Acoust Soc Am 82(4):1429–1435
77. Tjøtta JN, Tjøtta S. Interaction of sound waves. Part III: two real beams. The Journal of the Acoustical Society of America, 1988, 83(2): 487–495.
78. Berntsen J, Tjøtta JN, Tjøtta S (1989) Interaction of sound waves. Part IV: scattering of sound by sound. J Acoust Soc Am 86(5):1968–1983
79. Darvennes CM, Hamilton MF, Tjøtta JN et al (1991) Effects of absorption on the nonlinear interaction of sound beams. J Acoust Soc Am 89(3):1028–1036
80. Tjøtta JN, TenCate JA, Tjøtta S (1991) Effects of boundary conditions on the nonlinear interaction of sound beams. J Acoust Soc Am 89(3):1037–1049

81. Ji P, Yang J, Tian J (2008) A fast calculation method for the scattered sound fields from two sound beams using the Gaussian-Beam expansion technique. Jpn J Appl Phys 47(5):4253–4256
82. Ji P, Yang J, Gan WS (2009) The investigation of localized sound generation using two ultrasound beams. IEEE Trans Ultrason Ferroelectr Freq Control 56(6):1282–1287
83. Ji P (2008) Theoretical and experimental study of multi-beam sound field. Ph.D. Thesis of the graduate school of Chinese Academy of Sciences
84. Ye C, Kuang Z, Ji P et al (2008) Generation of audible sound from two ultrasonic beams with dummy head. Jpn J Appl Phys 47(5):4333–4335
85. Garner G (2011) Design of optimal directional parametric acoustic arrays in air. Ph.D. Dissertation of North Carolina State University
86. Shi C (2013) Investigation of the steerable parametric loudspeaker based on phased array techniques. Ph.D. Thesis of Nanyang Technological University
87. Li S (2011) Progress in the application of parametric arrays in underwater acoustics engineering. Techn Acoust 30(1):9–16
88. Helmholtz HLF (2009) On the sensations of tone as a physiological basis for the theory of music, 3rd edn. Cambridge: Cambridge University Press
89. Moffett MB, Westervelt PJ, Beyer RT (1970) Large-amplitude pulse propagation-a transient effect. J Acoust Soc Am 47:1473–1474
90. Moffett MB, Westervelt PJ, Beyer RT (1971) Large-amplitude pulse propagation-a transient effect. II. J Acoust Soc Am 49(1):339–343
91. Zheng X, Huang T, Wang C (2008) Calculation and simulation of parametric source sound field of circular plane emitter in air. J Huazhong Univ Sci Technology (Natural Science Edition) 36(2):112–114
92. Kite TD, Post JT, Hamilton MF (1998) Parametric array in air: distortion reduction by preprocessing. In: Proceedings of the 16th International Congress on Acoustics and the 135th Meeting of the Acoustical Society of America, vol 2, pp 1091–1092
93. Bennett MB, Blackstock DT (1975) Parametric array in air. J Acoust Soc Am 57(3):562–568
94. Yoneyama M, Fujimoto JI, Kawamo Y et al (1983) The audio spotlight: an application of nonlinear interaction of sound waves to a new type of loudspeaker design. J Acoust Soc Am 73(5):1532–1536
95. Kamakura T, Yoneyama M, Ikegaya K (1984) Development of a parametric loudspeaker for practical use. In: 10th International Symposium on Nonlinear Acoustics, Kobe, pp 147–150
96. Kamakura T, Aoki K, Kumamoto Y (1991) Suitable modulation of the carrier ultrasound for a parametric loudspeaker. Acustica 73(4):215–217
97. Merklinger HM (1975) Improved efficiency in the parametric transmitting array. J Acoust Soc Am 58(4):784–787
98. Kamakura T, Tani M, Kumamoto Y et al (1994) Parametric sound radiation from a rectangular aperture source. Acta Acustica united with Acustica 80(4):332–338
99. Havelock DI, Brammer AJ (2000) Directional loudspeakers using sound beams. J Audio Eng Soc 48(10):908–916
100. Sha K, Yang J, Gan WS (2003) A complex virtual source approach for calculating the diffraction beam field generated by a rectangular planar source. IEEE Trans Ultrason Ferroelectr Freq Control 50(7):890–897
101. Wen J, Breazeale M (1988) A diffraction beam field expressed as the superposition of Gaussian beams. J Acoust Soc Am 83(5):1752–1756
102. Dong W, Wu Q (1999) Audio sound reproduction based on nonlinear interaction of acoustic waves. J Audio Eng Soc 47(7–8):602–606
103. Garner G, Steer MB (2012) A cascaded second-order approach to computing third-order scattering of noncollinear acoustic beams. Appl Acoust 73(12):1220–1230
104. Johnson SJ, Steer MB (2014) An efficient approach to computing third- order scattering of sound by sound with application to parametric arrays. IEEE Trans Ultrason Ferroelectr Freq Control 61(10):1729–1741

105. Asahi K, Nakayama M, Nishiura T (2020) Audible sound enhancement based on carrier with periodic square-wave for parametric loudspeaker. In: IEEE 9th Global Conference on Consumer Electronics (GCCE), pp 12–15
106. Huang K (1983) Various characteristics of transient signal parametric array. J Marine Science and Applicaton 4(2):79–89
107. Lou G (1988) Research on transient signal phase-controlled parametric array. J Marine Science and Application 9(1):63–72
108. Liu G (2015) Research on the demodulation technology of the secondary wave of the transient signal parametric emission array. Master's Thesis of Harbin Engineering University
109. Li Z, Shi L, Fang E (2016) Acoustic field characteristics of parametric array formed by transient signals. J Naval Aeronaut Astronaut Univ 31(6):607–613
110. Li Z (2015) Study on the nonlinear acoustic field characteristics of parametric arrays. Ph.D. Thesis of Harbin Engineering University
111. Wygant IO, Kupnik M, Yaralioglu G et al (2007) 6F-4 50-kHz Capacitive micromachined ultrasonic transducers for generating highly directional sound with parametric arrays. In: Proceedings of IEEE Ultrasonics Symposium, pp 519–522
112. Wygant IO, Kupnik M, Windsor JC et al (2009) 50 kHz capacitive micromachined ultrasonic transducers for generation of highly directional sound with parametric arrays. IEEE Trans Ultrasonics, Ferroelectrics Freq Control 56(1):193–203
113. Hwang Y, Ahn H, Nguyen DN et al (2018) An underwater parametric array source transducer composed of PZT/thin-polymer composite. Sens Actuators, A 279:601–616
114. Nakano Y, Fukumori T, Nakayama M et al (2018) A study on audible low-frequency sound emphasis based on multiplexed double sideband modulation in parametric loudspeaker. Proc INTER-NOISE NOISE-CON Congr Conf 258(5):2054–2065
115. Iwasaki T, Ikefuji D, Nakayama M et al (2013) Parametric loudspeaker for speech signal based on the combination of amplitude and frequency modulations. J Acoust Soc Am 133(5):3302–3302
116. Konabe R, Matsui T, Ikefuji D et al (2014) A design of reflective audio spot with parabolic reflector for sound pressure improvement on separating emission of carrier and sideband waves. Proc INTER-NOISE NOISE-CON Congr Conf 249(5):2890–2896
117. Ogami Y, Fukumori T, Nakayama M et al (2017) Enhanced audio spot based on multiple carrier waves with extra parametric loudspeaker. Proc INTER-NOISE NOISE-CON Congr Conf 255(4):3233–3240
118. Smith BV (1971) An experimental study of a parametric end-fire array. J Sound Vib 14(1):7–21
119. Farias F, Abdulla W (2015) On Rayleigh distance and absorption length of parametric loudspeakers. In: Asia-Pacific Signal and Information Processing Association Annual Summit and Conference (APSIPA), pp 1262–1265
120. Li S, Yi Y, Zhang J (2020) The parametric source method for measuring characteristics of underwater acoustic materials in a pressure vessel. Chin J Acoust 39(1):83–93
121. Zhong J, Kirby R, Qiu X (2020) A spherical expansion for audio sounds generated by a circular parametric array loudspeaker. J Acoust Soc Am 147(5):3502–3510
122. Zhong J, Kirby R, Qiu X (2020) A non-paraxial model for the audio sound behind a non-baffled parametric array loudspeaker. J Acoust Soc Am 147(3):1577–1580
123. Zhong J, Wang S, Kirby R et al (2020) Insertion loss of a thin partition for audio sounds generated by a parametric array loudspeaker. J Acoust Soc Am 148(1):226–235
124. Zhong J, Wang S, Kirby R et al (2020) Reflection of audio sounds generated by a parametric array loudspeaker. J Acoust Soc Am 148(4):2327–2336
125. Gan WS, Yang J, Kamakura T (2012) A review of parametric acoustic array in air. Appl Acoust 73(12):1211–1219
126. Yang J, Tian J, Gan WS (2014) Parametric loudspeaker: from theory to applications. In: The 21st International Congress on Sound and Vibration, Beijing, pp 1–18
127. Boullosa RR, Pérez-López A, Dorantes-Escamilla R et al (2016) An airborne parametric array. Appl Acoust 112:116–122

128. Kim Y, Kim S (2002) Novel preprocessing technique to improve harmonic distortion in airborne parametric array. In: 6th International Conference on Signal Processing, vol 2, pp 1815–1818
129. Zhu L, Florencio D (2015) 3D numerical modeling of parametric speaker using finite-difference time-domain. In: IEEE International Conference on Acoustics, Speech and Signal Processing (ICASSP), pp 5982–5986
130. Kamakura T, Tasahide Y, Ikegaya K (1985) Studies for the realization of parametric loudspeaker. J Acoust Soc Japan 41(6):378–385
131. Croft JJ, Norris JO (2002) Theory, history, and the advancement of parametric loudspeaker. White paper of American Technology Corporation
132. Croft JJ, Spencer ME, Norris JO (2003) Modulator processing for a parametric speaker system. Patent US 20080063214A1
133. Wu M, Yang J (2016) Low complexity single sideband modulation technique for parametric array loudspeaker. Acta Acustica 41(5):650–656
134. Tan EL, Gan WS, Ji P et al (2008) Distortion analysis and reduction for the parametric array. In: Proceedings of the 124th Audio Engineering Society Convention
135. Sakai S, Kamakura T (2008) Dynamic single sideband modulation for realizing parametric loudspeaker. In: The 18th International Symposium on Nonlinear Acoustics, Stockholm, Sweden, pp 613–616
136. Shi C, Kajikawa Y (2014) A comparative study of preprocessing methods in the parametric loudspeaker. In: Asia-Pacific Signal and Information Processing Association Annual Summit and Conference (APSIPA), pp 1–5
137. Ikefuji D, Nakayama M, Nishiura T et al (2013) Weighted double sideband modulation toward high quality audible sound on parametric loudspeaker. In: IEEE International Conference on Acoustics, Speech and Signal Processing (ICASSP), pp 843–847
138. Ji P, Tan EL, Gan WS et al (2011) A comparative analysis of preprocessing methods for the parametric loudspeaker based on the Khokhlov-Zabolotskaya-Kuznetsov equation for speech reproduction. IEEE Trans Audio Speech Lang Process 19(4):937–946
139. Nakayama M, Nishiura T (2017) Synchronized amplitude-and-frequency modulation for a parametric loudspeaker. In: Asia-Pacific Signal and Information Processing Association Annual Summit and Conference (APSIPA), pp 130–135
140. Hatano Y, Shi C, Kajikawa Y (2017) Compensation for nonlinear distortion of the frequency modulation-based parametric array loudspeaker. IEEE/ACM Trans Audio, Speech, Lang Process 25(8):1709–1717
141. Lee KCM, Yang J, Gan WS et al (2002) Modeling nonlinearity of air with Volterra kernels for use in a parametric array loudspeaker. Presented at the 112th Audio Engineering Society Convention, Munich
142. Ji W, Gan WS (2012) Identification of a parametric loudspeaker system using an adaptive Volterra filter. Appl Acoust 73(12):1251–1262
143. Mu Y, Ji P, Ji W et al (2013) On the linearization of a parametric loudspeaker system by using third-order inverse Volterra filters. In: Proceedings of IEEE 1st China Summit and International Conference on Signal and Information Processing, pp 570–574
144. Mu Y, Ji P, Ji W et al (2014) Modeling and compensation for the distortion of parametric loudspeakers using a one-dimension Volterra filter. IEEE/ACM Trans Audio, Speech, Lang Process 22(12):2169–2181
145. Mu Y, Ji P, Ji W et al (2014) Multipoint linearization of parametric loudspeakers using Volterra filters. In: Asia-Pacific Signal and Information Processing Association Annual Summit and Conference (APSIPA)
146. Mu Y, Ji P, Wu M et al (2014) Research on intermodulation distortion compensation of parametric array loudspeaker based on Volterra filter. J Appl Acoust 33(4):283–292
147. Shi C, Kajikawa Y (2015) Ultrasound-to-ultrasound Volterra filter identification of the parametric array loudspeaker. In: IEEE International Conference on Digital Signal Processing (DSP), pp 1–4

148. Shi C, Kajikawa Y (2015) Identification of the parametric array loudspeaker with a Volterra filter using the sparse NLMS algorithm. In: IEEE International Conference on Acoustics, Speech and Signal Processing (ICASSP), pp 3372–3376
149. Shi C, Mu H, Gan WS (2013) A psychoacoustical preprocessing technique for virtual bass enhancement of the parametric loudspeaker. In: IEEE International Conference on Acoustics, Speech and Signal Processing (ICASSP), pp 31–35
150. Woodward B, Cook J, Goodson A et al (1994) A phase steered parametric array for sub-bottom profiling. In: Proceedings of the 6th International Conference on Electronic Engineering in Oceanography, pp 77–82
151. Olszewski D, Prasetyo F, Linhard K (2005) Steerable highly directional audio beam loudspeaker. INTERSPEECH, Lisbon, Portugal
152. Yang J, Gan WS, Er MH et al (2006) Steering of directional sound beams. Patent US 7146011, 2006-12-05
153. Tan KS, Gan WS, Yang J et al (2003) Constant beamwidth beamformer for difference frequency in parametric array. In: Proceedings of International Conference on Multimedia and Expo (ICME), pp 593–596
154. Yang J, Gan WS, Tan KS et al (2005) Acoustic beamforming of a parametric speaker comprising ultrasonic transducers. Sens Actuators, A 125(1):91–99
155. Yang J, Tan KS, Gan WS et al (2005) Beamwidth control in parametric acoustic array. Jpn J Appl Phys 44(9A):6817–6819
156. Gan WS, Yang J, Tan KS et al (2006) A digital beamsteerer for difference frequency in a parametric array. IEEE Trans Audio Speech Lang Process 14(3):1018–1025
157. Takeoka S, Yamasaki Y (2010) Acoustic projector using directivity controllable parametric loudspeaker array. In: Proceedings of 20th International Congress on Acoustics (ICA), pp 921–925
158. Shi C, Gan WS (2011) Grating lobe elimination in steerable parametric loudspeaker. IEEE Trans Ultrason Ferroelectr Freq Control 58(2):437–450
159. Shi C, Gan WS (2012) Product directivity models for parametric loudspeakers. J Acoust Soc Am 131(3):1938–1945
160. Shi C, Kajikawa Y (2015) A convolution model for computing the far-field directivity of a parametric loudspeaker array. J Acoust Soc Am 137(2):777–784
161. Shi C, Kajikawa Y, Gan WS (2014) An overview of directivity control methods of the parametric array loudspeaker. APSIPA Trans Signal Inform Process 3:1–30
162. Shi C, Kajikawa Y, Gan WS (2015) Generating dual beams from a single steerable parametric loudspeaker. Appl Acoust 99:43–50
163. Shi C, Nomura H, Kamakura T et al (2014) Development of a steerable stereophonic parametric loudspeaker. In: Asia-Pacific Signal and Information Processing Association Annual Summit and Conference (APSIPA), pp 1–5
164. Wu S, Wu M, Huang C et al (2012) FPGA-based implementation of steerable parametric loudspeaker using fractional delay filter. Appl Acoust 73(12):1271–1281
165. Luan G, Zhang J, Wang R (2005) Piezoelectric transducer and transducer array. Peking University Press, Beijing
166. Hong W, Toda M (1999) Curved PVDF airborne transducer. IEEE Trans Ultrason Ferroelectr Freq Control 46(6):1375–1386
167. Toda M, Tosima S (2000) Theory of curved, clamped, piezoelectric film, air-borne transducers. IEEE Trans Ultrason Ferroelectr Freq Control 47(6):1421–1431
168. Toda M (2002) Cylindrical PVDF film transmitters and receivers for air ultrasound. IEEE Trans Ultrason Ferroelectr Freq Control 49(5):626–634
169. Toda M, Thompson M (2007) Temperature dependence of high frequency parameters of PVDF for length mode ultrasonic air transducers. IEEE Sens 484–487
170. Toda M (2008) Excitation of acoustic waves from cylindrical polyvinylidene fluoride (PVDF) film confined in a concentric wall. IEEE Trans Ultrason Ferroelectr Freq Control 55(7):1653–1659

171. Toda M (2002) New type of matching layer for air-coupled ultrasonic transducers. IEEE Trans Ultrason Ferroelectr Freq Control 49(7):972–979
172. Toda M (2005) New type of acoustic filter using periodic polymer layers for measuring audio signal components excited by amplitude-modulated high-intensity ultrasonic waves. J Audio Eng Soc 53(10):930–941
173. Haller MI, Khuri-Yakub BT (1996) A surface micromachined electrostatic ultrasonic air transducer. IEEE Trans Ultrason Ferroelectr Freq Control 43(1):1–6
174. Jin XC, Ladabaum I, Khuri-Yakub BT (1998) Surface micromachined capacitive ultrasonic transducers. IEEE Trans Ultrason Ferroelectr Freq Control 45(3):678–690
175. Noble RA, Jones ADR, Robertson TJ et al (2001) Novel, wide bandwidth, micromachined ultrasonic transducers. IEEE Trans Ultrason Ferroelectr Freq Control 48(6):1495–1507
176. Oralkan O, Ergun AS, Johnson JA et al (2002) Capacitive micromachined ultrasonic transducers: next-generation arrays for acoustic imaging? IEEE Trans Ultrason Ferroelectr Freq Control 49(11):1596–1610
177. Erguri AS, Huang Y, Zhuang X et al (2005) Capacitive micromachined ultrasonic transducers: fabrication technology. IEEE Trans Ultrason Ferroelectr Freq Control 52(12):2242–2258
178. Zhang X, Zhang H, Li D et al (2021) Design of a hexagonal air-coupled capacitive micromachined ultrasonic transducer for air parametric array. Nanotechnol Precis Eng 4(1):013004
179. Percin G, Khuri-Yakub BT (2002) Piezoelectrically actuated flextensional micromachined ultrasound transducers. I. Theory. IEEE Trans Ultrason Ferroelectr Freq Control 49(5):573–584
180. Percin G, Khuri-Yakub BT (2002) Piezoelectrically actuated flextensional micromachined ultrasound transducers. II. Fabrication and experiments. IEEE Trans Ultrason Ferroelectr Freq Control 49(5):585–595
181. Akasheh F, Myers T, Fraser JD et al (2004) Development of piezoelectric micromachined ultrasonic transducers. Sens Actuators, A 111(2–3):275–287
182. Muralt P, Lafermann N, Baborowski J et al (2005) Piezoelectric micromachined ultrasonic transducers based on PZT thin films. IEEE Trans Ultrason Ferroelectr Freq Control 52(12):2276–2288
183. Wang Z, Zhu W, Zhu H et al (2005) Fabrication and characterization of piezoelectric micromachined ultrasonic transducers with thick composite PZT films. IEEE Trans Ultrason Ferroelectr Freq Control 52(12):2289–2297
184. Lee L, Kang D, Moon W (2009) A micro-machined source transducer for a parametric array in air. J Acoust Soc Am 125(4):1879–1893
185. Je Y, Lee H, Been K et al (2015) A micromachined efficient parametric array loudspeaker with a wide radiation frequency band. J Acoust Soc Am 137(4):1732–1743
186. Ahn A, Been K, Kim ID et al (2019) A critical step to using a parametric array loudspeaker in mobile devices. Sensors 19(20):1–24
187. Moffett MB, Henriquez TA (1982) Hydrophone nonlinearity measurements. J Acoust Soc Am 72(1):1–6
188. Kamakura T, Sakai S, Nomura H et al (2008) Parametric audible sounds by phase-cancellation excitation of primary waves. J Acoust Soc Am 123(5):3694–3694
189. Hara Y, Maekawa Y, Nomura H et al (2010) Parametric sources designed for reducing carrier ultrasounds. In: Proceedings of the symposium ultrasonic electronics, Tokyo, Japan
190. Ji P, Liu W, Wu S et al (2012) An alternative method to measure the on-axis differencerequency sound in a parametric loudspeaker without using an acoustic filter. Appl Acoust 73(12):1244–1250
191. Ye C, Kuang Z, Wu M et al (2011) Development of an acoustic filter for parametric loudspeaker in air. Jpn J Appl Phys 50(7):913–919
192. Ji P, Hu W, Yang J (2016) Development of an acoustic filter for parametric loudspeaker using phononic crystals. Ultrasonics 67:160–167
193. Ji P, Yang J (2016) A preliminary experimental study of parameter's effects on parametric loudspeaker. Presented at the IEEE international ultrasonics symposium (IUS), Tours, France

194. Ji P, Yang J (2019) An experimental investigation about parameters 'effects on spurious sound in parametric loudspeaker. Appl Acoust 148:67–74
195. Nakashima Y, Yoshimura T, Naka N et al (2006) Prototype of mobile super directional loudspeaker. NTT DoCoMo Tech J 8(1):25–32
196. Nakashima Y, Yoshimura T, Ohya T (2005) Prototype of parametric array loudspeaker on mobile phone and its acoustical characteristics. In: Proceedings of the 118th Audio Engineering Society Convention
197. Nakadai K, Tsujino H (2005) Towards new human-humanoid communication: listening during speaking by using ultrasonic directional speaker. In: Proceedings of the IEEE International Conference on Robotics and Automation (ICRA), Barcelona, Spain, pp 1483–1488
198. Johannes R, Beh JW, Gan WS et al (2010) Investigation of 3D audio rendering with parametric array loudspeakers. In: Proceedings of the 128th Audio Engineering Society Convention
199. Phanomchoeng G, Rajamani R, Hourdos J (2010) Directional sound for long-distance auditory warnings from a highway construction work zone. IEEE Trans Veh Technol 59(5):2266–2276
200. Castagnéde B, Moussatov A, Lafarge D et al (2008) Low frequency in situ metrology of absorption and dispersion of sound absorbing porous materials based on high power ultrasonic non-linearly demodulated waves. Appl Acoust 69(7):634–648
201. Castagnède B, Saeid M, Moussatov A et al (2006) Reflection and transmission at normal incidence onto air-saturated porous materials and direct measurements based on parametric demodulated ultrasonic waves. Ultrasonics 44(2):221–229
202. Haupt RW, Rolt KD (2005) Standoff acoustic laser technique to locate buried land mines. Lincoln Lab J 15(1):3–22
203. Calicchia P, Simone SD, Marcoberardino LD et al (2012) Near to far-field characterization of a parametric loudspeaker and its application in non-destructive detection of detachments in panel paintings. Appl Acoust 73(12):1296–1302
204. Gibson D, Smith M, Scales J et al (2012) Producing an intense collimated beam of sound via a nonlinear ultrasonic array. J Appl Phys 111(12):124910
205. Achanta A, Mckenna J, Rudd K et al (2005) Non-linear acoustic concealed weapons detection. Mater Eval 63(12):1195–1202
206. Kaduchak G, Sinha DN, Lizon DC et al (2000) A non-contact technique for evaluation of elastic structures at large stand-off distances: applications to classification of fluids in steel vessels. Ultrasonics 37(8):531–536
207. Sayin U, Guasch O (2013) Directivity control and efficiency of parametric loudspeakers with horns. J Acoust Soc Am 134(2):EL153–EL157
208. Arnela M, Guasch O, Sánchez-Martín P et al (2018) Construction of an omnidirectional parametric loudspeaker consisting in a spherical distribution of ultrasound transducers. Sensors 18(12):4317
209. Zheng X (2006) High directionality audio sound wave theory and control research. Ph.D. Thesis of Huazhong University of Science and Technology
210. Zhao T (2014) Parametric array difference frequency sound field distribution characteristics research. Master's Thesis of Harbin Engineering University
211. Zhao Y (2009) Multi-channel parametric array signal source and amplifier circuit design. Master's Thesis of Harbin Engineering University
212. Li X (2018) Parametric array loudspeaker signal predistortion processing and modulation method research. Master's Thesis of University of Electronic Science and Technology
213. Sun N (2016) Parametric loudspeaker system design. Master's Thesis of Nanjing University
214. Nie X (2009) Acoustic parametric array and its testing technology research. Master's Thesis of National University of Defense Technology
215. Ji M (2005) Audio signal directional propagation technology research. Master's Thesis of Northwestern Polytechnical University
216. Wang L (2011) Audio directional loudspeaker signal processing software system design and implementation. Master's Thesis of University of Electronic Science and Technology
217. Lin J (2010) Sound wave directional loudspeaker difference frequency sound field directionality research. Master's Thesis of Fudan University

218. Wang Y (2013) Acoustic parametric array pretreatment algorithm research and implementation. Master's Thesis of Chongqing Jiaotong University
219. Liu S (2008) Acoustic parametric array technology research. Master's Thesis of National University of Defense Technology
220. Song Y (2017) Acoustic parametric array wind speed measurement system design. Master's Thesis of University of Electronic Science and Technology
221. Zhao Z (2017) Research on underwater buried target detection methods. Master's Thesis of University of Electronic Science and Technology of China
222. Wang Y, Long B, Yang C et al (2019) A study on the acoustic testing and evaluation of a parametric array bird driving device. China Measuremnt &Test 45(9):156–160
223. Liu W, Yang J (2010) A simple and accurate method for calculating the Gaussian beam expansion coefficients. Chin Phys Lett 27(12):105–108
224. Yang J, Ji P, Gan WS (2007) Distortion reduction for the parametric loudspeaker. In: 19th International Congress on Acoustics (ICA), Madrid, Spain
225. Ji P, Gan WS, Tan EL et al (2010) Performance analysis on recursive single-sideband amplitude modulation for parametric loudspeakers. In: IEEE International Conference on Multimedia and Expo (ICME), pp 748–753
226. Kuang Z, Ye C, Yang J (2008) Research on the directivity of new planar PVDF ultrasonic transducers. In: Proceedings of the national acoustics conference, pp 540–541
227. Ye C, Kuang Z, Ji W et al (2007) Experimental study on multi-path high directivity audio source. Technical Acoustics 26(5):1026–1027
228. Ye C, Kuang Z, Ma D et al (2007) Impedance matching of high directivity audio source. In: Proceedings of national environmental acoustics conference, pp 441–443
229. Zhang W, Kuang Z, Ye C et al (2010) Impedance matching of PVDF film transducer in high directivity audio source. Audio Engineering 34(4):30–32
230. Kuang Z, Ye C, Wu M et al (2010) A new method for field measurement of material absorption characteristics. Acta Acustica 35(2):162–168
231. Ye C, Wu M, Yang J (2011) Research on active noise control using parametric array loudspeaker. Audio Eng 35(3):61–63
232. Wu S, Wu M, Yang J (2013) Research on parametric array loudspeaker in pipeline noise control. J Appl Acoust 32(6):439–445
233. Chen M, Huang D, Xu S et al (2007) An approximate realization method of the square root signal processing algorithm of audio directional loudspeaker. In: IEEE International Conference on Mechatronics and Automation (ICMA), Harbin, China
234. Xu L, Cao J, Huang D (2005) Design and characterization of a PVDF ultrasonic acoustic transducer applied in audio beam loudspeaker. In: IEEE International Conference on Mechatronics and Automation, Niagara Falls, Canada
235. Li X, Xu L, Chen M et al (2006) Implementation of audio directional algorithm based on ADSP-BF533. Microcomput Inform 22(14):176–178
236. Liu S, Yao W, Xu L (2007) Application of DSP in new audio directional system. Audio Engineering 3:27–30
237. Zhu H (2007) Design and implementation of directional sound source based on DSP. Master's Thesis of Shandong University of Science and Technology
238. Tian X (2011) Design of directional loudspeaker based on TMS320VC5509A. Master's Thesis of Shandong University of Science and Technology
239. Yang Y, Wen Z, Feng H et al (2008) System design and test of ultra-directional loudspeaker. Tech Acoust 27(1):66–70
240. Cao J, Xu L, Li H (2019) Design and test of audio directional PVDF ultrasonic sensor. Exp Sci Technol 3(3):1–4
241. Chen M, Xu L, Huang D et al (2007) Resonant frequency adjustable PVDF film sound frequency directional transducer. Piezoelectr Acoustoopt 29(1):53–55
242. Berktay HO (1972) Near-field effects in parametric end-fire arrays. J Sound Vib 20(2):135–143
243. Berktay HO, Leahy DJ (1974) Farfield performance of parametric transmitters. J Acoust Soc Am 55(3):539–546

244. Yang D, Hu S, Shi J et al (2018) The effect of seawater relaxation on the propagation of finite amplitude sound waves. J Harbin Eng Univ 39(2):193–199
245. Schulkin M, Marsh HW (1962) Sound absorption in sea water. J Acoust Soc Am 34(6):864–865
246. Evans LB, Bass HE, Sutherland LC (1972) Atmospheric absorption of sound: theoretical predictions. J Acoust Soc Am 51(5B):1565–1575
247. Bass HE, Sutherland LC, Zuckerwar AJ (1990) Atmospheric absorption of sound: update. J Acoust Soc Am 88(4):2019–2021
248. Bass HE, Sutherland LC, Zuckerwar AJ et al (1995) Atmospheric absorption of sound: further developments. J Acoust Soc Am 97(1):680–683
249. Garrett GS, Tjøtta JN, Rolleigh RL et al (1984) Reflection of parametric radiation from a finite planar target. J Acoust Soc Am 75(5):1462–1472
250. Huijssen J, Bouakaz A, Verweij MD et al (2003) Simulations of the nonlinear acoustic pressure field without using the parabolic approximation. IEEE Symp Ultrason 2:1851–1854
251. Huijssen J, Verweij MD (2010) An iterative method for the computation of nonlinear, wide-angle, pulsed acoustic fields of medical diagnostic transducers. J Acoust Soc Am 127(1):33–44
252. Kamakura T, Nomura H, Clement GT (2013) Application of the split-step Padé approach to nonlinear field predictions. Ultrasonics 53(2):432–438
253. Zeng X, McGough RJ (2008) Evaluation of the angular spectrum approach for simulations of near-field pressures. J Acoust Soc Am 123(1):68–76
254. Zeng X, McGough RJ (2009) Optimal simulations of ultrasonic fields produced by large thermal therapy arrays using the angular spectrum approach. J Acoust Soc Am 125(5):2967–2977
255. Alais P, Hennion PY (1979) Study by a Fourier method of the nonlinear interaction of two acoustic radiations in an absorbing fluid. Particular case of parametric emission. Acta Acustica united with Acustica 43(1):1–11
256. JingY, Tao M, Clement GT (2011) Evaluation of a wave-vector-frequency-domain method for nonlinear wave propagation. J Acoust Soc Am 129(1):32–46
257. Landsberger BJ, Hamilton MF (2001) Second-harmonic generation in sound beams reflected from, and transmitted through, immersed elastic solids. J Acoust Soc Am 109(2):488–500
258. Frijlink ME, Kaupang H, Varslot T et al (2008) Abersim: a simulation program for 3D nonlinear acoustic wave propagation for arbitrary pulses and arbitrary transducer geometries. IEEE Ultrason Symp 1282–1285
259. Du Y, Jensen JA (2013) Fast simulation of non-linear pulsed ultrasound fields using an angular spectrum approach. Ultrasonics 53(2):588–594
260. Jensen JA (1996) Field: a program for simulating ultrasound systems. Med Biol Eng Compu 34:351–353
261. Cervenka M, Bednarik M (2013) Non-paraxial model for a parametric acoustic array. J Acoust Soc Am 134(2):933–938
262. Haigh AA, Treeby BE, McCreath EC (2012) Ultrasound simulation on the cell broad-band engine using the Westervelt equation. In: International Conference on Algorithms and Architectures for Parallel Processing. Springer, Berlin, pp 241–252
263. Bakhvalov NS, Zhileikin YM, Zabolotskaya EA et al (1979) Harmonic generation in sound beams. Sov Phys Acoust 25:101–106
264. Bakhvalov NS, Zhileikin YM, Zabolotskaya EA et al (1976) Nonlinear propagation of a sound beam in a nondissipative medium. Sov Phys Acoust 22:272–274
265. Bakhvalov NS, Zhileikin YM, Zabolotskaya EA (1980) Nonlinear propagation of sound beams with a uniform amplitude distribution. Sov Phys Acoust 26:95–100
266. Bakhvalov NS, Zhileikin YM, Zabolotskaya EA (1979) Parametric interaction of sound beams. Sov Phys Acoust 25:280–283
267. Bakhvalov NS, Zhileikin YM, Zabolotskaya EA et al (1978) Propagation of finite amplitude sound beams in a dissipative medium. Sov Phys Acoust 24:271–275
268. Godunov SK (1959) A finite difference method for the numerical computation and discontinuous solutions of the equations of fluid dynamics. Matematicheskii Sbornik 47:271–306

269. Zhileikin YM, Zhuravleva TM, Rudenko OV (1980) Nonlinear effects in the propagation of high-frequency sound waves in tubes. Sov Phys Acoust 26:32–34
270. McKendree FS (1981) A numerical solution of the second-order nonlinear acoustic wave equation in one and three dimensions. Doctoral Dissertation of The Pennsylvania State University
271. Frøysa KE, Tjøtta JN, Berntsen J (1993) Finite amplitude effects in sound beams: pure tone and pulsed excitation. In: Hobæk H (ed) Asymptotic modelling in fluid mechanics. World Scientific, Singapore, pp 233–238
272. Hamilton MF, Tjøtta JN, Tjøtta S (1985) Nonlinear effects in the farfield of a directive sound source. J Acoust Soc Am 78(1):202–216
273. Tjøtta JN, Tjøtta S, Vefring EH (1990) Propagation and interaction of two collinear finite amplitude sound beams. J Acoust Soc Am 88(6):2859–2870
274. Tjøtta JN, Tjøtta S, Vefring EH (1991) Effects of focusing on the nonlinear interaction between two collinear finite amplitude sound beams. J Acoust Soc Am 89(3):1017–1027
275. Kamakura T, Hamada N, Aoki K et al (1989) Nonlinearly generated spectral components in the nearfield of a directive sound source. J Acoust Soc Am 85(6):2331–2337
276. Nachef S, Cathignol D, Tjøtta JN et al (1995) Investigation of a high intensity sound beam from a plane transducer. Experimental and theoretical results. J Acoust Soc Am 98(4):2303–2323
277. Hart TS, Hamilton MF (1988) Nonlinear effects in focused sound beams. J Acoust Soc Am 84(4):1488–1496
278. Averkiou M, Hamilton M (1995) Measurements of harmonic generation in a focused finite-amplitude sound beam. J Acoust Soc Am 98(6):3439–3442
279. Baker AC, Berg AM, Sahin A et al (1995) The nonlinear pressure field of plane, rectangular apertures: experimental and theoretical results. J Acoust Soc Am 97(6):3510–3517
280. Kamakura T, Tani M, Kumamoto Y et al (1992) Harmonic generation in finite amplitude sound beams from a rectangular aperture source. J Acoust Soc Am 91(6):3144–3151
281. Lee YS, Hamilton MF (1995) Time-domain modeling of pulsed finite-amplitude sound beams. J Acoust Soc Am 97(2):906–917
282. Averkiou MA, Cleveland RO (1999) Modeling of an electrohydraulic lithotripter with the KZK equation. J Acoust Soc Am 106(1):102–112
283. Averkiou MA, Lee YS, Hamilton MF (1993) Self-demodulation of amplitude-and frequency-modulated pulses in a thermoviscous fluid. J Acoust Soc Am 94(5):2876–2883
284. Averkiou MA, Hamilton MF (1997) Nonlinear distortion of short pulses radiated by plane and focused circular pistons. J Acoust Soc Am 102(5):2539–2548
285. Kim W, Sparrow VW (2002) Audio application of the parametric array-Implementation through a numerical model. In: Proceedings of the 113th Audio Engineering Society Convention
286. Yang X, Cleveland RO (2005) Time domain simulation of nonlinear acoustic beams generated by rectangular pistons with application to harmonic imaging. J Acoust Soc Am 117(1):113–123
287. Christopher T, Parker KJ (1991) New approaches to nonlinear diffractive field propagation. J Acoust Soc Am 90(1):488–499
288. Christopher T, Parker KJ (1991) New approaches to the linear propagation of acoustic fields. J Acoust Soc Am 90(1):507–521
289. Kamakura T (2004) Two model equations for describing nonlinear sound beams. Jpn J Appl Phys 43(5B):2808–2812
290. Christopher T (1994) Modeling the dornier HM3 lithotripter. J Acoust Soc Am 96(5):3088–3095
291. Christopher T (1998) Experimental investigation of finite amplitude distortion-based, second harmonic pulse echo ultrasonic imaging. IEEE Trans Ultrason Ferroelectr Freq Control 45(1):158–162
292. Sapozhnikov OA, Khokhlova VA, Cathignol D (2004) Nonlinear waveform distortion and shock formation in the near field of a continuous wave piston source. J Acoust Soc Am 115(5):1982–1987

293. Khokhlova VA, Souchon R, Tavakkoli J et al (2001) Numerical modeling of finite-amplitude sound beams: shock formation in the near field of a cw plane piston source. J Acoust Soc Am 110(1):95–108
294. Zemp RJ, Tavakkoli J, Cobbold RSC (2003) Modeling of nonlinear ultrasound propagation in tissue from array transducers. J Acoust Soc Am 113(1):139–152
295. Tavakkoli J, Cathignol D, Souchon R et al (1998) Modeling of pulsed finite-amplitude focused sound beams in time domain. J Acoust Soc Am 104(4):2061–2072
296. Ma Y, Ma Q, Zhang D et al (2006) Theoretical and experimental study on superharmonic sound field and its bio-tissue imaging. Acta Acustica 31(5):433–437
297. Tao C, Ma J, Zhu Z et al (2004) Study on shock wave formation of nonlinear ultrasound beam. Acta Acustica 29(2):115–121
298. Li S (2004) Study on the near-field sound field of rectangular transducer parametric array. In: Proceedings of the national acoustics conference, pp 358–359
299. Du H (2007) Research on several nonlinear problems in biomedical ultrasound. Ph.D. Thesis of University of Science and Technology of China
300. Zhang Z, Zeng X (2008) Time-domain finite difference method simulation of nonlinear propagation of strong acoustic waves in air. Proceedings of the 2008 national acoustics conference, pp 28–29
301. Howard CQ, Hansen CH, Zander AC (2005) A review of current ultrasound exposure limits. J Occup Health Saf Aust NZ 21(3):253–257
302. Lee S, Katsuura T, Towatari-Umeno T et al (2010) The effects of parametric speaker sound on salivary hormones and a subjective evaluation. Neuro Endocrinol Lett 31(4):524–529
303. Cavanagh E, Cook BD (1980) Gaussian-Laguerre description of ultrasonic fields—numerical example: circular piston. J Acoust Soc Am 67(4):1136–1140
304. Cook BD, Arnoult WJ (1976) Gaussian-Laguerre/Hermite formulation for the nearfield of an ultrasonic transducer. J Acoust Soc Am 59(1):9–11
305. Huang D, Breazeale MA (1999) A Gaussian finite-element method for description of sound diffraction. J Acoust Soc Am 106:1771–1781
306. Wen JJ, Breazeale MA (1990) Computer optimization of the Gaussian beam description of an ultrasonic field. Comput Acoust Scattering, Gaussian Beams Aeroacoust 2:181–196
307. Ding D, Xu J (2006) The Gaussian beam expansion applied to Fresnel field integrals. IEEE Trans Ultrason Ferroelectr Freq Control 53(1):246–250
308. Ding D, Zhang Y (2004) Notes on the Gaussian beam expansion. J Acoust Soc Am 116(3):1401–1405
309. Ding D, Zhang Y (2004) A simple calculation approach for the second harmonic sound beam generated by an arbitrary distribution source. Chin Phys Lett 21(3):503–506
310. Ding D, Shui Y, Lin J et al (1996) A simple calculation approach for the second harmonic sound field generated by an arbitrary axial-symmetric source. J Acoust Soc Am 100(2):727–733
311. Ding D (2004) A simplified algorithm for second-order sound beams with arbitrary source distribution and geometry (L). J Acoust Soc Am 115(1):35–37
312. Ding D (2000) A simplified algorithm for the second-order sound fields. J Acoust Soc Am 108(6):2759–2764
313. Ding D, Zhang Y, Liu J (2003) Some extensions of the Gaussian beam expansion: radiation fields of the rectangular and the elliptical transducer. J Acoust Soc Am 113(6):3043–3048
314. Ding D, Tong X, He P (2005) Supplementary notes on the Gaussian beam expansion. J Acoust Soc Am 118(2):608–611
315. Ding D, Lu Z (1996) A simplified algorithm for the piston sound field. Acta Acustica 21(4):421–428
316. Ding D, Lin J, Shui Y et al (1993) An analytical description of the sound field of a piston-type ultrasonic transducer. Acta Acustica 18(4):249–255
317. Yang J, Sha K, Gan WS et al (2004) Nonlinear wave propagation for a parametric loudspeaker. IEICE Trans Fundam Electron Commun Comput Sci 87(9):2395–2400
318. Yang J, Sha K, Gan WS et al (2005) Modeling of finite-amplitude sound beams: second order fields generated by a parametric loudspeaker. IEEE Trans Ultrason Ferroelectr Freq Control 52(4):610–618

319. Kim HJ, Schmerr LW, Sedov A (2006) Generation of the basis sets for multi-Gaussian ultrasonic beam models—an overview. J Acoust Soc Am 119(4):1971–1978
320. Schmerr LW (1998) Fundamentals of ultrasonic nondestructive evaluation. Springer, New York
321. Kim HJ, Song SJ, Schmerr LW (2006) An ultrasonic measurement model using a multi-Gaussian beam model for a rectangular transducer. Ultrasonics 44:969–974
322. Huang R, Schmerr LW, Sedov A (2005) Multi-Gaussian ultrasonic beam modeling for multiple curved interfaces—an ABCD matrix approach. Res Nondestr Eval 16(4):143–174
323. Huang R, Schmerr LW, Sedov A (2007) Multi-Gaussian beam modeling for multilayered anisotropic media, I: modeling fundamentals. Res Nondestr Eval 18(4):193–220
324. Huang R, Schmerr LW, Sedov A (2007) Multi-gaussian beam modeling for multilayered anisotropic media, II: numerical examples of slowness surface and geometry effects. Res Nondestr Eval 21(4):221–240
325. Huang R, Schmerr LW, Sedov A (2008) Modeling the radiation of ultrasonic phased-array transducers with Gaussian beams. IEEE Trans Ultrason Ferroelectr Freq Control 55(12):2692–2702
326. Spies M (1999) Transducer field modeling in anisotropic media by superposition of Gaussian basis functions. J Acoust Soc Am 105:633–638
327. Spies M (2007) Ultrasonic field modeling for immersed components using Gaussian beam superposition. Ultrasonics 46:138–147
328. Prange MD, Shenoy RG (1996) A fast Gaussian beam description of ultrasonic fields based on Prony's method. Ultrasonics 34:117–119
329. Schmerr LW, Lopez-Sanchez A, Sedov A (2010) K-space Prony's method for generating the basis functions of multi-Gaussian beam models. Ultrasonics 50(6):600–605
330. Yu J, Zhang D, Liu X et al (2007) Theoretical and experimental study of the nonlinear acoustic field of a conical PVDF focusing transducer. Acta Physica Sinica 56(10):5909–5914
331. Liu X, Gong X, Lu Y (2011) Study on temperature rise caused by piston focusing transducer. J Biomed Eng China 30(6):914–919
332. Zhao X, Gang T, Zhang B (2008) Acoustic field calculation of phased array transducer based on non-paraxial approximation multi-Gaussian beam model. Acta Acustica 33(5):475–480
333. Zhao X, Gang T (2009) Nonparaxial multi-Gaussian beam models and measurement models for phased array transducers. Ultrasonics 49(1):126–130
334. Jeong H (2009) Time reversal study of ultrasonic waves for anisotropic solids using a Gaussian beam model. NDT&E Int 42:210–214
335. Liu W (2011) Calculation of directional beams—Gaussian beam expansion method and three-dimensional time-domain finite difference algorithm. Ph.D. Thesis of the graduate school of the Chinese Academy of Sciences
336. Liu W, Yang J, Tian J (2011) Three-dimensional time-domain modeling of finite amplitude waves. Acta Acustica 36(4):349–357
337. Lee YS (1993) Numerical solution of the KZK equation for pulsed finite amplitude sound beams in thermoviscous fluids. Doctoral Dissertation of The University of Texas
338. Engquist B, Majda A (1977) Absorbing boundary conditions for numerical simulation of waves. Proc Natl Acad Sci 74(5):1765–1766
339. Liu Q, Tao J (1997) The perfectly matched layer for acoustic waves in absorptive media. J Acoust Soc Am 102(4):2072–2082
340. Taflove A (1995) Computational electrodynamics: the finite-difference time domain method. Artech House, Boston
341. Ames WF (1992) Numerical methods for partial differential equations, 3rd edn. Academic, New York
342. Pompei FJ (1999) The use of airborne ultrasonics for generating audible sound beams. J Audio Eng Soc 47(9):726–731
343. Agazzi O, Messerschmitt DG, Hodges D (1982) Nonlinear echo cancellation of data signals. IEEE Trans Commun 30(11):2421–2433

344. Casar-Corredera J, Garcia-Otero M, Figueiras-Vidal A (1985) Data echo nonlinear cancellation. In: IEEE International Conference on Acoustics, Speech, and Signal Processing (ICASSP), pp 1245–1248
345. Sicuranza GL, Bucconi A, Mitri P (1984) Adaptive echo cancellation with nonlinear digital filters. In: IEEE International Conference on Acoustics, Speech, and Signal Processing (ICASSP), pp 130–133
346. Hu R, Ahmed HM (1994) Echo cancellation in high speed data transmission systems using adaptive layered bilinear filters. IEEE Trans Commun 42(234):655–663
347. Stenger A, Trautmann L, Rabenstein R (1999) Nonlinear acoustic echo cancellation with 2nd order adaptive Volterra filters. In: IEEE International Conference on Acoustics, Speech and Signal Processing (ICASSP), pp 877–880
348. Guérin A, Faucon G, Bouquin-Jeannes L (2003) Nonlinear acoustic echo cancellation based on Volterra filters. IEEE Trans Speech Audio Process 11(6):672–683
349. Benedetto S, Biglieri E, Daffara R (1979) Modeling and performance evaluation of nonlinear satellite links: a Volterra series approach. IEEE Trans Aerosp Electron Syst 4:494–507
350. Gutierrez A, Ryan WE (1995) Performance of adaptive Volterra equalizers on nonlinear satellite channels. In: IEEE international Conference on Communications (ICC), pp 488–492
351. Beidas BF (2011) Intermodulation distortion in multicarrier satellite systems: analysis and turbo Volterra equalization. IEEE Trans Commun 59(6):1580–1590
352. Lin WH (2001) Adaptive third-order Volterra satellite channel equalizer. Master's Thesis of National Sun Yat-sen University
353. Tan L, Jiang J (2001) Adaptive Volterra filters for active control of nonlinear noise processes. IEEE Trans Signal Process 49(8):1667–1676
354. Sicuranza GL, Carini A (2004) Filtered-X affine projection algorithm for multichannel active noise control using second-order Volterra filters. IEEE Signal Process Lett 11(11):853–857
355. Tan L, Jiang J (2009) Adaptive second-order Volterra filtered-X RLS algorithms with sequential and partial updates for nonlinear active noise control. In: IEEE Conference on Industrial Electronics and Applications (ICIEA), pp 1625–1630
356. Kumar MK (2010) Filtered-x affine projection algorithm in active noise control using Volterra filter. Master's Thesis of Thapar University
357. Hunter I, Korenberg M (1986) The identification of nonlinear biological systems: Wiener and Hammerstein cascade models. Biol Cybern 55:135–144
358. Korenberg MJ, Hunter I (1986) The identification of nonlinear biological systems: LNL cascade models. Biol Cybern 55(2–3):125–134
359. Marmarelis PZ, Marmarelis VZ (1978) Analysis of physiological systems: the white-noise approach. Springer, Berlin
360. Volterra V (1887) On the functions that depend on other functions. Rend. R. Academy of Lincei 2 Sem, pp 97–105
361. Frechet M (1910) On continuous functionals. Annales Scientifiques de L'Ecole Normale Superieure 27:193–216
362. Lesiak C, Krener AJ (1978) The existence and uniqueness of Volterra series for nonlinear systems. IEEE Trans Autom Control 23(6):1090–1095
363. Sandberg IW (1983) The mathematical foundations of associated expansions for mildly nonlinear systems. IEEE Trans Circ Syst 30(7):441–455
364. Boyd S, Chua LO (1985) Fading memory and the problem of approximating nonlinear operators with Volterra series. IEEE Trans Circ Syst 32(11):1150–1161
365. Brockett RW (1976) Volterra series and geometric control theory. IEEE Trans Automatica 12(2):167–176
366. Balakrishnan A (1969) Computing algorithms for detection problems. In: Proceedings of the symposium on computer processing in communications, pp 603–615
367. Nowak RD, Vanveen BD (1994) Random and pseudorandom inputs for Volterra filter identification. IEEE Trans Signal Process 42(8):2124–2135
368. Pearson RK, Ogunnaike BA, Doyle FJ (1996) Identification of structurally constrained second-order Volterra models. IEEE Trans Signal Process 44(11):2837–2846

369. Schetzen M (1980) The Volterra and Wiener theories of nonlinear systems. Krieger Publishing Company
370. Boyd S, Tang YS, Chua LO (1983) Measuring Volterra kernels. IEEE Trans Circ Syst 30(8):571–577
371. Chua LO, Liao Y (1991) Measuring Volterra kernels III: How to estimate the highest significant order. Int J Circ Theory Appl 19(2):189–209
372. Chua LO, Liao Y (1989) Measuring Volterra kernels (II). Int J Circ Theory Appl 17(2):151–190
373. Tsujikawa M, Shiozaki T, Kajikawa Y et al (2000) Identification and elimination of second-order nonlinear distortion of loudspeaker systems using Volterra filter. In: Proceedings of IEEE International Symposium on Circuits and Systems, pp 249–252
374. Bicken G, Carey GF, Stearman RO (2002) Frequency domain kernel estimation for 2nd- order Volterra models using random multi-tone excitation. VLSI Design 15(4):701–713
375. Jones JCP (2007) Simplified computation of the Volterra frequency response functions of non-linear systems. Mech Syst Signal Process 21(3):1452–1468
376. Bjorsell N, Suchanek P, Handel P et al (2008) Measuring Volterra kernels of analog-to-digital converters using a stepped three-tone scan. IEEE Trans Instrum Meas 57(4):666–671
377. Koh T, Powers EJ (1985) 2nd-order Volterra filtering and its application to nonlinear system identification. IEEE Trans Acoust Speech Signal Process 33(6):1445–1455
378. Takahama Y, Kajikawa Y, Nomura Y (2002) A formulation of the convergence property for the second-order adaptive Volterra filter using NLMS algorithm. Electron Commun Japan Part III-Fundam Electron Sci 85(10):40–50
379. Lee JS, Mathews VJM (1993) A fast recursive least squares adaptive second order Volterra filter and its performance analysis. IEEE Trans Signal Process 41(3):1087–1402
380. Liang J, Ji B, Zhang J et al (2008) Recursive least squares-like algorithms for the adaptive second-order lattice notch filter. Digital Signal Process 18(3):291–306
381. Ozden MT, Kayran AH, Panayirci EP (1996) Adaptive Volterra filtering with complete lattice orthogonalization. IEEE Trans Signal Process 44(8):2092–2098
382. Syed MA, Mathews VJ (1993) QR-Decomposition based algorithms for adaptive Volterra filtering. IEEE Trans Circ Syst I: Fundam Theory Appl 40(6):372–382
383. Davila CE, Welch AJ, Rylander HG (1987) A 2nd-order adaptive Volterra filter with rapid convergence. IEEE Trans Acoust Speech Signal Process 35(9):1259–1263
384. Kajikawa YK, Nomura Y (1998) The summational projection algorithm for the adaptive Volterra filter. In: European Signal Processing Conference, pp 745–748
385. Kajikawa Y (2000) The adaptive Volterra filter: its present and future. Electron Commun Japan Part III-Fundam Electron Sci 83(12):51–61
386. Sarti A, Pupolin SP (1992) Recursive techniques for the synthesis of a pth-order inverse of a Volterra system. Eur Trans Telecommun 3(4):315–322
387. Nowak RD, VanVeen BD (1997) Volterra filter equalization: a fixed point approach. IEEE Trans Signal Process 45(2):377–388
388. Lashkari K (2004) High quality sound from small loudspeakers using the exact inverse. In: Conference Record of the 38th Asilomar Conference on Signals, Systems and Computers, pp 430–434
389. Pearson R, Ogunnaike BA (2002) Identification and control using Volterra models. Springer, Berlin
390. Ogunfunmi T (2007) Adaptive nonlinear system identification: the Volterra and wiener model approaches. Springer Science and Business Media, New York
391. Rugh WJ (1981) Nonlinear system theory. Johns Hopkins University Press, Baltimore
392. Bernardini R (2000) A fast algorithm for general Volterra filtering. IEEE Trans Commun 48(11):1853–1864
393. Kinoshita S, Kajikawa Y, Nomura Y (2001) Volterra filters using multirate signal processing and their application to loudspeaker systems. In: IEEE International Conference on Acoustics, Speech, and Signal Processing (ICASSP), pp 3497–3500

394. Ishikawa T, Nakashima K, Kajikawa Y et al (2000) A consideration on elimination of nonlinear distortion of the loudspeaker system by using digital Volterra filter. Electron Commun Japan Part III-Fundam Electron Sci 83(2):110–118
395. Frank WA (1995) An efficient approximation to the quadratic Volterra filter and its application in real-time loudspeaker linearization. IEEE Trans Signal Process 45(1):97–113
396. Kafka VS, Appel U (1999) An approximation to the Volterra series with multiple linear ARMA filters. In: IEEE EURASIP workshop on nonlinear signal and image processing, pp 675–678
397. Korenberg MJ, Hunter IW (1986) The identification of nonlinear biological-systems—Inl cascade models. Biol Cybern 55(2–3):125–134
398. Marmarelis VZ (1993) Identification of nonlinear biological-systems using Laguerre expansions of kernels. Ann Biomed Eng 21(6):573–589
399. Farina A (2000) Simultaneous measurement of impulse response and distortion with a swept-sine technique. In: Proceedings of the 108th Audio Engineering Society Convention
400. Farina A, Bellini A, Armelloni E (2001) NonLinear convolution: a new approach for the auralization of distortion systems. In: Proceedings of the 110th Audio Engineering Society Convention
401. Farina A (2007) Real-time auralization employing a not-linear, not-time-invariant convolver. In: Proceedings of the 123th Audio Engineering Society Convention
402. Novak A, Simon L, Lotton P (2010) Analysis, synthesis, and classification of nonlinear systems using synchronized swept-sine method for audio effects. EURASIP J Adv Signal Process, p 793816
403. Novak A, Simon L, Kadlec F et al (2010) Nonlinear system identification using exponential swept-sine signal. IEEE Trans Instrum Meas 59(8):2220–2229
404. Rebillat M, Hennequin R, Corteel É et al (2010) Prediction of harmonic distortion generated by electro-dynamic loudspeakers using cascade of Hammerstein models. In: Proceedings of the 128th Audio Engineering Society Convention
405. Rebillat M, Hennequin R, Corteel E et al (2011) Identification of cascade of Hammerstein models for the description of nonlinearities in vibrating devices. J Sound Vib 330(5):1018–1038
406. Tronchin L (2012) The emulation of nonlinear time-invariant audio systems with memory by means of Volterra series. J Audio Eng Soc 60(12):984–996
407. Tronchin L (2012) Nonlinear convolution applied to acoustic signals. In: Proceedings of the 16th WSEAS International Conference on Circuits and Systems, Kos Island, Greece
408. Sniedovich M (2010) Dynamic programming: foundations and principles. CRC Press, Boca Roton
409. Wu M, Ye C, Wu S et al (2010) Effects on the performance of parametric loudspeaker caused by non ideal ultrasound transducer. Japan J Appl Phys 49(7):07HE19-1-4
410. Shi C, Kajikawa Y (2016) Effect of the ultrasonic emitter on the distortion performance of the parametric array loudspeaker. Appl Acoust 112:108–115
411. Parks TW, Burrus CS (1987) Digital filter design. Wiley, New York
412. Mu Y (2014) Research on nonlinear distortion compensation of loudspeaker system. Ph.D. Thesis of the graduate school of the Chinese Academy of Sciences
413. Kadlec F, Novak A, Simon L et al (2008) Measurement and analysis of non-linear distortion of electro-acoustic and audio systems. In: The 15th International Congress on Sound and Vibration
414. Park T, Kim S, Sung KM (2011) System identification for acoustic echo cancellation using stepped sine method related to FFT size. In: Proceedings of the 130th Audio Engineering Society Convention
415. Germain FG (2011) A nonlinear analysis framework for electronic synthesizer circuits. Master's Thesis of McGill University
416. Schmitz T, Embrechts JJ (2014) Improvement in non-linear guitar loudspeaker sound reproduction. In: International Conference on Systems, Signals and Image Processing

417. Schmitz T, Embrechts JJ (2013) Nonlinear guitar loudspeaker simulation. In: Proceedings of the 134th Audio Engineering Society Convention
418. Potirakis S, Tatlas NA, Zafeiropoulos N et al (2013) Assessment of military intercom headsets for maximum voice reproduction level in high noise conditions. Appl Acoust 74(6):870–881
419. Sanderg IW (1992) Uniform approximation with doubly finite Volterra series. IEEE Trans Signal Process 40(6):1438–1442
420. Mathews VJ, Sicuranza GL (2000) Polynomial signal processing. Wiley, New York
421. Lucas BG, Tjøtta JN, Muir TG (1983) Field of a parametric focusing source. J Acoust Soc Am 73(6):1966–1971
422. Sugahara A, Lee H, Sakamoto S et al (2019) Measurements of acoustic impedance of porous materials using a parametric loudspeaker with phononic crystals and phase-cancellation method. Appl Acoust 152:54–62
423. Ju HS, Kim YH (2010) Near-field characteristics of the parametric loudspeaker using ultrasonic transducers. Appl Acoust 71(9):793–800
424. Sigalas MM, Economou EN (1992) Elastic and acoustic wave band structure. J Sound Vib 158(2):377–382
425. Kushwaha MS, Halevi P, Dobrzynski L et al (1993) Acoustic band structure of periodic elastic composites. Phys Rev Lett 71(13):2022–2025
426. Liu Z, Zhang Z, Mao Y et al (2000) Locally resonant sonic materials. Science 289(5485):1734–1736
427. Hirsekorn M, Delsanto PP, Batra NK et al (2004) Modelling and simulation of acoustic wave propagation in locally resonant sonic materials. Ultrasonics 42(1–9):231–235
428. Maekawa Z (1968) Noise reduction by screens. Appl Acoust 1(3):157–173
429. Kamakura T, Nomura H, Clement GT (2015) Linear and nonlinear ultrasound fields formed by planar sources with random pressure distributions. Acoust Sci Technol 36(3):208–215
430. Minami S, Kuroda J, Oikawa Y (2018) Individual difference of ultrasonic transducers for parametric array loudspeaker. In: IEEE international Conference on Acoustics, Speech and Signal Processing (ICASSP), Calgary, Alberta, Canada, pp 486–490
431. Blackstock DT (2000) Fundamentals of physical acoustics. Wiley, New York
432. Murata Manufacturing Co L. Ultrasonic sensor [Online]. Available: http://www.murata.com.cn/products/sensor/index.html
433. Gutierrez R, Valls J (2011) Low cost hardware implementation of logarithm approximation. IEEE Trans Very Large Scale Integr Syst 19(12):2326–2330
434. Xu L, Gu L, Chen X et al (2009) Research on analysis and optimization techniques of MBE. Comput Eng Appl 45(33):87–89
435. Zhang H, Li H (2009) Optimization of WMA open source decoding program based on S3C2410. Comput Eng Des 30(1):13–15, 34
436. Johnson DH, Dudgeon DE (1993) Array signal processing. Prentice Hall, New Jersey
437. Shi C, Bai R (2019) Design of a constant beamwidth beamformer for the parametric array loudspeaker. In: 23rd International Congress on Acoustics (ICA), Aachen, Germany
438. Gou J, Liang J, Bai R et al (2020) Beampattern design for the parametric array loudspeaker using mixed Gaussian directivity method. Proc INTER-NOISE NOISE-CON Congr Conf 261(3):3665–3672
439. Treeby BE, Cox BT (2010) Modeling power law absorption and dispersion for acoustic propagation using the fractional Laplacian. J Acoust Soc Am 127(5):2741–2748
440. Treeby BE, Cox BT (2011) A k-space Green's function solution for acoustic initial value problems in homogeneous media with power law absorption. J Acoust Soc Am 129(6):3652–3660
441. Murphy P, Krukowski A, Tarczynski A (1997) An efficient fractional sample delayer for digital beam steering. In: IEEE International Conference on Acoustics, Speech, and Signal Processing (ICASSP), pp 2245–2248
442. Vesma J, Saramaki T (1997) Optimization and efficient implementation of FIR filters with adjustable fractional delay. IEEE Int Symp Circ Syst (ISCAS) 4:2256–2259

443. Proakis JG, Manolakis DG (1996) Digital signal processing, 3rd edn. Prentice-Hall, Englewood Cliffs, NJ
444. Harris FJ (1978) On the use of windows for harmonic analysis with the discrete Fourier transform. Proc IEEE 66(1):51–83
445. Burrus CS, Soewito AW, Gopinath RA (1992) Least squared error FIR filter design with transition bands. IEEE Trans Signal Process 40(6):1327–1340
446. Selesnick IW, Lang M, Burrus CS (1996) Constrained least square design of FIR filters without specified transition bands. IEEE Trans Signal Process 44(8):1879–1892
447. Jerri AJ (1977) The Shannon sampling theorem—its various extensions and applications: a tutorial review. Proc IEEE 65(11):1565–1596
448. Schafer RW, Rabiner LR (1973) A digital signal processing approach to interpolation. Proc IEEE 61(6):692–702
449. Strube HW (1975) Sampled−data representation of a nonuniform lossless tube of continuously variable length. J Acoust Soc Am 57(1):256–257
450. Laine UK (1988) Digital modelling of a variable length acoustic tube. The proceedings of the Nordic acoustical meeting, Tampere, Finland
451. Liu GS, Wei CH (1990) Programmable fractional sample delay filter with Lagrange interpolation. Electron Lett 26(19):1608–1610
452. Pyfer M, Ansari R (1987) The design and application of optimal FIR fractional-slope phase filters. In: IEEE International Conference on Acoustics, Speech, and Signal Processing (ICASSP), pp 896–899
453. Oetken G (1979) A new approach for the design of digital interpolating filters. IEEE Trans Acoust Speech Signal Process 27(6):637–643
454. Papoulis A (1977) Signal analysis. McGraw-Hill, New York
455. Adams JW (1987) Alternate approach to digital phase shift filters. In: 1st IASTED International Symposium on Signal Processing and its Applications, vol 1, pp 160–165
456. Farrow CW (1988) A continuously variable digital delay element. In: Proceedings of IEEE International Symposium on Circuits and Systems, pp 2641–2645
457. Shi Y (1998) Modeling of acoustic waves for linear phased arrays. Master's Thesis of Massachusetts Institute of Technology
458. Brooks LA, Zander AC, Hansen CH (2005) Investigation into the feasibility of using a parametric array control source in an active noise control system. Proc Acoust 39–45
459. Kidner MRF, Petersen C, Zander AC et al (2006) Feasibility study of localised active noise control using an audio spotlight and virtual sensors. Proc Acoust 55–61
460. Tanaka N, Tanaka M (2010) Active noise control using a parametric array loudspeaker. In: The 17th International Congress on Sound and Vibration, Cairo, Egypt
461. Tanaka N, Tanaka M (2010) Active noise control using a steerable parametric array loudspeaker. J Acoust Soc Am 127:3526
462. Tanaka N, Tanaka M (2011) Mathematically trivial control of sound using a parametric beam focusing source. J Acoust Soc Am 129(1):165–172
463. Komatsuzaki T, Iwata Y (2011) Active noise control using high-directional parametric loudspeaker. J Environ Eng 6(1):140–149
464. Saeid M, Castagnede B, Moussatov A et al (2004) Application of nonlinearly demodulated acoustic signals for the measurement of the acoustical coefficient of reflection for air saturated porous materials. Comptes Rendus Mecanique 332(10):849–858
465. Lueg P (1936) Process of silencing sound oscillations. Patent DE 655508
466. Lueg P (1936) Process of silencing sound oscillations. Patent US 2043416
467. Olson HF, May EG (1953) Electronic sound absorber. J Acoust Soc Am 25(6):1130–1136
468. Olson HF (1956) Electronic control of noise, vibration and reverberation. J Acoust Soc Am 28(5):996–972
469. Nelson PA, Elliott SJ (1992) Active control of sound. Academic Press, London
470. Elliott SJ (2001) Signal processing for active control. Academic Press, London
471. Chen K (2003) Active noise control. National Defense Industry Press, Beijing
472. Kuo SM, Morgan DR (1999) Active noise control: a tutorial review. Proc IEEE 87(6):943–973

473. Kuo SM, Morgan DR (1996) Active noise control systems: algorithms and DSP implementations. Wiley, New York
474. Tian J (1992) Theoretical analysis of the energy mechanism of active noise reduction in monopole secondary sound source pipelines. Acta Acustica 17(5):369–374
475. Snyder SD, Hansen CH (1989) Active noise control in ducts: some physical insights. J Acoust Soc Am 86(1):184–194
476. Snyder SD, Tanaka N (1993) To absorb or not to absorb: control source power output in active noise control systems. J Acoust Soc Am 94(1):185–195
477. Elliott SJ, Joseph P, Nelson PA et al (1991) Power output minimization and power absorption in the active control of sound. J Acoust Soc Am 90(5):2501–2512
478. Nocke C, Mellert V (2002) Brief review on in situ measurement techniques of impedance or absorption. The proceedings of the forum acusticum, Sevilla, Spain
479. Allard J, Champoux Y (1989) In situ two-microphone technique for the measurement of the acoustic surface impedance of materials. Noise Control Eng J 32(1):15–23
480. Ye C (2011) Theoretical and experimental study of parametric array loudspeakers. Ph.D. Thesis of the graduate school of the Chinese Academy of Sciences
481. Hou H, Yang J, Li H et al (2000) Theoretical study on the radiation impedance law of sound sources in active noise reduction. J Northwestern Polytech Univ 18(4):557–561
482. Hou H, Yang J, Li H et al (2001) The law of electroacoustic impedance and its application in active noise reduction systems. J Northwestern Polytech Univ 19(2):246–250
483. Guo J, Pan J, Bao C (1997) Actively created quiet zones by multiple control sources in free space. J Acoust Soc Am 101(3):1492–1501
484. Guo J, Pan J (1997) Further investigation on actively created quiet zones by multiple control sources in free space. J Acoust Soc Am 102(5):3050–3053
485. Guo J, Pan J (1999) Actively created quiet zones for broadband noise using multiple control sources and error microphones. J Acoust Soc Am 105(4):2294–2303
486. Guo J, Pan J (1998) Effects of reflective ground on the actively created quiet zones. J Acoust Soc Am 103(2):944–952
487. Tseng WK, Rafaely B, Elliott SJ (2000) Local active sound control using 2-norm and infinity-norm pressure minimization. J Sound Vib 234(3):427–439
488. Wu S(2013) Research on the application of parametric array loudspeakers in sound field control. Ph.D. Thesis of the graduate school of the Chinese Academy of Sciences
489. Glover J (1977) Adaptive noise canceling applied to sinusoidal interferences. IEEE Trans Acoust Speech Signal Process 25(6):484–491
490. Kuang Z (2011) Study on the characteristics of sound absorption coefficient of materials measured by parametric loudspeaker array. Ph.D. Thesis of the graduate school of the Chinese Academy of Sciences
491. Chung J, Blaser D (1980) Transfer function method of measuring acoustic intensity in a duct system with flow. J Acoust Soc Am 68:1570–1577
492. Chung J, Blaser D (1980) Transfer function method of measuring induct acoustic properties. II-Experiment. J Acoust Soc Am 68(3):914–921
493. (1998) Acoustics-Determination of sound absorption coefficient and impedance in impedance tubes-Part 2 transfer-function method (ISO 10534–2)
494. Boulanger P, Waters-Fuller T, Attenborough K et al (1997) Models and measurements of sound propagation from a point source over mixed impedance ground. J Acoust Soc Am 102:1432–1442
495. Nocke C (2000) In-situ acoustic impedance measurement using a free-field transfer function method. Appl Acoust 59(3):253–264

The manufacturer's authorised representative in the EU is Springer Nature Customer Service Centre GmbH, Europaplatz 3, 69115 Heidelberg, Germany. If you have any concerns regarding our products, please contact ProductSafety@springernature.com

Printed and bound by CPI Group (UK) Ltd, Croydon, CR0 4YY
01/10/2026
02225937-0015